Further Research in Food Processing Safety and Quality from Field to Fork

Further Research in Food Processing Safety and Quality from Field to Fork

Guest Editors

Tamara Lazarević-Pašti
Nebojša Potkonjak

Basel • Beijing • Wuhan • Barcelona • Belgrade • Novi Sad • Cluj • Manchester

Guest Editors

Tamara Lazarević-Pašti
VINČA Institute of Nuclear
Sciences—National Institute
of the Republic of Serbia
University of Belgrade
Belgrade
Serbia

Nebojša Potkonjak
Department of Physical
Chemistry
VINČA Institute of Nuclear
Sciences—National Institute
of the Republic of Serbia
University of Belgrade
Belgrade
Serbia

Editorial Office
MDPI AG
Grosspeteranlage 5
4052 Basel, Switzerland

This is a reprint of the Special Issue, published open access by the journal *Foods* (ISSN 2304-8158), freely accessible at: https://www.mdpi.com/journal/foods/special_issues/food_processing_contamination_farm_Fork.

For citation purposes, cite each article independently as indicated on the article page online and as indicated below:

Lastname, A.A.; Lastname, B.B. Article Title. *Journal Name* **Year**, *Volume Number*, Page Range.

ISBN 978-3-7258-3541-6 (Hbk)
ISBN 978-3-7258-3542-3 (PDF)
https://doi.org/10.3390/books978-3-7258-3542-3

Cover image courtesy of Tamara Lazarević-Pašti

© 2025 by the authors. Articles in this book are Open Access and distributed under the Creative Commons Attribution (CC BY) license. The book as a whole is distributed by MDPI under the terms and conditions of the Creative Commons Attribution-NonCommercial-NoDerivs (CC BY-NC-ND) license (https://creativecommons.org/licenses/by-nc-nd/4.0/).

Contents

Tamara Lazarević-Pašti and Nebojša Potkonjak
Further Research in Food Processing Safety and Quality from Field to Fork
Reprinted from: *Foods* **2025**, *14*, 546, https://doi.org/10.3390/foods14040546 1

Ammar Al-Hamry, Tianqi Lu, Jing Bai, Anurag Adiraju, Tharun K. Ega, Igor A. Pašti and Olfa Kanoun
Layer-by-Layer Deposited Multi-Modal PDAC/rGO Composite-Based Sensors
Reprinted from: *Foods* **2023**, *12*, 268, https://doi.org/10.3390/foods12020268 4

Tamara Tasić, Vedran Milanković, Katarina Batalović, Stefan Breitenbach, Christoph Unterweger, Christian Fürst, et al.
Application of Viscose-Based Porous Carbon Fibers in Food Processing—Malathion and Chlorpyrifos Removal
Reprinted from: *Foods* **2023**, *12*, 2362, https://doi.org/10.3390/foods12122362 21

Vedran Milanković, Tamara Tasić, Milica Pejčić, Igor Pašti and Tamara Lazarević-Pašti
Spent Coffee Grounds as an Adsorbent for Malathion and Chlorpyrifos—Kinetics, Thermodynamics, and Eco-Neurotoxicity
Reprinted from: *Foods* **2023**, *12*, 2397, https://doi.org/10.3390/foods12122397 41

Andreja Leskovac and Sandra Petrović
Pesticide Use and Degradation Strategies: Food Safety, Challenges and Perspectives
Reprinted from: *Foods* **2023**, *12*, 2709, https://doi.org/10.3390/foods12142709 61

Federica Flamminii, Silvia Minetti, Adriano Mollica, Angelo Cichelli and Lorenzo Cerretani
The Effect of Washing, Blanching and Frozen Storage on Pesticide Residue in Spinach
Reprinted from: *Foods* **2023**, *12*, 2806, https://doi.org/10.3390/foods12142806 85

Ana Jocić, Slađana Marić, Danijela Tekić, Jasmina Mušović, Jelena Milićević, Sanja Živković and Aleksandra Dimitrijević
Green Pre-Treatment Strategy Using Ionic Liquid-Based Aqueous Two-Phase Systems for Pesticide Determination in Strawberry Samples
Reprinted from: *Foods* **2024**, *13*, 4106, https://doi.org/10.3390/foods13244106 102

Valeria Paucar-Quishpe, Darío Cepeda-Bastidas, Richar Rodríguez-Hidalgo, Ximena Pérez-Otáñez, Cecilia Perez, Sandra Enríquez, et al.
Evaluating the Human Risks of Consumption of Foods of Bovine Origin with Ivermectin Residues in Ecuador
Reprinted from: *Foods* **2024**, *13*, 3470, https://doi.org/10.3390/foods13213470 121

Alejandra Cantoral, Sonia Collado-López, Larissa Betanzos-Robledo, Héctor Lamadrid-Figueroa, Betzabeth A. García-Martínez, Camilo Ríos, et al.
Dietary Risk Assessment of Cadmium Exposure Through Commonly Consumed Foodstuffs in Mexico
Reprinted from: *Foods* **2024**, *13*, 3649, https://doi.org/10.3390/foods13223649 135

Alice N. Mafe and Dietrich Büsselberg
Mycotoxins in Food: Cancer Risks and Strategies for Control
Reprinted from: *Foods* **2024**, *13*, 3502, https://doi.org/10.3390/foods13213502 151

Alice N. Mafe and Dietrich Büsselberg
Impact of Metabolites from Foodborne Pathogens on Cancer
Reprinted from: *Foods* **2024**, *13*, 3886, https://doi.org/10.3390/foods13233886 176

Aleksandra Savić, Jelena Mutić, Milica Lučić, Jelena Vesković, Andrijana Miletić and Antonije Onjia
Ultrasound-Assisted Extraction Followed by Inductively Coupled Plasma Mass Spectrometry and Multivariate Profiling of Rare Earth Elements in Coffee
Reprinted from: *Foods* **2025**, *14*, 275, https://doi.org/10.3390/foods14020275 219

Milica Lučić, Nebojša Potkonjak, Ivana Sredović Ignjatović, Steva Lević, Zora Dajić-Stevanović, Stefan Kolašinac, et al.
Influence of Ultrasonic and Chemical Pretreatments on Quality Attributes of Dried Pepper (*Capsicum annuum*)
Reprinted from: *Foods* **2023**, *12*, 2468, https://doi.org/10.3390/foods12132468 235

Anna Augustyńska-Prejsnar, Paweł Hanus, Małgorzata Ormian, Miroslava Kačániová, Zofia Sokołowicz and Jadwiga Topczewska
The Effect of Temperature and Storage Duration on the Quality and Attributes of the Breast Meat of Hens after Their Laying Periods
Reprinted from: *Foods* **2023**, *12*, 4340, https://doi.org/10.3390/foods12234340 251

Aleksandra Dimitrijević, Slađana Marić, Ana Jocić, Danijela Tekić, Jasmina Mušović and Joana S. Amaral
Green Extraction Strategy Using Bio-Based Aqueous Biphasic Systems for Polyphenol Valorization from Grape By-Product
Reprinted from: *Foods* **2024**, *13*, 954, https://doi.org/10.3390/foods13060954 266

Anna Pęksa, Agnieszka Tajner-Czopek, Artur Gryszkin, Joanna Miedzianka, Elżbieta Rytel and Szymon Wolny
Assessment of the Content of Glycoalkaloids in Potato Snacks Made from Colored Potatoes, Resulting from the Action of Organic Acids and Thermal Processing
Reprinted from: *Foods* **2024**, *13*, 1712, https://doi.org/10.3390/foods13111712 287

Editorial

Further Research in Food Processing Safety and Quality from Field to Fork

Tamara Lazarević-Pašti * and Nebojša Potkonjak

VINČA Institute of Nuclear Sciences—National Institute of the Republic of Serbia, University of Belgrade, Mike Petrovica Alasa 12–14, 11000 Belgrade, Serbia; npotkonjak@vin.bg.ac.rs
* Correspondence: tamara@vin.bg.ac.rs

This Editorial refers to the Special Issue "Further Research in Food Processing Safety and Quality from Field to Fork", which addresses critical aspects of ensuring the safety and quality of food throughout the entire production chain. Modern food processing technologies enhance product quality and contribute to reducing waste, maximizing the use of byproducts, and promoting sustainable practices. Advances in green extraction, pesticide residue assessment, and novel sensor technologies are crucial for undertaking contemporary food safety challenges. These approaches align with global efforts to reduce environmental impact, ensure regulatory compliance, and improve consumer trust in food systems. This Special Issue highlights the importance of innovative technologies and strategies for improving food safety and minimizing risks to public health from agricultural practices to final processing stages.

The presented collection of research focuses on investigations aimed at the following:

(i) Advanced methodologies for detecting and mitigating contaminants, such as pesticides, mycotoxins, and heavy metals, in food products;

(ii) Evaluate the impact of pretreatment, processing, and storage conditions on food quality and safety attributes;

(iii) Promote the valorization of agro-industrial byproducts through green and sustainable extraction techniques;

(iv) Explore innovative materials and sensors for real-time monitoring and quality assurance in food systems.

This collection includes 15 thoroughly reviewed manuscripts, each offering valuable insights into diverse aspects of food processing safety and quality. The contributions collectively underline the necessity of integrating advanced technologies, interdisciplinary approaches, and sustainability principles to ensure safe and high-quality food from field to fork.

Al-Hamry et al. [1] developed multi-modal sensors based on PDAC/rGO films for detecting dimethoate, demonstrating their real-time efficiency in food safety monitoring. In the field of pesticide removal, Tasić et al. [2] investigated viscose-derived activated carbon fibers for the adsorption of malathion and chlorpyrifos from liquid food samples, highlighting their high efficiency, selectivity, and reusability. Similarly, Milanković et al. [3] explored using spent coffee grounds as a sustainable adsorbent for organophosphate pesticide removal, showing promising adsorption capacities and safe application in food and water treatment.

Aside from detection and removal, regulatory aspects and degradation strategies are also examined. Leskovac and Petrović [4] reviewed pesticide contamination concerns, focusing on organophosphates, addressing regulation gaps, and highlighting the need for

Received: 29 January 2025
Revised: 5 February 2025
Accepted: 5 February 2025
Published: 7 February 2025

Citation: Lazarević-Pašti, T.; Potkonjak, N. Further Research in Food Processing Safety and Quality from Field to Fork. *Foods* **2025**, *14*, 546. https://doi.org/10.3390/foods14040546

Copyright: © 2025 by the authors. Licensee MDPI, Basel, Switzerland. This article is an open access article distributed under the terms and conditions of the Creative Commons Attribution (CC BY) license (https://creativecommons.org/licenses/by/4.0/).

global harmonization of food safety standards. Flamminii et al. [5] studied the impact of washing, blanching, freezing, and frozen storage on pesticide residues in spinach, identifying optimal processing conditions to minimize contamination. Lastly, Jocić et al. [6] proposed a green pretreatment strategy using ionic liquid-based aqueous two-phase systems for pesticide extraction from strawberries, achieving high extraction efficiencies while maintaining sustainability.

While pesticide-related food safety remains critical, other contaminants such as toxic elements, pharmaceutical residues, and pathogens also pose significant health risks. Several studies have explored these concerns, focusing on their occurrence, impact, and potential mitigation strategies. Paucar-Quishpe et al. [7] assessed human exposure to ivermectin residues in bovine-derived foods, revealing that while contamination levels in meat were negligible, milk samples exceeded safety limits, highlighting the need for stricter regulations and monitoring. Cantoral et al. [8] investigated dietary cadmium exposure in Mexico, identifying high contamination levels in certain vegetables and cocoa products, with children exceeding the tolerable weekly intake, emphasizing the necessity of food safety policies and surveillance programs.

Besides toxic element contamination, microbial hazards and their health effects are also explored. Mafe and Büsselberg [9] reviewed the carcinogenic risks of mycotoxins in food, detailing their sources, health implications, and current detection and control strategies while also calling for technological advancements to enhance food safety. Similarly, Mafe and Büsselberg [10] examined foodborne pathogens and their metabolites, particularly their role in cancer development, highlighting detection challenges and the importance of innovative control strategies.

Finally, Savić et al. [11] developed an advanced analytical method for determining rare earth elements in coffee, demonstrating their presence in varying concentrations but concluding that they pose a negligible health risk to consumers.

Beyond contamination concerns, food processing and storage methods play a crucial role in determining food products' safety, quality, and nutritional value. Lučić et al. [12] examined how ultrasound and chemical pretreatments affect the quality of dried red pepper, revealing that ultrasonic treatment altered morphology. At the same time, potassium metabisulfite was more effective than citric acid in preserving color, antioxidant activity, and overall quality. Augustyńska-Prejsnar et al. [13] studied the impact of temperature and storage duration on post-laying hen meat, finding that lower temperatures (2 °C) better maintained microbiological and sensory quality, prolonging shelf life compared to storage at 6 °C.

Innovative extraction methods for bioactive compounds are also explored. Dimitrijević et al. [14] developed a green extraction method for polyphenols from grape byproducts using aqueous biphasic systems, achieving high extraction efficiency (~99% for resveratrol and 78% for gallic acid), demonstrating the potential for sustainable polyphenol recovery.

Pęksa et al. [15] investigated glycoalkaloid content in potato snacks made from colored potatoes, assessing the effects of organic acid soaking and thermal processing. Malic acid and the purple-fleshed variety were most effective in reducing α-chaconine levels, with all processing methods significantly lowering toxic glycoalkaloid concentrations.

In conclusion, the studies presented in this Special Issue provide substantial contributions to the field of food safety, quality, and sustainability. Each study emphasizes innovative methods for assessing food contaminants, such as pesticides, mycotoxins, and heavy metals, as well as the impact of foodborne pathogens and their metabolites. From developing efficient extraction techniques, including ultrasound-assisted extraction and ionic liquid-based systems, to evaluating potential health risks associated with food contaminants, the research highlights the importance of optimizing analytical methods and ensuring consumer safety. Furthermore, these studies emphasize the role of advanced technologies in improving food monitoring systems and enhancing the overall sustainability of food production. Future research should focus on refining these methods, particularly in sensor development, analytical precision, and risk assessment, to further enhance food safety, regulatory compliance, and environmental sustainability.

Author Contributions: Conceptualization, T.L.-P.; validation, T.L.-P. and N.P.; formal analysis, T.L.-P. and N.P.; data curation, T.L.-P. and N.P.; writing—original draft preparation, T.L.-P. and N.P.; writing—review and editing, T.L.-P. and N.P.; visualization, T.L.-P. and N.P.; supervision, T.L.-P.; project administration, T.L.-P. All authors have read and agreed to the published version of the manuscript.

Funding: The authors acknowledge the support provided by the Serbian Ministry of Science, Technological Development and Innovations (contract number: 451-03-66/2024-03/200017).

Data Availability Statement: Not applicable.

Conflicts of Interest: The authors declare no conflicts of interest.

References

1. Al-Hamry, A.; Lu, T.; Bai, J.; Adiraju, A.; Ega, T.; Pašti, I.; Kanoun, O. Layer-by-Layer Deposited Multi-Modal PDAC/rGO Composite-Based Sensors. *Foods* **2023**, *12*, 268. [CrossRef] [PubMed]
2. Tasić, T.; Milanković, V.; Batalović, K.; Breitenbach, S.; Unterweger, C.; Fürst, C.; Pašti, I.; Lazarević-Pašti, T. Application of Viscose-Based Porous Carbon Fibers in Food Processing—Malathion and Chlorpyrifos Removal. *Foods* **2023**, *12*, 2362. [CrossRef] [PubMed]
3. Milanković, V.; Tasić, T.; Pejčić, M.; Pašti, I.; Lazarević-Pašti, T. Spent Coffee Grounds as an Adsorbent for Malathion and Chlorpyrifos—Kinetics, Thermodynamics, and Eco-Neurotoxicity. *Foods* **2023**, *12*, 2397. [CrossRef] [PubMed]
4. Leskovac, A.; Petrović, S. Pesticide Use and Degradation Strategies: Food Safety, Challenges and Perspectives. *Foods* **2023**, *12*, 2709. [CrossRef]
5. Flamminii, F.; Minetti, S.; Mollica, A.; Cichelli, A.; Cerretani, L. The Effect of Washing, Blanching and Frozen Storage on Pesticide Residue in Spinach. *Foods* **2023**, *12*, 2806. [CrossRef] [PubMed]
6. Jocić, A.; Marić, S.; Tekić, D.; Mušović, J.; Milićević, J.; Živković, S.; Dimitrijević, A. Green Pretreatment Strategy Using Ionic Liquid-Based Aqueous Two-Phase Systems for Pesticide Determination in Strawberry Samples. *Foods* **2024**, *13*, 4106. [CrossRef]
7. Paucar-Quishpe, V.; Cepeda-Bastidas, D.; Rodríguez-Hidalgo, R.; Pérez-Otáñez, X.; Perez, C.; Enríquez, S.; Guzman, E.; Ulcuango, F.; Grijalva, J.; Vanwambeke, S.; et al. Evaluating the Human Risks of Consumption of Foods of Bovine Origin with Ivermectin Residues in Ecuador. *Foods* **2024**, *13*, 3470. [CrossRef]
8. Cantoral, A.; Collado-López, S.; Betanzos-Robledo, L.; Lamadrid-Figueroa, H.; García-Martínez, B.; Ríos, C.; Díaz-Ruiz, A.; Mariscal-Moreno, R.; Téllez-Rojo, M. Dietary Risk Assessment of Cadmium Exposure Through Commonly Consumed Foodstuffs in Mexico. *Foods* **2024**, *13*, 3649. [CrossRef] [PubMed]
9. Mafe, A.; Büsselberg, D. Mycotoxins in Food: Cancer Risks and Strategies for Control. *Foods* **2024**, *13*, 3502. [CrossRef] [PubMed]
10. Mafe, A.; Büsselberg, D. Impact of Metabolites from Foodborne Pathogens on Cancer. *Foods* **2024**, *13*, 3886. [CrossRef]
11. Savić, A.; Mutić, J.; Lučić, M.; Vesković, J.; Miletić, A.; Onjia, A. Ultrasound-Assisted Extraction Followed by Inductively Coupled Plasma Mass Spectrometry and Multivariate Profiling of Rare Earth Elements in Coffee. *Foods* **2025**, *14*, 275. [CrossRef]
12. Lučić, M.; Potkonjak, N.; Sredović Ignjatović, I.; Lević, S.; Dajić-Stevanović, Z.; Kolašinac, S.; Belović, M.; Torbica, A.; Zlatanović, I.; Pavlović, V.; et al. Influence of Ultrasonic and Chemical Pretreatments on Quality Attributes of Dried Pepper (*Capsicum annuum*). *Foods* **2023**, *12*, 2468. [CrossRef] [PubMed]
13. Augustyńska-Prejsnar, A.; Hanus, P.; Ormian, M.; Kačániová, M.; Sokołowicz, Z.; Topczewska, J. The Effect of Temperature and Storage Duration on the Quality and Attributes of the Breast Meat of Hens after Their Laying Periods. *Foods* **2023**, *12*, 4340. [CrossRef]
14. Dimitrijević, A.; Marić, S.; Jocić, A.; Tekić, D.; Mušović, J.; Amaral, J. Green Extraction Strategy Using Bio-Based Aqueous Biphasic Systems for Polyphenol Valorization from Grape Byproduct. *Foods* **2024**, *13*, 954. [CrossRef]
15. Pęksa, A.; Tajner-Czopek, A.; Gryszkin, A.; Miedzianka, J.; Rytel, E.; Wolny, S. Assessment of the Content of Glycoalkaloids in Potato Snacks Made from Colored Potatoes, Resulting from the Action of Organic Acids and Thermal Processing. *Foods* **2024**, *13*, 1712. [CrossRef]

Disclaimer/Publisher's Note: The statements, opinions and data contained in all publications are solely those of the individual author(s) and contributor(s) and not of MDPI and/or the editor(s). MDPI and/or the editor(s) disclaim responsibility for any injury to people or property resulting from any ideas, methods, instructions or products referred to in the content.

Article

Layer-by-Layer Deposited Multi-Modal PDAC/rGO Composite-Based Sensors

Ammar Al-Hamry [1], Tianqi Lu [1], Jing Bai [1], Anurag Adiraju [1], Tharun K. Ega [1], Igor A. Pašti [2] and Olfa Kanoun [1,*]

[1] Measurement and Sensor Technology, Department of Electrical Engineering and Information Technology, Chemnitz University of Technology, 09107 Chemnitz, Germany
[2] University of Belgrade—Faculty of Physical Chemistry, Studentski trg 12-16, 11158 Belgrade, Serbia
* Correspondence: olfa.kanoun@etit.tu-chemnitz.de

Abstract: Different environmental parameters, such as temperature and humidity, aggravate food spoilage, and different volatile organic compounds (VOCs) are released based on the extent of spoilage. In addition, a lack of efficient monitoring of the dosage of pesticides leads to crop failure. This could lead to the loss of food resources and food production with harmful contaminants and a short lifetime. For this reason, precise monitoring of different environmental parameters and contaminations during food processing and storage is a key factor for maintaining its safety and nutritional value. Thus, developing reliable, efficient, cost-effective sensor devices for these purposes is of utmost importance. This paper shows that Poly-(diallyl-dimethyl ammonium chloride)/reduced Graphene oxide (PDAC/rGO) films produced by a simple Layer-by-Layer deposition can be effectively used to monitor temperature, relative humidity, and the presence of volatile organic compounds as indicators for spoilage odors. At the same time, they show potential for electrochemical detection of organophosphate pesticide dimethoate. By monitoring the resistance/impedance changes during temperature and relative humidity variations or upon the exposure of PDAC/rGO films to methanol, good linear responses were obtained in the temperature range of 10–100 °C, 15–95% relative humidity, and 35 ppm–55 ppm of methanol. Moreover, linearity in the electrochemical detection of dimethoate is shown for the concentrations in the order of 10^2 μmol dm^{-3}. The analytical response to different external stimuli and analytes depends on the number of layers deposited, affecting sensors' sensitivity, response and recovery time, and long-term stability. The presented results could serve as a starting point for developing advanced multi-modal sensors and sensor arrays with high potential for analytical applications in food safety and quality monitoring.

Keywords: multi-modal sensor; poly(diallyl dimethyl ammonium chloride); reduced graphene oxide; layer-by-layer deposition; temperature sensor; relative humidity sensor; volatile organic compounds; electrochemical senor

1. Introduction

Efficient food processing, manufacturing, and monitoring processes are some of the essential strategies in battling the worldwide rising famine epidemic, especially during the current economic crisis that also hit the food market. Based on statistical data, 11.3% of the World's population is hit by famine, with a quarter of this number being in sub-Saharan Africa and over 500 million people in Asia [1]. In order to overcome this challenge effectively, it is necessary to implement monitoring rules for food production and processing in order to reduce the loss of valuable resources. Specifically, while many different, mutually interconnected factors, such as politics, religion, poverty, and inequality, contribute to global famine, significant amounts of food are lost in the post-harvesting and post-production period (as discussed further below), and these resources could be redirected to famine-affected regions. Another vital problem that impacts healthy food production is growing pollution, where many toxic substances such as pesticides end up in the final products, risking in that way human health.

Temperature and odor control in food processing is one of the key aspects during production, as any abnormal values of both parameters are strongly related to microbial growth and the quality of the final products [2,3]. In fact, there are pieces of evidence that in developed countries, nearly 30% of food is lost in the post-harvesting period [4]. Furthermore, as storage and processing conditions can influence food quality, it is also very important to consider monitoring methods, which could be performed by analyzing volatile organic compounds (VOCs) emitted, alcohols being the most common [5]. However, all these aspects are mutually connected as temperature and moisture affect the growth of microorganisms, which in turn cause food decomposition and emission of VOCs. Finally, monitoring pollutant residues in food, pesticides being of the major concern, is of crucial importance as they impose direct health risks [6].

In this regard, developing novel, cost-effective, and efficient sensors for monitoring environmental conditions in food processing and storage is essential, as well as the development of sensing platforms for monitoring different pollutants in food. For this reason, the research in sensing technologies is very active. For example, different solutions have been found, such as temperature sensors [7], humidity sensors [8], VOCs sensors [9–11], metal ions [12,13], as well as for pesticides and other contaminants [14,15]. In all these applications, sensitive materials based on conductive polymers and carbon materials play very important roles as they are cheap, affordable, and environmentally benign. Nevertheless, for any of the desired applications, a sensor has to fulfill several requirements, including good linearity, high sensitivity, low hysteresis, rapid response and recovery time, and selectivity.

Composites of carbon materials and conductive polymers have found their place in sensor technologies. Additional impetus for such composite-based sensors was the discovery of novel low-dimensional carbon nanostructures, primarily graphene, known for many exceptional properties and different applications [16–19]. Another important aspect is the development of novel sensor production strategies so that materials with controllable properties, supramolecular structure, and miniaturization can be achieved, such as different printing and deposition techniques, including Layer-by-Layer (LbL) deposition techniques [20,21]. There are many examples of various LbL-based sensors in the literature. The process is based on the self-assembly of oppositely charged layers, like carboxylated single-walled carbon nanotube (SWNT) self-assembly with a polycation, poly(diallyl dimethyl ammonium chloride) (PDDA) used for pH sensing [22]. As polycations, different conductive polymers can be used, like polyaniline (PANI) [23,24], poly(dimethyldiallyl ammonium chloride) (PDDAC) [25], polyethylene glycol (PEG) [26], and as carbon material graphene oxide (GO) [27], CVD graphene [28], different nanotubular carbon forms [22,26], and others. These sensors have been applied for temperature sensing [29], humidity measurements [25,30], VOCs sensing [26,31], and also as electrochemical sensors [32]. In addition, these carbon-based nanomaterials are commonly demonstrated as multi-modal sensitive materials [33].

Different polymer/carbon combinations can be utilized for measuring different parameters, but the question is whether it is possible to reach composite formulations with more than one sensor application. This issue is highly important for rationalizing and economizing sensor production while not compromising on the sensors' performances and applicability. Based on the above discussion, GO is a good candidate for different sensor applications with facile methods of preparation and coating. With the LbL technique, tuned film thickness, sensitivity, stability, and adhesion can be obtained [31,34]. The main aim of this research is to investigate and develop sensors as a proof of concept to monitor different environmental parameters, gases, and pesticides, which can be used for efficient food monitoring during processing, storage, and other applications. Here we present LbL-manufactured PDAC/rGO composite-based sensors, which can be used as temperature, relative humidity, and VOC sensors, while they also show potential for use in electrochemical sensors of pesticides. Sensors' properties are affected by the number of deposited PDAC/rGO layers and temperature treatment, and, depending on the selected application, the sensor architecture can additionally be optimized to obtain the best possible sensor performance. The sensor properties of PDAC/rGO layers were investigated for their response to; temperature in the range of 10–100 °C, relative humidity (15–95%), methanol VOC (25–55 ppm), and dimethoate pesticide (100–700 $\mu M/dm^3$). Optical properties were investigated by Raman spectroscopy, atomic force spectroscopy, and scanning electron microscopy. PDAC/rGO sensors were demonstrated to differentiate the changes in food and beverage.

2. Materials and Methods

2.1. Materials Preparation and Sensor Fabrication

GO oxide solution (0.4 wt.%) was purchased from Graphenea (San Sebastián, Spain) and used without further purification. For the LbL preparation of the sensor electrodes, it was diluted using high-purity deionized water to 0.1 wt.%. The PDAC used in this work was purchased from Sigma-Aldrich (Taufkirchen, Germany). It is a light-yellow viscous liquid with a solution concentration of 20 wt.%. For the LbL procedure, it was diluted to the concentration of 1 wt.% and used as such for all the experiments described from now on.

Sensor electrodes were prepared using the LbL procedure, employing screen-printed silver interdigitated electrode (IDE) on Kapton HN substrate, with dimensions of 15 × 4 mm (see Figure S1a). Before the LbL process, the IDEs were cleaned with isopropanol and dried with nitrogen to remove dust contamination on the electrode film. To coat only the active area on the IDE electrodes, the undesired area around the silver IDE section, and on the back of the film completely using Oramask Film 810, purchased from Orafol GmbH, Oranienburg, Germany. Next, the silver IDE was fixed on the NEMESYS pump precision device (Cetoni GmBH, Korbußen, Germany), and the electrode was immersed in the PDAC solution for 5 min. After this step, IDE was rinsed in deionized water for 30 s, then taken out and slowly dried with nitrogen gas. In the next step, IDE was immersed in the GO solution for 5 min, after which the rinsing and drying step was repeated identically as previously described [35]. This way, a single bi-layer is formed, and the electrode is marked as $(PDAC/GO)_1$. The above process is repeated N times, and such electrodes are designated as $(PDAC/GO)_N$. Thus, N gives the number of deposited PDAC/GO bi-layers. The LbL process is schematically presented in Figure S1b.

After the LbL procedure, the $(PDAC/GO)_N$ electrodes were thermally reduced on a hot plate to obtain PDAC/rGO-NL sensor electrodes. The temperature was chosen by thermal reduction of the $(PDAC/GO)_4$ electrode. The resistance of the obtained PDAC/rGO-4L sensor electrode reached a plateau after 200 °C, so temperature of 200 °C was subsequently used to reduce all the sensors reported here.

2.2. Sensor Physical and Chemical Characterization

Sensor electrodes were characterized using Scanning Electron Microscopy with Energy-Dispersive X-ray Spectroscopy (SEM-EDX), Raman spectroscopy, and Atomic Force Microscopy (AFM).

Raman spectra, excited with a diode-pumped solid-state high brightness laser (excitation wavelength 532 nm), were collected on a DXR Raman microscope (Thermo Scientific, Waltham, MA, USA) equipped with an Olympus optical microscope and a CCD detector. The laser beam was focused on the sample using an objective magnification of 10×. The scattered light was analyzed by the spectrograph with a 900 lines mm^{-1} grating. Laser power on the sample was kept at 1 mW to prevent thermal degradation of the samples. The AFM analysis was performed using Agilent 5600LS (Keysight, Santa Rosa, CA, USA) in tapping mode. SEM-EDX characterization was performed using a Phenom ProX (Phenom, Eindhoven, the Netherlands).

2.3. Sensor Performance

2.3.1. Temperature Measurement

For the static temperature measurements, a fabricated sensor was placed on the heating plate. To prevent uneven or incomplete sensor heating due to the upward bending of the sensor material, a Kapton film is placed over the sensor, followed by a metal plate to ensure that the sensor is in full contact with the heating plate. Sensor resistance was measured using Keithley Sourcemeter 2602.

A preheated silicon oil tank (thermostat C12 CS, Lauda, Lauda-Königshofen, Germany) was used to determine sensor response time accurately. The sensors were fixed to a metal bracket controlled by an air pump and connected to the Keysight Sourcemeter. When the oil temperature is stable at 80 °C, the air pump pneumatic plunge arm inserts the sensor into the oil for an accurate response time measurement. The sensors were removed from the oil two minutes later, and the recovery time was measured.

2.3.2. Humidity Measurement

The relative humidity (RH) measurement system was composed of 5 parts: a computer with a LabVIEW program, measurement device, gas control system, Arduino UNO board with SHT85 RH-Temperature sensor, and test chamber (see Figure S2). The reference sensor was used to measure and control the actual humidity in the chamber. In this experiment, Agilent LCR Meter 4284A was used as the measuring device. In addition, two FLOW BUS (Bronkhorst, Veenendaal, the Netherlands) were used for the gas control system. The block and connection diagram is shown in Figure S2.

2.3.3. VOCs Measurement

Owlstone V-OVG (Owlstone, Connecticut, USA) was used as a gas-generating system. The VOCs gas (methanol) was mixed with nitrogen and introduced into the chamber containing the sensor electrode. The sensor response was measured using Keysight DAQ973A and processed using LabVIEW 2021 software.

2.3.4. Electrochemical Measurement

Electrochemical measurements were carried out in an all-glass one-compartment electrochemical cell. As working electrodes, modified Ag electrodes were used, while Saturated Calomel Electrode (SCE) and a wide Pt foil were used as a reference and counter electrode, respectively. As a supporting electrolyte 1 mol/dm^3 KNO$_3$ was used. Experiments were performed using Gamry Interface 1010E Potentiostat/Galvanostat/ZRA (Gamry instrument, Warminster, PA, USA). Measurements were performed in a quiescent solution. Cyclic voltammetry was investigated in the potential range -0.30 to $+0.40$ V vs. SCE. Working electrodes were produced by LbL procedure as described, but on continuous screen-printed Ag strips so that the performance of (PDAC/rGO-NL)@Ag is measured. In this case, Ag served as a substrate and a current collector. Analyte, organophosphate pesticide dimethoate, was injected stepwise in the electrolyte, and the voltammetric response of the sensor electrodes was detected.

3. Results and Discussion

3.1. Sensor Physical Properties

SEM-EDX analysis was performed in order to confirm the deposition of GO and PDAC on the silver IDEs. As we reported previously [35], high transparency and low concentration of the GO sheets and PDAC prevented direct observation of deposited layers on the silver substrate. However, the EDX analysis confirmed the presence of carbon and oxygen (in the ratio 2:1), nitrogen which is present in PDAC, and the underlying Ag. This clearly confirms an effective functionalization of the silver electrode by PDAC/GO layers. Nevertheless, more direct pieces of evidence of the presence of GO and PDAC in the as-deposited films come from Raman spectroscopy. Based on the Raman spectra taken for the composites reduced at 100 °C it can be seen that some inhomogeneity is present for one and two layers, which disappear already for 4 L. Raman spectra clearly show the D and G bands of graphene, while for the samples with one and two layers, the characteristic bands of PDAC are also visible (see Figure S3).

Following the reduction of produced (PDAC/GO)$_N$ composites, the sensors' resistance decreased as the reduction temperature was raised to 200 °C, after which a slight increase was seen (Figure 1a). The initial resistance of (PDAC/GO)$_4$ is 286.9 kΩ. The initial resistance decreases rapidly when the reduction temperature is higher than 180 °C. As the reduction temperature rises to 220 °C, the initial resistance was found to be 3.8 kΩ for the PDAC/rGO-4L sensor. When the reduction temperature continued to rise, the initial resistance increased slightly. For this reason, sensors described starting in Section 3.2 have been produced with a reduction temperature of 200 °C. The primary effect of reduction temperature on film resistance is sought in removing oxygen-containing functional groups from GO. The oxygen-containing functional groups break the ultra-long distance conjugated large π-bonds on the graphene surface, reducing the electron migration rate. The removal of oxygen-containing functional groups, although generating defects, caused the electron migration rate to rebound [10,36]. The formation of defects during the reduction was confirmed here using Raman spectroscopy. For the PDAC/rGO-4L sensor, it is clear that the intensity of the D band, associated with the presence of defects [37], increases when the composite is reduced at 200 °C (see Figures 1b and S3).

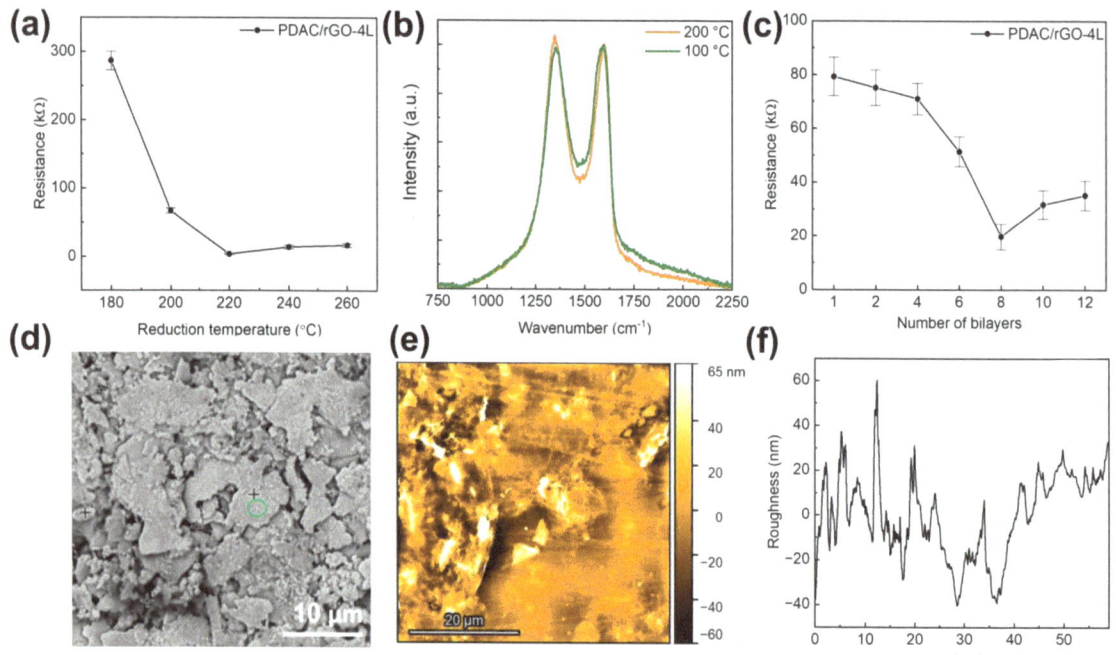

Figure 1. (**a**) Initial resistance of the proposed PDAC/rGO-4L sensors as a function of reduction temperature; (**b**) Raman spectra of PDAC/rGO-4L composite electrodes reduced at 100 °C and 200 °C (spectra were normalized so that the intensity of the G band is set to 1); (**c**) Initial resistance of the proposed PDAC/rGO sensors as a function of the number of deposited bi-layers; (**d**) Low-magnification SEM image showing overall morphology of the sensor; (**e**) AFM image showing overall topology of PDAC/rGO-2L sensor; (**f**) Roughness profile of the PDAC/rGO-2L sensor from AFM measurements.

When the number of film layers is less than or equal to four layers, although the initial resistance of the sensor decreases with the increase of the number of layers, the decrease is not large. The initial resistance of the sensor decreases from 79.3 kΩ at one layer to 71 kΩ at four layers, a decrease of approximately 10%. When the number of film layers exceeds four, the initial resistance begins to drop significantly. When the number of layers is eight, the initial resistance of the sensor reaches the lowest value, 19.8 kΩ. Starting at that point, as the number of layers of the sensor continues to rise, the initial resistance of the sensor begins to rise again. The increase in the film conductivity with the rising number of layers is a consequence of the increasing number of channels across the layers, causing an increase in electron mobility. Sarker and Hong [38] reported a similar result: the sheet resistances of the multilayer films decreased exponentially as the number of bi-layers increased, reaching a minimum resistance for 15 bilayers.

The SEM analysis was performed on the Ag part of the interdigitated structure covered by PDAC/rGO layers to ensure good conductivity of the analysis spot (see Figure 1d). It was only possible to observe the overall morphology with Ag flakes, but not PDAC/rGO composite, while the EDX analysis indicated the presence of the bi-layers. This suggests very thin deposits are formed, which was confirmed by the AFM analysis (see Figures 1e and 2f). It was found that the thickness of the composite in the case of the PDAC/rGO-2L sensor is very low. However, it can be assumed that as the number of layers with PDAC/rGO gradually increases, the film formation becomes more and more effective. On the other hand, the roughness of the PDAC/rGO-12L sensor composite film is still more than twice that of PDAC/rGO-2L. The RMS roughness of PDAC/rGO-2L was found to be 30.87 nm, and that of PDAC/rGO-12L was 64.27 nm.

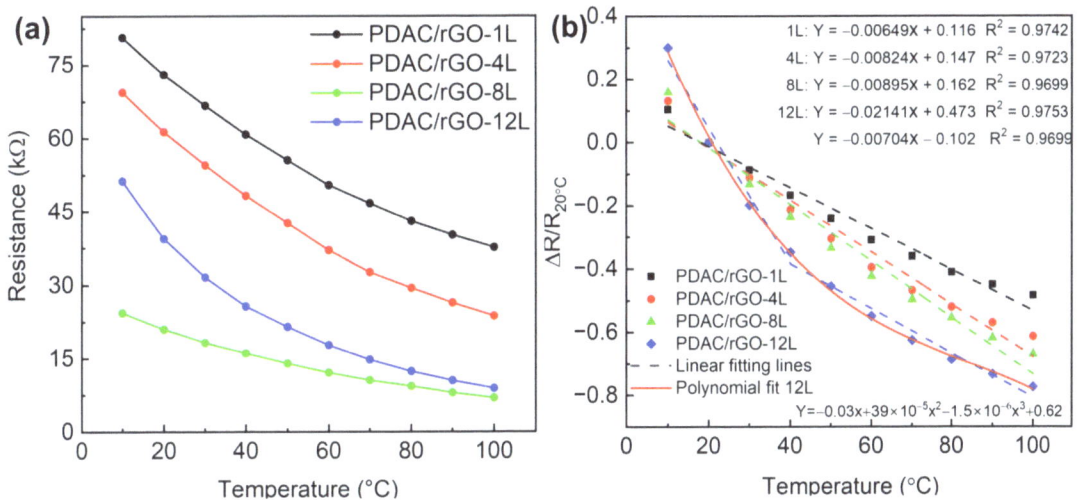

Figure 2. Static thermal properties of PDAC/rGO sensors: (**a**) Resistance of PDAC/rGO sensors and (**b**) sensitivity of PDAC/rGO.

3.2. PDAC/rGO-Composites for Temperature Measurements

First, the response and quantitative relationship between the number of sensor layers and temperature response was established. Figure 2a shows the response curves of 1-layer, 4-layer, 8-layer, and 12-layer sensors to temperature changes in the range of 10 °C–100 °C. The resistance value at room temperature, i.e., 20 °C, was chosen as the initial resistance value ($R_{20\,°C}$) so that the response is calculated as $\Delta R/R_{20\,°C}$. The calibration curves of the PDAC/rGO sensors are shown in Figure 2b. The sensitivity of the PDAC/rGO-1L, -4L, and -8L were 48.43%, 61.26%, and 66.80% at 100 °C, respectively. The sensitivity of the PDAC/rGO-12L sensor increased from 19.83% at 30 °C to 77.31% at 100 °C. This is because, as the number of sensor layers increases, the rGO surface can ionize more electrons from more oxygen-containing functional groups, leading to an increase in sensor sensitivity. The sensitivity curve for the different numbers of bi-layers is provided in Figure S5a. The temperature coefficient of resistance (TCR = $\Delta R/R_0\ 1/\Delta T$), as extracted from the linear fitting, are -0.65, -0.85, and $-0.89\%/°C$ for PDAC/rGO-1L, -4L, and -8L, respectively. For PDAC/rGO-12L, it is $-2.1\%/°C$ in the range (10–40 °C) and $-0.7\%/°C$ in the range (40–100 °C).

In the next step, the response and recovery time of the sensors are addressed. Figure 3a shows the dynamic thermal properties of PDAC/rGO sensors in the oil tank preheated to 80 °C, and Figure S4b gives the response time curve of PDAC/rGO sensors. The PDAC/rGO sensor response time increases as the number of layers increases. When the number of sensor layers is less than 6, the temperature response time is less than 1 s (0.78 s and 0.79 s for the one-layer and eight bi-layer-sensor, respectively). When the number of sensor layers is more than six, the PDAC/rGO sensor's response time is 2 s. To determine the recovery times, the sensors were heated from room temperature (20 °C) onto a hot plate to 100 °C. After 900 s of complete heating (when sensor response remained stable), the sensors were quickly removed from the hot plate and placed in a room-temperature environment, providing curves for determining the recovery time (see Figure 3b). The recovery time of PDAC/rGO sensors is about 24.87 s–35.80 s (see Figure S4c).

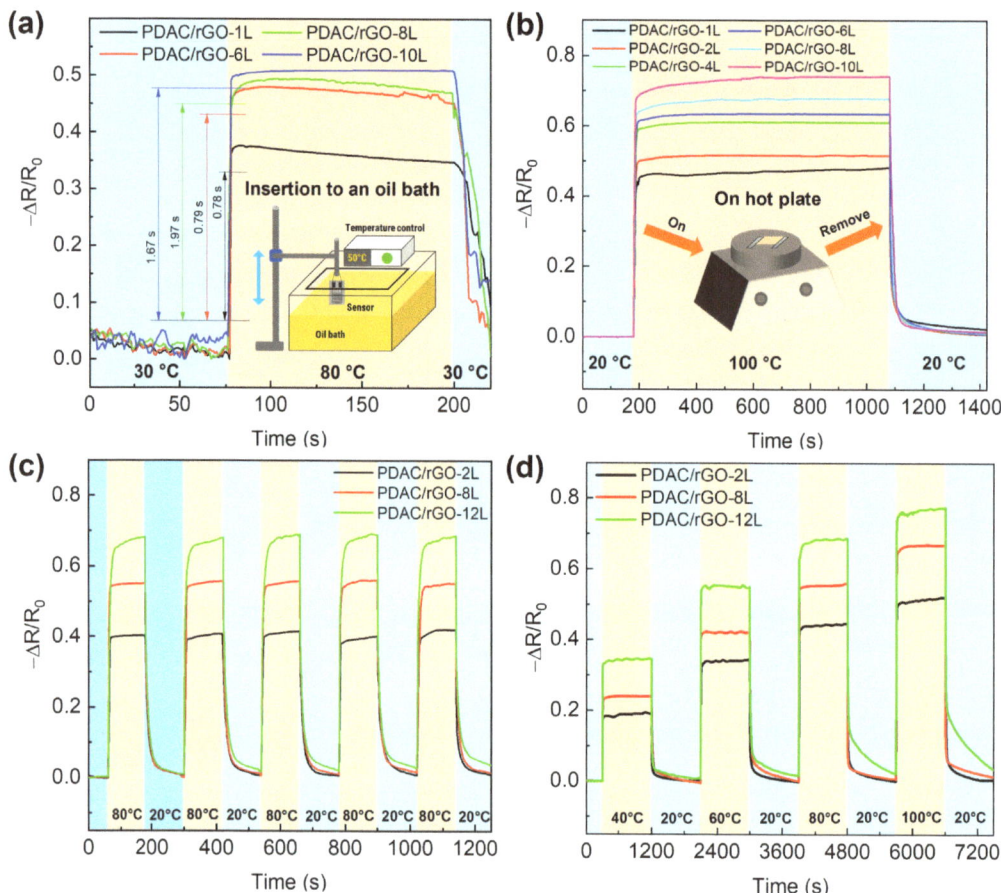

Figure 3. Dynamic thermal properties of PDAC/rGO sensors: (**a**) Response curves of PDAC/rGO sensors plunged abruptly in hot oil bath, and (**b**) recovery curves of PDAC/rGO sensors for different numbers of layers removed suddenly from hot plate; (**c**) repeatability and (**d**) step curves of thermal properties of PDAC/rGO sensors upon exposure to different heating–cooling programs.

In order to explore the response of different sensors to rapid temperature changes, the repeatability curve is measured. The measurements were carried out on a hot plate. The sensors were quickly placed on top of the hot plate at 80 °C and then at room temperature of 20 °C for 120 s only in each state for five consecutive sets of rapid cycling experiments. This protocol resulted in the repeatability curve shown in Figure 3c. PDAC/rGO-12L sensor was essentially equilibrated, and the 2L and 8L sensors were fully equilibrated. The recoverability of the 2L and 8L sensors also remains largely unchanged, but that of the 12L sensor decreases as the number of cycles increases.

In the second set of experiments, successive stages of heating at different temperatures and recovery of the sensor are tested. In these experiments, 40 °C, 60 °C, 80 °C, and 100 °C are chosen as the test temperatures. First, the sensor was placed on a hot plate from room temperature (20 °C) to the lowest test temperature of 40 °C and subjected to temperature response measurements for 900 s. After 900 s, the sensor is moved to room temperature for a recovery test for 900 s. The step curve is then repeated by placing the sensor on the hot plate at 60 °C, 80 °C, and 100 °C to obtain the step curve of the sensor, as shown in Figure 3d.

The calibration curves of PDAC/rGO sensors to temperature changes in these two sets of experiments were consistent. The PDAC/rGO sensors with the same number of layers have practically the same response time to temperature and basically do not change with the temperature. Further, the recovery time of 2L and 8L PDAC/rGO sensors to temperature remains almost unchanged, but the recoverability is slightly reduced. The PDAC/rGO-12L sensor has a significant downward trend in temperature measurement recovery time and recoverability due to the slow cooling process and heat capacity of the substrate.

The long-term stability measurements were performed by assessing sensor sensitivity at 40 °C, 60 °C, 80 °C, and 100 °C. The measurement interval was five days, and seven measurements were taken in total over one month under the same ambient conditions. Over one month period, the long-term stability of the PDAC/rGO-8L sensor decreased slightly (see Figure S4d). At the start date, the sensitivities of the PDAC/rGO sensor at 40 °C and 100 °C are 23.98% and 66.76%, respectively. By day 30 was reached, the sensitivity dropped to 23.4% at 40 °C and 64.44% at 100 °C. At 40 °C, the sensitivities of the PDAC/rGO sensor decrease by 0.27 and 0.58 at 15 and 30 days, respectively. When the test temperature reaches 100 °C, the sensitivities decrease by 1.12% on day 15 and 2.32% on day 30. Based on the obtained results, it can be safely concluded that the stability of the PDAC/rGO sensor is excellent.

3.3. PDAC/rGO-Composites for Relative Humidity Measurements

For the relative humidity measurements, the range of 15–95 RH% is investigated (see Figure 4a). The impedance measured in pure nitrogen gas was used to obtain the initial impedance value (Z_0) for the construction of calibration curves. The impedance was measured for relative humidity in the chamber set to 15%, 35%, 55%, 75%, 85%, and 95%. The impedance values of the sensors gradually increased as the humidity rose. PDAC/rGO sensors showed good humidity sensor characteristics due to the properties of rGO, where water molecules penetrate the interlayer between rGO flakes, causing the increase of tunneling effect and, thus, a decrease in the conductivity. Even at high RH%, there was no formation of water film due to hydrophobicity of rGO [39]. Expansion of the polymer-rGO composite film may as well play a role caused by water adsorption. The effect is the opposite as the relative humidity decreases. Figure 4b shows the humidity calibration curve for PDAC/rGO-NL sensors. The sensor's sensitivity increased monotonically with the number of layers (see Figure S5a, Supplementary Information) and reached a maximum of 39.56% PDAC/rGO-12L sensor at 95% RH.

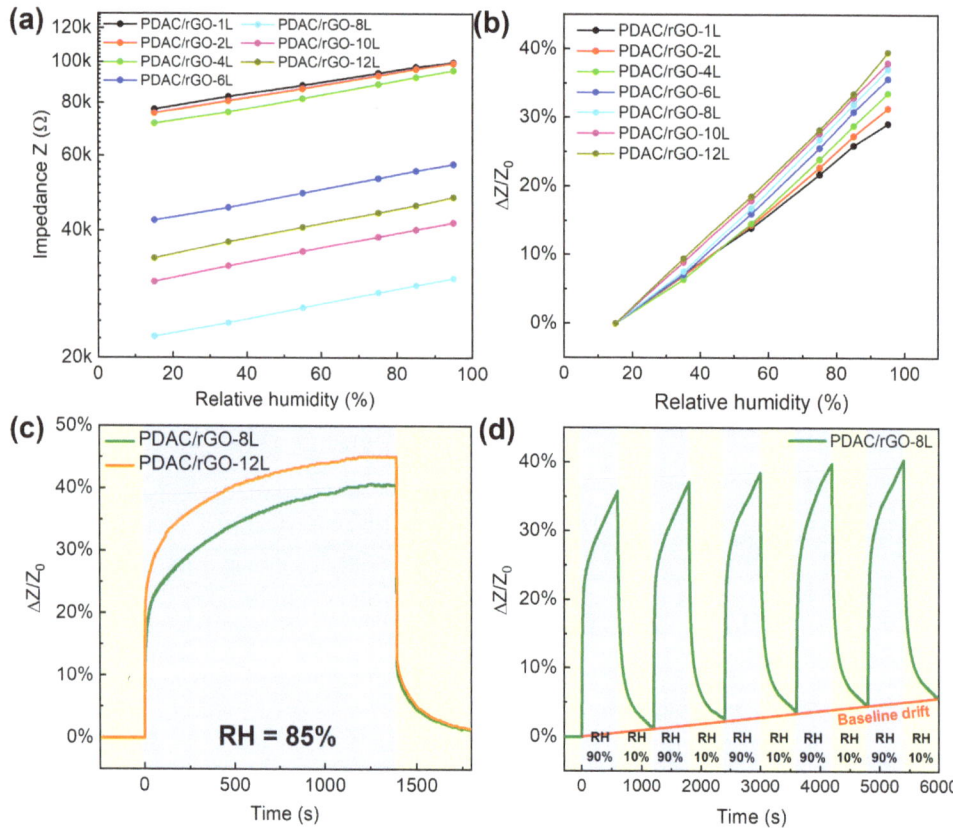

Figure 4. Static and dynamic characteristics of PDAC/rGO humidity sensors: (**a**) Impedance of PDAC/rGO sensors; (**b**) sensitivity of PDAC/rGO sensors; (**c**) response curves of PDAC/rGO-8L and PDAC/rGO-12L sensors; (**d**) repeatability of the humidity characteristic of PDAC/rGO-8L sensor.

Once the calibration curves have been established, the humidity sensor is tested more deeply. Figure 4c presents the response curve of PDAC/rGO-8L and PDAC/rGO-12L sensors. In terms of recovery, PDAC/rGO sensors showed recovery times of around 120 s.

PDAC/rGO-8L was chosen to investigate the repeatability of the sensor. At the beginning of the measurement, the sensor was placed in a chamber at 10% RH, and when stabilized, the humidity of the environment chamber was changed to 90% RH. After 600 s, the relative humidity is changed back to 10%, and this process was repeated five times (see Figure 4d). An obvious drift in the sensor response was seen due to the incomplete removal of adsorbed water during the cycling.

Finally, the long-term stability curve of the humidity response of the sensors is assessed using the PDAC/rGO-8L sensor. The measurement intervals and experiment duration were the same as in the case of temperature measurements, and sensor sensitivity for 15%, 35%, 55%, 75%, and 95% relative humidity was checked over one month (see Figure S5b). At a relative humidity of 15%, the initial sensitivity of the sensor was 5.882%, which decreased by 0.47% on day 15 and by 0.912% on day 30. At a relative humidity of 95%, the initial sensitivity was 46.03% and dropped by 1.009% and 1.987% for day 15 and day 30, respectively. This undoubtedly indicates the good stability of the sensors at different humidity levels.

3.4. PDAC/rGO-Composites for VOCs Measurements

The VOC response of the PDAC/rGO sensor is tested using the VOC generator as the VOC gas generator and methanol as the gas source, ranging from 25 ppm to 55 ppm. The sensor resistance value when the methanol gas concentration is 25 ppm is used as the initial resistance of the sensor (see Figure 5a). The response curve of the PDAC/rGO sensor to the methanol gas concentration and the calibration curves are shown in Figure 5b. Using the quadratic functions to link sensor response to the methanol concentration, it can be seen that sensitivity increases with methanol concentration.

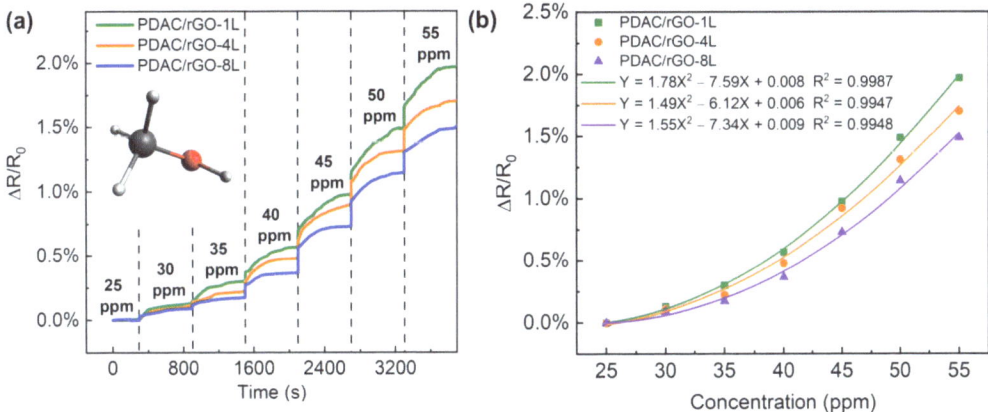

Figure 5. Response curve of the PDAC/rGO sensor to methanol gas (**a**) step response with time and (**b**) calibration curve.

The response of the PDAC/rGO sensor is positively correlated to the concentration of methanol. In contrast to the case of the relative humidity measurements, the highest sensitivity was observed for PDAC/rGO-1L sensor (1.97%) and the lowest for PDAC/rGO-8L, so the sensitivity dropped as the number of bi-layers increased. When the methanol concentration was below 40 ppm, the sensitivity changes for all three layers of PDAC/rGO sensors gradually increased and were very unstable, and when the methanol concentration was higher than 40 ppm, the change in sensitivity for all three sensors remained rather stable. After the methanol concentration exceeded 40 ppm, the sensitivity change of the three sensors was basically linear with the concentration change. For each 5 ppm increase in the methanol concentration, the sensitivity of the PDAC/rGO-1L, PDAC/rGO-4L, and PDAC/rGO-8L increases by approximately 0.5%, 0.4%, and 0.35%, respectively. Since a larger range of gas concentrations can be allowed by the diffusion tube in the VOC generator, the sample flow can also be expanded to a larger value so that the gas in the chamber can also quickly reach the measured concentration when the higher gas concentrations are required. The response times of the sensors, therefore, remain essentially the same.

3.5. PDAC/rGO-Composites as Electrochemical Sensors for Organophosphate Pesticides

In addition to using LbL-produced PDAC/rGO layers for temperature monitoring, relative humidity, and VOC measurements, here we also briefly communicate that the same fabrication approach can be used for the electrochemical detection of pesticides. While the method is not fully optimized, the proof-of-concept is clear and provides added value to the multi-modal PDAC/rGO-based sensors. As shown previously [40], the electrochemistry of dimethoate at (PDAC/rGO-NL)@Ag is interesting, showing clear anodic and cathodic peak corresponding to dimethoate oxidation and reduction of the oxidation products (see Figure 6a). Depending on the number of the deposited bi-layers, the response is different, and for 1, 2, and 4 bi-layers, we found that the highest response is for the (PDAC/rGO-1L)@Ag electrode (see Figure 6b). When cyclic voltammograms are recorded, linearity in the response, taken as anodic or cathodic peak current versus concentration of dimethoate, was confirmed (see Figure 6c). The highest sensitivity and the best linearity were observed when the difference between anodic and cathodic peak current was plotted as the function of dimethoate concentrations. While the sensor is not fully optimized for the electrochemical detection of dimethoate, and the linearity range is, to this point, confirmed for relatively high dimethoate concentrations (order of 10^2 µmol dm^{-3}), it is clear that PDAC/rGO composite layers can also be used for electrochemical applications.

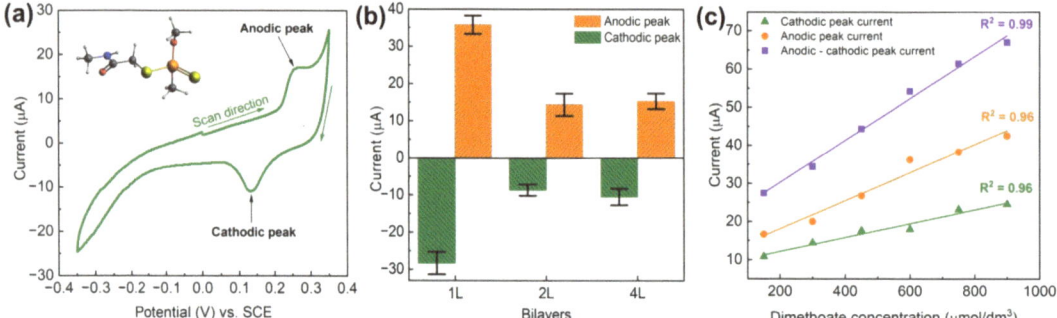

Figure 6. (**a**) Cyclic voltammetry of dimethoate at (PDAC/rGO-1L)@Ag, dimethoate concentration 100 µmol dm^{-3}, inset shows the molecular structure of dimethoate; (**b**) voltammetric response of (PDAC/rGO-NL)@Ag electrodes for dimethoate as a function of the number of layers, dimethoate concentration of 200 µmol dm^{-3}; (**c**) response of the (PDAC/rGO-1L)@Ag electrode in as the function of dimethoate concentration.

3.6. Critical Evaluation of Sensors' Properties

Here we have demonstrated the multi-modal nature of the LbL-produced PDAC/rGO composite-based sensors for temperature monitoring, relative humidity measurements, VOCs measurements, and electrochemical detection of dimethoate. The behavior of produced sensors varies depending on the different operational conditions, while the number of deposited layers also has a noticeable impact on the materials' performance. Considering temperature measurements, the resistance decreases with the increase in temperature and vice versa. Although the increase in temperature intensifies the irregular motion of molecules and causes the mobility of free electrons to drop slightly, the number of free electrons increases faster with the increase of temperature, so the resistance of material at high-temperature decreases. It was proved that the second effect becomes dominant when the temperature exceeds 120 °C. In combination with rGO, PDAC does not play a dominant role in response to temperature changes and only affects the initial conductivity, as reported in [41]. The temperature response dependence of the proposed sensor is not only determined by the nature of the insulating polymer but also by the intrinsic properties of the reduced graphene oxide-based component. Thermal treatment at 200 °C used here caused the reduction of a large number of oxygen-containing functional groups in parental GO. Thus, produced rGO exhibits a p-type semiconductor behavior (having a negative temperature coefficient) whose resistance decreases with increasing temperature, as reported in [42,43]. The mechanism can be explained as Arrhenius-like temperature dependence of resistance, indicating a band gap dominating transport behavior [44].

Opposite to the case of the increasing temperature, there is a positive response to the increasing humidity (increasing impedance). The resistance of the proposed sensor increases with humidity as the resistive response of the insulating polymer is related to the degree of moisture absorption of the material. Cavallo et al. [45] demonstrated that at an initial environment of high relative humidity (65–90%), the polymer swells due to the continuous absorption of water. This process increases the distance between its molecular chains, hindering the charge-hopping process and reducing the electrical conductivity.

Moreover, the increase in the number of deposited bi-layers increases the sensitivity of the relative humidity measurements (see Figure S5a). Namely, as the number of sensor layers increases, the PDAC/rGO film absorbs more moisture and expands, therefore increasing the response. For the PDAC/rGO sensor, the more layers there are, the more active cations can bind to water and, therefore, the higher the sensitivity [46].

In the cases of both temperature and relative humidity measurements, the sensitivity decayed slightly over one month of monitoring. In the case of humidity measurements, the sensitivity decreases practically in a linear fashion, and the rate is essentially constant over time. The performance of the sensors decayed faster for higher relative humidity. The PDAC/rGO-8L sensor has shown the largest sensitivity drop of 1.987% (absolute value) at 95% relative humidity in a long-term stability test, so the humidity response stability can be considered excellent.

In the case of the VOCs (methanol) detection, a positive sensor response is seen, i.e., increasing resistance with methanol concentration. The effective response of PDAC/GO films to methanol is a combination of PDAC and rGO contributions. The mechanism of the response of rGO to VOC has been explained by filling the defects on the rGO surface, resulting in interrupted long-range charge transport and increased resistance [10]. We note that studied sensors responded well to rising methanol concentration, but it was not the case with acetone. Methanol is more likely to swell the PDAC/rGO sensor surface and expand the distance between adjacent rGO domains, thereby increasing sensitivity. The effectiveness of the response of PDAC to methanol and the low response to acetone was previously confirmed by Al-Hamry et al. [47], who tested the PDAC/rGO-8L sensor at 2000 ppm with a sensitivity of about 15%. It is rather interesting to note that the sensor with only one bi-layer is the most sensitive in the case of methanol detection. This is likely because methanol cannot penetrate through thicker LbL films that remain partially unaffected by its presence and thus buffer the sensor response.

Finally, considering the electrochemical detection of dimethoate, the sensor behavior is rather promising, but it cannot compete yet with some state-of-the-art sensors based on aptamers or molecularly imprinted polymers [15,48,49]. However, we note that the first logical step towards improving the linear range and reducing the limit of detection is the application of more advanced electrochemical techniques, like square wave voltammetry or differential pulse voltammetry. On the other hand, it is also important to observe that the most efficient electrochemical sensor is the one with one deposited bi-layer. This might indicate that the electrochemical reaction occurs at the interface between Ag and the deposited bi-layers. For thicker layers, it is difficult for dimethoate, which is a much larger molecule compared to methanol and water (see Figure 6a, inset), to reach the interface and undergo electrochemical transformation. Another important point is to note that the electrodes for electrochemical testing cannot be reused. This result suggests that harsh electrochemical conditions cause irreversible changes in the electrode structure, while there is also a possibility that the reaction products remain on the electrode and block active sites for electrochemical reactions of dimethoate.

To put the presented results in a broader context, we compare the performance of the presented PDAC/rGO-NL sensors to those previously reported in the literature (see Table 1). Taking the multi-modal aspect of our proposed sensors into account, they stand hand to hand with sensors specifically tailored for different applications.

Table 1. The performance comparison of previously reported LbL-manufactured sensors and those reported in this work (RH—relative humidity). For this comparison, the sensors with carbon-based nanomaterials were considered.

Polymer	Nanomaterial	Target	Linear Range	Sensitivity	Response Time	Reference
PEDOT:PSS	Graphene	Temperature	33–45 °C	0.06 %/°C	20 s	[50]
Polyaniline	Graphene		25–80 °C	1.2 %/°C	-	[51]
PEDOT:PSS	CNT		30–80 °C	0.64 %/°C	4.8 s	[52]
PDAC	rGO		10–100 °C	0.7–2.1 %/°C	0.78 s	This work *, **
S-PANI	-	Humidity	50–90%	60% (90% RH)	15–27 s	[23]
PDDA	GO		11–97% RH	8.69–37.43%	108–147 s	[30]
PANI	GO		11–97% RH	20 Hz/% RH	5–13 s	[24]
PDDAC	GO		11–97% RH	25.4 Hz/% RH	1–7 s	[25]
PDAC	rGO		15–95% RH	46% (100% RH)	~10 s	This work *
PEG	MWCNT	VOCs	1–60 μM	-	-	[32]
PEG	MWCNT		10–1000 ppm	0.06 %/ppm	110 s	[26]
PDAC	rGO		35–55 ppm	0.1 %/ppm	<10 s	This work **

* 12 bi-layers; ** 1 bi-layer.

Based on the comparison with the literature data and the results provided in Section 3, there is certainly some space for improving the proposed sensors. However, a general evaluation of the performance reached to this point is presented in Figure 7. The temperature measurement performance of the presented sensors is appreciable, but in some other applications, further improvements are essential. For example, there is a significant drift in the humidity response characteristic, while the reuse for electrochemical applications and the corresponding sensitivity/response has to be significantly improved.

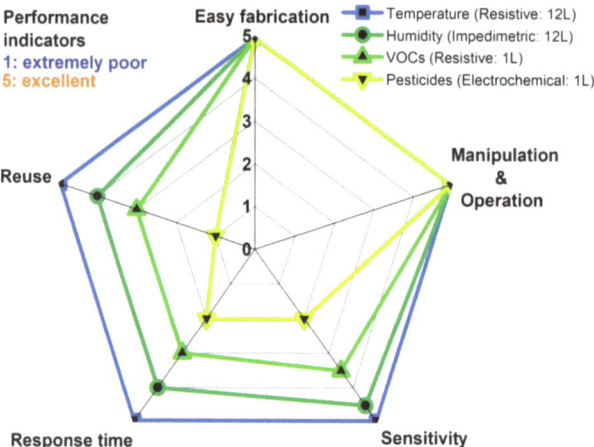

Figure 7. Spider diagram assessing the overall performance of PDAC/rGO-NL sensors for different applications presented in this work.

Finally, to demonstrate the direct applicability of PDAC/rGO sensors in food monitoring, we show the overall response of the PDAC/rGO to VOCs from two beverage samples (wine and coffee) and two meat samples (grounded beef and pork) in Figure 8. The experiment was performed as presented in Figure 8a. The sample was placed in a glass container for 15 min to fill the headspace with volatile gas. The measurement chamber with the proposed sensor was first filled with a clean and dry airflow, and then the headspace of the sample container was directed through a switching valve. Both airflows were controlled by the flow bus controller. The information on the real tested samples is shown in Table S1.

The sensor gave specific responses to each of the samples (see Figure 8b), which is not due to humidity, as the response to ground beef is the highest. Moreover, the response and recovery time are different for each sample, suggesting that the sensor is effectively responding to total VOCs from a specific sample. This study shows the ability of the sensor to be used in different cases for food quality monitoring which can be extended to give complex information about the food state and quality. Nevertheless, the calibration of sensors in different controlled environments needs to be performed and integrated with machine learning techniques so as to obtain an accurate and reliable fingerprint of food and beverages.

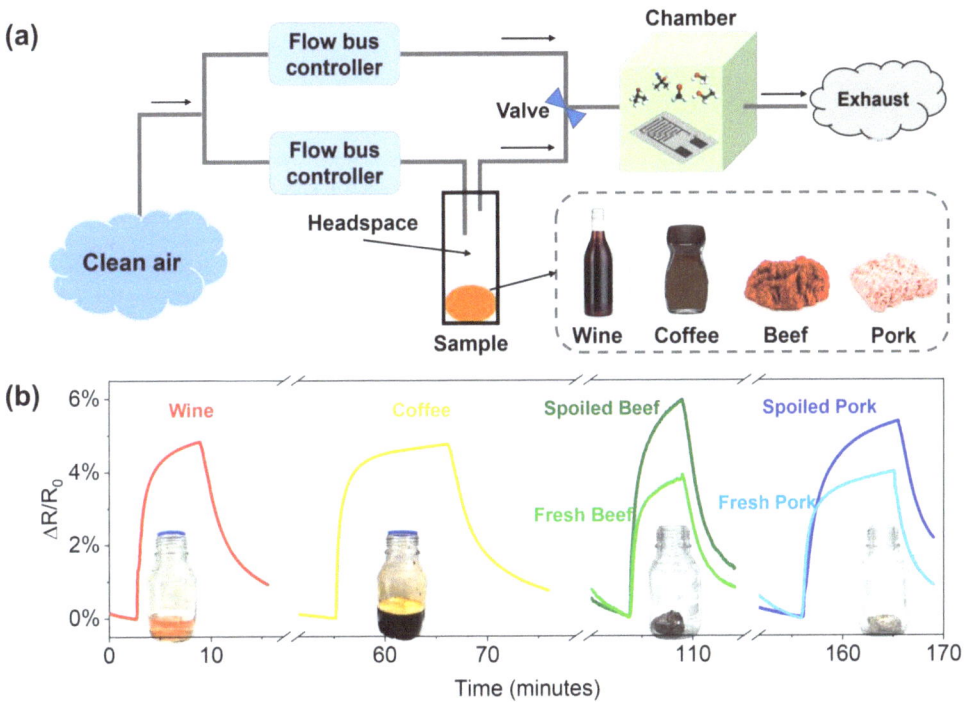

Figure 8. (a) Schematic representation of the total VOC measurement system for real samples; (b) the responses of the PDAC/rGO sensor to VOCs from beverage and meat samples.

4. Conclusions

Taking the importance of precise monitoring of different environmental parameters and contaminants in food processing, developing sensitive and cost-effective sensors is of utmost importance. In this paper, we have shown that PDAC/rGO composites produced by an LbL procedure can be effectively used for temperature and relative humidity monitoring, VOC detection, and electrochemical measurements of organophosphate pesticide dimethoate. In the case of temperature monitoring, the proposed sensors show relatively short response times in the range of 0.78–1.6 s and recovery times in the range of 24–35 s, while the sensitivity increases with the number of bilayers deposited on the electrode. In the case of relative humidity measurement, the effect of the number of deposited bi-layers is similar, but by increasing the number of layers, the sensitivity changes are not significant. The long-term stability of the proposed sensors for temperature and relative humidity measurements is excellent, as the responses changed to a small extent over one month of testing. Further, the methanol gas sensitivity of the PDAC/rGO sensor decreases with the number of deposited PDAC/rGO bi-layers. Good linearity is observed for methanol concentrations above 35 ppm. The PDAC/rGO-1L sensor measured a maximum sensitivity of 1.97% at a methanol concentration of 55 ppm. PDAC/rGO-1L is also the most sensitive for the electrochemical detection of dimethoate, although further improvements in this direction are absolutely necessary. With the numerous applications of LbL self-assembly layers in the field of sensors, further advancements are expected, particularly toward developing multi-modal sensors and sensor arrays. Such multi-sensors could have a tremendous impact on monitoring food safety and quality, as demonstrated by sensing total VOCs from two beverages (coffee and wine) and meat samples (pork and beef). Because of their demonstrated capabilities, sensor arrays based on LbL-PDAC/rGO could be developed to detect different physical and chemical phenomena to serve as an electronic nose or tongue where artificial intelligence can be utilized to extract desired information.

Supplementary Materials: The following supporting information can be downloaded at: https://www.mdpi.com/article/10.3390/foods12020268/s1, Figure S1. (a) Schematic representation of the screen-printed Ag electrode used for the LbL deposition of PDAC/GO bi-layers; (b) Protocol for the LbL deposition of the PDAC/GO bi-layers and (c) photograph of the LbL PDAC/GO-based sensor after thermal reduction; Figure S2. The block and connection diagram of the humidity measurement system; Figure S3. Raman spectra of PDAC/rGO composite electrodes reduced at 100 °C with (a) 1 bi-layer; (b) 2 bi-layers; (c) 4 bi-layers. Yellow and green spectra are taken at two different spots of the sample (marked Spot A and Spot B); Figure S4. Temperature measurement (a) sensitivity curve as the function of the number of the PDAC/rGO layers; (b) response time as the function of the number of the PDAC/rGO layers; (c) recovery time as the function of the number of the PDAC/rGO layers; (d) Long-term stability curves for the temperature measurements using PDAC/rGO-8L sensor; Figure S5. (a) Sensitivity of the response as response curve of different layers for the relative humidity measurements; (b) Long-term stability curves for the relative humidity measurements using PDAC/rGO-8L sensor; Table S1. Information of the detected real samples.

Author Contributions: Conceptualization, A.A.-H., I.A.P. and O.K.; methodology, A.A.-H., I.A.P., J.B., T.K.E. and T.L.; validation, A.A.-H., I.A.P., T.L. and A.A.; formal analysis, A.A.-H., I.A.P. and T.L.; investigation, A.A.-H., J.B., T.K.E., T.L. and I.A.P.; resources, A.A.-H., T.L. and I.A.P.; data curation, A.A.-H., T.L., A.A. and I.A.P.; writing—original draft preparation, A.A.-H., J.B., T.L. and I.A.P.; writing—review and editing, All; visualization, A.A.-H., J.B., T.L. and I.A.P.; supervision, O.K. and A.A.-H.; project administration, A.A.-H., I.A.P. and O.K.; funding acquisition, O.K. All authors have read and agreed to the published version of the manuscript.

Funding: This research was funded by Deutsche Forschungsgemeinschaft (DFG, German Research Foundation) within the projects "Photosens" (KA 1663/12) and "Dualsens" (KA 1663/13-1).

Data Availability Statement: The data are available upon request to the corresponding author.

Acknowledgments: Authors would like to thank Danica Bajuk Bogdanović (the University of Belgrade—Faculty of Physical Chemistry) for the Raman analysis of the sensor electrodes. We also thank Leonardo Paterno (Instituto de Química, Universidade de Brasília, Brasilia, Brazil) for the valuable discussion and advice during the LbL deposition work.

Conflicts of Interest: The authors declare no conflict of interest.

References

1. The State of Food Insecurity in the World 2014—Food and Agricultural Organization of the United Nations. Available online: https://www.fao.org/3/i4030e/i4030e.pdf (accessed on 22 November 2022).
2. Vera Zambrano, M.; Dutta, B.; Mercer, D.G.; MacLean, H.L.; Touchie, M.F. Assessment of Moisture Content Measurement Methods of Dried Food Products in Small-Scale Operations in Developing Countries: A Review. *Trends Food Sci. Technol.* **2019**, *88*, 484–496. [CrossRef]
3. Bourdichon, F.; Betts, R.; Dufour, C.; Fanning, S.; Farber, J.; McClure, P.; Stavropoulou, D.A.; Wemmenhove, E.; Zwietering, M.H.; Winkler, A. Processing Environment Monitoring in Low Moisture Food Production Facilities: Are We Looking for the Right Microorganisms? *Int. J. Food Microbiol.* **2021**, *356*, 109351. [CrossRef] [PubMed]
4. Global Food Losses and Food Waste—Food and Agricultural Organization of the United Nations. Available online: http://www.fao.org/docrep/014/mb060e/mb060e00.pdf (accessed on 22 November 2022).
5. Cui, Y.; Zhang, H.; Zhang, J.; Lv, B.; Xie, B. The Emission of Volatile Organic Compounds during the Initial Decomposition Stage of Food Waste and Its Relationship with the Bacterial Community. *Environ. Technol. Innov.* **2022**, *27*, 102443. [CrossRef]

6. Narenderan, S.T.; Meyyanathan, S.N.; Babu, B. Review of Pesticide Residue Analysis in Fruits and Vegetables. Pre-Treatment, Extraction and Detection Techniques. *Food Res. Int.* **2020**, *133*, 109141. [CrossRef] [PubMed]
7. Arman Kuzubasoglu, B.; Kursun Bahadir, S. Flexible Temperature Sensors: A Review. *Sens. Actuators A Phys.* **2020**, *315*, 112282. [CrossRef]
8. Lee, C.-Y.; Lee, G.-B. Humidity Sensors: A Review. *Sens. Lett.* **2005**, *3*, 1–15. [CrossRef]
9. Galstyan, V.; D'Arco, A.; Di Fabrizio, M.; Poli, N.; Lupi, S.; Comini, E. Detection of Volatile Organic Compounds: From Chemical Gas Sensors to Terahertz Spectroscopy. *Rev. Anal. Chem.* **2021**, *40*, 33–57. [CrossRef]
10. Lu, T.; Al-Hamry, A.; Rosolen, J.M.; Hu, Z.; Hao, J.; Wang, Y.; Adiraju, A.; Yu, T.; Matsubara, E.Y.; Kanoun, O. Flexible Impedimetric Electronic Nose for High-Accurate Determination of Individual Volatile Organic Compounds by Tuning the Graphene Sensitive Properties. *Chemosensors* **2021**, *9*, 360. [CrossRef]
11. Trigona, C.; Lu, T.; Talbi, M.; Baglio, S.; Al-Hamry, A.; Garcia-Martinez, C.; Kanoun, O. MEMS Based on Chitosan—Tetrasulfonated Copper Phthalocyanine Composite for Detection of Ethanol Vapor in Air. In Proceedings of the 2022 IEEE International Workshop on Metrology for Industry 4.0 & IoT (MetroInd4.0&IoT), Trento, Italy, 7 June 2022; pp. 240–244. [CrossRef]
12. Lu, T.; Al-Hamry, A.; Hao, J.; Liu, Y.; Qu, Y.; Kanoun, O. Machine Learning-Based Multi-Level Fusion Framework for a Hybrid Voltammetric and Impedimetric Metal Ions Electronic Tongue. *Chemosensors* **2022**, *10*, 474. [CrossRef]
13. Lu, T.; Al-Hamry, A.; Talbi, M.; Zhang, J.; Adiraju, A.; Hou, M.; Kanoun, O. Functionalized PEDOT:PSS Based Sensor Array for Determination of Metallic Ions in Smart Agriculture. In Proceedings of the 2021 6th International Conference on Nanotechnology for Instrumentation and Measurement (NanofIM), Opole, Poland, 25 November 2021; pp. 1–4. [CrossRef]
14. Xiang, H.; Cai, Q.; Li, Y.; Zhang, Z.; Cao, L.; Li, K.; Yang, H. Sensors Applied for the Detection of Pesticides and Heavy Metals in Freshwaters. *J. Sens.* **2020**, *2020*, 8503491. [CrossRef]
15. Kanoun, O.; Lazarević-Pašti, T.; Pašti, I.; Nasraoui, S.; Talbi, M.; Brahem, A.; Adiraju, A.; Sheremet, E.; Rodriguez, R.D.; Ben Ali, M.; et al. A Review of Nanocomposite-Modified Electrochemical Sensors for Water Quality Monitoring. *Sensors* **2021**, *21*, 4131. [CrossRef]
16. Basu, J.; Basu, J.K.; Bhattacharyya, T.K. The Evolution of Graphene-Based Electronic Devices. *Int. J. Smart Nano Mater.* **2010**, *1*, 201–223. [CrossRef]
17. Mukherjee, R.; Thomas, A.V.; Datta, D.; Singh, E.; Li, J.; Eksik, O.; Shenoy, V.B.; Koratkar, N. Defect-Induced Plating of Lithium Metal within Porous Graphene Networks. *Nat. Commun.* **2014**, *5*, 3710. [CrossRef]
18. Mahmood, N.; Zhang, C.; Yin, H.; Hou, Y. Graphene-Based Nanocomposites for Energy Storage and Conversion in Lithium Batteries, Supercapacitors and Fuel Cells. *J. Mater. Chem. A* **2014**, *2*, 15–32. [CrossRef]
19. Mukherjee, R.; Thomas, A.V.; Krishnamurthy, A.; Koratkar, N. Photothermally Reduced Graphene as High-Power Anodes for Lithium-Ion Batteries. *ACS Nano* **2012**, *6*, 7867–7878. [CrossRef]
20. Richardson, J.J.; Cui, J.; Björnmalm, M.; Braunger, J.A.; Ejima, H.; Caruso, F. Innovation in Layer-by-Layer Assembly. *Chem. Rev.* **2016**, *116*, 14828–14867. [CrossRef]
21. Lipton, J.; Weng, G.-M.; Röhr, J.A.; Wang, H.; Taylor, A.D. Layer-by-Layer Assembly of Two-Dimensional Materials: Meticulous Control on the Nanoscale. *Matter* **2020**, *2*, 1148–1165. [CrossRef]
22. Lee, D.; Cui, T. Layer-by-Layer Nano Self-Assembly of PH Sensors Based on Polyelectrolytes and Carboxylated Carbon Nanotubes. *ECS Trans.* **2008**, *16*, 3–9. [CrossRef]
23. Nohria, R.; Khillan, R.K.; Su, Y.; Dikshit, R.; Lvov, Y.; Varahramyan, K. Humidity Sensor Based on Ultrathin Polyaniline Film Deposited Using Layer-by-Layer Nano-Assembly. *Sens. Actuators B Chem.* **2006**, *114*, 218–222. [CrossRef]
24. Zhang, D.; Wang, D.; Li, P.; Zhou, X.; Zong, X.; Dong, G. Facile Fabrication of High-Performance QCM Humidity Sensor Based on Layer-by-Layer Self-Assembled Polyaniline/Graphene Oxide Nanocomposite Film. *Sens. Actuators B Chem.* **2018**, *255*, 1869–1877. [CrossRef]
25. Ren, X.; Zhang, D.; Wang, D.; Li, Z.; Liu, S. Quartz Crystal Microbalance Sensor for Humidity Sensing Based on Layer-by-Layer Self-Assembled PDDAC/Graphene Oxide Film. *IEEE Sens. J.* **2018**, *18*, 9471–9476. [CrossRef]
26. Liu, Z.; Yang, T.; Dong, Y.; Wang, X. A Room Temperature VOCs Gas Sensor Based on a Layer by Layer Multi-Walled Carbon Nanotubes/Poly-Ethylene Glycol Composite. *Sensors* **2018**, *18*, 3113. [CrossRef] [PubMed]
27. Su, P.-G.; Shieh, H.-C. Flexible NO2 Sensors Fabricated by Layer-by-Layer Covalent Anchoring and in Situ Reduction of Graphene Oxide. *Sens. Actuators B Chem.* **2014**, *190*, 865–872. [CrossRef]
28. Zhang, D.; Jiang, C.; Sun, Y. Room-Temperature High-Performance Ammonia Gas Sensor Based on Layer-by-Layer Self-Assembled Molybdenum Disulfide/Zinc Oxide Nanocomposite Film. *J. Alloys Compd.* **2017**, *698*, 476–483. [CrossRef]
29. Ke, F.; Song, F.; Zhang, H.; Xu, J.; Wang, H.; Chen, Y. Layer-by-Layer Assembly for All-Graphene Coated Conductive Fibers toward Superior Temperature Sensitivity and Humidity Independence. *Compos. Part B Eng.* **2020**, *200*, 108253. [CrossRef]
30. Zhang, D.; Tong, J.; Xia, B. Humidity-Sensing Properties of Chemically Reduced Graphene Oxide/Polymer Nanocomposite Film Sensor Based on Layer-by-Layer Nano Self-Assembly. *Sens. Actuators B Chem.* **2014**, *197*, 66–72. [CrossRef]
31. Gross, M.A.; Sales, M.J.A.; Soler, M.A.G.; Pereira-da-Silva, M.A.; da Silva, M.F.P.; Paterno, L.G. Reduced Graphene Oxide Multilayers for Gas and Liquid Phases Chemical Sensing. *RSC Adv.* **2014**, *4*, 17917. [CrossRef]
32. Liu, S.; Yan, J.; He, G.; Zhong, D.; Chen, J.; Shi, L.; Zhou, X.; Jiang, H. Layer-by-Layer Assembled Multilayer Films of Reduced Graphene Oxide/Gold Nanoparticles for the Electrochemical Detection of Dopamine. *J. Electroanal. Chem.* **2012**, *672*, 40–44. [CrossRef]

33. Yang, R.; Zhang, W.; Tiwari, N.; Yan, H.; Li, T.; Cheng, H. Multimodal Sensors with Decoupled Sensing Mechanisms. *Adv. Sci.* **2022**, *9*, 2202470. [CrossRef]
34. Brahem, A.; Al-Hamry, A.; Gross, M.A.; Paterno, L.G.; Ali, M.B.; Kanoun, O. Stability Enhancement of Laser-Scribed Reduced Graphene Oxide Electrodes Functionalized by Iron Oxide/Reduced Graphene Oxide Nanocomposites for Nitrite Sensors. *J. Compos. Sci.* **2022**, *6*, 221. [CrossRef]
35. Ega, T.K.; Al-Hamry, A.; Kanoun, O.; Lazarevic-Pasti, T.; Bogdanovic, D.B.; Pasti, I.A.; Rodriguez, R.D.; Sheremet, E.; Paterno, L.G. Detection of Dimethoate Pesticide Using Layer by Layer Deposition of PDAC/GO on Ag Electrode. In Proceedings of the 2019 16th International Multi-Conference on Systems, Signals & Devices (SSD), Istanbul, Turkey, 21–24 March 2019; pp. 621–625.
36. Al-Hamry, A.; Kang, H.; Sowade, E.; Dzhagan, V.; Rodriguez, R.D.; Müller, C.; Zahn, D.R.T.; Baumann, R.R.; Kanoun, O. Tuning the Reduction and Conductivity of Solution-Processed Graphene Oxide by Intense Pulsed Light. *Carbon* **2016**, *102*, 236–244. [CrossRef]
37. Ferrari, A.C.; Robertson, J. Interpretation of Raman Spectra of Disordered and Amorphous Carbon. *Phys. Rev. B* **2000**, *61*, 14095–14107. [CrossRef]
38. Sarker, A.K.; Hong, J.-D. Layer-by-Layer Self-Assembled Multilayer Films Composed of Graphene/Polyaniline Bilayers: High-Energy Electrode Materials for Supercapacitors. *Langmuir* **2012**, *28*, 12637–12646. [CrossRef]
39. Al-Hamry, A.; Kanoun, O. Humidity sensitivity investigation of reduced graphene oxide by impedance spectroscopy. In *Impedance Spectroscopy*; Kanoun, O., Ed.; De Gruyter: Berlin, Germany, 2018; pp. 49–52. ISBN 978-3-11-055892-0.
40. Al-Hamry, A.; Ega, T.K.; Pasti, I.A.; Bajuk-Bogdanovic, D.; Lazarevic-Pasti, T.; Rodriguez, R.D.; Sheremet, E.; Kanoun, O. Electrochemical Sensor Based on Reduced Graphene Oxide/PDAC for Dimethoate Pesticide Detection. In Proceedings of the 2019 5th International Conference on Nanotechnology for Instrumentation and Measurement (NanofIM), Sfax, Tunisia, 30 October 2019; pp. 1–7. [CrossRef]
41. Jung, H.; Park, J. Adherable Temperature Sensor to a Porous Structure of Fiber via Reduced Graphene Oxide and Poly(Diallyldimethylammonium Chloride) Complex. *J. Micromech. Microeng.* **2022**, *32*, 045001. [CrossRef]
42. Liu, G.; Tan, Q.; Kou, H.; Zhang, L.; Wang, J.; Lv, W.; Dong, H.; Xiong, J. A Flexible Temperature Sensor Based on Reduced Graphene Oxide for Robot Skin Used in Internet of Things. *Sensors* **2018**, *18*, 1400. [CrossRef]
43. Zhang, C.; Chen, Y.; Li, H.; Tian, R.; Liu, H. Facile Fabrication of Three-Dimensional Lightweight RGO/PPy Nanotube/Fe$_3$O$_4$ Aerogel with Excellent Electromagnetic Wave Absorption Properties. *ACS Omega* **2018**, *3*, 5735–5743. [CrossRef]
44. Muchharla, B.; Narayanan, T.N.; Balakrishnan, K.; Ajayan, P.M.; Talapatra, S. Temperature Dependent Electrical Transport of Disordered Reduced Graphene Oxide. *2D Mater.* **2014**, *1*, 011008. [CrossRef]
45. Cavallo, P.; Acevedo, D.F.; Fuertes, M.C.; Soler-Illia, G.J.A.A.; Barbero, C.A. Understanding the Sensing Mechanism of Polyaniline Resistive Sensors. Effect of Humidity on Sensing of Organic Volatiles. *Sens. Actuators B Chem.* **2015**, *210*, 574–580. [CrossRef]
46. Liu, L.; Ye, X.; Wu, K.; Zhou, Z.; Lee, D.; Cui, T. Humidity Sensitivity of Carbon Nanotube and Poly (Dimethyldiallylammonium Chloride) Composite Films. *IEEE Sens. J.* **2009**, *9*, 1308–1314. [CrossRef]
47. Al-Hamry, A.; Zubkova, T.; Baumann, R.; Kanoun, O.; Paterno, L.G.; Errachid, A. Sensitivity of Layer-by-Layer Deposited GO/PDAC to Volatile Organic Compounds. In *Proceedings of the Tagungsband*; AMA Service GmbH: Nürnberg, Germany, 2019; pp. 584–587. [CrossRef]
48. Aghoutane, Y.; Diouf, A.; Österlund, L.; Bouchikhi, B.; El Bari, N. Development of a Molecularly Imprinted Polymer Electrochemical Sensor and Its Application for Sensitive Detection and Determination of Malathion in Olive Fruits and Oils. *Bioelectrochemistry* **2020**, *132*, 107404. [CrossRef]
49. Fei, A.; Liu, Q.; Huan, J.; Qian, J.; Dong, X.; Qiu, B.; Mao, H.; Wang, K. Label-Free Impedimetric Aptasensor for Detection of Femtomole Level Acetamiprid Using Gold Nanoparticles Decorated Multiwalled Carbon Nanotube-Reduced Graphene Oxide Nanoribbon Composites. *Biosens. Bioelectron.* **2015**, *70*, 122–129. [CrossRef] [PubMed]
50. Vuorinen, T.; Niittynen, J.; Kankkunen, T.; Kraft, T.M.; Mäntysalo, M. Inkjet-Printed Graphene/PEDOT:PSS Temperature Sensors on a Skin-Conformable Polyurethane Substrate. *Sci. Rep.* **2016**, *6*, 35289. [CrossRef] [PubMed]
51. Pan, J.; Liu, S.; Zhang, H.; Lu, J. A Flexible Temperature Sensor Array with Polyaniline/Graphene–Polyvinyl Butyral Thin Film. *Sensors* **2019**, *19*, 4105. [CrossRef] [PubMed]
52. Ozioko, O.; Kumaresan, Y.; Dahiya, R. Carbon Nanotube/PEDOT: PSS Composite-Based Flexible Temperature Sensor with Enhanced Response and Recovery Time. In Proceedings of the 2020 IEEE International Conference on Flexible and Printable Sensors and Systems (FLEPS), Manchester, UK, 16 August 2020; pp. 1–4. [CrossRef]

Disclaimer/Publisher's Note: The statements, opinions and data contained in all publications are solely those of the individual author(s) and contributor(s) and not of MDPI and/or the editor(s). MDPI and/or the editor(s) disclaim responsibility for any injury to people or property resulting from any ideas, methods, instructions or products referred to in the content.

Article

Application of Viscose-Based Porous Carbon Fibers in Food Processing—Malathion and Chlorpyrifos Removal

Tamara Tasić [1], Vedran Milanković [1], Katarina Batalović [1], Stefan Breitenbach [2,3], Christoph Unterweger [2], Christian Fürst [2], Igor A. Pašti [4,*] and Tamara Lazarević-Pašti [1]

[1] VINČA Institute of Nuclear Sciences—National Institute of the Republic of Serbia, University of Belgrade, Mike Petrovica Alasa 12–14, 11000 Belgrade, Serbia; tamara.tasic@vin.bg.ac.rs (T.T.); vedran.milankovic@vin.bg.ac.rs (V.M.); kciric@vin.bg.ac.rs (K.B.); tamara@vin.bg.ac.rs (T.L.-P.)

[2] Wood K Plus—Kompetenzzentrum Holz GmbH, Altenberger Strasse 69, 4040 Linz, Austria; s.breitenbach@wood-kplus.at (S.B.); c.unterweger@wood-kplus.at (C.U.); c.fuerst@wood-kplus.at (C.F.)

[3] Institute of Chemical Technology of Inorganic Materials (TIM), Johannes Kepler University Linz, Altenberger Strasse 69, 4040 Linz, Austria

[4] Faculty of Physical Chemistry, University of Belgrade, Studentski Trg 12–16, 11158 Belgrade, Serbia

* Correspondence: igor@ffh.bg.ac.rs

Abstract: The increasing usage of pesticides to boost food production inevitably leads to their presence in food samples, requiring the development of efficient methods for their removal. Here, we show that carefully tuned viscose-derived activated carbon fibers can be used for malathion and chlorpyrifos removal from liquid samples, even in complex matrices such as lemon juice and mint ethanol extract. Adsorbents were produced using the Design of Experiments protocol for varying activation conditions (carbonization at 850 °C; activation temperature between 670 and 870 °C; activation time from 30 to 180 min; and CO_2 flow rate from 10 to 80 L h^{-1}) and characterized in terms of physical and chemical properties (SEM, EDX, BET, FTIR). Pesticide adsorption kinetics and thermodynamics were then addressed. It was shown that some of the developed adsorbents are also capable of the selective removal of chlorpyrifos in the presence of malathion. The selected materials were not affected by complex matrices of real samples. Moreover, the adsorbent can be regenerated at least five times without pronounced performance losses. We suggest that the adsorptive removal of food contaminants can effectively improve food safety and quality, unlike other methods currently in use, which negatively affect the nutritional value of food products. Finally, data-based models trained on well-characterized materials libraries can direct the synthesis of novel adsorbents for the desired application in food processing.

Keywords: biomass; biowaste; activated carbon materials; pesticides; organophosphates; adsorptive removal; properties–performance relations

1. Introduction

With an increasing global population, the demand for food is also rising [1]. Farmers and producers must find ways to enhance food production to meet current needs. This can be accomplished through various techniques, such as improving crop yields, diversifying crop varieties, reducing food waste and loss, and improving agricultural methods [2,3].

The Farm to Fork Strategy is an economic, environmental, and social policy initiative to transform and modernize the EU's food system [4]. It will enable the EU to become a more sustainable, resilient, and competitive food system while improving the quality and safety of the food we consume and reducing our environmental footprint. Some of Farm to Fork's key focal points are promoting sustainable consumption, stimulating the production of more nutritious and safe food, investing in research and innovation, and stimulating the shift toward ecological farming. One of the core objectives of the Farm to Fork Strategy is to reduce the use and risk of chemical pesticides and other synthetic substances in food production. To achieve this, the EU promotes more sustainable alternatives to chemical pesticides, such as integrated pest management, biocontrols, and organic farming. The Farm to Fork strategy also aims to ensure that any chemical pesticides used on food crops are in line with EU standards, which include an increased focus on the safety and protection of the environment.

Pesticides are chemical substances that kill or control insects, weeds, fungi, bacteria, and other organisms that can damage crops or transmit diseases to humans [5]. They vary in their mode of action, toxicity, persistence, and environmental impact. Using pesticides in food plant production is necessary for controlling pests, weeds, and diseases that can adversely affect plant yields. Organophosphate pesticides (OPs), such as malathion and chlorpyrifos, are commonly used to control pests on fruit crops. They are used on various fruits, such as apples, oranges, lemons, limes, peaches, nectarines, bananas, grapes, and watermelons. The use of OPs is highly regulated due to their ability to inhibit the activity of acetylcholinesterase (AChE) [6–9]. AChE is an important enzyme in the nervous system, and disruption of its activity leads to many health issues, even death [10]. Malathion and chlorpyrifos are used worldwide and, hence, are often found in food samples. While chlorpyrifos is well known to be highly poisonous [11], malathion is misleadingly considered moderately toxic. These estimations refer to the acute toxicity of the mentioned pesticides, while recent research indicates that these compounds are potentially neurotoxic, even in small amounts [6,7,12].

Food processing is any process or method used to transform raw ingredients into safe, edible, and shelf-stable food products [13]. Fruit and spice extracts are popular ingredients in the food industry [14]. They are made by extracting juice, essential oils, pigments, and other components from fresh fruits, such as lemon or mint. These extracts are used to enhance the taste, color, and texture of food products and provide antioxidants and other health benefits [15]. As such, they are often found in food products such as juices, yogurts, ice creams, gums, and baked goods. Because the plants used for extract production are often exposed to pesticides, the final products are also expected to contain their toxic residues.

The use of pesticides has been debated for decades, as many people are concerned about the potential long-term effects of these chemicals on our health and the environment [16]. Pesticide use is necessary for food production to protect crops from pests and diseases. However, the extensive use of pesticides can lead to pesticide accumulation, which can be detrimental to human health due to toxicity [17]. Because we still cannot avoid using them, the solution to the problem may be to act on the level of food processing [18].

Pesticides can be removed from samples using chemical, physical, or biological methods [19]. As an additional step in food processing, adsorption has the best potential. In terms of pesticide remediation, adsorption is a process of removing pesticide molecules from various samples by attaching them to an adsorbent material [20]. The adsorbent materials most commonly used for pesticide removal include activated carbon materials [20,21], mesoporous monetite [22], porous metal–organic frameworks [23], mineral surfaces [24], organohydrotalcite [25], zeolites [26], materials from the graphene family [27], metal nanoparticles [28], and others. Activated carbon materials have an excellent potential for removing pesticides during food processing. They are highly porous, cheap, available, and easy to use. In addition, they are mostly non-toxic, so their use is safe [29]. Moreover, carbon materials produced from biomass represent a sustainable choice for pesticide removal. The activated carbon materials can be produced by the pyrolysis of different biomass materials, such as wood, coconut husks, seed hulls, and other organic waste materials [30–32]. The activated carbon material has a high surface area, which allows it to absorb and remove many organic contaminants effectively.

Viscose fibers are a type of fiber made from cellulose that are commonly studied as a precursor for producing activated carbon materials. These fibers can be used in a variety of applications, including the removal of several pollutants. Plens et al. used activated carbon materials derived from viscose fibers for nitrogen (NOx) and sulfur (SOx) adsorption [33]. Bhati et al. were able to use carbonized viscose fiber for the effective removal of iodine and

CCl$_4$ [34]. It was also successfully used for the removal of heavy metals [35], dyes [36], and wastewater treatment [37]. In our previous paper, we used viscose fibers impregnated with diammonium hydrogen phosphate to efficiently remove different OPs from an aqueous solution [20,21,38]. A well-defined series of materials with gradually changing properties, such as pore size distribution, pore volume, and chemical composition, allowed for the analysis of the interconnection between the physico-chemical properties of adsorbents and their performance for OP removal.

Methods currently used during food processing with the aim of removing present contaminants include various chemical and physical techniques. Bleaching, UV radiation, and chemical treatment are commonly applied but severely affect the nutritional values of the products. We aimed to show an alternative filtration method for pesticide removal without the essential nutrient reduction.

This paper investigates the potential of using viscose-based porous carbon fibers as adsorbents for malathion (aliphatic OP) and chlorpyrifos (aromatic OP) removal considering fundamental and practical aspects. The main focus is on the investigated materials' application in food processing—the removal of malathion and chlorpyrifos residues from lemon and mint extracts. First, the series of carbon materials used as adsorbents were synthesized using the Design of Experiments (DoE) protocol [39], characterized in terms of morphology, chemical composition, and textural properties, and then the kinetics and thermodynamics of malathion and chlorpyrifos adsorption from aqueous solutions were addressed. Next, the materials performance was linked to the synthesis conditions and their properties using principal component analysis and principal component regression. Then, the practical applicability of the presented series of materials was addressed, considering food safety and sustainability. Finally, the feasibility and potential benefits of using the adsorptive removal of OPs in food processing are discussed.

2. Materials and Methods

2.1. Material Synthesis

Viscose fibers (1.7 dtex, 38 mm) provided by Lenzing AG (Lenzing, Austria) were used as the precursor. They were washed thoroughly with distilled water before use. The viscose fibers were centrifuged with a spin dryer for 15 min before being dried for 24 h at 90 °C in a drying cabinet. The residual moisture was determined with a moisture analyzer (MX-50, A&D Company, Tokyo, Japan) at 105 °C until the mass remained constant. The fibers were only used if the residual moisture was below 4.5%.

The carbonization of the precursor was carried out by loading 100–400 g into a chamber furnace (HTK 8 W, Carbolite Gero GmbH, Neuhausen, Germany). After the evacuation of the chamber, a nitrogen flow atmosphere of 250 L h^{-1} was established. The sample was heated to 850 °C at a 1.0 °C/min heating rate and held isothermal for 30 min. It was then cooled to room temperature under a nitrogen atmosphere.

Parametric space for carbonized viscose fibers activation was generated using the DoE approach. The three independent variables, the activation temperature, activation time, and CO$_2$ flow rate, were systematically varied each at three different levels using a Central Composite Design (CCD). As no significant differences in the carbon yield and porosity parameters could be measured, no replicates of the center point Run10 are included in this study (see Table 1 for details). The levels of the three variables were chosen based on preliminary tests. A temperature of 870 °C was found to be the highest suitable activation temperature, as further increase resulted in the complete consumption of the samples due to the fast kinetics of the activation reaction. An activation time of 105 min and a CO$_2$ flow rate of 45 L h^{-1} were found suitable in preliminary tests and used as center points in our CCD. The lower and upper limits were chosen in order to ensure the start of the activation process at the lower limits and avoid the complete consumption of the sample at the upper limits. Activation was performed in a rotary kiln (RSR-B 120/500/11, Nabertherm GmbH, Lilienthal, Germany). For the activation, 10 g of the sample was placed in the middle of the quartz glass reactor. Prior to use, N$_2$ was used to purge any air in the setup at a flow rate of 100 L h^{-1}. Subsequently, the materials were heated from room temperature to the desired final activation temperature under an N$_2$ flow rate of 50 L h^{-1}. The sample was kept isothermally under the N$_2$ flow for 30 min to ensure that the temperature was uniform throughout the reaction chamber. Afterward, the N$_2$ flow was terminated and replaced by CO$_2$ at the desired flow rate for a specified amount of time (Table 1). The activation process was completed by the termination of the CO$_2$ flow and restarting of the N$_2$ flow (50 L h^{-1}) until the kiln was cooled to room temperature.

Table 1. Experimental parameters for the activation of studied series of carbon materials. Carbonization was conducted at 850 °C with a heating rate of 1 °C min^{-1}. Missing sample numbers correspond to repetitions of center point Run10.

Sample	Activation Temperature	Activation Time	CO$_2$ Flow Rate
Run1	870 °C	180 min	80 L h^{-1}
Run2	670 °C	180 min	10 L h^{-1}
Run3	870 °C	30 min	80 L h^{-1}
Run5	870 °C	30 min	10 L h^{-1}
Run6	670 °C	30 min	80 L h^{-1}
Run7	670 °C	30 min	10 L h^{-1}
Run8	870 °C	180 min	10 L h^{-1}
Run9	670 °C	180 min	80 L h^{-1}
Run10	770 °C	105 min	45 L h^{-1}
Run12	770 °C	180 min	45 L h^{-1}
Run13	770 °C	105 min	80 L h^{-1}
Run15	670 °C	105 min	45 L h^{-1}
Run16	870 °C	105 min	45 L h^{-1}
Run17	770 °C	105 min	10 L h^{-1}

2.2. Material Characterization

A scanning electron microscope (SEM) PhenomProX (Thermo Fisher Scientific, Waltham, MA, USA) was used to investigate the samples' morphology and elemental composition using Energy-Dispersive X-ray Analysis (EDX).

N$_2$ isothermal adsorption (-196.15 °C) on a gas sorption system (Autosorb-iQ, Anton Paar QuantaTec Inc., Graz, Austria) was employed to analyze the specific surface area and textural structure of the activated carbon samples. Before the analysis, the samples were de-gassed for at least 2 h at 200 °C. The specific surface area was calculated using the method of Brunauer–Emmett–Teller (BET) [40], while the non-local density functional theory (NLDFT) was applied for derived pore size distribution (PSD) calculations.

A Nicolet iS20 FT-IR spectrophotometer (Thermo Fisher Scientific, Waltham, MA, USA) was used for the FTIR spectra recording. The applied wavenumber range was from 4000 to 500 cm^{-1} with 64 scans and 4 cm^{-1} resolution.

2.3. Adsorption Experiments

Adsorption experiments were carried out in batch (stationary conditions) and filter (dynamic conditions). For stationary analysis, prepared activated carbon fibers were first dispersed in double-distilled water. To provide the targeted concentration of adsorbent and OPs (malathion and chlorpyrifos), the desired amount of 10^{-1} mol dm^{-3} OP stock solution (Pestanal, Sigma Aldrich, Søborg, Denmark) was added. A laboratory shaker (Orbital Shaker-Incubator ES-20, Grant-Bio, Cambridgeshire, UK) was used for the adsorbent + OP mixture shaking and incubation at 25 °C for desired times (from 1 to 60 min). The adsorbent + OP mixture was centrifuged for 10 min at 14,500× g after the incubation. Next, the nylon filter (pore size 220 nm KX Syringe Filter, Kinesis, Cole Parmer, St. Neots, UK) was used for the supernatant filtration. The filtrate was subjected to ultra-performance liquid chromatography (UPLC) analysis, as described below, in order to determine the OP concentration after the adsorption. Modified commercial filters were used to analyze the OP adsorption onto the investigated materials under dynamic conditions. A total amount of 1 mg of each material was dispersed in 1 mL of deionized water and injected into the commercial nylon filter (pore size 220 nm KX Syringe Filter, Kinesis, Cole Parmer, St. Neots, UK). Compressed air was used for excess water removal from the modified filter. Next, 1 mL of the desired final concentration OP solution was injected through the modified filter with a flow rate of 1 mL min^{-1}. The filtrate was subjected to UPLC analysis, as described below. Modified filters were disassembled after the experiments to check for the uniformity of the carbon material layer. Uniform distribution of the adsorbent over the nylon membrane was observed in all cases.

The adsorption efficiency was calculated as Uptake = 100% × ($C_0 - C_{eq}$)/C_0 (C_0—the starting concentration of OPs) for stationary or dynamic experiments. UPLC was used to determine the concentration of OPs in filtrates after adsorption (C_{eq}). To confirm that there was no OP degradation during batch experiments, control experiments were performed in identical ways but without carbon materials.

A Waters ACQUITY UPLC system with a photodiode array (PDA) detector, controlled by Empower software, was used for OP analysis. An ACQUITY UPLC™ BEH C18 column (1.7 μm, 100 mm × 2.1 mm, Waters GmbH, Eschborn, Germany) was employed under isocratic conditions with 10% acetonitrile in water (v/v) as mobile phase A, and pure acetonitrile as mobile phase B. The eluent flow rate was 0.2 mL min^{-1}, and the injection volume was 5 μL in all cases. For malathion, the mobile-phase composition was 40% A and 60% B, while for chlorpyrifos, it was 20% A and 80% B. Both OPs were detected at 200 nm. Under these experimental conditions, the retention times were 3.07 ± 0.05 min and 2.53 ± 0.05 min for malathion and chlorpyrifos, respectively. The limit of detection of the method used was 1×10^{-7} mol dm^{-3}.

2.4. Principal Component Analysis

Principal component analysis (PCA) and principal component (PC) linear regression were performed using Scikit-learn built-in functions. Different levels/ranges of the considered input variables were scaled using the StandardScaler function in order to prepare the data for the statistical analysis.

2.5. Real-Samples Analysis

For real samples, lemon juice and mint extract were used. Because malathion is mainly used for lemon crop treatment, and chlorpyrifos is used for mint field preservation, we chose these extracts as reals samples. Juice pressed from 1 lemon (75 g) was diluted with 500 mL of tap water (pH = 4.5) and spiked with malathion to achieve the desired concentrations, which was followed by filtration. For mint extract preparation, 7 g of mint leaves (*Mentha spicata*) was mixed with 45 mL of 50% ethanol and left for 72 h at room temperature. After that, the extract was filtered and diluted with 200 mL of 50% ethanol (pH = 6.0). Finally, the mint extract was spiked with the desired amount of chlorpyrifos and consecutively filtered through a nylon filter. Prepared solutions were used in real-samples investigation.

2.6. Adsorbent Regeneration

The adsorbent regeneration filter was achieved by washing the modified filter with 5 mL of absolute ethanol for 1 min.

3. Results and Discussion

3.1. Material Properties

To investigate the samples' morphology, we used SEM. The SEM micrographs of samples Run1, Run6, and Run8 are presented in Figure 1 (from a1 to b3). The presented micrographs show that the morphology of all the investigated samples is the same and reflects the morphology of precursor viscose fibers, in agreement with our previous findings [20,41]. Namely, in the mentioned works, activated carbon fibers were produced by the carbonization of viscose fibers, either impregnated or without the impregnation step. In both cases, just like we find here, the morphology of the precursor was preserved. Thus, while passing through the carbonization and activation steps, as described in Table 1, fibers keep the morphology of the precursor (see Ref. [42] for SEM micrograph of the precursor), while only a shrinkage of the fiber diameter can be observed.

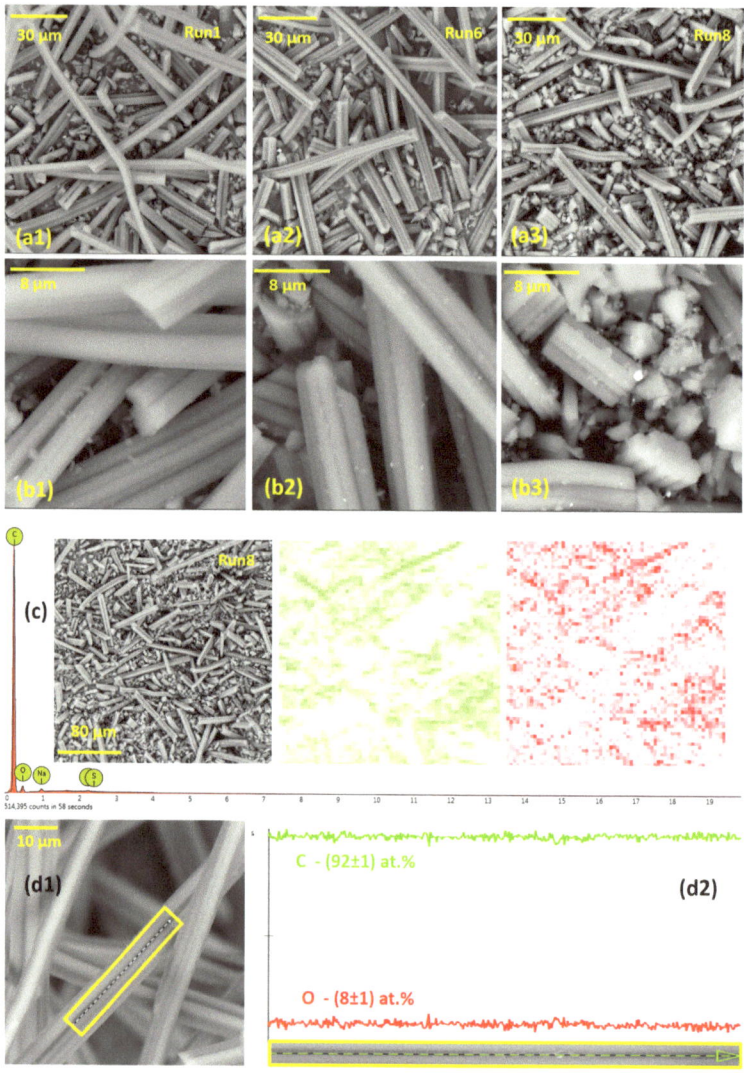

Figure 1. SEM images of samples Run1, Run6, and Run8 at two different magnifications and the EDX map of the Run8 sample with two major elements (C and O) (from (**a1**–**b3**)). EDX spectra (**c**) show only the additional presence of Na and S in the samples. Small debris observed on the fibers are due to the milling procedure, as explained in [43]. (**d1**,**d2**) show the selected region for the line EDX scan along a single fiber of the Run1 sample and the corresponding C and O distribution (Na and S are below 1 at.%) (resolution 512 pixels; three consecutive scans were averaged).

Upon the carbonization and activation, the fiber widths are around 8 µm, while the flower-like cross-sections (see Figure 1(b3)) are the result of the spinning process during precursor production. The length of the fibers after the milling step is 100–150 µm, while there is a lot of smaller debris of a few microns in length. While the presented finding might seem trivial, it is essential for the further analysis of the material performance. Namely, for the entire series of designed adsorbents, the morphology is not a parameter that could cause intra-series differences in the adsorption performance.

Using EDX, the chemical composition of the produced activated carbon fibers was determined (Table 2). C was found as a major element in all the samples, with roughly 7 at.% of oxygen and traces of Na and S, likely originating from the precursor. It is not surprising that the chemical composition does not show a significant intra-series variation, as the carbonization temperature was the same for all the samples, while the maximum activation temperature was also very close to the carbonization temperature. In all the samples, uniform elemental distribution was observed (Figure 1c), which is not surprising, as the precursor has a uniform chemical composition, while there was no impregnation agent with phase separation that could occur during the carbonization/activation. The overall elemental distribution is not only uniform at low magnification (Figure 1c), but also the elemental distribution along a single fiber (Figure 1(d1,d2)), with the variation under 1 at.% along approx. 35 µm of the fiber length. The main elements (carbon and oxygen) are expected to influence the adsorption process primarily through the specific interactions with the studied pesticides. Considering that chlorpyrifos has an aromatic moiety, it can not only interact with the sp^2 domains (graphitic) on the carbon surface, but also with oxygen functional groups via dipole interactions. Malathion is a polar aliphatic molecule, and so the interaction with hydrophobic sp^2 domains in carbon structures is not likely.

Table 2. Elemental composition of studied carbon materials obtained using EDX (average of four individual spot measurements).

Sample	Carbon (at.%)	Oxygen (at.%)	Sodium (at.%)	Sulfur (at.%)
Run1	91.88	7.60	0.45	0.07
Run2	92.55	7.18	0.21	0.06
Run3	92.13	7.58	0.27	0.02
Run5	92.39	7.28	0.28	0.05
Run6	93.39	6.31	0.26	0.05
Run7	93.63	6.11	0.23	0.03
Run8	92.24	7.38	0.33	0.05
Run9	92.29	7.38	0.27	0.06
Run10	93.08	6.65	0.26	0.01
Run12	91.83	7.91	0.22	0.05
Run13	92.60	7.14	0.23	0.03
Run15	93.99	5.79	0.20	0.02
Run16	92.35	7.33	0.30	0.02
Run17	93.13	6.65	0.21	0.02

Figure 2a presents N_2 adsorption isotherms, while Figure 2b gives the derived pore size distributions (PSDs) for the entire series of samples. The textural properties of the samples are summarized in Table 3, including the total pore volume (V_{tot}), average pore diameter (d_{mean}), and specific surface using the BET method (S_{BET}). The produced activated carbon fibers are dominantly microporous, with only one sample containing pores entering the mesopore domain (i.e., pores with diameters above 2 nm) (Run1, Figure 2b). The considered properties vary by one order of magnitude, suggesting the successful tuning of textural properties by the DoE protocol. Based on the obtained results, the activation temperature, followed by the activation time, play the key role in obtaining samples with higher S_{BET}.

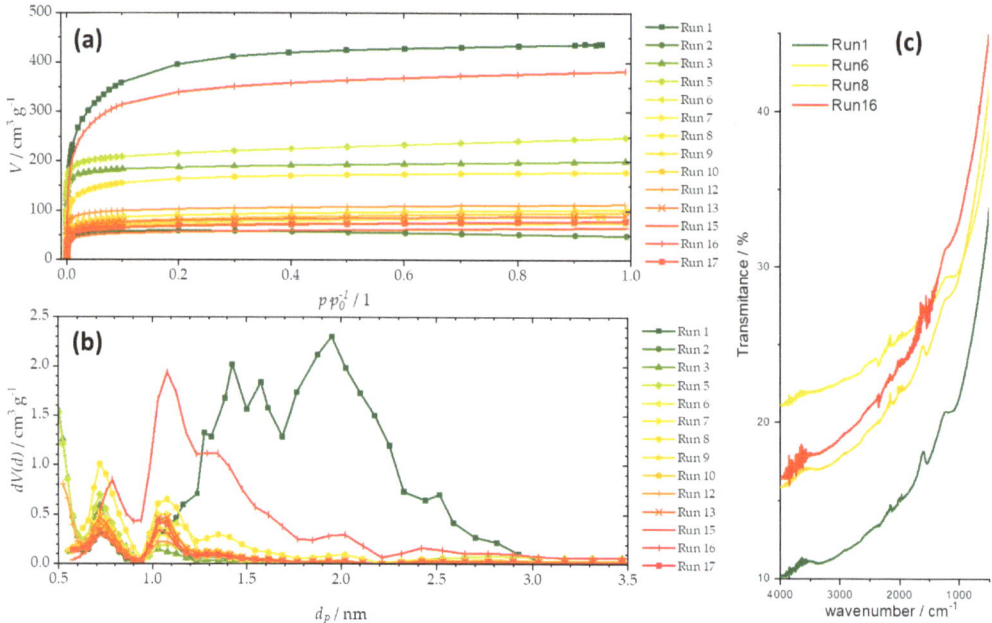

Figure 2. (**a**) Adsorption isotherms, (**b**) pore size distributions, and (**c**) ATR-FTIR spectra of selected carbon materials (FTIR spectra have a high degree of similarity, so the spectra are not given for the entire series).

Table 3. Textural properties of studied carbons: V_{tot}—total pore volume; d_{mean}—average pore diameter; S_{BET}—specific surface using BET method.

Sample	V_{tot}/cm^3 g^{-1}	d_{mean}/nm	S_{BET}/m^2 g^{-1}
Run1	0.547	1.951	1791
Run2	0.074	1.029	227
Run3	0.247	0.479	659
Run5	0.307	0.479	746
Run6	0.108	0.718	259
Run7	0.122	0.718	309
Run8	0.212	0.718	372
Run9	0.090	0.718	250
Run10	0.115	1.077	277
Run12	0.135	0.524	353
Run13	0.106	0.718	270
Run15	0.078	1.029	192
Run16	0.452	0.718	1141
Run17	0.097	0.574	232

Considering that the carbonization temperature was the same for all the samples, surface functional groups were also similar for all the samples. Functional groups were probed using ATR-FTIR, and largely featureless spectra were obtained with only a few characteristic bands (Figure 2c). This finding is also in line with the relatively small variations in the oxygen elemental content found using EDX. From the presented FTIR spectra, it can be seen that the bands do not differ significantly in their positions or intensities. Only two bands stand out: those in the spectral ranges of 1606–1633 cm^{-1} and 1242–1287 cm^{-1}. The first mentioned band could be associated with in-plane vibrations of the sp^2 hybridized C=C bonds, while the second originates from C–OH vibrations [44,45]. We note that, as previously discussed [42], the structural disorder in all the samples was the same (as derived from Raman spectroscopy data), as it was dependent only on the temperature at which the samples were carbonized, which is common for all the samples in the series.

3.2. Adsorption of OPs

3.2.1. Adsorption Kinetics

The malathion and chlorpyrifos adsorption kinetics was investigated in batch adsorption experiments, and the data were processed by fitting experimental data points into equations corresponding to two frequently used kinetic models—pseudo-first-order (Equation (1)) and pseudo-second-order (Equation (2)) kinetics [46]:

$$q_t = q_e \left(1 - e^{-k_1 t}\right) \tag{1}$$

$$q_t = \frac{k_2 q_e^2 t}{1 + k_2 q_e t} \tag{2}$$

In the equations above, q_t and q_e are the adsorbed amounts of OPs in a given moment of time and the equilibrium adsorption capacity, respectively. The rate constants are k_1 and k_2 for the pseudo-first-order and pseudo-second-order kinetics, respectively.

The experimental data and corresponding fits are presented in Figure 3. The obtained equilibrium adsorption capacities and rate constants are summarized in Supplementary Information, Tables S1 and S2. There is a striking difference between malathion and chlorpyrifos adsorption in terms of the adsorption rate, as chlorpyrifos adsorption is much faster than malathion adsorption. Moreover, for malathion, several activated carbon fibers in the studied series performed quite well (Run1, Run8, Run16, and partially Run3), while Run 6 showed a low equilibrium adsorption capacity but high malathion adsorption kinetics. Considering the textural properties shown in Figure 2, it seems that a high adsorption capacity and fast malathion adsorption kinetics require pores larger than 1.2 nm, as this is the range in which the best-performing samples (Run1, Run8, Run16) significantly differ from all the others. In contrast, all the studied adsorbents collect chlorpyrifos exceptionally fast, and equilibrium is reached practically within 10 min of contact. Such fast adsorption made the measurements of q_t practically impossible for $t < 1$ min, making the use of other kinetic models, such as the inter-particle diffusion model, inappropriate. Namely, there is a lack of experimental points that could cover different time domains corresponding to different processes dominating adsorption.

Figure 4 assembles the obtained equilibrium adsorption capacities for a malathion concentration of 5×10^{-4} mol dm^{-3} and chlorpyrifos concentrations of 5×10^{-5} mol dm^{-3} and 5×10^{-4} mol dm^{-3}. As can be seen, the studied adsorbents perform rather well for chlorpyrifos, irrespective of the concentrations. However, for malathion, only three samples have adsorption capacities similar to chlorpyrifos. These are the samples Run1, Run8, and Run16—those with higher V_{tot}, d_{mean}, and S_{BET}. Pore diameter seems quite important. For example, Run5 has a higher V_{tot} and S_{BET} than Run8 (Table 3), but a much lower d_{mean}. Run7–9, Run13, and Run16 have the same d_{mean}, but Run8 and Run16 have the largest V_{tot} among these samples and, thus, performed the best.

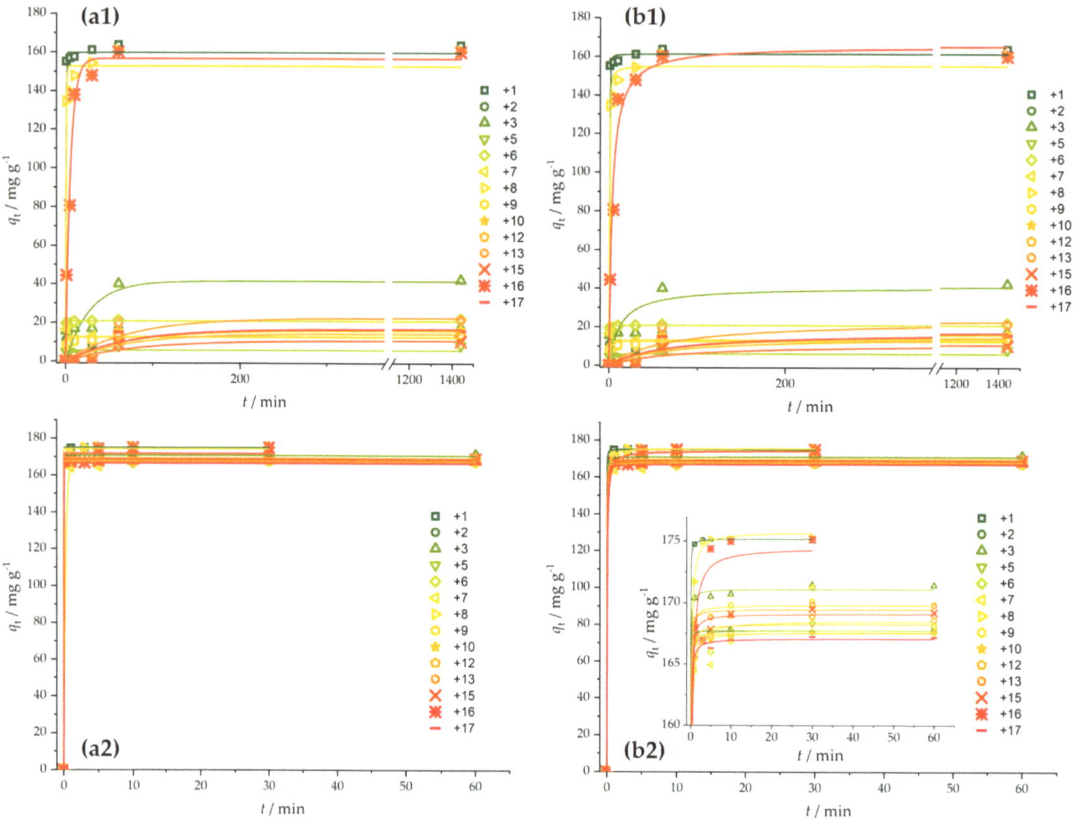

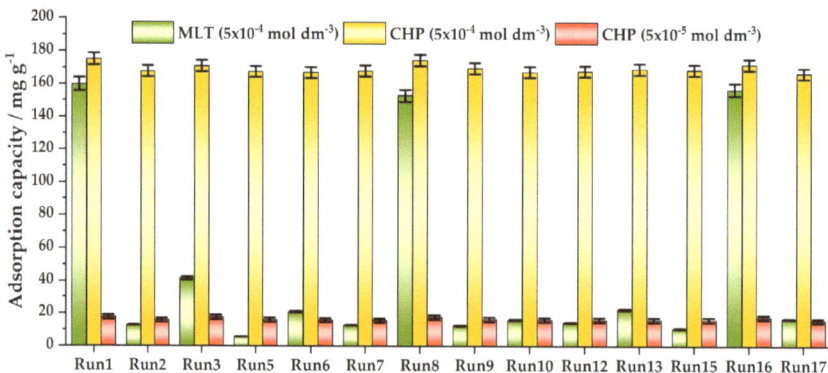

Figure 3. Kinetics of malathion (**a1,b1**) and chlorpyrifos (**a2,b2**): (**a**) pseudo-first reaction order, (**b**) pseudo-second reaction order. Pesticide concentrations were 5×10^{-4} mol dm^{-3}, and adsorbent loading was 1 mg mL^{-1}. Experimental uncertainties for presented data points are within 5%.

Figure 4. Adsorption capacities for malathion and chlorpyrifos for adsorbent loading of 1 mg mL^{-1}.

3.2.2. Adsorption Isotherms

The thermodynamics of the malathion and chlorpyrifos adsorption process was investigated by constructing adsorption isotherms and fitting experimental data into several adsorption isotherm models, which reveal different aspects of the adsorption process. We used the Freundlich (Equation (3)), Langmuir (Equation (4)), Temkin (Equation (5)), and Dubinin–Radushkevich (Equation (6)) models. The experimental data points and the corresponding fits are presented in Figure 5, while the fitted parameters are summarized in Supplementary Information, Tables S3–S6.

$$q_e = K_F C_e^{1/n} \tag{3}$$

$$q_e = \frac{q_{max} K_L C_e}{1 + K_L C_e} \tag{4}$$

$$q_e = \frac{RT}{b_T} \ln K_T C_e \tag{5}$$

$$q_e = q_{DR} e^{-K_{DR} \varepsilon^2} \tag{6}$$

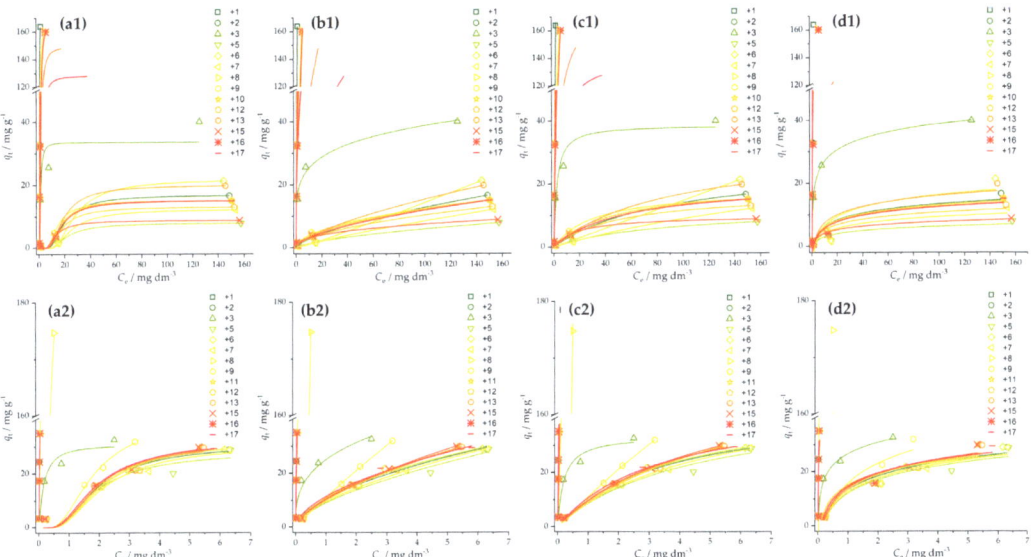

Figure 5. Adsorption isotherms for malathion (**a1,b1,c1,d1**) and chlorpyrifos (**a2,b2,c2,d2**): (**a**) Dubinin–Radushkevich, (**b**) Freundlich, (**c**) Langmuir, and (**d**) Temkin isotherms. Adsorbent loading was 1 mg mL^{-1}. Experimental uncertainties for presented data points are within 5%.

The used parameters are as follows: q_e (mg g^{-1})—equilibrium adsorption capacity; C_e (mg dm^{-3})—equilibrium adsorbate concentration; K_F (mg g^{-1} (mg dm^{-3})$^{1/n}$) and n—Freundlich constants; K_L (dm^3 mg^{-1}) and q_{max} (mg g^{-1})—Langmuir constant and theoretical maximum adsorption capacity of the monolayer, respectively; b_T (J g mol^{-1} mg^{-1}) and K_T (dm^3 mg^{-1})—Temkin isotherm constants; q_{DR}—maximum adsorption capacity; K_{DR} (mol^2 J^{-2})—constant associated with the mean free adsorption energy per mole of adsorbent, $\varepsilon = RT \times \ln(1 + 1/C_e)$.

Several adsorbents show a high affinity towards both malathion and chlorpyrifos, making the construction of complete adsorption isotherms difficult due to limitations in chlorpyrifos solubility, while the fitting in these cases is not highly reliable. Similar to the difference in the adsorption kinetics, the thermodynamics of malathion and chlorpyrifos adsorption is quite different. The n value of the Freundlich isotherm model is greater than 1 in all cases, indicating that adsorption is a favorable process. Constants indicating an affinity towards adsorption, those from the Temkin, Langmuir, and Freundlich adsorption isotherms, suggest that chlorpyrifos adsorption is more favored than malathion adsorption. The

Freundlich isotherm model describes most of the experimental results. Thus, the adsorption is likely physisorption, with a heterogeneous surface of carbon materials. Moreover, based on the Dubinin–Radushkevich model (i.e., the calculated adsorption free energy per mole of adsorbent ($E = (2K_{DR})^{-1/2} < 8$ kJ mol^{-1})) (Table S6, Supplementary Information), it can be seen that the adsorption process is physisorption, which means that there was no chemical bond formation between the OPs and investigated activated carbon fibers.

3.3. PCA Analysis

PCA and PC regression were further used to allow feature selection and to study the influence of the synthesis conditions and material properties on the adsorption capacity for both malathion and chlorpyrifos. We note that multiple regression was considered in the analysis of the data during our work, but because a high correlation of various input features was found in the PCA, we found regression based on PC components better suited for this analysis. Therefore, the reported discussion of the additive effects is based on PC regression analysis. The choice of the levels of variables related to the material synthesis conditions is explained in Section 2.1, while all the other variables (material properties and adsorption performance) were determined experimentally.

As a starting point, 10 input features were considered—those assembled in Tables 1–3. After scaling the data and performing PCA, the first three principal components were shown to account for over 80% of the variance (Supplementary Materials, Figure S1). However, a high correlation of multiple features was found in PC1. Therefore, further consideration was carried out by dividing the input features into two sets: the first contained the synthesis conditions (activation temperature, gas flow rate, and activation time) and material surface (S_{BET}), while the other set included all the material properties. Figure 6 shows the variance contribution of PCs and the heatmap plot of the variable input contribution to each PC for both cases considered.

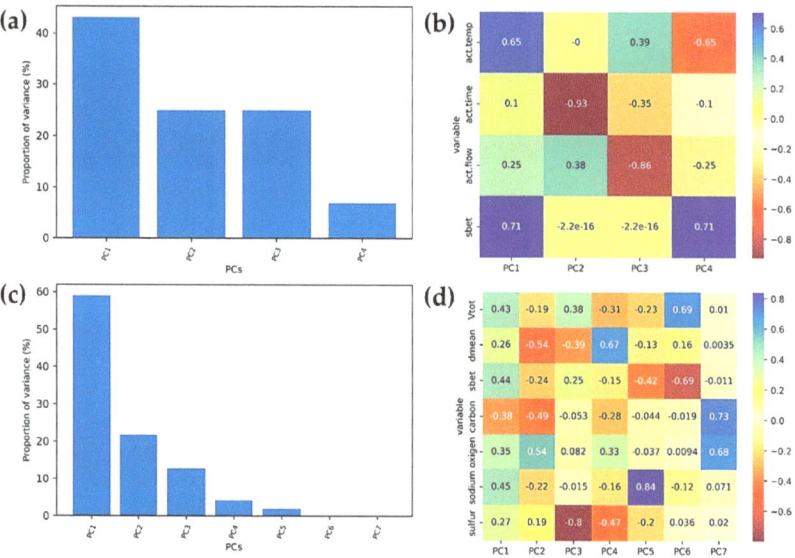

Figure 6. (**a**,**c**) PC variance proportion and (**b**,**d**) heatmap plot of input feature contributions in the PCs for two considered cases. PCA results for case 1 (**a**,**b**)—synthesis conditions are the main input features, and case 2 (**c**,**d**)—material properties are the input features.

Based on the PCA, we can see that, in the first case, in which the synthesis conditions are considered, the activation temperature and sample surface size are primary and correlated features in PC1, and the activation time is the main contributor to PC2. In case 2, in which seven material properties are considered as the input features, there is still a correlation of multiple input features in PC1, while PC2 is significantly correlated to the oxygen concentration and negatively correlated to the carbon concentration and pore diameter.

Linear regression was used to consider the performance of the PCs in both cases for the prediction of the malathion and chlorpyrifos adsorption capacities (Figure 4; pesticide concentrations: 5×10^{-4} mol dm^{-3}). We see that, in both cases, we obtained a better performance for chlorpyrifos. Taking material properties as a starting point for PCA (case 2) outperforms other studied approaches and provides significantly lower mean square errors (MSEs) (Table 4). Thus, linking the synthesis conditions and/or material properties to the material performance for OP removal seems to be a plausible way to rationalize synthetic routes for obtaining high-performance adsorbents. We note that the presented results should be considered as a starting point for building up powerful predictive models, which we plan to establish with the growth of our materials library.

Table 4. Mean square errors for linear regression model trained with various numbers of PCs from two cases described in the text: case 1, in which synthesis conditions and S_{BET} are taken as input features for PCA, and case 2, in which 7 material properties are considered as input features for PC (q_{MLT}—malathion adsorption capacity, q_{CHP}—chlorpyrifos adsorption capacity).

	Case 1		Case 2	
	MSE(q_{MLT}) /mg g^{-1}	MSE(q_{CHP}) /mg g^{-1}	MSE(q_{MLT}) /mg g^{-1}	MSE(q_{CHP}) /mg g^{-1}
PC1	63.35	2.33	43.77	0.94
PC1-2	56.5	2.26	27.26	0.59
PC1-3	56.19	3.06	187	6.65

3.4. Selective Removal of Pesticides—Adsorption from Mixtures

In the presented series of adsorbents, there are distinct differences in the adsorption performance, where some materials adsorb both malathion and chlorpyrifos, while all materials perform very well for chlorpyrifos removal. Considering that the used OPs are structurally different, the question is whether the adsorption of one OP affects the adsorption of the other one. To investigate the performance of activated carbon fibers towards removing malathion and chlorpyrifos in mixtures, we studied the removal of OPs from the mixture containing 5×10^{-5} mol dm^{-3} of malathion and chlorpyrifos, using the adsorbent dose of 1 mg mL^{-1}. The results are summarized in Table S7, Supplementary Information. Figure 7 presents the comparison of two selected samples, Run1 and Run3, towards the removal of malathion and chlorpyrifos from the mixtures. Sample Run1 performs well for both OPs (Figure 4 and Table S2, Supporting Information; adsorption capacity for chlorpyrifos 175 mg g^{-1}, and for malathion 165 mg g^{-1}, while sample Run3 performs well for chlorpyrifos and much lower for malathion (Figure 4 and Table S2, Supporting Information; adsorption capacity for chlorpyrifos 171 mg g^{-1}, and for malathion 41 mg g^{-1}). As can be seen, Run1 practically completely removes both pesticides (100% uptake, Table S7). Run3 removes chlorpyrifos to a high degree (although less than Run1, uptake is 96.9%), but only 20% of malathion (Table S7). We believe that this is a combination of two factors—the different specific surface areas (Table 3) and different adsorption kinetics of these two pesticides (Figure 3). The sample Run1 has a much larger specific surface and five times larger average pore diameter than Run3, allowing the adsorption of both pesticides due to the availability of a number of adsorption sites, while Run3 has a lower specific surface and much narrower pores. As chlorpyrifos adsorption is faster, it quickly occupies the available surface so that malathion cannot adsorb. Considering, for example, the fitting of the adsorption isotherms using the Freundlich model, the affinity of adsorbents is significantly higher towards chlorpyrifos (Table S3, Supplementary Information) compared to malathion. This leads to the conclusion that once chlorpyrifos is adsorbed onto the surface, it cannot be displaced by malathion. We note that the adsorption performance towards individual pesticides (Figure 4) reflects well on their adsorption performance in the mixture of pesticides (Table S7, Supplementary Information). Thus, it seems that by tailoring the synthetic conditions, which ultimately reflect the material properties, it is possible to obtain adsorbents that selectively remove certain compounds. Naturally, in the case of OPs, the target is to remove as many contaminants as possible, irrespective of their chemical structures. However, considering a broader perspective, particularly in the case of food processing and food safety, it is vital that contaminants are removed while essential nutrients remain in the treated samples, and that adsorbents can operate in complex matrices. This issue is highly relevant, as many contaminants, such as OPs, can be transferred from the original sources into food products. For liquid samples, adsorption seems to be an easy additional (filtration) step to be added in the processing stage, which could significantly improve the quality and safety of the final products if the selective removal of contaminants is possible.

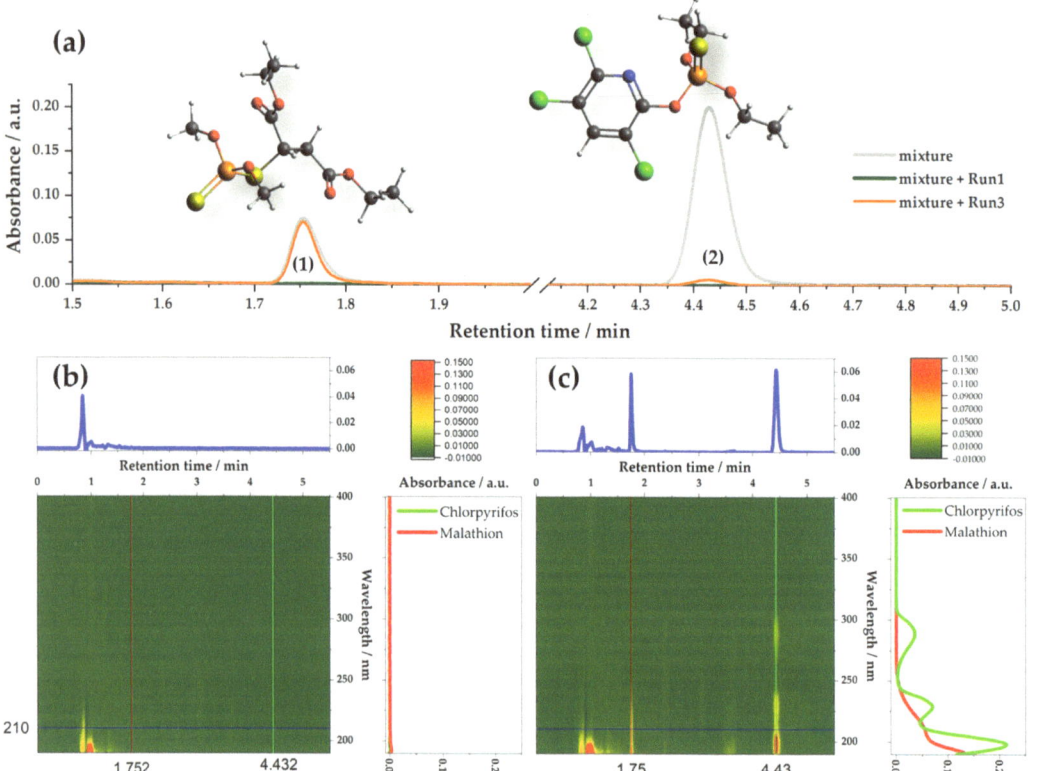

Figure 7. (**a**) UPLC chromatogram of the malathion + chlorpyrifos mixture (5×10^{-5} mol dm^{-3}) before and after the adsorptive removal using Run1 and Run3 samples: adsorbent loading was 1 mg mL^{-1}, and wavelength was 200 nm; peak 1 corresponds to malathion, peak 2 corresponds to chlorpyrifos; (**b**) PDA signal of the malathion + chlorpyrifos mixture after the adsorptive removal of pesticides using the Run1 sample, with extracted chromatogram at 210 nm (at which malathion and chlorpyrifos have the same absorption coefficients), and spectra for retention times 1.75 and 4.43 min; (**c**) the same as for (**b**) but with the sample Run3 as adsorbent.

The advantages of using adsorption are clear over aggressive chemical or physical treatments to remove contaminants, which could negatively impact the final product's nutritional value. Food processing nowadays includes various chemical and physical methods of food treatment using bleach, chlorine dioxide [47], ultraviolet (UV) radiation [48], and peracetic acid [49]. Even though their effects on cleansing food are proven, side effects of their use are evident. Stout et al. showed that the use of bleaching agents leads to the degradation of vitamins and carotenoids in whey protein [50], meaning that the application of bleach in food processing changes the chemical composition of food itself. Chlorine dioxide has similar disinfectant properties to bleach, but it is very unsafe for use in large amounts because it can damage red blood cells and the lining of the gastrointestinal tract. UV radiation has a high oxidation power, and it could cause changes in food taste and discolor it. Furthermore, Li et al. demonstrated that by exposing malathion to UV radiation, it degrades to its more toxic form—malaoxon—making the problem larger [51]. Peracetic acid is corrosive to the eyes, skin, and respiratory tract, making it unsafe to use. Moreover, using it in food processing could affect the content of antioxidant compounds in the food, as its disinfectant activity is based on the release of active oxygen [52].

As the treatment of food with carbon materials does not affect the food itself, only the OP on its surface, and does not cause the formation of more toxic products, it is safe to consider its use during food processing instead of the ones listed above. Moreover, activated carbon has a high potential for application in food packaging, aiming to maintain food quality and control food safety [53].

In the next step, we show that the presented activated carbon fibers can effectively remove the studied OPs from realistic food samples.

3.5. Application of Investigated Materials in Food Processing—Filtration Step for Pesticide Removal

In this section, we demonstrate that the presented activated carbon fibers can be used for the removal of chlorpyrifos and malathion from real samples using filtration. In particular, we focused on removing malathion from lemon juice, as lemons are frequently treated by malathion. In addition, we analyzed the removal of chlorpyrifos from mint ethanol extract, in which chlorpyrifos could be found due to the relatively high solubility in ethanol compared to water.

3.5.1. Applications in Real Samples

To investigate the performance of the best-performing materials in the series for the real-sample treatment, lemon juice and mint extract spiked with 5×10^{-5} mol dm^{-3} malathion and chlorpyrifos, respectively, were treated with the materials under the dynamic conditions, using modified nylon filters, as described in Section 2.3. For comparison, the adsorption of malathion and chlorpyrifos from deionized water and 50% ethanol, respectively, were investigated under identical conditions. The results are summarized in Table 5.

Table 5. Typical uptakes for 5×10^{-5} mol dm^{-3} malathion in deionized water and lemon juice and 5×10^{-5} mol dm^{-3} chlorpyrifos in 50% ethanol and mint ethanol extract. Presented results were obtained when 0.5 mg mL^{-1} of adsorbents were used under dynamic conditions at 20 °C.

Material	Uptake (%)			
	Malathion in Deionized Water	Malathion in Lemon Juice	Chlorpyrifos in 50% Ethanol	Chlorpyrifos in Mint Ethanol Extract
Run1	81 ± 4	73 ± 5	90 ± 3	84 ± 5
Run8	64 ± 4	66 ± 4	51 ± 4	53 ± 3
Run16	47 ± 3	39 ± 4	34 ± 3	31 ± 3

From Table 5, it can be seen that the investigated materials show comparable uptakes in real samples, as in water and ethanol. The influence of the matrix of the real sample in the analyzed samples was not pronounced and did not interfere with the performance of the investigated adsorbents.

3.5.2. Material Regeneration and Reuse

To investigate the possibility of material regeneration, samples Run1, Run8, and Run16 were used for the adsorption under the dynamic conditions for malathion and chlorpyrifos removal from real samples, as described above. After the first round of adsorption experiments, filters modified with material were washed with 5 mL of absolute ethanol for 1 min. Subsequently, filters were used again under the same experimental conditions. This cycle was repeated five times, and the results are presented in Figure 8.

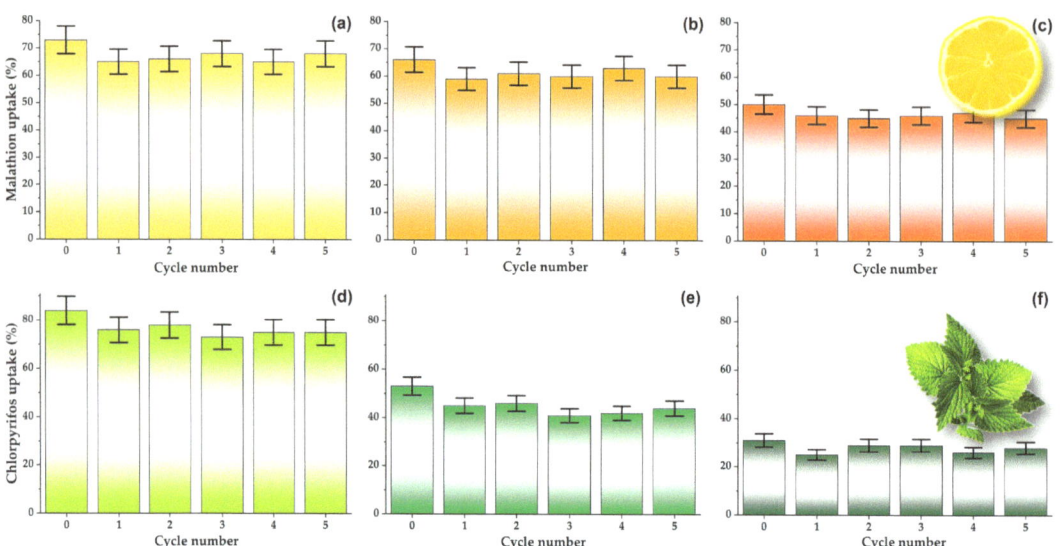

Figure 8. Regeneration of 0.5 mg mL^{-1} of Run1 (**a,d**), Run8 (**b,e**), and Run16 (**c,f**) after the adsorption of 5×10^{-5} mol dm^{-3} malathion in lemon juice (**a,b,c**) and 5×10^{-5} mol dm^{-3} chlorpyrifos in mint ethanol extract (**d,e,f**).

From the data presented in Figure 8, it can be concluded that the investigated material can be successfully regenerated for at least five cycles under the given experimental conditions. Namely, after a small drop in performance between the initial adsorption and the first cycle, the malathion and chlorpyrifos uptakes remained nearly constant in the subsequent five regeneration–adsorption cycles. The presented data strongly emphasize the potential economic aspect of carbon-based filter application in food processing.

4. Conclusions

In the present work, a series of viscose-derived activated carbon fibers was produced using the Design of Experiments protocol to set the parametric space for the activation step, allowing for the systematic tuning of the material properties. As a result, highly effective adsorbents for malathion and chlorpyrifos removal were obtained. Both malathion and chlorpyrifos were found to be physisorbed on the studied activated carbon fibers. All the studied materials performed very well for chlorpyrifos removal, giving adsorption capacities above 150 mg g^{-1}. The materials with higher surface areas, pore volumes, and larger pore diameters all activated at higher temperatures (870 °C) and also performed as excellent adsorbents for malathion. Moreover, a higher affinity of adsorbents towards chlorpyrifos also resulted in fast adsorption kinetics, where equilibrium was achieved within 10 min of contact with OP-containing solutions. By carefully tuning the carbon materials' properties, it was possible to selectively remove chlorpyrifos in the presence of malathion. These results have very important implications for further developments. Namely, the problems of complex matrices in food processing can be alleviated by proper material design, allowing the selective adsorptive removal of contaminants while not affecting the nutritive value of food like traditional aggressive chemical and physical treatments. As an example, here, we have demonstrated that the produced carbons can be used for malathion removal from lemon juice and chlorpyrifos removal from mint ethanol extracts, followed by several regeneration cycles, without apparent adsorption performance loss. Thus, precisely tailored carbon materials could be used to process liquid samples used in the food industry to remove contaminants and improve food safety and quality effectively. Considering future perspectives, the present work shows that by using different data-based models, such as PCA and PC regression, it is possible to link the material synthesis conditions and/or material properties to the adsorption performance. Thus, in principle, it is possible to devise models of different complexity that could provide optimal synthetic routes for carbon material for the adsorptive removal of targeted pollutant(s). It is easy to perceive the importance and the possibilities of such an approach; for example, using geographic and environmental data to tackle pollutants that are problematic for a given geographic region. To build a model that can be fed to machine learning models or artificial intelligence, it is essential to generate a sufficient amount of highly reliable data for the training, which requires a systematic approach to material synthesis, careful and in-depth characterization, and standardized protocols for material performance assessment.

Supplementary Materials: The following supporting information can be downloaded at https://www.mdpi.com/article/10.3390/foods12122362/s1, Table S1: Kinetics parameters for malathion (5×10^{-4} mol dm^{-3}) and chlorpyrifos (5×10^{-4} mol dm^{-3}) for pseudo-first-order kinetics, adsorbent dose 1 mg mL^{-1}; Table S2: Kinetics parameters for malathion (5×10^{-4} mol dm^{-3}) and chlorpyrifos (5×10^{-4} mol dm^{-3}) for pseudo-second-order kinetics, adsorbent dose 1 mg mL^{-1}; Table S3: Parameters for malathion and chlorpyrifos adsorption using Freundlich adsorption isotherm, adsorbent dose 1 mg mL^{-1}; Table S4: Parameters for malathion and chlorpyrifos adsorption using Langmuir adsorption isotherm, adsorbent dose 1 mg mL^{-1}; Table S5: Parameters for malathion and chlorpyrifos adsorption using Temkin adsorption isotherm, adsorbent dose 1 mg mL^{-1}; Table S6: Parameters for malathion and chlorpyrifos adsorption using Dubinin–Raduskevich adsorption isotherm, adsorbent dose 1 mg mL^{-1} (MLT—malathion, CHP—chlopryrifos); Figure S1: PC variance proportion (left) and heatmap plot of input feature contributions in the PCs (right) for the case of 10 input variables; Table S7: Adsorptive removal of malathion and chlorpyrifos in the mixture (5×10^{-5} mol dm^{-3} of each pesticide; 30 min equilibration time; 25 °C; adsorbent dose 1 mg mL^{-1}) given as pesticide uptake.

Author Contributions: Conceptualization, I.A.P. and T.L.-P.; methodology, I.A.P., S.B., C.U., C.F., K.B. and T.L.-P.; software, I.A.P.; formal analysis, V.M., T.T., K.B., T.L.-P. and S.B.; investigation, V.M., T.T., S.B. and T.L.-P.; resources, I.A.P., C.F. and T.L.-P.; writing—original draft preparation, I.A.P., V.M., T.T., K.B. and T.L.-P.; writing—review and editing, I.A.P., S.B., C.U., C.F. and T.L.-P. All authors have read and agreed to the published version of the manuscript.

Funding: S.B., C.U. and C.F. wish to thank the European Regional Development Fund (EFRE) and the province of Upper Austria for financial support for this study through the program IWB 2014–2020 (project BioCarb-K). T.L.P., V.M., K.B. and T.T. acknowledge the support provided by the Serbian Ministry of Education, Science and Technological Development (contract number: 451-03-47/2023-01/200017). I.A.P. acknowledges the support provided by the Serbian Ministry of Education, Science and Technological Development (contract number: 451-03-47/2023-01/200146).

Data Availability Statement: The data used to support the findings of this study can be made available by the corresponding author upon request.

Conflicts of Interest: The authors declare no conflict of interest.

References

1. Valin, H.; Sands, R.D.; van der Mensbrugghe, D.; Nelson, G.C.; Ahammad, H.; Blanc, E.; Bodirsky, B.; Fujimori, S.; Hasegawa, T.; Havlik, P.; et al. The future of food demand: Understanding differences in global economic models. *Agric. Econ.* **2014**, *45*, 51–67. [CrossRef]
2. Ponisio, L.C.; Ehrlich, P.R. Diversification, Yield and a New Agricultural Revolution: Problems and Prospects. *Sustainability* **2016**, *8*, 1118. [CrossRef]

3. Mzyece, A.; Ng'ombe, J.N. Does Crop Diversification Involve a Trade-Off Between Technical Efficiency and Income Stability for Rural Farmers? Evidence from Zambia. *Agronomy* **2020**, *10*, 1875. [CrossRef]
4. Farm to Fork Strategy. Available online: https://food.ec.europa.eu/horizontal-topics/farm-fork-strategy_en (accessed on 15 April 2023).
5. Bromilow, R.H. Pesticides. In *Encyclopedia of Soils in the Environment*; Hillel, D., Ed.; Elsevier: Oxford, UK, 2005; pp. 188–195.
6. Richendrfer, H.; Creton, R. Chlorpyrifos and malathion have opposite effects on behaviors and brain size that are not correlated to changes in AChE activity. *Neurotoxicology* **2015**, *49*, 50–58. [CrossRef]
7. Elmorsy, E.; Al-Ghafari, A.; Al Doghaither, H.; Salama, M.; Carter, W.G. An Investigation of the Neurotoxic Effects of Malathion, Chlorpyrifos, and Paraquat to Different Brain Regions. *Brain Sci.* **2022**, *12*, 975. [CrossRef] [PubMed]
8. Lazarević-Pašti, T.; Čolović, M.; Savić, J.; Momić, T.; Vasić, V. Oxidation of diazinon and malathion by myeloperoxidase. *Pestic. Biochem. Physiol.* **2011**, *100*, 140–144. [CrossRef]
9. Lazarević-Pašti, T.; Nastasijević, B.; Vasić, V. Oxidation of chlorpyrifos, azinphos-methyl and phorate by myeloperoxidase. *Pestic. Biochem. Physiol.* **2011**, *101*, 220–226. [CrossRef]
10. Mitić, M.; Lazarević-Pašti, T. Does the application of acetylcholinesterase inhibitors in the treatment of Alzheimer's disease lead to depression? *Expert Opin. Drug Metab. Toxicol.* **2021**, *17*, 841–856. [CrossRef]
11. Ubaid Ur Rahman, H.; Asghar, W.; Nazir, W.; Sandhu, M.A.; Ahmed, A.; Khalid, N. A comprehensive review on chlorpyrifos toxicity with special reference to endocrine disruption: Evidence of mechanisms, exposures and mitigation strategies. *Sci. Total Environ.* **2021**, *755*, 142649. [CrossRef]
12. Sparling, D.W.; Fellers, G. Comparative toxicity of chlorpyrifos, diazinon, malathion and their oxon derivatives to larval *Rana boylii*. *Environ. Pollut.* **2007**, *147*, 535–539. [CrossRef] [PubMed]
13. MacDonald, R.; Reitmeier, C. Chapter 6—Food Processing. In *Understanding Food Systems*; MacDonald, R., Reitmeier, C., Eds.; Academic Press: Cambridge, MA, USA, 2017; pp. 179–225.
14. Bolouri, P.; Salami, R.; Kouhi, S.; Kordi, M.; Asgari Lajayer, B.; Hadian, J.; Astatkie, T. Applications of Essential Oils and Plant Extracts in Different Industries. *Molecules* **2022**, *27*, 8999. [CrossRef]
15. Proestos, C. The Benefits of Plant Extracts for Human Health. *Foods* **2020**, *9*, 1653. [CrossRef]
16. Singh, S.; Sharma, S.; Sarma, S.J.; Misra, K.; Brar, S.K. Chapter 10—Pesticides in Water. In *Handbook of Water Purity and Quality*, 2nd ed.; Ahuja, S., Ed.; Academic Press: Amsterdam, The Netherlands, 2021; pp. 231–253.
17. Colosio, C.; Rubino, F.M.; Moretto, A. Pesticides. In *International Encyclopedia of Public Health*, 2nd ed.; Quah, S.R., Ed.; Academic Press: Oxford, UK, 2017; pp. 454–462.
18. Alamu, E.O.; Mooya, A. Chapter 10—Food Processing Technologies and Value Addition for Improved Food Safety and Security. In *Smart Technologies for Sustainable Smallholder Agriculture*; Nhamo, N., Chikoye, D., Gondwe, T., Eds.; Academic Press: Cambridge, MA, USA, 2017; pp. 201–210.
19. Savić, J.Z.; Petrović, S.Ž.; Leskovac, A.R.; Lazarević Pašti, T.D.; Nastasijević, B.J.; Tanović, B.B.; Gašić, S.M.; Vasić, V.M. UV-C light irradiation enhances toxic effects of chlorpyrifos and its formulations. *Food Chem.* **2019**, *271*, 469–478. [CrossRef] [PubMed]
20. Jocić, A.; Breitenbach, S.; Bajuk-Bogdanović, D.; Pašti, I.A.; Unterweger, C.; Fürst, C.; Lazarević-Pašti, T. Viscose-Derived Activated Carbons Fibers as Highly Efficient Adsorbents for Dimethoate Removal from Water. *Molecules* **2022**, *27*, 1477. [CrossRef] [PubMed]
21. Jocić, A.; Breitenbach, S.; Pašti, I.A.; Unterweger, C.; Fürst, C.; Lazarević-Pašti, T. Viscose-derived activated carbons as adsorbents for malathion, dimethoate, and chlorpyrifos—Screening, trends, and analysis. *Environ. Sci. Pollut. Res.* **2022**, *29*, 35138–35149. [CrossRef]
22. Mirković, M.M.; Pašti, T.D.L.; Došen, A.M.; Čebela, M.Ž.; Rosić, A.A.; Matović, B.Z.; Babić, B.M. Adsorption of malathion on mesoporous monetite obtained by mechanochemical treatment of brushite. *RSC Adv.* **2016**, *6*, 12219–12225. [CrossRef]
23. Bondžić, A.M.; Lazarević Pašti, T.D.; Pašti, I.A.; Bondžić, B.P.; Momčilović, M.D.; Loosen, A.; Parac-Vogt, T.N. Synergistic Effect of Sorption and Hydrolysis by NU-1000 Nanostructures for Removal and Detoxification of Chlorpyrifos. *ACS Appl. Nano Mater.* **2022**, *5*, 3312–3324. [CrossRef]
24. Masini, J.C.; Abate, G. Guidelines to Study the Adsorption of Pesticides onto Clay Minerals Aiming at a Straightforward Evaluation of Their Removal Performance. *Minerals* **2021**, *11*, 1282. [CrossRef]
25. Bruna, F.; Pavlovic, I.; Barriga, C.; Cornejo, J.; Ulibarri, M.A. Adsorption of pesticides Carbetamide and Metamitron on organohydrotalcite. *Appl. Clay Sci.* **2006**, *33*, 116–124. [CrossRef]
26. Milojević-Rakić, M.; Janošević, A.; Krstić, J.; Nedić Vasiljević, B.; Dondur, V.; Ćirić-Marjanović, G. Polyaniline and its composites with zeolite ZSM-5 for efficient removal of glyphosate from aqueous solution. *Microporous Mesoporous Mater.* **2013**, *180*, 141–155. [CrossRef]
27. Lazarević-Pašti, T.; Anićijević, V.; Baljozović, M.; Anićijević, D.V.; Gutić, S.; Vasić, V.; Skorodumova, N.V.; Pašti, I.A. The impact of the structure of graphene-based materials on the removal of organophosphorus pesticides from water. *Environ. Sci. Nano* **2018**, *5*, 1482–1494. [CrossRef]
28. Momić, T.; Pašti, T.L.; Bogdanović, U.; Vodnik, V.; Mraković, A.; Rakočević, Z.; Pavlović, V.B.; Vasić, V. Adsorption of Organophosphate Pesticide Dimethoate on Gold Nanospheres and Nanorods. *J. Nanomater.* **2016**, *2016*, 8910271. [CrossRef]

29. Zieliński, B.; Miądlicki, P.; Przepiórski, J. Development of activated carbon for removal of pesticides from water: Case study. *Sci. Rep.* **2022**, *12*, 20869. [CrossRef] [PubMed]
30. Mohammad, S.G.; Ahmed, S.M.; Amr, A.E.-G.E.; Kamel, A.H. Porous Activated Carbon from Lignocellulosic Agricultural Waste for the Removal of Acetampirid Pesticide from Aqueous Solutions. *Molecules* **2020**, *25*, 2339. [CrossRef]
31. Kalsoom; Khan, S.; Ullah, R.; Adil, M.; Waheed, A.; Khan, K.A.; Ghramh, H.A.; Alharby, H.F.; Alzahrani, Y.M.; Alghamdi, S.A.; et al. Adsorption of Pesticides Using Wood-Derived Biochar and Granular Activated Carbon in a Fixed-Bed Column System. *Water* **2022**, *14*, 2937. [CrossRef]
32. Wiśniewska, M.; Pawlak, N.; Sternik, D.; Pietrzak, R.; Nowicki, P.P. Production of Activated Carbons from Food/Storage Waste. *Materials* **2023**, *16*, 1349. [CrossRef] [PubMed]
33. Plens, A.C.O.; Monaro, D.L.G.; Coutinho, A.R. Adsorption of SO_X and NO_X in activated viscose fibers. *An. Acad. Bras. Ciênc.* **2015**, *87*, 1149–1160. [CrossRef] [PubMed]
34. Bhati, S.; Mahur, J.; Dixit, S.; Choubey, O. Surface and adsorption properties of activated carbon fabric prepared from cellulosic polymer: Mixed activation method. *Bull. Korean Chem. Soc.* **2013**, *34*, 569–573. [CrossRef]
35. Sun, Y.; Cheng, B.; Wu, X.; Meng, K.; Duan, S.; Tao, T.; Min, X.; Huang, Z.; Fang, M.; Ding, H. Fabrication and Heavy Metals Adsorption Performance of Viscose Based Activated Carbon Fibers. *SSRN Electron. J.* **2022**, preprint. [CrossRef]
36. Liu, Q.-X.; Zhou, Y. R.; Wang, M.; Zhang, Q.; Ji, T.; Chen, T.-Y.; Yu, D.-C. Adsorption of methylene blue from aqueous solution onto viscose-based activated carbon fiber felts: Kinetics and equilibrium studies. *Adsorpt. Sci. Technol.* **2019**, *37*, 312–332. [CrossRef]
37. Liu, F.; Zhou, L.; Tao, L.; Qian, L.; Yu, G.; Deng, S. Adsorption behavior and mechanism of Au(III) on caffeic acid functionalized viscose staple fibers. *Chemosphere* **2020**, *253*, 126704. [CrossRef]
38. Lazarević-Pašti, T.; Jocić, A.; Milanković, V.; Tasić, T.; Potkonjak, N.; Breitenbach, S.; Unterweger, C.; Fürst, C.; Pašti, I.A. Kinetics of Dimethoate, Malathion, and Chlorpyrifos Adsorption on Cellulose-Derived Activated Carbons–Linking Performance to the Physicochemical Properties. *Preprints.org* **2022**, 2022110449. Available online: https://www.researchgate.net/publication/365709749_Kinetics_of_Dimethoate_Malathion_and_Chlorpyrifos_Adsorption_on_Cellulose-derived_Activated_Carbons_-_Linking_Performance_to_the_Physicochemical_Properties (accessed on 8 May 2023).
39. Breitenbach, S.; Unterweger, C.; Lumetzberger, A.; Duchoslav, J.; Stifter, D.; Hassel, A.W.; Fürst, C. Viscose-based porous carbon fibers: Improving yield and porosity through optimization of the carbonization process by design of experiment. *J. Porous Mater.* **2021**, *28*, 727–739. [CrossRef]
40. Osterrieth, J.W.M.; Rampersad, J.; Madden, D.; Rampal, N.; Skoric, L.; Connolly, B.; Allendorf, M.D.; Stavila, V.; Snider, J.L.; Ameloot, R.; et al. How Reproducible are Surface Areas Calculated from the BET Equation? *Adv. Mater.* **2022**, *34*, 2201502. [CrossRef] [PubMed]
41. Breitenbach, S.; Gavrilov, N.; Pašti, I.; Unterweger, C.; Duchoslav, J.; Stifter, D.; Hassel, A.W.; Fürst, C. Biomass-Derived Carbons as Versatile Materials for Energy-Related Applications: Capacitive Properties vs. Oxygen Reduction Reaction Catalysis. *C* **2021**, *7*, 55. [CrossRef]
42. Breitenbach, S.; Lumetzberger, A.; Hobisch, M.A.; Unterweger, C.; Spirk, S.; Stifter, D.; Fürst, C.; Hassel, A.W. Supercapacitor Electrodes from Viscose-Based Activated Carbon Fibers: Significant Yield and Performance Improvement Using Diammonium Hydrogen Phosphate as Impregnating Agent. *C* **2020**, *6*, 17. [CrossRef]
43. Anićijević, V.; Tasić, T.; Milanković, V.; Breitenbach, S.; Unterweger, C.; Fürst, C.; Bajuk-Bogdanović, D.; Pašti, I.A.; Lazarević-Pašti, T. How Well Do Our Adsorbents Actually Perform?—The Case of Dimethoate Removal Using Viscose Fiber-Derived Carbons. *Int. J. Environ. Res. Public Health* **2023**, *20*, 4553. [CrossRef] [PubMed]
44. Dun, W.; Guijian, L.; Ruoyu, S.; Xiang, F. Investigation of Structural Characteristics of Thermally Metamorphosed Coal by FTIR Spectroscopy and X-ray Diffraction. *Energy Fuels* **2013**, *27*, 5823–5830. [CrossRef]
45. Norberg, I.; Nordström, Y.; Drougge, R.; Gellerstedt, G.; Sjöholm, E. A new method for stabilizing softwood kraft lignin fibers for carbon fiber production. *J. Appl. Polym. Sci.* **2013**, *128*, 3824–3830. [CrossRef]
46. Revellame, E.D.; Fortela, D.L.; Sharp, W.; Hernandez, R.; Zappi, M.E. Adsorption kinetic modeling using pseudo-first order and pseudo-second order rate laws: A review. *Clean. Eng. Technol.* **2020**, *1*, 100032. [CrossRef]
47. Fukayama, M.Y.; Tan, H.; Wheeler, W.B.; Wei, C.I. Reactions of aqueous chlorine and chlorine dioxide with model food compounds. *Environ. Health Perspect.* **1986**, *69*, 267–274. [CrossRef] [PubMed]
48. Ramesh, T.; Nayak, B.; Amirbahman, A.; Tripp, C.P.; Mukhopadhyay, S. Application of ultraviolet light assisted titanium dioxide photocatalysis for food safety: A review. *Innov. Food Sci. Emerg. Technol.* **2016**, *38*, 105–115. [CrossRef]
49. Zoellner, C.; Aguayo-Acosta, A.; Siddiqui, M.W.; Dávila-Aviña, J.E. Chapter 2—Peracetic Acid in Disinfection of Fruits and Vegetables. In *Postharvest Disinfection of Fruits and Vegetables*; Siddiqui, M.W., Ed.; Academic Press: Cambridge, MA, USA, 2018; pp. 53–66.
50. Stout, M.A.; Park, C.W.; Drake, M.A. The effect of bleaching agents on the degradation of vitamins and carotenoids in spray-dried whey protein concentrate. *J. Dairy Sci.* **2017**, *100*, 7922–7932. [CrossRef] [PubMed]
51. Li, W.; Zhao, Y.; Yan, X.; Duan, J.; Saint, C.P.; Beecham, S. Transformation pathway and toxicity assessment of malathion in aqueous solution during UV photolysis and photocatalysis. *Chemosphere* **2019**, *234*, 204–214. [CrossRef] [PubMed]

52. González-Aguilar, G.; Ayala-Zavala, J.F.; Chaidez-Quiroz, C.; Heredia, J.B.; Campo, N.C.-D. Peroxyacetic Acid. In *Decontamination of Fresh and Minimally Processed Produce*; Wiley-Blackwell, John Wiley & Sons: Hoboken, NJ, USA, 2012; pp. 215–223.
53. Chaemsanit, S.; Matan, N.; Matan, N. Activated Carbon for Food Packaging Application: Review. *Walailak J. Sci. Technol. (WJST)* **2018**, *15*, 255–271. [CrossRef]

Disclaimer/Publisher's Note: The statements, opinions and data contained in all publications are solely those of the individual author(s) and contributor(s) and not of MDPI and/or the editor(s). MDPI and/or the editor(s) disclaim responsibility for any injury to people or property resulting from any ideas, methods, instructions or products referred to in the content.

Article

Spent Coffee Grounds as an Adsorbent for Malathion and Chlorpyrifos—Kinetics, Thermodynamics, and Eco-Neurotoxicity

Vedran Milanković [1], Tamara Tasić [1], Milica Pejčić [1], Igor Pašti [2] and Tamara Lazarević-Pašti [1,*]

[1] VINČA Institute of Nuclear Sciences—National Institute of the Republic of Serbia, University of Belgrade, Mike Petrovica Alasa 12–14, 11000 Belgrade, Serbia; vedran.milankovic@vin.bg.ac.rs (V.M.); tamara.tasic@vin.bg.ac.rs (T.T.); milica.pejcic@vin.bg.ac.rs (M.P.)

[2] Faculty of Physical Chemistry, University of Belgrade, Studentski Trg 12–16, 11158 Belgrade, Serbia; igor@ffh.bg.ac.rs

* Correspondence: tamara@vin.bg.ac.rs

Citation: Milanković, V.; Tasić, T.; Pejčić, M.; Pašti, I.; Lazarević-Pašti, T. Spent Coffee Grounds as an Adsorbent for Malathion and Chlorpyrifos—Kinetics, Thermodynamics, and Eco-Neurotoxicity. *Foods* **2023**, *12*, 2397. https://doi.org/10.3390/foods12122397

Academic Editor: Thierry Noguer

Received: 18 May 2023
Revised: 14 June 2023
Accepted: 15 June 2023
Published: 16 June 2023

Copyright: © 2023 by the authors. Licensee MDPI, Basel, Switzerland. This article is an open access article distributed under the terms and conditions of the Creative Commons Attribution (CC BY) license (https://creativecommons.org/licenses/by/4.0/).

Abstract: Coffee is one of the most popular beverages, with around 10.5 million tons manufactured annually. The same amount of spent coffee grounds (SCGs) might harm the environment if disposed of carelessly. On the other hand, pesticide contamination in food and biowaste is a rising problem. Because pesticides are hazardous and can cause serious health consequences, it is critical to understand how they interact with food biowaste materials. However, it is also a question if biowaste can be used to remediate rising pesticide residues in the environment. This study investigated the interactions of SCGs with the organophosphate pesticides malathion (MLT) and chlorpyrifos (CHP) and addressed the possibility of using SCGs as adsorbents for the removal of these pesticides from water and fruit extracts. The kinetics of MLT and CHP adsorption on SCGs fits well with the pseudo-first-order kinetic model. The Langmuir isotherm model best describes the adsorption process, giving the maximal adsorption capacity for MLT as 7.16 mg g^{-1} and 7.00 mg g^{-1} for CHP. Based on the thermodynamic analysis, it can be deduced that MLT adsorption on SCGs is exothermic, while CHP adsorption is an endothermic process. The adsorption efficiency of MLT and CHP using SCGs in a complicated matrix of fruit extracts remained constant. The neurotoxicity results showed that no more toxic products were formed during adsorption, indicating that SCGs are a safe-to-use adsorbent for pesticide removal in water and fruit extracts.

Keywords: biowaste; spent coffee grounds; organophosphates; pesticides; food processing; adsorption

1. Introduction

Food biowaste represents an organic waste material derived from food processing, preparation, and serving activities, including agricultural by-products and kitchen scraps [1]. It is becoming a significant global issue, as the components of biowaste release CO_2 and other greenhouse gases in the process of breaking down, in addition to the economic and moral factors of throwing food away. Up to 30% of food biowaste globally is leftovers from restaurants, grocery stores, and households [2].

Coffee is one of the most popular beverages, with around 10.5 million tons produced yearly [3]. However, the same amount of spent coffee grounds (SCGs) could negatively affect the environment if disposed of improperly. When exposed to different environmental influences, coffee grounds release nitrogen and carbon dioxide into the atmosphere [4,5]. Nitrogen reacts with oxygen to form smog and ozone, which are both detrimental to the environment, while carbon dioxide is a greenhouse gas [6]. If disposed of in water, spent coffee grounds could increase acidity, thus decreasing oxygenation. These factors could lead to an increase in algae and bacteria growth [7]. To reduce the negative environmental impact of spent coffee grounds, they should be composted, reused, or recycled instead of thrown away.

SCGs have recently been gaining attention for their potential uses in many fields, including gardening, energy production, and environmental protection. Once recycled, they can play essential roles in many industries, especially energy production, agroecology, and sustainable practices [8,9]. Spent coffee grounds are composed mainly of cellulose, lignin, and carbohydrates [10,11]. These properties make it a suitable adsorbent material that can be used to remove heavy metals, dyes, pesticides, and other pollutants from water, while Pujol et al. [10] particularly emphasized SCGs as a potential sorbent for hydrophobic pollutants.

Pesticides are one of the most widely used substances around the world. Although they are essential in controlling pests and protecting crops, they can also cause significant harm to human health and the environment if used excessively or carelessly [12]. Organophosphate pesticides (OPs) are widely used and efficient but very harmful for different species. OPs' toxicity is connected to their ability to inhibit acetylcholinesterase (AChE) activity. AChE is an enzyme with a key role in neurotransmission in animals. Its inhibition severely affects humans and other non-targeted species, such as bees [13,14]. Therefore, developing effective methods to remove OPs from the environment is essential. Except for the central (thio)phosphate moiety common for all OPs, they can have very different structures. For this work, we considered two OPs, chlorpyrifos and malathion. CHP has an aromatic moiety, while MLT is an aliphatic molecule. Thus, the parallel investigation of these two pesticides allows us to better understanding of the role of the adsorbent structure and chemistry in the adsorption of structurally different OPs.

Chlorpyrifos (CHP) is an organophosphate pesticide commonly used to control pests on fruits and vegetables, such as apples, oranges, lemons, limes, peaches, nectarines, bananas, grapes, tomatoes, peppers, and strawberries [15]. It is typically applied through aerial spraying or ground application and is highly regulated due to its toxicity [16]. Using chlorpyrifos in mint cultivation can help protect the mint crop against aphids, whiteflies, and spider mites [17]. China produces more than 30% of the world's supply of chlorpyrifos, followed by the United States and India. In Europe, the highest production of chlorpyrifos is located in France and the Netherlands. It is highly regulated due to its potential toxicity. Improper use of chlorpyrifos can lead to health problems in humans and other animals, so careful selection and application of this pesticide are necessary.

Malathion (MLT) is an insecticide that is widely used in lemon cultivation [18,19]. Like other organophosphate pesticides, malathion exerts its effects by affecting the nervous system of the pests [20]. Although it is moderately toxic to humans, it is essential to use malathion responsibly, as overuse of this pesticide can harm other animals' health and the environment.

The presence of pesticides in food and biowaste materials is a growing concern. Due to their toxic nature and potential health impacts, it is crucial to understand how pesticides interact with food biowaste materials. Adsorption is one of the most widely studied processes for remediation by which a pollutant, such as a pesticide, binds to a surface and is subsequently removed. This surface binding can be either physical or chemical [21]. Physical adsorption occurs when adsorbate molecules are attracted to the surface of the adsorbent material without forming chemical bonds [22]. Chemical adsorption, on the other hand, occurs when adsorbate molecules chemically react with the surface of the adsorbent material [21]. One of the most significant benefits is that adsorption reduces pollutants entering soil, water, and air systems. Moreover, it is cost-effective and does not require complex technologies or processes [23].

Several factors influence the adsorption of pesticides on food biowaste materials. Some of them are the surface properties of biowaste, such as surface functional groups, the type of biowaste material, the pH of the solution, and the chemical and physical properties of the pesticide [24]. Adsorption effectiveness also depends on the pesticide concentration in the solution, the temperature, and the time in contact with the biowaste material [25,26].

The adsorption of pesticides on food biowaste materials has many important implications, leading to the possible accumulation of pesticides or indicating the potential applicability of biowaste in pesticide remediation.

This study investigated the potential of using SCGs as an adsorbent for malathion and chlorpyrifos removal from water and fruit extracts. Our goal was to use SCGs as received, with minimal additional treatment, and without an intense use of energy and/or chemicals and release of greenhouse gases. In this way, we aimed to investigate the possible accumulation of OPs in waste SCGs and assess the direct reuse of SCGs in compliance with contemporary environmental protection strategies [27]. First, the results of the physicochemical characterization of SCGs are presented. Then, CHP and MLT adsorption on SCGs in water is analyzed. The kinetics and thermodynamics of MLT and CHP adsorption on SCGs are discussed. The adsorption of OP from realistic samples (plant extract) on SCGs is also addressed to determine the possibility of using SCGs as an adsorbent and detect if the complex matrixes influence the process. In addition, the eco-neurotoxicity of the spiked fruit extracts was monitored during the remediation process to determine if more toxic products were formed, such as pesticides' oxo-forms. Finally, this method's feasibility and potential benefits in food processing are discussed. At the time of submitting this article, to the best of our knowledge, no research articles were dedicated to the investigation of remediation of MLT and CHP using SCGs as a sustainable material. Moreover, in this study, the eco-neurotoxicity testing of the samples before and after the adsorption process was performed for the first time.

2. Materials and Methods

2.1. Adsorbent Preparation

Coffee (purchased from the local market, 80% Arabica and 20% Robusta) was brewed (treated with boiling water) and left for 2 h at room temperature until it became to cold. Next, the coffee grounds were separated with filtration and left to dry at room temperature for 24 h. Then, in order to release the leftover moisture, the obtained spent coffee grounds were dried in the oven at 80 °C for 1 h and ground for 15 min, using an agate stone mortar. Finally, 100 mg of the material was rinsed with 50 mL of HCl, NaOH (Centrochem, Stara Pazova, Srbija), and H_2O and eventually dispersed in 50 mL 50% EtOH (J.T. Baker, Phillipsburg, NJ, USA), obtaining the stock dispersion concentration of 2 mg mL^{-1}.

2.2. Adsorbent Characterization

The fractional sieving method was used to determine the particle size distribution in the SCG sample (sieve sizes 200, 100, and 63 μm; ROTH, Karlsruhe, Germany). For the investigation of samples' morphology and elemental composition, Scanning Electron Microscopy (SEM) and Energy Dispersive X-Ray Analysis (EDX) were performed using a scanning electron microscope PhenomProX (Thermo Fisher Scientific, Waltham, MA, USA). Fourier-transform infrared (FTIR) spectra were recorded using a Nicolet iS20 FT-IR spectrophotometer (Thermo Fisher Scientific, Waltham, MA, USA). The applied wavenumber range was from 4000 to 500 cm^{-1}, with 64 scans and 4 cm^{-1}. A TA Instruments SDT 2960 thermoanalytical device (TA Instruments, Inc. New Castle, DE, USA) was used for thermogravimetric analysis (TGA). The analysis was performed with a heating rate of 10 °C min^{-1} up to the temperature of 900 °C and under purging helium gas (Messer, Belgrade, Serbia). For the temperature-programmed desorption (TPD) analysis, the SCG sample was heated in a vacuum (starting pressure 1×10^{-7} mbar), with a constant heating rate of 10 °C min^{-1}, from room temperature to 1000 °C (on the heater). Desorbed gaseous products were detected using a quadrupole mass spectrometer EXTORR XT300 (Extorr Inc., New Kensington, PA, USA).

2.3. Adsorption Experiments

A mixture of 1 mL SCG stock dispersion and the designated amounts of chlorpyrifos and malathion (Pestanal, Sigma Aldrich, Søborg, Denmark) stock solutions (made in

50 vol.% ethanol in water) was made to deliver the targeted concentration of adsorbent and OP. After that, the mixtures were put in a laboratory shaker and left for the specified period. Subsequently, they underwent centrifugation at 14,500 rpm, with their supernatant filtered through a nylon membrane. Ultra-Performance Liquid Chromatography (UPLC) analysis was then conducted to determine the concentrations of CHP and MLT with a Waters ACQUITY UPLC system and a Photodiode array (PDA) detector managed by Empower 3 software. An ACQUITY UPLC™ BEH C18 column (1.7 µm, 100 mm × 2.1 mm) was used under isocratic conditions of 10% acetonitrile (J.T. Baker, Phillipsburg, NJ, USA) in water (v/v) for mobile phase A and pure acetonitrile for mobile phase B. The eluent flow rate was 0.2 mL min^{-1} in all cases, with an injection volume of 5 µL. The mobile phase used for chlorpyrifos contained 20% A and 80% B, and 40% A and 60% B for malathion. The retention time for malathion was 3.2 min, and for chlorpyrifos, it was 2.7 min. Both OPs were detected at 200 nm. Control experiments were performed identically but without an adsorbent.

2.4. Plant Extracts Preparation

Lemon juice and mint extract were used as food samples. Lemon juice was made by squeezing one lemon (75 g) and diluting it with 500 mL of tap water (pH = 4.5) before adding MLT to the desired concentration. Next, the sample was filtered. The mint extract was prepared by mixing 7 g of *Mentha spicata* leaves with 45 mL of 50% ethanol and leaving it for 72 h at room temperature; this extract was then filtered and diluted with 200 mL of 50% ethanol (pH = 6.0). CHP was then added in the required amount, and the resulting solution was further filtered through a nylon filter. The prepared extracts were used for sample analysis.

2.5. Eco-Neurotoxicity Assessment

The physiological effects of the treated solutions were analyzed using AChE inhibition measurements. By employing modified Ellman's procedure [28,29], 2.5 IU commercially purified AChE (Sigma Aldrich, Taufkirchen, Germany) from an electric eel was exposed to the OP solutions in 50 mM phosphate buffer, pH = 8.0, at 37 °C in a 0.650 mL final volume. The combination of acetylcholine-iodide (AChI, Sigma Aldrich, Taufkirchen, Germany) and DTNB (Sigma Aldrich, Taufkirchen, Germany) as a chromogenic reagent triggered the enzymatic reaction. The reaction was allowed to proceed for 8 min before being stopped with 10% sodium dodecyl sulfate (SDS). Thiocholine, the reaction product, reacts with DTNB, forming 5-thio-2-nitrobenzoate, whose optical adsorption was then read at 412 nm. The enzyme concentration was kept constant and set to produce an optimal spectrophotometric signal. The physiological effects were quantified as the AChE inhibition, which is given as follows:

$$I = 100 \times \frac{A_0 - A}{A_0} \quad (1)$$

where A_0 and A stand for the AChE activity in the absence of OP and the one measured after exposure to a given OP.

3. Results and Discussion

3.1. Physicochemical Characterization of SCGs

3.1.1. Particle Sizes, Morphology, and Chemical Composition

The particle size distribution, determined by the fractional sieving method (Figure 1a), shows that the largest fraction of SCG particles is in the 100–200 µm range. Particle sizes in the order of 10^2 µm generally agree with previous reports of SCG properties, but the SCG particles reported here are somewhat smaller than in earlier reports [30,31], likely due to the grounding step we performed. We note that particle sizes of the same order of magnitude are frequently reported for biobased carbon materials used as adsorbents (see, for example, [32]). However, carbon materials typically have large specific surface areas (SSA) due to the developed internal pore structure. SSA is usually determined by gas adsorption measurements, but in our case, a reliable determination of SSA was not possible, and this is suspected to be due to very low SSA. Indeed, one of the scarce literature reports that claims the SSA of SCGs indicated the SSA in the range from 0.19 to 2.3 m^2 g^{-1} [33].

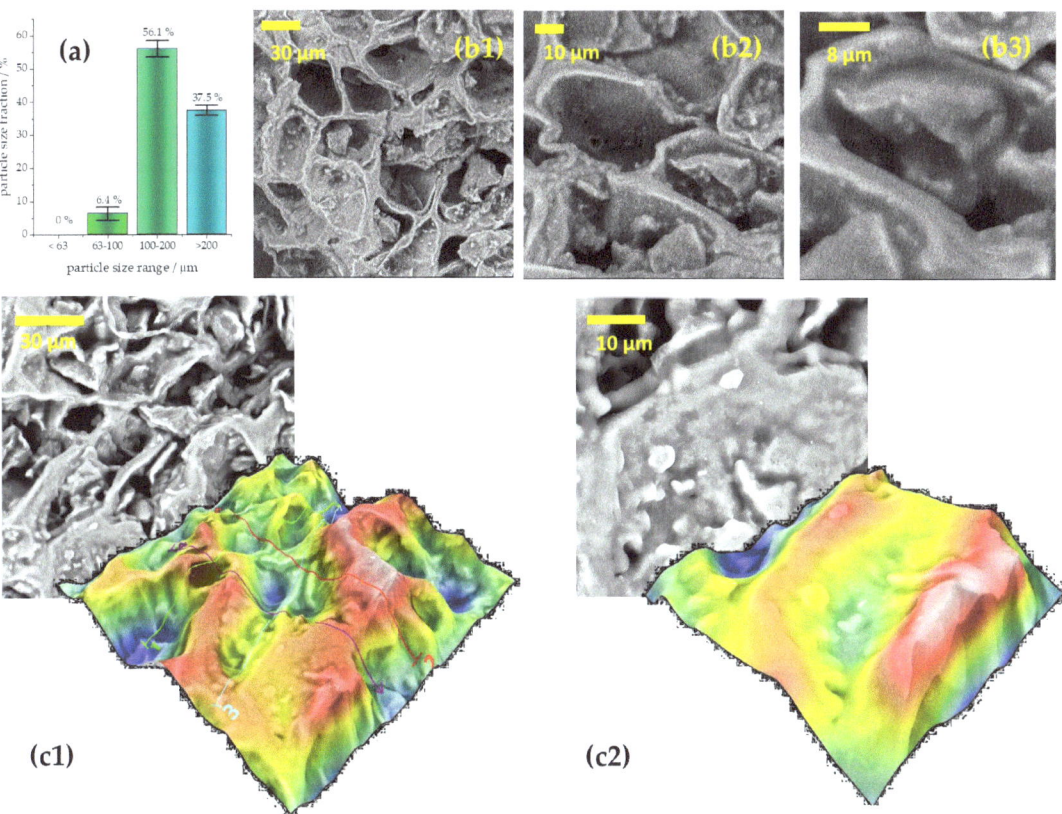

Figure 1. (**a**) Particle size distribution obtained by fractional sieving; (**b1–b3**) SEM micrographs of SCGs at magnifications ×2000, ×5000, and ×10,000 (from left to right); (**c1,c2**) 3D surface reconstruction of SCG powder under magnifications ×2000 and ×5000.

To investigate the morphology of SCGs, SEM was used. In addition, the chemical composition of the SCGs was investigated using EDX. The results are presented in Figure 1. The SEM micrograph showed that SCGs have a sponge-like, porous, heterogeneous structure (Figure 1(b1–b3)). Similar morphologies and large pores with diameters of approx. 30 μm were reported previously [34,35]. It can be seen that the cavities are randomly distributed on the materials' surface, with different size openings. SEM images were taken without deposition of the conductive layer on the SCGs sample, resulting in charge accumulation at higher magnifications. However, smooth pore walls can be observed without bright spots, indicating heavy elements in the sample. The surface 3D SEM reconstruction confirms these findings (Figure 1(c1,c2)). At a lower magnification (×2000), a rough surface profile is seen, with macropores entering the interior of the particles and contributing to the overall internal pore structure. However, higher magnification at the pore wall (×5000) indicates a rather smooth surface.

The EDX analysis showed that C, O, and N are present in high percentages in SCGs, namely 62.6 at. %, 28.1 at. %, and 9.0 at. %, respectively. Besides those mentioned, Mg, P, Ca, K, and S are present in traces (Figure 2, top row), constituting less than 0.3 at. % of the SCG sample. These elements are uniformly distributed in the sample, as indicated by the EDX maps (Figure 2, bottom row). The EDX analysis agrees with the literature and elemental composition of SCGs, as the other research reported a similar elemental composition with a different order of trace elements from source to source [36,37]. Here, we found that the trace elements' composition (in at. %) follows the order K (0.09) > S = Mg (0.05) > P = Ca (0.02). We note that a previously low N content (<2%) was reported for SCGs [10], but using the elemental analysis, which also gives the H content, thus effectively reducing the contents of the other elements in the samples; meanwhile, here, H cannot be determined using EDX.

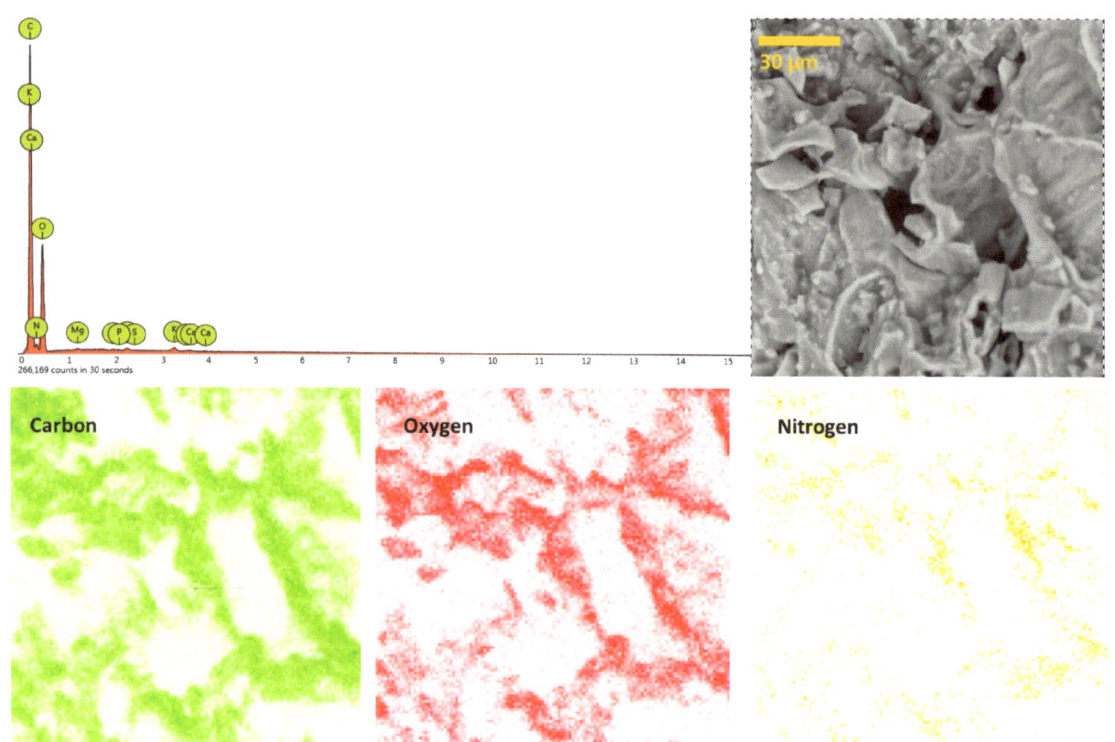

Figure 2. (**Top row**) EDX spectra of the indicated region (taken at a magnification of ×2000, 15 keV); (**bottom row**) EDX maps of carbon, oxygen, and nitrogen (64 × 64 pixels).

3.1.2. Surface Functional Groups of SCGs

The FTIR spectrum of the SCGs is shown in Figure 3a. The obtained FTIR spectrum correlates well with the literature and assumed compounds in SCGs [36–38]. The bands present in the FTIR spectrum of the SCGs are assigned to a specific vibration. The broad band on 3307 cm^{-1} was attributed to the stretching of the O-H group due to hydrogen bonding. Bands at 3010 cm^{-1} and 2924 cm^{-1} indicated the presence of C-H bonds where C is sp^2 and sp^3 hybridized, respectively. The sharp band on 1745 cm^{-1} was assigned to C=O stretch vibration, while the 1635 cm^{-1} band was attributed to C=N stretch vibration. The presence of a condensed aromatic system is assumed due to bands on 1521 cm^{-1} and 1441 cm^{-1} representing aromatic skeletal stretch and aromatic vibrations coupled with aromatic C-H in-plane vibrations, respectively. It is a clear indication of ligning presence, as found in Ref. [10]. An aliphatic C-H stretch vibration was assigned to the band on 1376 cm^{-1}. The deformation vibration of the C-N and C-O bands in secondary alcohols and C-N are assigned to bands on 1237 cm^{-1} and 1152 cm^{-1}. The band on 1028 cm^{-1} was attributed to the coupling of aromatic C-H in-plane deformation vibrations and C-O stretch vibrations in primary alcohols.

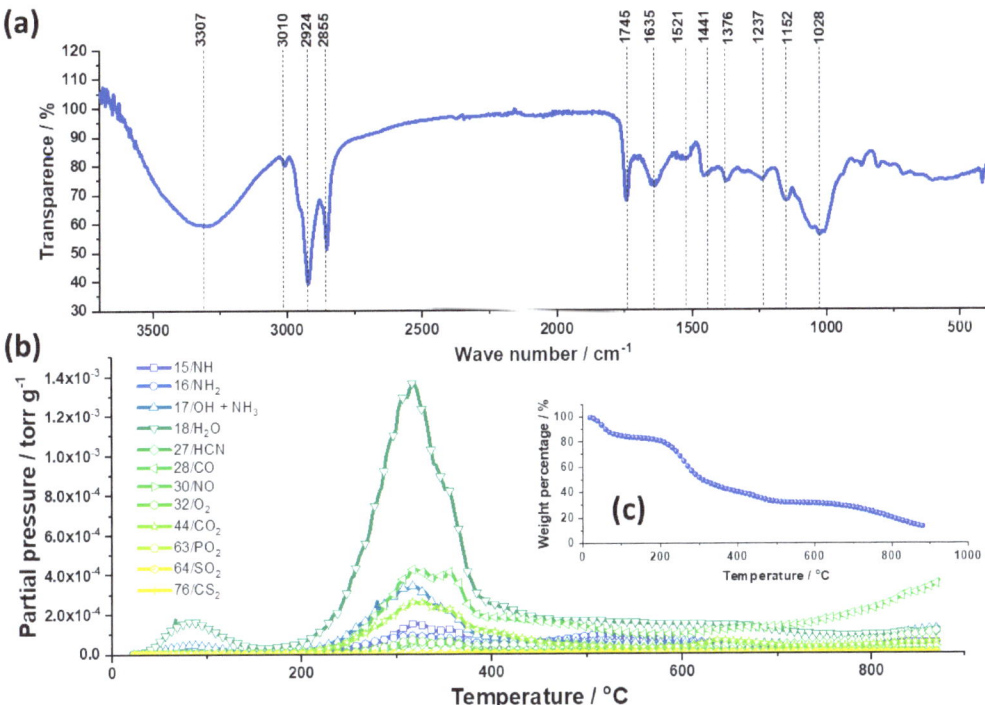

Figure 3. (**a**) ATR-FTIR spectrum of SCGs. (**b**) TPD curves of most common functional groups and molecules in SCGs. Inset (**c**) gives the TGA curve of SCGs under a helium atmosphere.

The TGA curve (Figure 3c) shows that by increasing the temperature from 20 to 900 °C in the helium atmosphere, SCGs go through three significant weight losses. In the first stage, from 20 to 85 °C, SCGs lose about 15% of their weight. The second stage starts at 200 and ends at 500 °C. It is the stage that represents carbonization, and during this stage, SCGs lose 48% of their initial weight. The third stage, during which SCGs lose another 15% of their initial weight, starts at 670 °C and lasts up to 900 °C. At the end of the heating process, 12% of the initial SCGs' weight is left.

The TPD curves for the most common functional groups and molecules desorbed during the heating process of SCGs are shown in Figure 3b. The same trend that TGA showed can be seen on these curves. In the correlation of TGA and TPD, it can be concluded that, during the first stage (20–85 °C), SCGs release physisorbed water. During the phase of SCG carbonization (200–500 °C), the weight loss is primarily due to the release of chemisorbed water, CO, and CO_2 molecules. Besides these, it can be noticed that all of the investigated functional groups and molecules are desorbed in this stage. During the third stage (670–900 °C), additional CO and CO_2 molecules are released, leading to the final weight loss. The evolved functional groups agree well with the elemental content found by EDX and also with the FTIR spectra of SCGs. A high concentration of H is expected in the sample, in line with Ref. [10], originating from physisorbed and strongly bound water, OH and NH_2 groups, and terminal C-H bonds.

3.2. Adsorption of Chlorpyrifos and Malathion on Spent Coffee Grounds

3.2.1. Kinetics of Adsorption

To investigate the kinetic parameters of the adsorption process, 1 mg mL^{-1} of SCG suspension was incubated with CHP and MLT in the concentration 5×10^{-5} mol dm^{-3} for various time intervals (from 1 min to 24 h) at 25 °C. The concentration of adsorbed pesticides was calculated as a difference between their initial concentration (C_0) and the equilibrium concentration (C_e) measured with UPLC after removing the adsorbent. The kinetics of CHP and MLT adsorption onto SCGs were analyzed using the nonlinear pseudo-first, pseudo-second, Elovich, and intraparticle diffusion models. The corresponding equations are listed in Table 1.

Table 1. Kinetic models' equations.

Kinetic Model	Equation
Pseudo-first-order model	$q_t = q_e \left(1 - e^{-k_1 t}\right)$
Pseudo-second-order model	$q_t = \frac{q_e^2 k_2 t}{1 + q_e k_2 t}$
Elovich model	$q_t = \frac{1}{\beta} \ln(1 + \alpha \beta t)$
Intraparticle diffusion model	$q_t = k_{id} t^{0.5} + C$

In Table 1, parameter q_t represents the amount of adsorbed OP at time t (mg g^{-1}), while q_e is the amount of adsorbed OP at equilibrium (mg g^{-1}). Parameter k_1 stands for the adsorption rate of pseudo-first order (min^{-1}), k_2 is the adsorption rate of pseudo-second order (g mg^{-1} min^{-1}), α is the initial adsorption rate in the Elovich model (mg g^{-1} min^{-1}), and β is desorption constant in the Elovich model (g mg^{-1}). Parameter k_{id} represents the adsorption rate constant of the intraparticle diffusion model (mg g^{-1} min$^{-0.5}$), and C is connected to a boundary layer (mg g^{-1}).

The nonlinear forms of the mentioned kinetic models and the intraparticle diffusion model fittings are shown in Figure 4. The kinetic parameters and corresponding R^2, χ^2, and Root Mean Square Error (RMSE) values are given in Table 2.

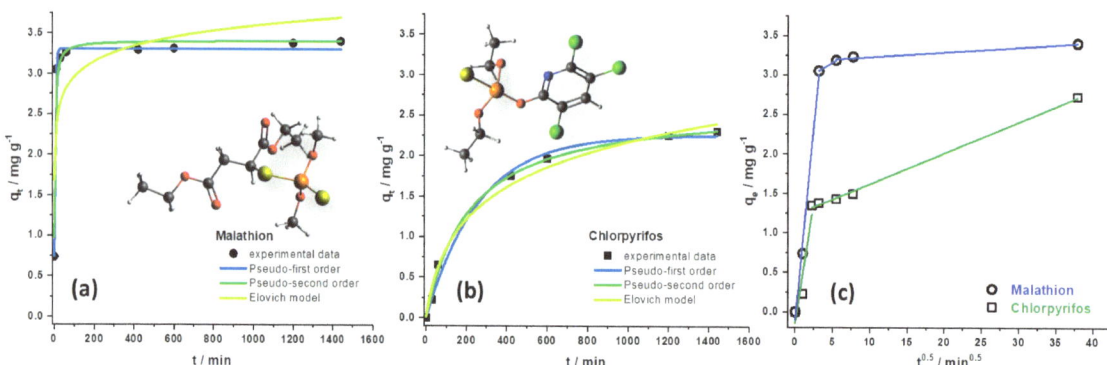

Figure 4. Graphical representation of kinetic models for the adsorption of MLT (**a**) and CHP (**b**) on SCGs. (**c**) Fitting experimental data into the intraparticle diffusion kinetic model for MLT and CHP. Insets present the optimized molecular structures of MLT and CHP (color legend for atoms: dark gray—C; red—O; yellow—S; orange—P; blue—N; green—Cl; and white—H).

Table 2. Kinetic parameters for MLT and CHP adsorption on SCGs.

Parameter	MLT	CHP
Pseudo-first order		
q_e (mg g^{-1})	3.31 ± 0.03	2.26 ± 0.07
k_1 (min^{-1})	0.25 ± 0.02	0.0038 ± 0.0005
χ^2	0.004	0.006
R^2	0.997	0.994
RMSE	0.0714	0.0964
Pseudo-second order		
q_e (mg g^{-1})	3.41 ± 0.08	2.66 ± 0.08
k_2 (mg min^{-1} g^{-1})	0.12 ± 0.03	0.0018 ± 0.0003
χ^2	0.037	0.008
R^2	0.970	0.993
RMSE	0.1720	0.0565
Elovich model		
α (mg g^{-1} min^{-1})	2.97 ± 0.23	0.018 ± 0.005
β (g mg^{-1})	1.47 ± 0.17	1.55 ± 0.18
χ^2	0.215	0.011
R^2	0.859	0.990
RMSE	0.5948	0.0579
Intraparticle diffusion model		
part I		
C (mg g^{-1})	0.00 ± 0.16	0.00 ± 0.28
K_{id} (mg g^{-1} min$^{-0.5}$)	0.98 ± 0.08	0.61 ± 0.19
R^2	0.987	0.921
part II		
C (mg g^{-1})	2.86 ± 0.01	1.23 ± 0.02
K_{id} (mg g^{-1} min$^{-0.5}$)	0.060 ± 0.001	0.039 ± 0.002
R^2	–	0.997
part III		
C (mg g^{-1})	3.17 ± 0.02	–
K_{id} (mg g^{-1} min$^{-0.5}$)	0.0062 ± 0.0009	–
R^2	0.963	–

By observing the results presented in Figure 4, it can be concluded that the adsorption equilibrium is reached after 10 min for MLT and 1440 min for CHP. In addition, it can be supposed that in the case of both MLT and CHP adsorption, the pseudo-first-order kinetic model fitted the experimental data better compared to the pseudo-second and Elovich models, according to the R^2 and χ^2 values. The calculated adsorption capacities of SCGs under experimental conditions are 3.31 mg g^{-1} for MLT and 2.26 mg g^{-1} for CHP. A higher k_1 value for MLT adsorption indicates a higher adsorption rate than that of CHP.

The value α, representing the initial adsorption rate, and the value β, the desorption constant, obtained from the Elovich model constant also indicate that the adsorption of CHP is noticeably slower than the adsorption of MLT on SCGs, as the α value is lower than the β value in the case of CHP adsorption.

By plotting q_t against $t^{0.5}$, three linear stages for MLT and two linear stages for CHP could be clearly observed in the intraparticle kinetic model (Figure 3c). The breakpoints for MLT adsorption onto SCGs are 10 min and 60 min, and for CHP, 10 min. The first rapid stage is attributed to the diffusion of MLT and CHP molecules through the solution to the external surface of the SCGs [39]. The second stage represents intraparticle diffusion, in which MLT and CHP molecules diffuse into the pores of the SCGs. The third stage seen for MLT is attributed to the final equilibrium stage. The K_{id} values decrease by one order of magnitude after the breakpoints, indicating slower adsorption in every subsequent stage. The C value, representing the boundary layer, increases, suggesting that the boundary layer has a high significance in the adsorption of MLT and CHP onto SCGs. It can be concluded that the three mentioned processes control the rate of molecule adsorption for MLT and two for CHP, but for each time range during the adsorption process, one process at a time determines the adsorption kinetics. We believe that differences in the molecular structure of MLT and CHP, the first being aliphatic, while the other one possessing aromatic moiety (Figure 4, insets), renders drastically different adsorption kinetics. As CHP is a less polar molecule, also reflected in much lower solubility compared to MLT, this property likely reduces its adsorption rate on the solvated surface of SCGs. It also slows down its diffusion into pores, as stage two in the intraparticle diffusion model for CHP extends over 1000 min without reaching the final equilibration stage.

3.2.2. Adsorption Isotherms

To further study the adsorption process, 1 mg mL^{-1} of SCGs was incubated with MLT and CHP in the concentration range from 5×10^{-6} to 1×10^{-4} mol dm^{-3} for 24 h at 25, 30, and 35 °C. The obtained data were fitted with several nonlinear isotherm models (Freundlich, Langmuir, Temkin, and Dubinin–Radushkevich (DR)). The corresponding equations for used isotherms are given in Table 3.

Table 3. Adsorption isotherms' equations.

Model	Nonlinear Form
Freundlich isotherm	$q_t = K_F C_e^{\frac{1}{n}}$
Langmuir isotherm	$q_t = \frac{q_{max} K_L C_e}{1 + K_L C_e}$
Temkin isotherm	$q_t = \frac{RT}{b_T} \ln K_T C_e$
Dubinin–Radushkevich isotherm	$q_t = q_{DR} e^{-K_{DR} \varepsilon^2}$

In Table 3, the q_t parameter represents the amount of adsorbed OP for a given C_e (mg g^{-1}), C_e is the equilibrium concentration (mg dm^{-3}), and K_F is the Freundlich isotherm constant related to the adsorption capacity ((mg g^{-1})(dm^3 g^{-1})$^{1/n}$). Parameter n stands for the so-called adsorption intensity, q_{max} represents the maximum amount of adsorbed OP (mg g^{-1}), and K_L is the Langmuir isotherm constant (dm^3 mg^{-1}). R is the universal gas constant (8314 J K^{-1} mol^{-1}), T is the temperature (K), and b_T represents the Temkin isotherm constant (J g mol^{-1} mg^{-1}), and K_T is the Temkin isotherm equilibrium binding constant (dm^3 mg^{-1}). At the same time, q_{DR} is the theoretical isotherm saturation capacity (mg g^{-1}), while K_{DR} (mol^2 J^{-2}) is the Dubinin–Radushkevich isotherm constant. The Polanyi potential, ε (J mol^{-1}), is calculated from C_e, while K_{DR} is related to the adsorption mean free energy (E), as $E = 1/(2K_{DR})^{1/2}$.

Nonlinear forms fitting of all investigated isotherms are shown in Figure 5. The parameters obtained from the different models provide essential information on the adsorption mechanisms and the surface properties and affinities of SCGs towards MLT and CHP. Table 4 summarizes the parameters obtained from the nonlinear plots of each isotherm. The fitting of the models to the experimental data was evaluated by χ^2 and R^2. The R^2 values are the highest ($R^2 > 0.9$), while the χ^2 and RMSE values are the lowest for the Langmuir model, suggesting that that model could best describe the experimental data. Still, most of the used isotherm models fit with most experimental data, so the calculated parameters are relevant for discussion.

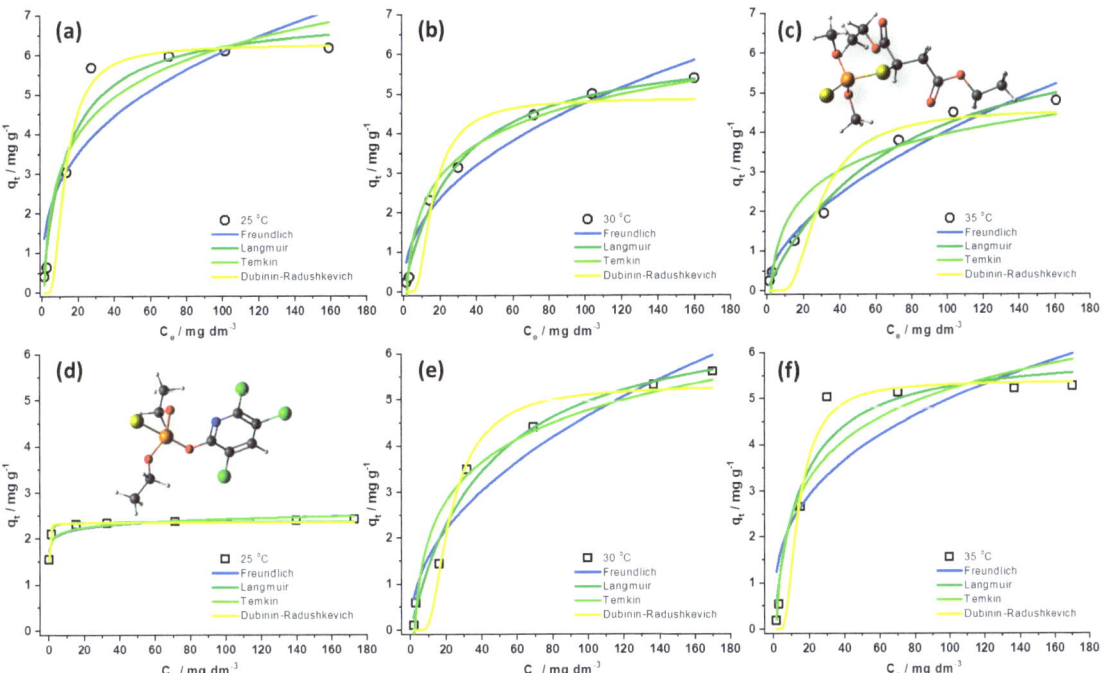

Figure 5. Top row: graphical representations of isotherm models for MLT adsorption on SCGs at (**a**) 25 °C, (**b**) 30 °C, and (**c**) 35 °C. Bottom row: graphical representations of isotherm models for CHP adsorption on SCGs at (**d**) 25 °C, (**e**) 30 °C, and (**f**) 35 °C.

The n-value of the Freundlich isotherm model for MLT adsorption is higher than 1, indicating that the adsorption is a favorable process. The decrease of the n-value with the increase in temperature indicates that the affinity of the adsorbent for MLT declines with the rising temperature. As mentioned, the Langmuir model showed the best agreement with the experimental results, indicating that the monolayer of MLT is adsorbed on an energetically homogenous surface without the interaction between adsorbed molecules. It indicates that all active sites are energetically equivalent and that equilibrium is reached when a monolayer is formed. Even though the temperature increase had almost no impact on the maximum amount (q_{max} = 7.16 mg g^{-1}) of MLT that could be adsorbed by SCGs, with the decrease of the K_L value, it can be concluded that the interaction between MLT and SCGs was weakened. It is also confirmed by the decrease of the E value obtained from the Dubinin–Radushkevich isotherm model. As E << 8000 J mol^{-1}, the proposed mechanism of MLT adsorption on SCGs is likely to be physisorption [40]. The Temkin isotherm model showed that the b_T values, which are related to adsorption heat, increase with the temperature. It could indicate that the molecules of MLT are more readily adsorbed onto the SCGs' surface, contrasting with the previous parameters obtained from the rest of the tested isotherms. However, since the χ^2 and RMSE values are higher for the Temkin isotherm than other investigated models, it is unsurprising that the findings provided different indications.

Table 4. Adsorption parameters for MLT and CHP adsorption on SCGs.

	MLT			CHP		
T (°C) →	25	30	35	25	30	35
Freundlich isotherm						
K_f ((dm^3 mg^{-1})$^{1/n}$)	1.27 ± 0.54	0.64 ± 0.19	0.33 ± 0.10	1.90 ± 0.09	0.54 ± 0.20	1.07 ± 0.51
n	2.94 ± 0.84	2.29 ± 0.34	1.83 ± 0.22	19.2 ± 4.1	2.13 ± 0.37	2.98 ± 0.95
χ^2	1.312	0.242	0.119	0.018	0.341	1.203
R^2	0.808	0.946	0.967	0.808	0.934	0.771
RMSE	1.1456	0.4919	0.3456	0.1348	0.5840	1.097
Langmuir isotherm						
K_L (dm^3 mg^{-1})	0.072 ± 0.023	0.035 ± 0.004	0.014 ± 0.002	9.23 ± 0.98	0.024 ± 0.006	0.07 ± 0.03
q_{max} (mg g^{-1})	7.06 ± 0.53	6.33 ± 0.21	7.16 ± 0.64	2.36 ± 0.03	7.00 ± 0.55	6.04 ± 0.52
χ^2	0.316	0.022	0.045	0.003	0.108	0.323
R^2	0.954	0.995	0.988	0.968	0.979	0.938
RMSE	0.5622	0.1485	0.2115	0.0550	0.3282	0.5687
Temkin isotherm						
K_T (dm^3 mg^{-1})	0.92 ± 0.37	0.61 ± 0.10	0.52 ± 0.23	365000 ± 1200	0.50 ± 0.15	0.76 ± 0.33
b_T (J g mol^{-1} mg^{-1})	1800 ± 210	2170 ± 120	2600 ± 370	16700 ± 240	2070 ± 210	2130 ± 280
χ^2	0.532	0.080	0.420	0.015	0.288	0.487
R^2	0.922	0.982	0.885	0.845	0.944	0.907
RMSE	0.7293	0.2832	0.6479	0.1212	0.5367	0.6982
Dubinin–Radushkevich isotherm						
q_{DR} (mg g^{-1})	6.26 ± 0.22	4.89 ± 0.30	4.60 ± 0.38	2.34 ± 0.04	5.31 ± 0.25	5.39 ± 0.20
K_{DR} (mol^2 J^{-2}) × 10^5	2.13 ± 0.41	3.15 ± 0.92	9.7 ± 3.4	0.0021 ± 0.0002	6.20 ± 1.20	2.31 ± 0.46
E (J mol^{-1})	153 ± 15	126 ± 19	72 ± 13	4850 ± 230	89.8 ± 8.7	147 ± 15
χ^2	0.155	0.273	0.313	0.007	0.158	0.120
R^2	0.977	0.939	0.914	0.928	0.969	0.977
RMSE	0.3937	0.5221	0.5597	0.0825	0.3978	0.3460

In the case of CHP adsorption, the n-values obtained from Freundlich isotherm models are also above 1, again indicating the favorability of the adsorption process. In addition, the n-value decreases with the temperature increase. The affinity of the adsorbent for CHP is declining with the rising temperature. Still, the n-values should be taken cautiously regarding the not-high R^2 values for all temperatures (Table 4). Similar to the results for MLT adsorption, the Langmuir isotherm model provides the best fit for adsorption at all temperatures. In theory, that indicates that CHP is adsorbed on an energetically homogenous surface without the interaction between adsorbed molecules in the form of the monolayer. Adsorption of CHP onto SCGs at 25 °C provides the highest K_L value (K_L = 9.23 dm^3 mg^{-1}). It is in agreement with the n-values from the Freundlich isotherm. The obtained qmax values rise with the increasing temperature, and the maximum adsorption capacity is calculated to be 7.00 mg g^{-1}. The energy of adsorption obtained from the DR isotherm model indicates that the adsorption mechanism is physisorption. Still, at the temperature of 25 °C, there is a strong indication that the chemisorption also occurs. The Temkin isotherm model showed that the b_T values decreased with the temperature increase, meaning that the readiness of the CHP molecule for adsorption on SCGs was decreasing, which agrees with the other calculated parameters.

3.2.3. Thermodynamic Analysis

The thermodynamic parameters of adsorption provide information about the interactions between the adsorbate molecules and the surface. The most important thermodynamic adsorption parameters are the enthalpy (ΔH), entropy (ΔS), and Gibbs free energy (ΔG) of the adsorption process. These parameters give an indication of the heat, randomness changes, and spontaneity of the process. However, we note that the exact values of the thermodynamic parameters strongly depend on the choice of the standard state, as discussed in Ref. [41]. Thus, we adopted the recommendations for the standard state by Chen et al. [41].

The mentioned thermodynamic parameters are linked through the Gibbs–Helmholtz equation:

$$\Delta G^0 = \Delta H^0 - T\Delta S^0 \quad (2)$$

To determine the ΔH^0 and ΔS^0 values to calculate ΔG^0, it is necessary to plot the Van't Hoff equation, assuming that $\Delta G^0 = -RT\ln K^0_{dist}$:

$$\ln K^0_{dist} = -\frac{\Delta H^0}{RT} + \frac{\Delta S^0}{R} \quad (3)$$

where the standard distribution coefficient is defined by the following:

$$K^0_{dist} = \frac{q_e}{C_e} \times \frac{C^0}{q^0} \quad (4)$$

The values of C^0 and q^0 define the standard state for OP in solution (1 mol dm^{-3}) and in the adsorbed state (1 mol kg^{-1}), rendering the standard distribution coefficient to dimensionless quantity. Figure 6a shows graphic representations of the amount of MLT and CHP dependence on temperature and Van't Hoff plot (Figure 6b). The extracted thermodynamic parameters for the adsorption of MLT and CHP onto SCGs are summarized in Table 5.

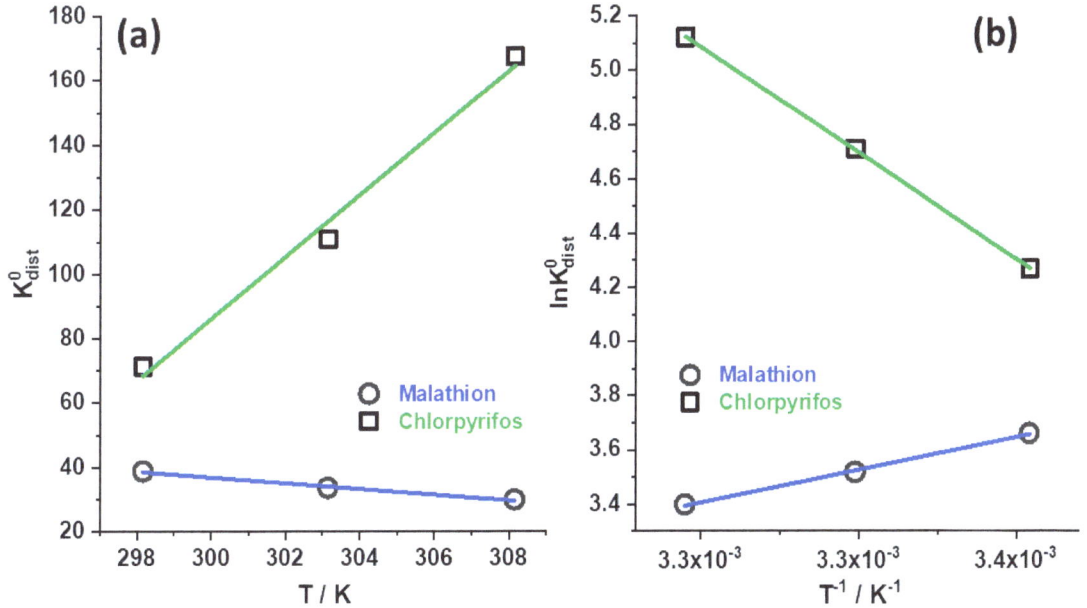

Figure 6. (a) Graphic representations of the amount of MLT (starting concentration 5×10^{-4} mol dm^{-3}) and CHP (starting concentration 1×10^{-4} mol dm^{-3}) dependence on temperature. (b) Van't Hoff plot.

Table 5. Thermodynamic parameters of MLT and CHP adsorption for a SCG loading of 1 mg mL^{-1}.

	MLT			CHP		
T (°C)	25	30	35	25	30	35
ΔH^0 (kJ mol^{-1})		−(19.9 ± 0.8)			65.3 ± 0.7	
ΔS^0 (J mol^{-1}K^{-1})		−(36.5 ± 2.8)			254.6 ± 2.2	
ΔG^0 (kJ mol^{-1})	−(9.1 ± 1.7)	−(8.8 ± 1.7)	−(8.7 ± 1.7)	−(10.6 ± 1.3)	−(11.9 ± 1.3)	−(13.1 ± 1.3)
R^2		0.996			0.9997	

Figure 6a shows that increasing the temperature decreases the distribution coefficient of MLT, while the distribution coefficient of CHP increases. Temperature dependence is linear in both cases, but the effect of temperature is more pronounced for CHP. The obtained thermodynamic parameters (Table 5) explain this behavior and indicate that the adsorption of MLT is an exothermic process and that the adsorption is followed by a small negative change of entropy. Thus, for MLT, the adsorption on SCGs is an enthalpy-driven process, compensating for the reduction in entropy, giving ΔG^0 negative, and increasing with temperature. On the other hand, the case of CHP adsorption onto SCGs is the opposite, as adsorption is an endothermic process, but the adsorption process is followed by a large positive entropy. Hence, CHP adsorption onto SCGs is an entropy-driven process. Again, the Gibbs free energy change is negative (slightly more negative than for MLT) and decreases with temperature, suggesting spontaneous adsorption, in line with experimental findings. Considering that a large fraction of the SCGs is constituted of lignin [10], we believe that the differences in thermodynamics of MLT and CHP adsorption can be explained by different types of interactions between SCGs and the studied OPs, keeping in mind that the adsorption process was classified as physisorption. We suggest that, in the case of MLT, dipole and dispersion interactions (Figure 7a) dominate the adsorption, thus becoming less prominent as the temperature increases. Such interactions can be placed between MLT molecules and different functional groups observed by FTIR (Figure 3a) and TPD. In the case of CHP, relatively strong π-π stacking between the aromatic ring of CHP and phenolic moieties of lignin can also take place (Figure 7b). However, in this case, desolvation of the surface has to take place, resulting in a positive change of the enthalpy and a large increase in entropy, resulting in negative Gibbs free energy change. As mentioned in Section 3.2.1., CHP and MLT show drastically different kinetic behavior, which was hypothesized to originate from different structures of these two pesticides. The argument mentioned above and the results regarding thermodynamic functions of the adsorption process support previous assumptions. In the case of CHP, it is proposed that desolvation of the SCG surface slows the adsorption down when compared to the MLT adsorption. As the dipole interactions are electrostatic, thus decaying with the distance but generally reaching longer ranges then π-π stacking (operative at distances below 4 Å [42]), it is assumed that MLT adsorption can proceed without significant slowdown due to desolvation. At the same time, the molecular forces capture MLT molecules at more considerable distances compared to CHP.

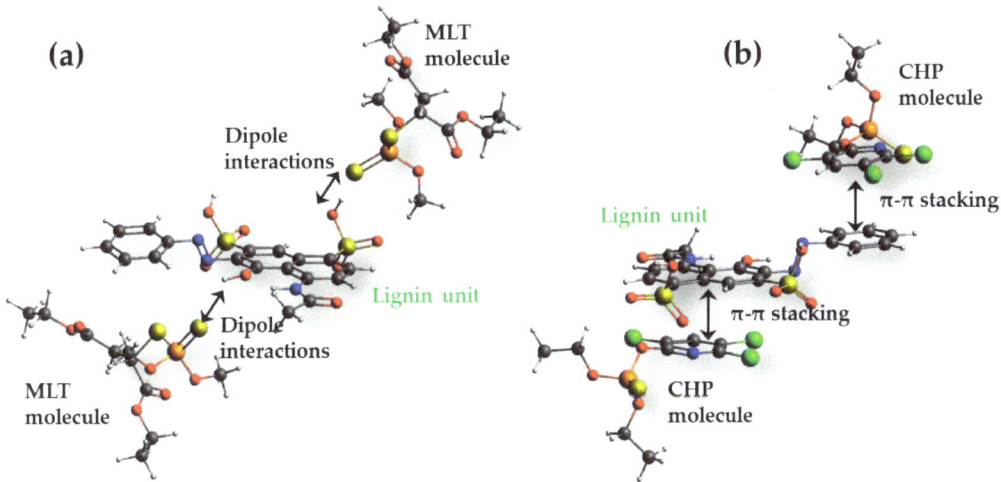

Figure 7. Schematic presentation of lignin monomer unit (structure data taken from [43] with (**a**) two MLT molecules via dipole interactions and (**b**) two CHP molecules via π-π stacking interactions. Structures were optimized by molecular mechanics in Avogadro [44] with an MMFF94 force field [45].

3.3. Application of SCGs in Food Processing—Plant Extracts Testing and Eco-Neurotoxicology Assessment

To assess the potential for SCGs' application in food processing, the material was tested as an adsorbent in plant extracts. As mentioned, MLT is used to treat lemon crops, and CHP is used in mint crops. Therefore, diluted lemon juice for MLT and mint extract for CHP were prepared as described in Section 2.4. The samples were spiked with respective OP in the concentration of 5×10^{-5} mol dm^{-3}. For comparison, the adsorptions of MLT and CHP from deionized water and 50% ethanol, respectively, were investigated under identical conditions. The results are shown in Table 6.

Table 6. OP concentrations and AChE inhibition in the tested samples (MLT in deionized water and lemon juice and CHP in 50% ethanol and mint ethanol extract) before and after the adsorption. Results were obtained when 1 mg mL^{-1} of adsorbent (SCG) was used at 25 °C.

Sample	[OP] (mol dm^{-3})		AChE Inhibition (% of Control)	
	Before the Adsorption	After the Adsorption	Before the Adsorption	After the Adsorption
MLT in deionized water	5.00×10^{-5}	3.48×10^{-5}	66 ± 5	45 ± 5
MLT in lemon juice	5.00×10^{-5}	3.50×10^{-5}	64 ± 5	50 ± 6
CHP in 50% ethanol	5.00×10^{-5}	4.00×10^{-5}	90 ± 4	81 ± 6
CHP in mint ethanol extract	5.00×10^{-5}	4.15×10^{-5}	92 ± 5	85 ± 5

Our primary goal was to reduce the toxicity of plant extract samples and confirm that no respective oxo-analogs are formed during adsorption. To estimate the eco-neurotoxicity of the samples before and after the adsorption of OPs on SCGs, the AChE inhibition was measured, as described in Section 2.5. AChE inhibition is an important benchmark for eco-neurotoxicity, as it indicates a general impact of contaminants on the species containing AChE in their nervous systems [46]. The results are also summarized in Table 6.

From the data presented in Table 6, it is obvious that the matrix effects in plant extracts tested in this work have a minor impact on the OPs' uptake. The amount of the adsorbed pesticides did not significantly change in the plant-extract analysis compared to deionized water and ethanol, indicating that SCGs as the adsorbent maintain a good performance and can be successfully used in plant extracts' treatment. Moreover, the AChE inhibition due to the contact with samples was reduced after the adsorption in all cases, indicating that no more toxic products, such as OPs' oxo-forms, were formed during this process. While the reduction of toxicity is not as prominent as in the case of some much more potent adsorbents, such as carbon materials [47,48], we note that SCGs' performance falls well within the range of previously reported biowaste adsorbents and surpasses many of them in terms of MLT and CHP adsorption capacity, as is discussed further. Moreover, an important advantage is that such an adsorbent is truly green and sustainable, as it does not require a carbonization step during which large amounts of CO_2 are released. In fact, with adsorption capacities in the order of 10^0 mg g^{-1}, SCGs perform quite similarly to carbons derived from biomass if the capacity is calculated versus raw material (precursor). Namely, during the carbonization and activation steps, a large amount of the precursor is lost, giving a yield in the order of 10^0–10^1% [49]. Thus, carbon materials with typical capacities for OPs' removal around 100 mg g^{-1} under high and realistic adsorbent loadings [50] have adsorption capacities in the order of 10^0 mg g^{-1} when referring to the raw precursor mass. Another critical point is the actual affinity of adsorbents towards contaminants. Here, the presented SCGs have a very low SSA (the direct measurement was not possible), but the adsorption capacities are in the order of 10^0 mg g^{-1}. Carbon-based adsorbents, on the other hand, have SSAs in the order of hundreds of m^2 g^{-1}. Thus, the affinity of the SCG surface towards MLT and CHP is quite similar to that of carbon materials, resulting from specific chemical composition and surface chemistry (Figure 7). Hence, as the adsorbent surface, not the adsorbent mass, is responsible for the adsorption process, the guidelines for improving SCGs' (mass normalize) adsorption capacities are straightforward, increasing SSA while maintaining surface chemistry unchanged. In the next section, in order to put the obtained results in a proper context, we compare the performance of the here-described SCGs with previously reported biowaste-based adsorbents for MLT and CHP.

3.4. Biowaste as Adsorbent for Different Pollutants—Evaluation of SCGs' Performance Compared to the Data Available in the Literature

SCGs have been studied for the adsorption of various water contaminants. The removal of Cd using SCGs was investigated by Patterer et al. [51] and Kim and Kim [52]. Patterer et al. concluded that the adsorption of Cd is an exothermic and spontaneous process and that the adsorption capacity is 4.484 mg g^{-1} [51]. In their research, Kim and Kim reported that the adsorption of Cd is irrelevant to the pH values of the solution in the range of 4–8, while the adsorption capacity of the SCGs was 19.32 mg g^{-1} [52]. Besides heavy metals, SCGs are shown to be excellent adsorbents of dyes. In their work, Block et al. found that non-treated SCGs are a promising adsorbent for methylene blue, removing 80% of the dye in 3 h of contact but adsorbing only 20% of methyl orange, making it a partly selective adsorbent [53].

Many sustainable materials can be effectively used as adsorbents for CHP. For example, Memon et al. successfully used walnut shell powder to remediate chlorpyrifos. The adsorption efficiency increased with the increase in temperature, similar to the here-reported behavior of SCGs. The used material had an adsorption capacity of 0.99 mmol g^{-1} (347 mg g^{-1}) [54]. Moreover, Rojas et al. investigated the adsorption of chlorpyrifos on sunflower seed shells, rice husk, composted sewage sludge, and soil and concluded that their adsorption capacities are 0.035 mg g^{-1}, 0.014 mg g^{-1}, 0.010 mg g^{-1}, and 0.02 mg g^{-1} respectively [55]; thus, two orders of magnitude below here reported capacities of SCGs towards CHP.

Apart from CHP, research has been conducted on MLT remediation, but to a much lesser extent. The biosorption of MLT from aqueous solutions was investigated using two types of herbal leaf powder by Yadamari et al. The adsorption capacity for *Achyranthes aspera* was 3.401 mg g^{-1}, and for *Phyllanthus niruri*, it was 2.664 mg g^{-1} [56]. Veličković et al. reported the adsorption capacity of malathion on pulverized river shellfish shells of 46.462 mg g^{-1}. The adsorption of malathion in the mentioned investigation was assessed as exothermic, similar to the results presented in this work [57]. A comparative analysis of the maximum adsorption capacity obtained in this work and the previously mentioned research available in the literature is given in Table 7.

Table 7. Maximum adsorption capacities for MLT and CHP adsorption for various biowastes and raw materials.

Material	OP	q_{max} (mg g^{-1})	Refs.
Walnut shell powder	CHP	347	[54]
Sunflower seed shells	CHP	0.035	[55]
Rice husk	CHP	0.014	
Composted sewage sludge	CHP	0.010	
Soil	CHP	0.020	
Achyranthes aspera leaves powder	MLT	3.401	[56]
Phyllanthus niruri leaves powder	MLT	2.664	
Pulverized river shellfish shell powder	MLT	46.462	[57]
Spent coffee grounds	MLT	7.157	This work
Spent coffee grounds	CHP	7.004	

From the presented data, it can be seen that SCGs are a reasonably good adsorbent for MLT and CHP remediation. Moreover, the obtained adsorption capacities are comparable with the previously reported MLT and CHP adsorption capacities on different types of biowastes and raw materials. Considering the discussion provided in the previous section regarding the performance of carbon-based adsorbents, losses of precursor mass, and significant CO_2 evolution during the carbonization process, we believe that SCGs have the potential for practical application in remediation processes, as they effectively stand side by side with apparently much more efficient adsorbents. However, we acknowledge that a detailed (socio-)techno-economical assessment is needed to correctly judge the possible use of SCGs in environmental remediation processes.

4. Conclusions

In this work, MLT and CHP adsorption on SCGs from aqueous solutions was analyzed. The adsorption kinetics was best described by the pseudo-first-order kinetic model for both MLT and CHP adsorption, while MLT adsorption was found to be significantly faster compared to CHP. The Langmuir isotherm model best describes the adsorption process for both pesticides, giving the maximal adsorption capacity of 7.16 mg g^{-1} for MLT and 7.00 mg g^{-1} for CHP; while based on the Dubinin–Radushkevich isotherm model, the adsorption process can be safely classified as physisorption. MLT adsorption on SCGs was found to be exothermic, whereas CHP adsorption is endothermic and entropy driven. The adsorption efficiency of MLT and CHP using SCGs in a complicated matrix of fruit extracts remained constant. The eco-neurotoxicity results showed that no more toxic products were formed during adsorption, indicating that SCGs are a safe-to-use adsorbent for pesticide removal. Their benefits as an adsorbent should be considered in line with the fact that their adsorption performance is similar to carbon materials if referred to the precursor mass used for carbon production while completely removing complicated steps involving carbonization and activation, resulting in significant CO_2 emissions. Considering that SCGs are globally available, their use in the remediation process also alleviates the problem of waste management, being particularly beneficial for less-developed countries, which, as a rule, suffer from extensive pesticide usage.

Author Contributions: Conceptualization, V.M. and T.L.-P.; methodology, V.M., M.P. and T.L.-P.; validation, I.P. and T.L.-P.; formal analysis, V.M., T.T. and M.P.; investigation, V.M., T.T. and M.P.; writing—original draft preparation, V.M., T.T. and M.P.; writing—review and editing, I.P. and T.L.-P.; visualization, V.M. and I.P.; supervision, I.P. and T.L.-P.; funding acquisition, T.L.-P. All authors have read and agreed to the published version of the manuscript.

Funding: The authors acknowledge the support provided by the Serbian Ministry of Science, Technological Development, and Innovations (contract numbers: 451-03-47/2023-01/200017 and 451-03-47/2023-01/200146).

Data Availability Statement: The data presented in this study are available on request from the corresponding author.

Acknowledgments: The authors would like to thank Nemanja Gavrilov for the SEM-EDX analysis and Slavko Mentus for the TGA analysis.

Conflicts of Interest: The authors declare no conflict of interest.

References

1. Mahro, B.; Timm, M. Potential of Biowaste from the Food Industry as a Biomass Resource. *Eng. Life Sci.* **2007**, *7*, 457–468. [CrossRef]
2. Carmona-Cabello, M.; García, I.L.; Sáez-Bastante, J.; Pinzi, S.; Koutinas, A.A.; Dorado, M.P. Food waste from restaurant sector–Characterization for biorefinery approach. *Bioresour. Technol.* **2020**, *301*, 122779. [CrossRef] [PubMed]
3. Available online: https://www.ico.org/#:~:text=Total%20production%20for%20coffee%20year (accessed on 15 April 2023).
4. Sermyagina, E.; Mendoza Martinez, C.L.; Nikku, M.; Vakkilainen, E. Spent coffee grounds and tea leaf residues: Characterization, evaluation of thermal reactivity and recovery of high-value compounds. *Biomass Bioenergy* **2021**, *150*, 106141. [CrossRef]
5. Plaza, M.G.; González, A.S.; Pevida, C.; Pis, J.J.; Rubiera, F. Valorisation of spent coffee grounds as CO_2 adsorbents for postcombustion capture applications. *Appl. Energy* **2012**, *99*, 272–279. [CrossRef]
6. Rani, B.; Singh, U.; Chuhan, A.; Sharma, D.; Maheshwari, R. Photochemical smog pollution and its mitigation measures. *J. Adv. Sci. Res.* **2011**, *2*, 28–33.
7. Xu, H.; Wang, W.; Liu, X.; Yuan, F.; Gao, Y. Antioxidative phenolics obtained from spent coffee grounds (*Coffea arabica* L.) by subcritical water extraction. *Ind. Crops Prod.* **2015**, *76*, 946–954. [CrossRef]
8. Kim, M.-S.; Min, H.-G.; Koo, N.; Park, J.; Lee, S.-H.; Bak, G.-I.; Kim, J.-G. The effectiveness of spent coffee grounds and its biochar on the amelioration of heavy metals-contaminated water and soil using chemical and biological assessments. *J. Environ. Manag.* **2014**, *146*, 124–130. [CrossRef]
9. Kondamudi, N.; Mohapatra, S.K.; Misra, M. Spent Coffee Grounds as a Versatile Source of Green Energy. *J. Agric. Food Chem.* **2008**, *56*, 11757–11760. [CrossRef]
10. Pujol, D.; Liu, C.; Gominho, J.; Olivella, M.À.; Fiol, N.; Villaescusa, I.; Pereira, H. The chemical composition of exhausted coffee waste. *Ind. Crops Prod.* **2013**, *50*, 423–429. [CrossRef]
11. Krishna Mohan, G.; Babu, A.; Kalpana, K.; Ravindhranath, K. Removal of chromium (VI) from water using adsorbent derived from spent coffee grounds. *Int. J. Environ. Sci. Technol.* **2017**, *16*, 101–112. [CrossRef]
12. Hossain, M.S.; Chowdhury, M.A.Z.; Pramanik, M.K.; Rahman, M.A.; Fakhruddin, A.N.M.; Alam, M.K. Determination of selected pesticides in water samples adjacent to agricultural fields and removal of organophosphorus insecticide chlorpyrifos using soil bacterial isolates. *Appl. Water Sci.* **2015**, *5*, 171–179. [CrossRef]
13. Colovic, M.B.; Krstic, D.Z.; Lazarevic-Pasti, T.D.; Bondzic, A.M.; Vasic, V.M. Acetylcholinesterase inhibitors: Pharmacology and toxicology. *Curr. Neuropharmacol.* **2013**, *11*, 315–335. [CrossRef] [PubMed]
14. Keifer, M.C.; Firestone, J. Neurotoxicity of Pesticides. *J. Agromed.* **2007**, *12*, 17–25. [CrossRef] [PubMed]
15. Zhang, X.; Starner, K.; Spurlock, F. Analysis of Chlorpyrifos Agricultural Use in Regions of Frequent Surface Water Detections in California, USA. *Bull. Environ. Contam. Toxicol.* **2012**, *89*, 978–984. [CrossRef]
16. Solomon, K.R.; Williams, W.M.; Mackay, D.; Purdy, J.; Giddings, J.M.; Giesy, J.P. Properties and Uses of Chlorpyrifos in the United States. In *Ecological Risk Assessment for Chlorpyrifos in Terrestrial and Aquatic Systems in the United States*; Giesy, J.P., Solomon, K.R., Eds.; Springer International Publishing: Cham, Switzerland, 2014; pp. 13–34.
17. El-Haoud, H.; Boufellous, M.; Mohammed, F.; Bengueddour, R. Risks of phytosanitary uses and residue analysis pesticides on green mint (*Mentha spicata* L.) in the province of Benslimane in Morocco. *Am. J. Innov. Res. Appl. Sci.* **2017**, *5*, 467–475.
18. Manrakhan, A.; Kotze, C.; Daneel, J.H.; Stephen, P.R.; Beck, R.R. Investigating a replacement for malathion in bait sprays for fruit fly control in South African citrus orchards. *Crop. Prot.* **2013**, *43*, 45–53. [CrossRef]
19. Mediouni Ben Jemâa, J.; Bachrouch, O.; Allimi, E.; Dhouibi, M. Field evaluation of Mediterranean fruit fly mass trapping with Tripack (R) as alternative to malathion bait-spraying in citrus orchards. *Span. J. Agric. Res.* **2010**, *8*, 400–408. [CrossRef]
20. Kori, R.K.; Singh, M.K.; Jain, A.K.; Yadav, R.S. Neurochemical and Behavioral Dysfunctions in Pesticide Exposed Farm Workers: A Clinical Outcome. *Indian J. Clin. Biochem.* **2018**, *33*, 372–381. [CrossRef]

21. Pasti, I.; Breitenbach, S.; Unterweger, C.; Fuerst, C. Carbon Materials as Adsorbents for Organophosphate Pesticides in Aqueous Media-Critical Overview. In *Organophosphates: Detection, Exposure and Occurrence. Volume 1: Impact on Health and the Natural Environment*; Nova Science Publishers, Inc.: New York, NY, USA, 2022; p. 251. ISBN 978-1-68507-652-8. [CrossRef]
22. Vidali, G.; Ihm, G.; Kim, H.-Y.; Cole, M.W. Potentials of physical adsorption. *Surf. Sci. Rep.* **1991**, *12*, 135–181. [CrossRef]
23. Foo, K.Y.; Hameed, B.H. Detoxification of pesticide waste via activated carbon adsorption process. *J. Hazard. Mater.* **2010**, *175*, 1–11. [CrossRef]
24. Lazarević-Pašti, T.; Anićijević, V.; Baljozović, M.; Anićijević, D.V.; Gutić, S.; Vasić, V.; Skorodumova, N.V.; Pašti, I.A. The impact of the structure of graphene-based materials on the removal of organophosphorus pesticides from water. *Environ. Sci. Nano* **2018**, *5*, 1482–1494. [CrossRef]
25. Yang, Y.; Chun, Y.; Sheng, G.; Huang, M. pH-dependence of pesticide adsorption by wheat-residue-derived black carbon. *Langmuir* **2004**, *20*, 6736–6741. [CrossRef] [PubMed]
26. Marczewski, A.W.; Seczkowska, M.; Deryło-Marczewska, A.; Blachnio, M. Adsorption equilibrium and kinetics of selected phenoxyacid pesticides on activated carbon: Effect of temperature. *Adsorption* **2016**, *22*, 777–790. [CrossRef]
27. Hickel, J. The contradiction of the sustainable development goals: Growth versus ecology on a finite planet. *Sustain. Dev.* **2019**, *27*, 873–884. [CrossRef]
28. Lazarević-Pašti, T.D.; Bondžić, A.M.; Pašti, I.A.; Vasić, V.M. Indirect electrochemical oxidation of organophosphorous pesticides for efficient detection via acetylcholinesterase test. *Pestic. Biochem. Physiol.* **2012**, *104*, 236–242. [CrossRef]
29. Ellman, G.L.; Courtney, K.D.; Andres, V.; Featherstone, R.M. A new and rapid colorimetric determination of acetylcholinesterase activity. *Biochem. Pharmacol.* **1961**, *7*, 88–95. [CrossRef] [PubMed]
30. Lauberts, M.; Mierina, I.; Pals, M.; Latheef, M.A.A.; Shishkin, A. Spent Coffee Grounds Valorization in Biorefinery Context to Obtain Valuable Products Using Different Extraction Approaches and Solvents. *Plants* **2023**, *12*, 30. [CrossRef]
31. Johnson, K.; Liu, Y.; Lu, M. A Review of Recent Advances in Spent Coffee Grounds Upcycle Technologies and Practices. *Front. Chem. Eng.* **2022**, *4*, 31. [CrossRef]
32. Saeidi, N.; Lotfollahi, M.N. Effects of powder activated carbon particle size on adsorption capacity and mechanical properties of the semi activated carbon fiber. *Fibers Polym.* **2015**, *16*, 543–549. [CrossRef]
33. Meshram, S.; Katiyar, D.; Asha, T.; Dewangan, G.P.; Joshi, A.N.; Thakur, R.S. Preparation and characterization of activated carbon from spent coffee grounds using NaOH and KCl as activating agents. *J. Indian Chem. Soc.* **2020**, *97*, 1115–1118. [CrossRef]
34. Safarik, I.; Horska, K.; Svobodova, B.; Safarikova, M. Magnetically modified spent coffee grounds for dyes removal. *Eur. Food Res. Technol.* **2012**, *234*, 345–350. [CrossRef]
35. Sertoli, L.; Carnier, R.; Abreu, C.A.d.; Coscione, A.R.; Melo, L.C.A. Coffee Waste Biochar: Characterization and Zinc Adsorption from Aqueous Solution. *Coffee Sci.* **2019**, *14*, 518–529. [CrossRef]
36. Chien, H.-W.; Kuo, C.-J.; Kao, L.-H.; Lin, G.-Y.; Chen, P.-Y. Polysaccharidic spent coffee grounds for silver nanoparticle immobilization as a green and highly efficient biocide. *Int. J. Biol. Macromol.* **2019**, *140*, 168–176. [CrossRef]
37. Nieber, K. The impact of coffee on health. *Planta Med.* **2017**, *83*, 1256–1263. [CrossRef] [PubMed]
38. Li, X.; Strezov, V.; Kan, T. Energy recovery potential analysis of spent coffee grounds pyrolysis products. *J. Anal. Appl. Pyrolysis* **2014**, *110*, 79–87. [CrossRef]
39. Cheung, A.; Szeto, Y.S.; McKay, G. Intraparticle Diffusion Process During Acid Dye Adsorption Onto Chitosan. *Bioresour. Technol.* **2007**, *98*, 2897–2904. [CrossRef]
40. Inglezakis, V.J.; Zorpas, A.A. Heat of adsorption, adsorption energy and activation energy in adsorption and ion exchange systems. *Desalination Water Treat.* **2012**, *39*, 149–157. [CrossRef]
41. Chen, T.; Da, T.; Ma, Y. Reasonable calculation of the thermodynamic parameters from adsorption equilibrium constant. *J. Mol. Liq.* **2021**, *322*, 114980. [CrossRef]
42. Alvarez, S. A cartography of the van der Waals territories. *Dalton Trans.* **2013**, *42*, 8617–8636. [CrossRef]
43. Available online: https://pubchem.ncbi.nlm.nih.gov/compound/lignin (accessed on 15 April 2023).
44. Hanwell, M.D.; Curtis, D.E.; Lonie, D.C.; Vandermeersch, T.; Zurek, E.; Hutchison, G.R. Avogadro: An advanced semantic chemical editor, visualization, and analysis platform. *J. Cheminformatics* **2012**, *4*, 17. [CrossRef]
45. Halgren, T.A. Merck molecular force field. I. Basis, form, scope, parameterization, and performance of MMFF94. *J. Comput. Chem.* **1996**, *17*, 490–519. [CrossRef]
46. Anićijević, V.J.; Petković, M.; Pašti, I.A.; Lazarević-Pašti, T.D. Decomposition of Dimethoate and Omethoate in Aqueous Solutions—Half-Life, Eco-Neurotoxicity Benchmarking, and Mechanism of Hydrolysis. *Water Air Soil Pollut.* **2022**, *233*, 390. [CrossRef]
47. Jocić, A.; Breitenbach, S.; Pašti, I.A.; Unterweger, C.; Fürst, C.; Lazarević-Pašti, T. Viscose-derived activated carbons as adsorbents for malathion, dimethoate, and chlorpyrifos—Screening, trends, and analysis. *Environ. Sci. Pollut. Res.* **2022**, *29*, 35138–35149. [CrossRef]
48. Jocić, A.; Breitenbach, S.; Bajuk-Bogdanović, D.; Pašti, I.A.; Unterweger, C.; Fürst, C.; Lazarević-Pašti, T. Viscose-Derived Activated Carbons Fibers as Highly Efficient Adsorbents for Dimethoate Removal from Water. *Molecules* **2022**, *27*, 1477. [CrossRef] [PubMed]
49. Breitenbach, S.; Unterweger, C.; Lumetzberger, A.; Duchoslav, J.; Stifter, D.; Hassel, A.W.; Fuerst, C. Viscose-based porous carbon fibers: Improving yield and porosity through optimization of the carbonization process by design of experiment. *J. Porous Mater.* **2021**, *28*, 727–739. [CrossRef]

50. Anićijević, V.; Tasić, T.; Milanković, V.; Breitenbach, S.; Unterweger, C.; Fürst, C.; Bajuk-Bogdanović, D.; Pašti, I.A.; Lazarević-Pašti, T. How Well Do Our Adsorbents Actually Perform?—The Case of Dimethoate Removal Using Viscose Fiber-Derived Carbons. *Int. J. Environ. Res. Public Health* **2023**, *20*, 4553. [CrossRef]
51. Patterer, M.S.; Bavasso, I.; Sambeth, J.E.; Medici, F. Cadmium Removal from Acqueous Solution by Adsorption on Spent Coffee Grounds. *Chem. Eng. Trans.* **2017**, *60*, 157. [CrossRef]
52. Kim, M.-S.; Kim, J.-G. Adsorption Characteristics of Spent Coffee Grounds as an Alternative Adsorbent for Cadmium in Solution. *Environments* **2020**, *7*, 24. [CrossRef]
53. Block, I.; Günter, C.; Duarte Rodrigues, A.; Paasch, S.; Hesemann, P.; Taubert, A. Carbon Adsorbents from Spent Coffee for Removal of Methylene Blue and Methyl Orange from Water. *Materials* **2021**, *14*, 3996. [CrossRef]
54. Memon, G.Z.; Moghal, M.; Memon, J.R.; Memon, N.N.; Bhanger, M. Adsorption of selected pesticides from aqueous solutions using cost effective walnut shells. *Adsorption* **2014**, *4*, 43–56. [CrossRef]
55. Rojas, R.; Morillo, J.; Usero, J.; Vanderlinden, E.; El Bakouri, H. Adsorption study of low-cost and locally available organic substances and a soil to remove pesticides from aqueous solutions. *J. Hydrol.* **2015**, *520*, 461–472. [CrossRef]
56. Yadamari, T.; Yakkala, K.; Battala, G.; Gurijala, R.N. Biosorption of malathion from aqueous solutions using herbal leaves powder. *Am. J. Anal. Chem.* **2011**, *2*, 37. [CrossRef]
57. Veličković, Z.S.; Vujičić, B.D.; Stojanović, V.N.; Stojisavljević, P.N.; Bajić, Z.J.; Đokić, V.R.; Ivanković, N.D.; Otrisal, P.P. Pulverized river shellfish shells as a cheap adsorbent for removing of malathion from water: Examination of the isotherms, kinetics, thermodynamics and optimization of the experimental conditions by the response surface method. *Vojnoteh. Glas./Mil. Tech. Cour.* **2021**, *69*, 871–904. [CrossRef]

Disclaimer/Publisher's Note: The statements, opinions and data contained in all publications are solely those of the individual author(s) and contributor(s) and not of MDPI and/or the editor(s). MDPI and/or the editor(s) disclaim responsibility for any injury to people or property resulting from any ideas, methods, instructions or products referred to in the content.

Review

Pesticide Use and Degradation Strategies: Food Safety, Challenges and Perspectives

Andreja Leskovac and Sandra Petrović *

Vinca Institute of Nuclear Sciences-National Institute of the Republic of Serbia, University of Belgrade, M. Petrovića Alasa 12-14, 11000 Belgrade, Serbia; andreja@vin.bg.ac.rs
* Correspondence: sandra@vin.bg.ac.rs

Abstract: While recognizing the gaps in pesticide regulations that impact consumer safety, public health concerns associated with pesticide contamination of foods are pointed out. The strategies and research directions proposed to prevent and/or reduce pesticide adverse effects on human health and the environment are discussed. Special attention is paid to organophosphate pesticides, as widely applied insecticides in agriculture, veterinary practices, and urban areas. Biotic and abiotic strategies for organophosphate pesticide degradation are discussed from a food safety perspective, indicating associated challenges and potential for further improvements. As food systems are endangered globally by unprecedented challenges, there is an urgent need to globally harmonize pesticide regulations and improve methodologies in the area of food safety to protect human health.

Keywords: organophosphate pesticides; pesticide use; biotic degradation strategy; abiotic degradation strategy; food safety

1. Introduction

As the world's population grows, the industrialization of agriculture and the expansion of livestock production to meet increasing food demand create opportunities and challenges for food safety. These challenges place more responsibility on food manufacturers and processors to ensure food safety, preventing food contamination before it reaches the consumer [1].

The continuous development of agriculture intensifies the application of pesticides globally to reduce crop yield losses and increase productivity and product quality [2]. Proximately 2 million tons of pesticides are currently applied to crops worldwide each year to increase productivity and reduce losses from pests and diseases [3]. According to the Food and Agricultural Organization (FAO) of the United Nations, the United States of America was the largest user of pesticides in 2020, while the next 10 largest pesticide users in the world are Brazil, China, Argentina, the Russian Federation, Canada, France, Australia, India, and Italy [4]. In the 2022 update, FAO reported that total pesticide use in China significantly decreased, moving China to third place in pesticide usage globally [4]. However, even though a plateau has been reached in recent years, total pesticide use has increased by approximately 50% compared to the 1990s [4]. The pesticide use by region and the top five largest pesticide users in the world are shown in Figure 1.

Pesticide regulatory systems established to protect humans and the environment vary from country to country [5]. This variability implies that each country can adopt regulations to define acceptable concentrations of particular pesticides in food and feed and restrict or prohibit the usage of particular pesticides due to their unacceptable health or environmental effects. The Joint Meeting on Pesticide Residues (JMPR) is an expert body established mutually by the FAO and the World Health Organization (WHO) that is responsible for establishing toxicological endpoints, such as acceptable daily intake (ADI) and acute reference dose (ARfD), based on experimental data. Additionally, the JMPR recommends the maximum concentrations of pesticide residues (maximum residue

Citation: Leskovac, A.; Petrović, S. Pesticide Use and Degradation Strategies: Food Safety, Challenges and Perspectives. *Foods* **2023**, *12*, 2709. https://doi.org/10.3390/foods12142709

Academic Editor: Roberto Romero-González

Received: 30 June 2023
Revised: 11 July 2023
Accepted: 13 July 2023
Published: 15 July 2023

Copyright: © 2023 by the authors. Licensee MDPI, Basel, Switzerland. This article is an open access article distributed under the terms and conditions of the Creative Commons Attribution (CC BY) license (https://creativecommons.org/licenses/by/4.0/).

levels, or MRLs) in food and feed to the Codex Committee on Pesticide Residues (CCPR) for consideration [6]. The recommended MRLs in food and feed that are considered safe for consumers were finally adopted by the Codex Alimentarius Commission. The MRLs, which provide a wide margin of safety based on good agricultural practice, are the most implemented standards regarding food safety [7]. However, regardless of the prevailing framework the Codex provides, the MRLs differ considerably across countries [8].

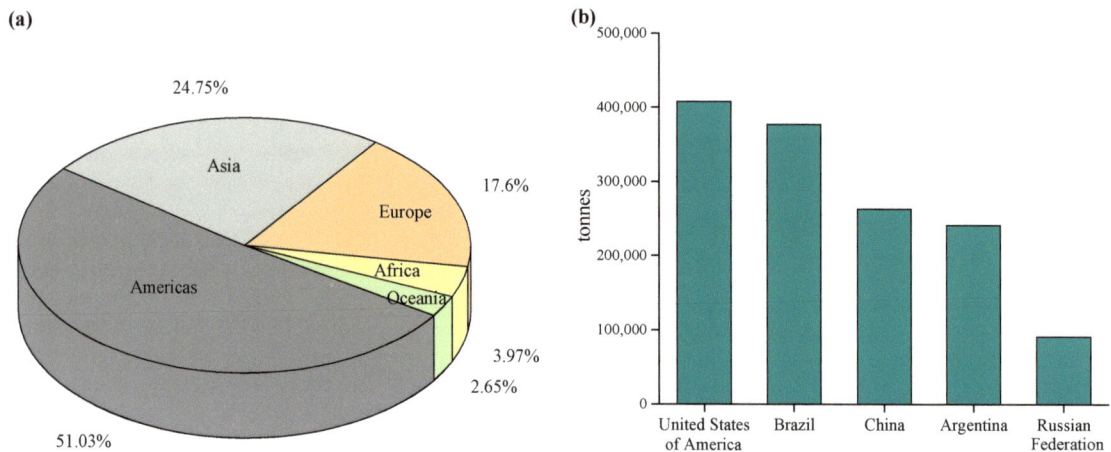

Figure 1. (a) Pesticides use by region and (b) the top five pesticide users in 2020.

In the EU, the European Commission regulation 396/2005 directly concerns public health, establishing a system of setting and monitoring the MRLs in food and feed [9]. In the USA, the Environmental Protection Agency (EPA) is responsible for the pesticide registration, regulations, and establishment of MRLs in food and feed following the Federal Insecticide, Fungicide, and Rodenticide Act (FIFRA) and the Federal Food, Drug, and Cosmetic Act (FFDCA) [10]. The US Department of Agriculture (USDA) and the Food and Drug Administration (FDA) are responsible for measuring and collecting data on pesticide residues in fruits, vegetables, grains, meat, and dairy products nationwide and in products imported from other countries. In China, the Ministry of Agriculture and Rural Affairs (MARA) is the main pesticide regulatory body responsible for pesticide registration and management [11]. In Brazil, pesticide regulations are supervised by the Ministry of Health through the National Sanitary Surveillance Agency (ANVISA), the Brazilian Institute for the Environment and Renewable Natural Resources (Ibama), and the Ministry of Agriculture, Livestock, and Food Supplies (MAPA) [12].

Of the total amount of pesticides used worldwide, organophosphate pesticides (OPs) account for approximately 33% [13]. As effective and broad-spectrum insecticides, they are extensively used worldwide in agriculture, homes, gardens, and veterinary practices [14]. In the last decade, over 100 organophosphorus compounds have been commercially used as insecticides to control pests in agricultural food commodities [15], of which the medium- or low-toxic OPs, such as dimethoate, phoxim, chlorpyrifos, and trichlorfon, are widely used [16]. Although numerous OPs are no longer approved in most developed countries, they are still in use in many developing countries, causing long-term negative effects on human health and the environment [7]. Acute and/or chronic exposure to OPs can occur directly from occupational and non-occupational use and indirectly through the consumption of pesticide residues that can remain in food and drinking water [17]. Pesticide residues and their metabolites can contaminate soils and water, enter the food chain, and, as a final point, display toxic effects, affecting human health [16,18,19].

The increased quantity and frequency of pesticide utilization worldwide consequently increased their impact on the environment and human health. The excessive use and misuse of pesticides, especially in developing countries, can cause environmental pollution and adverse human health effects in the long run. While recognizing gaps in pesticide regulations that impact consumer safety, public health concerns related to pesticide contamination of foods and recent strategies proposed to prevent and/or reduce their adverse effects on human health and the environment are discussed. Particular attention is paid to biotic and abiotic strategies used for OPs degradation, identifying challenges and potential for future improvements.

References for this manuscript, published by June 2023, were collected from scientific databases (PubMed, Science Direct, Scopus, Google Scholar, Taylor & Francis platform, and BioMed Central platform), using the keywords "organophosphate pesticides", "pesticide use", "food safety", "pesticide regulations", "human health", "environmental pollution", "pesticide detection", "biotic degradation", "abiotic degradation" and combinations thereof. The relevant, up-to-date peer-reviewed articles, published in English, addressing current pesticide degradation strategies and public health concerns related to pesticide exposure and food safety are included. In addition, grey literature relevant to the topic, including selected reports of government agencies and international organizations, is incorporated.

2. Public Health Concerns Related to Pesticide Exposure

One of the major issues related to food safety is the lack of globally harmonized pesticide legislation and safety standards [20]. Pesticide MRLs in foods and feeds significantly differ, especially among developed and developing countries. The differences in regulations also cause trade issues since many developing countries use unauthorized pesticides or different MRLs [20]. Also, the EU MRLs are more stringent than the Codex MRLs, raising concern about whether the Codex MRL values sufficiently protect consumer health [20]. Most developed countries established their own MRL policies, and for developing countries, meeting the MRL requirements of developed countries can be particularly challenging [21].

Pesticide poisoning and mortality occur mostly in developing countries and are usually associated with insufficient occupational safety standards and regulations, inadequate application, and poor labeling of pesticides [22]. Inadequate regulatory systems also result in the import of pesticides banned in developed countries, while a lack of awareness among farmers and inadequate personal protective equipment cause poor pesticide practices. According to a report by the European Parliament (2021) [7] and Pesticide Atlas Kenya Edition (2022) [23], many pesticides no longer authorized in the European Union are still allowed to be manufactured and exported in developing countries. For example, until its ban in 2020, chlorpyrifos was the most commonly used pesticide in food production in the EU [24]. However, it is still being applied in China, India, and many other countries of the Global South [24,25]. Brazil, the largest pesticide consumer in Latin America, approved 475 new pesticides in 2019, of which about a third contain active substances that have been banned or restricted in the EU [7]. In 2019, Brazil imported 14 hazardous compounds, including chlorpyrifos, fipronil, cyanamide, and propineb [23]. Kenya, as a major importer of banned pesticides mainly from the EU and China, has registered 51 active ingredients prohibited in the EU, such as trichlorfon, atrazine, fipronil, iprodione, acetochlorines, and 1,3-dichloropropene [23]. The United States also allows the production and export of domestically banned pesticides to low- and middle-income countries where they have been linked to significant adverse health effects on the local population [26]. In addition, food containing residues of banned pesticides is frequently reimported back to the countries that allow their production and export, contributing to a global pesticide exposure risk [23]. To address the gap in the regulations of pesticides that pose risks to human health and the environment, in 2020, the European Commission drafted a legislative initiative to prohibit the production and export of hazardous chemicals banned in the EU, which is expected to come into force in 2023 [23].

There are numerous reports indicating pesticide contamination of foods. For example, an earlier study from Ghana reported that chlorpyrifos, diazinon, deltamethrin, fenvalerate, and permethrin concentrations exceeded their respective EU MRLs in some ready-to-eat vegetable samples collected from different sites along the food chain [27]. Similar results were obtained in the study of 160 samples of commonly consumed fruits and vegetables collected from all supply chain stages (distribution, storage, and handling from farm to fork) in the Kampala Metropolitan Area, Uganda. In 95.6% of the samples, multiple pesticide residues were detected, of which 91.3% were organophosphates [28]. The analysis of 1183 bovine milk samples from different locations in India demonstrated that approximately 8% contained organochlorines, organophosphates (ethion, profenofos, chlorpyrifos), synthetic pyrethroids, and phenylpyrazole residues, exceeding the MRL values. Chlorpyrifos was the most common OP detected [29]. Moreover, the residues of hexachlorocyclohexane (HCH), dichloro-diphenyl trichloroethane (DDT), and endosulfan were also found in some of the milk samples, although their usage was restricted or banned [29]. A recent study from Egypt reported that approximately 40% of the pesticide residues detected in samples of vegetables and fruits from the market exceeded the permissible MRLs. The most frequently detected pesticides were insecticides; the results obtained for lambda-cyhalothrin, fipronil, dimethoate, and omethoate in spinach, zucchini, kaki, and strawberry, respectively, indicate they may cause acute or chronic poisoning when consumed in amounts equal to 0.1 or 0.2 kg per day [30]. Another study from Egypt reported the presence of multiple pesticide residues (cypermethrin, thiamethoxam, chlorpyrifos, and lambda-cyhalothrin) in strawberry and tomato-based products available on the market. It was found that 27% of the average pesticide residues in the tested samples exceeded the maximum residue levels (MRLs) [31]. A recent study from Algeria has revealed the contamination of honey samples with OPs (methyl parathion, coumaphos, and fenitrothion), exceeding the MRL (MRL 50 ng/g) [32].

As mentioned, pesticide MRLs in food imported from outside the EU are generally higher than in foods from EU countries [33]. However, an enhanced level of pesticide residues in foods was also reported in EU countries. For example, a previous study from Poland reported an exceedingly high presence of chlorpyrifos in all of the investigated fruits and vegetable peels and also a high level of methyl parathion, especially in the peel of potatoes and pulp of zucchini [34]. Recent research from the UK has shown that out of the total 33,911 analyzed samples from imported foods (including from EU countries), 50.2% contained detectable residues, and 3.3% of the total analyzed samples were above MRLs [35]. Also, the contamination of foodstuffs, such as honey, with OP residues was reported in studies conducted in Italy, Spain, Belgium, France, Germany, Switzerland, and from outside Europe, such as South America and North America [36]. A recent study on more than 200 cereal and legume samples from Italy, Eastern Europe, and some non-European countries has reported the presence of pesticide residues in the grain samples (contamination percentage of 7%), which were below the MRLs, while no pesticide was found in the analyzed legumes. The most abundant pesticides in cereal samples were cyfluthrin, deltamethrin, phenothrin, cypermethrin, fenvalerate, chlorpyrifos, and pirimiphos-methyl [37].

The latest EFSA annual report, considering the assessment of pesticide residue levels in foods on the European market in 2021, has shown that 96.1% of the samples analyzed were below the MRL, while 3.9% exceeded this level, of which 2.5% were non-compliant [38]. The MRL exceedance and non-compliance rates were lower than those reported in 2020 (the MRL exceedance rate of 5.1% and the non-compliance rate of 3.6%). However, samples imported from third countries showed a 5-fold MRL exceedance rate (10.3%) and non-compliance rate (6.4%) compared to the EU-derived samples, which showed 2.1% MRL exceedance and 1.3% non-compliance [38]. Given the safety margins incorporated into the ADI and ARfD, the MRL exceedance does not necessarily imply a risk to human health, so case-by-case assessments are required to determine whether dietary intakes exceed the health-based limits. The EFSA report shows that no consumer intake concern was identified in the chronic health risk assessment. However, out of the total samples analyzed under the acute assessment, 1.1% exceeded the health-based guidance values (HBGVs) in 29 pesticides out of the 190 analyzed [38].

As expected, food products in developed countries are systematically monitored for pesticide residues to ensure compliance with national legislation and consumer safety. In contrast, the monitoring of food in developing countries is often restricted; nevertheless, this issue is also reported in developed countries, as shown in the case of the US, where the FDA inspects only 1–2% of import shipments [20]. Therefore, an increasing public health concern associated with pesticide contamination of food is completely justified and points out the necessity to globally harmonize and standardize MRLs to ensure consistent and effective food safety regulations worldwide. Establishing uniform MRLs is a fundamental step that must be followed to prevent and avoid any health risks. In addition, the lack of consensus regarding MRLs undermines pesticide controls, so the continuous, internationally harmonized monitoring of foods to ensure consumer safety is required.

Over the past years, the main concern has been related to the potential risk of combined exposure to multiple pesticide residues in the diet and the dose addition of these compounds. According to the current regulations, the risk assessment of exposure to chemicals mainly relies on assessing individual substances and a few groups of substances that are expected to occur together [39]. The current methods used for human risk assessments assume that different components in mixtures act additively and behave as if they were dilutions of each other [39]. In this respect, the evaluation of exposure to multiple chemicals assumes that compounds with the same mechanism of toxicological action may have a cumulative effect that should be considered. In this regard, much effort has been put toward developing comprehensive frameworks dealing with human risk assessment of combined exposure to multiple chemicals [40,41]. As a result, methodologies developed enabled a grouping of chemicals into cumulative assessment groups (CAGs) based on their effects on target organs/systems and then with respect to their modes of action. Such methodologies have been developed only for multiple pesticide residues in food [42]. In 2021, EFSA published a report on a retrospective (2016–2018) cumulative risk assessment of dietary exposure to OPs ($n = 36$) and N-methyl carbamate insecticides ($n = 11$), which was conducted for chronic inhibition of erythrocyte acetylcholinesterase (AChE) [42]. It was concluded that cumulative exposure to pesticides, causing effects on the AChE, did not reach the threshold for regulatory consideration for any of the populations assessed [42].

However, the effects of combined exposure to multiple pesticide residues can be more complex due to their possible interactions. Scientific data about the possible synergistic effects of multiple pesticide residues as well as the effects of exposure to multiple residues that display different modes of action remains very limited [20,43]. Additionally, exposures to different chemicals may arise from separate sources, which should also be considered [44]. Consequently, these gaps in our knowledge may lead to an underestimation of the real health risk.

Recent nutritional recommendations to increase the consumption of fruit, vegetables, and whole grains may increase dietary pesticide intakes leading to severe cumulative toxicity and increased risk of various chronic illnesses, including cancer, respiratory, metabolic, reproductive, and neurologic disorders [1,45]. Urinary levels of pesticides or their metabolites are commonly used as biomarkers of human pesticide exposure [46]. Recently, the European Human Biomonitoring Initiative (HBM4EU) prioritized the collection of information on human exposure to pyrethroids pesticides, organophosphate pesticides (chlorpyrifos, dimethoate, and glyphosate), polyethoxylated tallow amine (additive in glyphosate formulations), and phenyl pyrazole insecticide (fipronil) for the period 2000–2022 [47]. However, as no proper urinary biomarkers existed for dimethoate and polyethoxylated tallow amine (POEA), the European human biomonitoring data on these substances was unavailable. The study results indicate extensive exposure to pyrethroids, chlorpyrifos, and glyphosate in the general European population, with noticeable geographical differences. The highest urinary levels for all the investigated pesticides were reported in Cyprus and Valencia (Spain) [47]. As for the OPs, the high detection rate of chlorpyrifos metabolite,

3,5,6-trichloro pyridine-2-phenol (TCP), was reported in most studies. However, as chlorpyrifos and chlorpyrifos-methyl have been banned in the EU since February 2020 [48], the exposure level in the general population is expected to have decreased. Recently, POEA exposure biomarkers have been identified; the first LC-MS/MS method for rapid analysis of 11 POEA homologues in human plasma was developed and validated using the plasma samples of glyphosate-poisoned patients [49].

Several studies have shown that organic food consumption may be one way to achieve a considerable reduction in dietary exposure to pesticides, including OPs, minimizing potential health risks [43,50–52]. Organic farming stipulates the non-use of synthetic fertilizers and most pesticides, leading to the absence or decrement of the concentration of pesticide residues in foods compared to conventional farming [45,52]. A recent study assessing the EU agricultural soils of organic and conventional farms reported that the pesticide residue levels in organic fields were 70–90% lower than in conventional ones [43]. However, although synthetic pesticides are not used in organic farming, pesticide residues can still be present in organic farming soils [53]. Furthermore, persistent compounds, such as DDT, remain at relatively high levels in organic fields, likely due to historical applications, despite being banned in many EU countries since the 1970s [54]. Therefore, to ensure minimal pesticide residue levels, transitioning to organic farming requires conversion transition periods adapted based on the initial residue mixtures and their residence time in the soil [43]. In addition, there is a severe research gap considering the effects of complex pesticide mixtures present in the soil-on-soil health and, consequently, on food quality and human health [43].

Several dietary intervention studies have shown that an organic diet significantly reduces urinary pesticide residue excretion compared to conventional food consumption [45]. However, these studies usually monitor a small number of selected pesticides and do not evaluate mineral- and plant-extract-based pesticides that are commonly used in organic farming. In addition, urinary pesticide residue excretion may result from both dietary and environmental pesticide exposure, and according to current knowledge, the relative contribution of these two sources to total chronic pesticide exposure is not possible to estimate [45].

The risk assessment of pesticide effects on human health and the environment is complex and considers the types and dosage of pesticides used, the periods and levels of exposure, and the environmental characteristics of the locality where pesticides are applied. In addition, although some toxic pesticides have been banned, they continue to be detected frequently in the environment due to their long degradation half-lives, thus contaminating the soil and water sources [55]. Therefore, although there is a requirement for pesticides to be produced, distributed, and used under regulations, due to their frequent applications, mistreatments, and heterogeneous regulatory limits, pesticides and their metabolites have been frequently detected in crops, agricultural soils, and water sources, posing a potential threat to human health [16,56]. Therefore, the cumulative risk assessment of the pesticide effects on human health should consider both the dietary and non-dietary routes of exposure and be regulated by an extensive legal framework harmonized globally to ensure and maintain food safety and security. Based on the above, developing and implementing improved strategies to protect human health and the environment is mandatory.

3. Strategies Aimed to Protect Human Health and the Environment from Pesticide Exposure

Due to the prominent scientific progress in chemistry, biology, and molecular biology, the approaches to protecting human health and the environment from pesticide exposure have continuously improved. These include searching for novel pesticides, developing methods for detecting and re-assessing the safety of the currently used pesticides, and developing methods for pesticide degradation into less toxic products. All of these approaches should meet the requirements of Integrated Pest Management (IPM), a strategy adopted by the EU in 2009 through Directive 2009/128/EC, also called the Sustainable Use of Pesticides

Directive (SUD) [57]. The IPM strategy focuses on managing pests through a combination of sustainable biological, physical, and other non-chemical methods minimizing the risk to human health and the environment associated with the use of chemical products. According to IPM, chemical pesticides should be applied only as a last resort. Instead, the use of competitive plant material (e.g., resistant cultivars and certified seed), non-chemical tools (e.g., seed coating, flaming, beneficial microorganisms, etc.), and novel cultivation techniques (e.g., intercropping, crop rotation and diversification, stale seedbed technique, etc.) should be applied [58]. However, although this concept was made obligatory in the EU in 2014, limited progress has been achieved thus far, and goals set by the SUD have mainly been left unaccomplished [59].

General strategies for the development of new, effective, environmentally friendly pesticides encompass (i) the development of pesticides that are rapidly degradable and less residual in the environment; (ii) the development of pesticides that are effective at extremely low doses; and (iii) the development of selective chemicals effective in the control of pests but not toxic against humans or non target species [60]. In the last decade, at least 105 pesticides (fungicides, insecticides, nematicides, acaricides, and herbicides) have been launched or are under development [60]. However, although most of them appear safe for humans and the environment, only a few products have been developed for practical use. Pesticide development has increasingly shifted from chemical to biological pesticides, including RNAi pesticides, abiotic stress control agents, genetically modified crops, and seeds, which are believed to affect the environment to a lesser extent than chemical pesticides [60]. It has been estimated that biological pesticides will equal chemical pesticides on the market by the 2050s [61]. Nevertheless, pesticides, including OPs, are frequently used and are expected to be continuously used in the future.

The detection of pesticides and re-assessment of the safety of the already-used pesticides envisage the development and implementation of new techniques with better reliability than existing ones. Novel techniques should enable a better prediction of the potential hazards of pesticides and, henceforward, contribute to reducing their adverse effects on human health and the environment. The simultaneous presence of different contaminants, including pesticides, in the same food (the so-called "cocktail effect") represents a significant aspect of food safety that requires comprehensive research [62]. The possible interactions between different chemical contaminants in food may result in partial detoxification (if antagonism or inhibition takes place) or in an increase in toxicity due to synergism or potentiation, even when each compound is present at a level below toxicity reference values [44,63]. Therefore, there is an urgency to develop approaches that will enable the evaluation of these effects before confirming an effective risk [62]. Concerning the mixture risk assessment (MRA), according to Regulation (EC) No 396/2005 [9], the dietary risk assessment of pesticide exposure should take into account cumulative and synergistic effects in the setting of pesticide MRLs when methods become available. In 2021, the EU Commission and European Food Safety Authority (EFSA) developed and adopted an Action Plan (2022–2030) under document SANTE/10178/2021. It focuses on assessing human health risks from dietary and non-dietary exposure to pesticides and accelerates work on developing methods for cumulative risk assessment and their subsequent implementation [64]. As proposed, methodologies enabling human health risk assessment of combined exposure to multiple chemicals (RACEMiC) will be implemented by 2030 [39]. Currently, the methods, data, and tools for dietary MRA for pesticides are mostly available; however, significant scientific gaps considering the non-dietary exposure to pesticide mixtures still exist [39]. In addition, according to regulations, the dose-addition assumption will still be applied for the combined toxicity of the chemicals unless evidence for antagonistic or synergistic interactions is available [39]. Therefore, there is an urgency to increase knowledge about potential interactions among multiple contaminants in food to evaluate the effective risk of exposure. In this respect, the development of new analytical procedures and microbiological methods for food safety control is necessary [62].

Numerous detection methods based on chemical, physical and biological parameters have been utilized so far to identify even trace amounts of the pesticides. Traditional methods for pesticide detection comprise instrumental techniques such as gas chromatography (GC), high-performance liquid chromatography (HPLC), or chromatographic methods coupled with mass spectrometry (MS) detectors, which provide profuse qualitative and quantitative information on the residues with high accuracy [65]. However, the main limitations of these methods are the time-consuming sample preparation, the requirement for highly trained technicians, and the expensive equipment [66]. In that terms, Near-infrared spectroscopy (NIR), as a simple, reliable, and cheap technique, could be helpful as it can be used to predict soil composition and absorption of OPs, such as chlorpyrifos, methyl parathion, and phoxim [67]. In addition, significant research has been devoted to developing microfluidic paper-based analytical devices (µPADs) as an inexpensive alternative to highly sophisticated instrumentation in analytical applications for food and water monitoring that can be used for continual testing, especially in developing countries [68]. Furthermore, the use of µPAD sensors is frequently associated with smartphone-based detection of pesticides [67]. For example, Sicard et al. developed a highly selective and sensitive µPAD sensor and a mobile application suitable for on-site colorimetric identification of paraoxon and malathion based on the inhibition of immobilized AChE [69].

The development of sensors (electrochemical sensors, optical sensors, including chemiluminescence, fluorescence, and colorimetric sensors, and piezoelectric sensors) and biosensors represents a novel strategy for monitoring pesticide contamination of food. Moreover, introducing nanomaterials into their structure improves the efficacy of sensors and biosensors as analytical tools for detecting pesticides [66,67]. Using biosensors in pesticide detection might enable proficient and precise analysis at a low cost. AChE-based biosensors have been commonly used to detect diazinon, dimethoate, dichlorvos, chlorpyrifos, malathion, methyl parathion, glyphosphate, and other OPs [66]. However, the main limitation of this method is the lack of stability and persistent need for a substrate for quantifying the pesticide level. For that reason, several methods based on AChE inhibition, including colorimetric and electrochemical assays, have been developed to improve the system's stability and enable a more efficient analysis with lower detection capabilities. Moreover, the immobilization of AChE with different nanocomposites has been considered a potent tool to increase the response of the biosensor in pesticide detection [70]. In that sense, numerous nanomaterials have been developed for detecting, degrading, and removing pesticides [65].

As discussed before, the food and feed might frequently be contaminated with more than one pesticide, so there is a growing need to develop and improve sensitive multi-residue detection methods. As comprehensively discussed by Jia et al. [71], two strategies for rapid multi-residue detection methods have been proposed; the first one, based on different recognition elements, comprises the use of antibodies, aptamers, and molecularly imprinted polymers (MIPs). The second strategy, based on the inherent characteristics of pesticides, encompasses the use of enzymatic inhibition-based sensors, NIR spectroscopy, and surface-enhanced Raman scattering (SERS) spectroscopy. In addition, numerous sensitive and reliable Liquid chromatography tandem mass spectrometry (LC-MS/MS) techniques and high-resolution mass spectrometry (LC-HRMS) techniques are developed and validated for simultaneous analysis of pesticides, veterinary drug residues, and other contaminants in foods [72–74]. Likewise, the method (CEN 15662) for pesticide residue analysis in foods proposed by European Committee for Standardization [75] encompasses a QuEChERS extraction followed by multi-residue determination on GC-MS/MS and LC-MS/MS. Furthermore, an advanced QuEChERS mega-method for simultaneous determination of at least 300 compounds, including pesticides, veterinary drugs, and environmental contaminants in matrices such as muscles of beef lamb, goat, and fish using LC-MS/MS and GC-MS/MS is also validated [76,77]. The proposed methods and their further improvements are expected to enable efficient and accurate multi-residue screening and detection.

4. Degradation Strategies for Organophosphate Pesticides

The broad-spectrum insecticidal activity, chemical stability, high efficiency, and low cost of production make OPs one of the predominant pesticides widely used in agriculture, veterinary practice, and urban areas. However, since they can seriously affect human health, their application has become one of the primary anthropogenic sources of environmental pollution.

Numerous pesticides are not easily degradable; they remain in the soil, leak into groundwater and surface water, and contaminate the surrounding ecosystem. In addition, depending on their chemical qualities, they can enter the organism and bioaccumulate in food chains [78]. For this reason, efforts are being made to find effective ways for their degradation into non-toxic or less toxic forms to reduce their impact on humans and restore the pristine environment.

Several biotic and abiotic degradation strategies have been developed and continuously improved to minimize human exposure to pesticides and their potential adverse environmental effects. Various biotic and abiotic techniques applied for OP degradation are presented in Figure 2. The biotic strategy includes transformation processes mediated by microorganisms, fungi, or plants, while the abiotic strategy implies the direct chemical or mechanical breakdown of pesticides into non-toxic forms. The types of transformation processes by which a pesticide will degrade depend on its structural affinity for specific types of transformation and the environmental conditions to which it is exposed [79].

4.1. Biotic Degradation Strategy

The biotic approach is based on the ability of microorganisms to convert hazardous contaminants into relatively simple and non-toxic compounds. Contaminants, including OPs, can be accumulated in the soil and agricultural runoff water through agricultural application, and their removal from the environment can be achieved by biodegradation and/or bioremediation. While biodegradation is a naturally occurring process, bioremediation is a man-made, engineered process whose efficiency may depend on moisture, temperature, redox conditions, organic matter, pH, and nutrients that influence chemical diffusion and microbial activity in the soil [80].

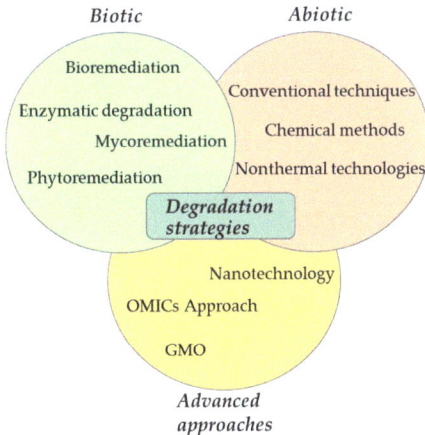

Figure 2. Biotic and abiotic degradation strategies applied for organophosphate pesticides.

Biodegradation is a process that involves the complete decomposition of an organic compound into its inorganic constituents. Due to its low cost, simple application, high effectiveness, and lack of secondary contamination, biodegradation is considered an effective tool for the remediation of pesticide contamination [81–83]. Generally, under optimal conditions, biodegradation represents the bioconversion of a substance (via a series of intermediates) into small, inert end products (mineralization) [84]. Advances in approaches for soil pesticide degradation, such as biostimulation and bioaugmentation, may enable effective OP detoxification. In contrast to bioattenuation, which occurs naturally without human intervention, biostimulation enables accelerated biodegradation by providing the right conditions for microorganisms in the soil. The optimum nutritional ratio of carbon, nitrogen, and phosphorus is crucial for biostimulation [85]. Land farming and composting are biostimulation activities that involve carbon sources, nutrients, and humidity control [86].

Nowadays, the remediation industry and scientific community are focusing on bioremediation systems that use bioaugmentation processes. Altered microorganisms, obtained from the environment or genetically modified in the laboratory, are often utilized in bioaugmentation to accelerate the detoxification and breakdown processes in contaminated environments [86]. However, bioremediation is restricted to biodegradable compounds since not all toxins in contaminated soils are substrates for microbial absorption.

In general, the biological removal of organophosphorus compounds has become the method of choice since many microorganisms have been found to have the metabolic pathways and enzymes necessary for the degradation of a variety of xenobiotic compounds, including OPs [87]. Microorganisms are the most significant candidates for biodegradation/bioremediation because of their straightforward cellular structure, small genome size, quick replication, rapid evolution, and adaptation to contaminated environments [88]. In addition, microorganisms can metabolize contaminants, including OPs, using them as nutrient and/or energy sources. Bacteria, fungi, actinomycetes, and algae are the most capable bio-transformers and pesticide degraders [84,89].

A wide range of microorganisms is reported to selectively hydrolyze a variety of organophosphorus contaminants, including the species of the genera *Arthrobacter* [90], *Enterobacter* [91], *Burkholderia* [90], *Pseudomonas* [92], *Serratia* [93], *Sphingobium* [87] and *Bacillus, Flavobacterium, Micrococcus,* and *Plesiomonas* [94], algae such as *Chlorella, Stichococcus,* and *Scenedesmus* [95,96], as well as fungi including *Penicillium oxalicum* [97], *Fusarium* sp. [98] and *Aspergillus sydowii* [99], *Cladosporium cladosporioides* [100], *Aspergillus niger* [101], *Aspergillus fumigates* [102], among others.

In 1973 the first bacteria, *Flavobacterium* sp., capable of degradation of organophosphorus compounds was discovered [103]. *Flavobacterium* can degrade almost all known P-O bonds through enzymatic hydrolysis. Next, the bacteria *Pseudomonas diminuta*, which acts similarly by cleaving P-O bonds by OP-degrading enzymes, was isolated in the United States in 1982 [104]. Many bacteria and fungi that may utilize organophosphorus compounds as a carbon, nitrogen, or phosphorus source have been discovered in subsequent years.

The microbial degradation of OPs involves complex processes of oxidation, reduction, hydrolysis, dealkylation, hydroxylation, alkylation, and ring cleavage, where hydrolysis usually represents the first stage in the degradation process, followed by continued degradation of the less hazardous compounds [105,106]. Usually, phosphorus is present as a phosphonate or a phosphate ester. As esters, they possess several hydrolysis-vulnerable sites. Microorganisms degrade OPs by hydrolyzing P-O alkyl and aryl bonds, which is a crucial step in detoxification [94,105,107]. The degradation rate varies and depends on microorganism species, their catalytic activities, and various environmental parameters such as temperature, pH, and sunlight availability [108]. In general, the action of bacteria is related to their genes and associated enzymes that hydrolyze and detoxify OPs [94]. Numerous functional genes have been reported to date, including *opd* (*opdE, opdA, opdC*), *amp* (*ampA*), *oph* (*ophB*), and *mpd* [109,110], while enzymes involved in the biodegradation and detoxification of OPs include cytochromes P450, phosphatase, esterase, hydrolase, and oxygenase [111]. Most studied OP degrading enzymes are hydrolases, such as organophosphorus hydrolase (OPH), organophosphorus acid anhydrolase (OPAA), and methyl parathion hydrolase (MPH) [105].

The OPHs, also termed phosphotriesterases and paraoxonase, are found in microorganisms, animals, and plants and are the most widely studied OP-degrading enzymes due to their catalytic efficiency [106]. OPHs are metalloenzymes that hydrolyze the triester linkage in organophosphate insecticides. These enzymes are encoded by the OP-degrading (*opd*) gene and have been initially found in *Sphingobium fuliginis* (*Flavobacterium* sp.) and *Brevundimonas diminuta* (*Pseudomonas diminuta*) [103,112]. The amino acid sequences of these enzymes are highly consistent; they share the same $(\alpha/\beta)_8$-barrel structural folds and an active site with two transition metal ions, such as zinc, iron, cobalt, or manganese [113]. Although the natural substrate of OPH remains unclear, it was verified that synthetic paraoxon is the best substrate for OPH. Additionally, OPH effectively hydrolyzes organophosphate triglyceride pesticides containing P-O bonds, such as paraoxon, parathion, and diazinon, but also hydrolyzes P-F, P-CN, and P-S bonds [114]. In addition, a variant of the OPH enzyme, OPDA encoded by the *opdA* gene, obtained from *Agrobacterium radiobacter* [115], can hydrolyze a wide range of OPs and G-type nerve agents like tabun, sarin, soman, and ethylsarine [116]. Moreover, OPDA is the only enzyme currently used commercially for bioremediation and pesticide decontamination of water sources [117,118].

OP-degrading enzyme OPAA, encoded by the *opaA* gene, was identified in *Alteromonas undina* and *Alteromonas haloplanktis* [119] as a member of the dipeptidase family with no enzyme or gene-sequence similarities with OPH or MPH. It hydrolyzes OPs and nerve agents G/V-series, acting on the P-O, P-F, P-S, and P-CN bonds [120].

Numerous methyl parathion degrading (*mpd*) genes were cloned to date, and phylogenetic analysis showed that they evolved apart from *opd* genes. The analysis also suggested that *mpd* and β-lactamase gene homologs, both members of the β-lactamase superfamily, are present in *Methylibium petroleiphilum*, *Azoarcus* sp., *Leptothrix cholodnii*, *Chromobacterium violaceum* and *Sinorhizobium meliloti* [121].

Different fungi are also involved in the remediation of contaminants in wastewater, soil, and organic wastes. The elimination of contaminants, including pesticides, through the utilization of fungi is known as mycoremediation [122,123]. Toxins and contaminants can be stored within fungal structures and serve as a carbon source upon enzymatic degradation [124]. In general, fungi have shown the potential to transform or degrade harmful pesticides into non-harmful or less harmful compounds through oxidation, decarboxylation, and enzymatic hydrolysis. Numerous fungi, including *Penicillium spp.*, *Fusarium spp.*, *Aspergillus flavus*, *Aspergillus niger*, *Trichoderma harzianum*, *Trametes versicolor*, *Pleurotus ostreatus*, *Lentinula edodes*, *Bjerkandera adusta*, *Rhizoctonia solani*, *Sporothrix cyanescens*, *Mortierella* are involved in this process [125–128]. Studies have shown that *Aspergillus oryzae*, *Aspergillus niger*, *Aspergillus flavus*, *Penicillium waksmanii*, *Acremonium* sp. participate in decomposing OPs, such as chlorpyrifos, malathion, parathion, and ethion [128,129]. For example, *Aspergillus oryzae* degrades malathion into β-monoacid and dicarboxylic acid by carboxylesterase activity, successively converting it into inorganic phosphate. Furthermore, *Candida cylindracea* and *Fusarium oxysporum* are also reported to degrade malathion by hydrolyses [130]. The *Penicillium waksmanii* is shown to degrade parathion into aminoparathion, which is less hazardous than the parental compounds [129]. Similarly, *Aspergillus flavus*, *Aspergillus niger*, and *Trichoderma harzianum* are found to degrade chlorpyrifos and endosulfan [127,128].

In general, mycoremediation can be considered a cost-effective and eco-friendly approach for the degradation of pesticides since fungi grow and survive in diverse agroecosystems. Also, numerous fungi have shown the potential to transform or degrade harmful pesticides into non-harmful or less harmful compounds. Therefore, mycoremediation is extremely advantageous in protecting the environment and living organisms, including humans.

Phytoremediation, also known as plant-assisted bioremediation, is a solar-powered technology that uses contaminant-scavenging plant species [131]. During this process, plants remove contaminants, including pesticides, from the environment through phytoextraction, phytodegradation, phytovolatilization, and rhizofiltration [132–134] and transform them into less dangerous ones [135]. The major plant-associated enzymes involved in pesticide phytoremediation are carboxylesterase, cytochrome P450, and glutathione

S-transferase [136,137]. Although phytoremediation is a cost-effective and valuable method for remediation, it has certain restrictions, such as the requirement for contaminants to be in the zone accessible for plant roots [138]. Furthermore, if pesticides are highly water soluble, the root system will not be able to reach them, and no degradation will occur. In addition, excessive pesticide concentrations can be hazardous to plants [123].

4.2. Abiotic Degradation Strategy

The abiotic strategy encompasses various approaches applied to reduce pesticide residue levels in foods and feeds to maximum permissible limits (MRLs) as prescribed by regulatory bodies, thus minimizing the risk of consumer exposure. However, there are many challenges in these processes, especially in terms of detecting the metabolites or intermediates of the pesticide degradation, which in some cases could be more toxic than the parent compounds. The complexity of the compounds present in the food matrix, the identification of numerous degradation products that may be formed by different pathways that are difficult to predict, and the lack of adequate commercial standards for the degraded products are just some of the challenges that occurred during the assessment of the toxicity of the pesticide degradation products [139]. As a consequence, numerous studies investigating pesticide reduction using different technologies have not implemented the screening and detection of degraded pesticide metabolites.

As shown in Figure 2, conventional techniques (chemical washing, peeling, drying, heating, etc.) and advanced approaches, such as chemical and nonthermal methods, have been employed in food processing to degrade pesticide residues.

Conventional methods, such as washing and cleaning, are of limited efficacy in pesticide removal due to the hydrophobicity of numerous pesticides. On the other hand, heat processing techniques, such as saucing, canning, blanching, and boiling, may significantly (but often not completely) reduce pesticide levels [31,140,141]. However, these techniques may be followed by a reduction of nutritional and taste-related features of foods and are not appropriate for vegetables and fruits that are consumed raw [142].

The chemical methods for the oxidative degradation of OPs mainly include the utilization of the various chemical oxidants, such as ferrate (VI), manganese dioxide, manganese dioxide charged with bisulfite, and nano-structured titania-iron mixed oxides, which provided encouraging results in environmental remediation, especially treatments of water [143–146]. However, limited effects of chemical techniques on pesticide removal rates are also reported [147].

The nonthermal technologies (cold plasma, high pressure processing, pulsed electric field, ultrasound, pulsed light, ultraviolet light, irradiation, oscillating magnetic field, ozonization, etc.) were developed to overcome the disadvantages of the conventional methods and to facilitate pesticide residue removal in fresh fruits and vegetable products and the environment as well [147,148].

As a feasible and relatively cheap technique, ozonated water washing has been frequently used for OPs (methyl-parathion, parathion, diazinon, chlorpyrifos) degradation in fruits and vegetables [142]. However, it was demonstrated that degradation products of pesticide residues upon ozone treatment (methyl paraoxon, paraoxon, and diazoxon) were more toxic than the parent compounds, which indicates the need for further processing using different technologies. In addition, this process requires higher treatment time and enhances the risk of oxidative degradation of bioactive substances in food commodities [149].

Gamma irradiation has also been used efficiently for the degradation of chlorfenvinphos, dimethoate, diazinon, and profenofos in the environment, especially in water [150]. Moreover, the extended treatment of chlorfenvinphos in tap water and groundwater further removed its degradation products [151]. However, these techniques are less feasible in food processing, especially in pesticide removal from rough-surfaced foods [142].

Ultrasonication is an unexplored area for pesticide removal from food commodities. This method was commonly performed in combination with ozone and UV irradiation treatments, enhancing their efficacy [142]. Recently, a novel advanced oxidation process

(AOP), i.e., the coupled free chlorine/ultrasound (FC/US) process, was utilized for the removal of dimethoate, trichlorfon, and carbofuran from lettuce, where removal efficiencies reached 86.7%, 79.8%, and 71.3%, respectively [152]. No noticeable damage to the quality of vegetables was observed after the FC/US process. However, when used solely, the efficacy of ultrasonic washing in pesticide degradation was variable and dependent on the surface morphology of the food commodities, with a variable impact on their nutritional properties [153].

High-pressure processing (HPP) is an environmentally friendly technology proposed as useful in reducing food contaminants such as pesticides and mycotoxins [154]. The efficiency of the HPP process depends on processing parameters, the chemical structure of the pesticide or mycotoxin, and the food matrix [154]. For example, its application successfully reduced chlorpyrifos levels in tomato samples under optimized conditions [155].

Cold plasma has been investigated for a wide range of applications, including its potential use for pesticide removal from agricultural commodities and wastewater [156]. The efficiency of plasma for pesticide reduction depends on several factors, such as the type of gas used in plasma and its flow rate, electrode distance, plasma power or voltage, exposure time, and others. These factors determine the amount of active species available for the pesticide reaction and subsequent degradation [142]. In the case of OPs, it has been shown that plasma species supplant the phosphoryl groups of pesticides, forming the phosphoxons, the unstable, toxic metabolites of the parent compound. It was also demonstrated that the toxicity of the formed metabolites largely depends on the chemical structure of the pesticides. A study analyzing the reduction of omethoate and dichlorvos in goji berry after gas barrier discharge plasma treatment reported increased toxicity of the products upon initial plasma treatment time of 0–6 min, after which it declined [157]. The study investigating chlorpyrifos degradation in tomatoes showed an 89.18% reduction after the 5 W plasma power for 6 min treatment [158]. The detection of the secondary metabolite, TCP, after the plasma treatment was also confirmed, but it was shown to be less toxic compared to the parent chlorpyrifos and its primary metabolite, chlorpyrifos-oxon. In addition, further intensified treatments led to the complete mineralization of the TCP metabolite [158]. Similar effects were reported upon plasma degradation of diazinon and phoxim in cucumber and table grapes, respectively [159,160]. Even though an encouraging rate of pesticide reduction was observed in the mentioned studies, some adverse effects, such as changes in the texture and total phenolic content of commodities, were also observed [158].

The pulsed electric field (PEF) methodology has been widely used for food preservation, reducing food contaminants, maintaining the nutritional values of the products, and removing pesticide residues from foods and wastewater [161]. For example, a study investigating the degradation of 16 pesticide residues in raw strawberries after treatment with PEF and boiling reported a removal efficacy of 92.9% and up to 91.2% when combined with ultrasonication [162]. A similar removal efficiency was reported for pyraclostrobin, chlorpyrifos ethyl, cyprodinil, malathion, and tau-fluvalinate in cherry juice after treatment with PEF (24.7 kV/cm, 655 µs) in combination with ozone and ultrasonication [163]. Nevertheless, additional studies on various food commodities are needed to support these findings.

In recent years, extensive research has been dedicated to the effects of UV light irradiation on the pesticide residues retained in fruits and vegetables and the environment. It was shown that pesticide degradation by photolysis depends on the type of pesticide residues, UV light sources, light intensity, and irradiation time [164]. Some previous studies analyzing photodegradation of OPs in the honey, including coumaphos, methyl parathion, and fenitrothion after 1 h with 250, 500, and 750 W/m^2 sunlight irradiation, have shown that coumaphos exhibited the best degradation performance (97.02% after 1 h) under 750 W/m^2 sunlight irradiation [165]. Recently, numerous studies have been conducted to investigate the degradation of pesticides using vacuum ultraviolet (V-UV) and ultraviolet light-C (UV-C) light sources. It was shown that V-UV was more effective than UV-C in the degradation of some fungicides under the same reaction conditions [166]. Likewise, it was reported that V-UV/UV used to remove some carbamate pesticides from the water was much more effective than UV, reaching the removal efficiency of at least 90% at a V-UV fluence of 12 mJ/cm^2 [167]. As most pesticides show absorption maxima at relatively short UV wavelengths, their photostability under the UV-C treatment and the toxicity of their photodegradation products should be assessed. A recent study investigating the toxicity of chlorpyrifos and its formulations (emulsifiable concentrate—EC and oil-in-water emulsion—EW) after UV-C irradiation showed that chlorpyrifos concentration decreased during UV-C irradiation. In contrast, the concentration of its product, chlorpyrifos-oxon, increased, reaching a maximal concentration after 17 min (EW) and 80 min (chlorpyrifos and EC) of irradiation, when subsequently decreased [168]. The same study demonstrated the pro-oxidative and genotoxic effects of their photodegradation products. Noteworthy, chlorpyrifos was more genotoxic compared to its formulations. Another study investigating the effects of UV-C irradiation of glyphosate in water showed at least a 90% reduction in glyphosate concentrations and the generation of less toxic degradation products, reducing the overall toxicity to aquatic organisms [169]. Several techniques using UV have been applied for pesticide removal from wastewater. For example, a study using pulsed light (PL) technology for the photodegradation of several OPs in water showed >50% pesticide removal in a very short time [170]. However, the toxicity of photodegradation products was not assessed.

Catalyst methodologies have also been used for pesticide removal, mainly for wastewater treatment and environmental remediation [171]. The photocatalytic process is based on pesticide residue breakdown by AOP, where photons degrade pesticides to CO_2, H_2O, and inorganic compounds. Usually, it comprises a catalyst such as TiO_2, which, combined with UV light, accelerates the degradation of pesticides, mainly in the soil and agricultural wastewater [171,172]. For example, UV-C/TiO_2 was efficiently used for the photocatalytic degradation of diazinon in water [173]. A similar photocatalytic activity was also reported for the $CoFe_2O_4$@TiO_2 nanocomposite used to degrade chlorpyrifos [174]. An interesting study showed that the rate of degradation of malathion by UV light alone was lower than that observed when photocatalytic treatments, such as UV/H_2O_2, UV/TiO_2, and UV/Fenton systems, were applied; however, in contrast to the photocatalytic processes, no increase in toxicity of the malathion aqueous solution after UV irradiation alone was observed [175]. Therefore, it is assumed that applications of photocatalytic oxidation in food processing can be limited due to the observed toxicity of the treated solutions [147].

In general, identifying pesticide degradation products and assessing their toxicity while employing different pesticide removal methodologies have to be carried out in order to ensure food safety and, thus, protect human health.

5. Future Perspectives on Pesticide Use and Management

One of the biggest challenges in the 21st century is how to feed the increasing population while decreasing the adverse consequences on the environment and human health that arose due to the continuous deployment of pesticides, fertilizers, and freshwater [176,177]. The European Green Deal and its Farm to Fork Strategy proposed targets to establish sustainable agriculture, considering nature conservation to ensure a fair, healthy, and environmentally friendly food system [178]. Under the Farm to Fork Strategy [178], the EC plans to take actions to reduce by 50% the use and risk of chemical pesticides, including the use of more hazardous pesticides, until 2030. To achieve this goal, the EC will revise the SUD [57] and promote alternative practices, such as IPM, to protect harvests from pests and diseases. In this respect, strong support should be given to farmers and accelerate the transition toward sustainable agriculture. In addition, the Farm to Fork Strategy encourages organic farming intending to have at least 25% of EU agricultural land under organic farming management by 2030 [178]. However, some issues must be resolved to meet the proposed targets. As discussed, the risk assessment considering pesticide cocktails found in the major agricultural systems must be implemented since their effects on soil health and, consequently, on food safety are still unknown [43]. In addition, the issue of legacy pesticides that can persist in the environment for several decades after they were banned must be resolved [179]. Since current EU policy [178] leaves these issues unaddressed, improved strategies encompassing innovative methods to test the effects of pesticide cocktails on soil health and targeting and remediation of legacy pesticides in the environment are urgently required [43,179]. Soil remediation as well as the establishment of rich above-ground plant systems, should be prioritized to alleviate the effects of current and legacy pesticides in soils [43]. Thus, besides reducing pesticide usage, to ensure food safety, it is necessary to implement novel approaches to detect pesticide residues, assess the real risk of combined exposure to multiple residues, and degrade them into non-toxic products to safeguard consumer health.

Considering pesticide screening and detection, the new procedures developed within the green chemistry framework should be prioritized [63]. In this regard, novel analytical techniques have been proposed based on nanosystems for accurate, green, and ultrasensitive detection of pesticide residues in food and the environment [180,181]. In addition, novel, smartly engineered nanomaterials, and advanced instrumentation should facilitate pesticide detection in complex food matrices and make it more sensitive, cost-effective, and less time-consuming [180].

Similarly, pesticide degradation strategies should rely on efficient, cost-effective, environmentally friendly techniques. The integration of nanotechnology and advanced materials can offer innovative approaches to pesticide degradation. Nanomaterials can be designed to enhance the efficiency of degradation processes through increased surface area, catalytic activity, and selectivity [179]. Moreover, green nanomaterials, produced using different parts of plants (seeds, fruit, leaves, and flower) or microorganisms (bacteria, algae, and fungi), are suggested to be biocompatible, biodegradable, cost-effective, eco-friendly, and efficient in environmental remediation [182]. However, there is a severe knowledge gap concerning identifying metabolites or intermediates of pesticide degradation, which in some cases could be more toxic than parental compounds. Therefore, further comprehensive studies on pesticide degradation products and their toxicity are required.

Another promising direction in pesticide degradation is developing and implementing advanced, more efficient bioremediation techniques using various bioinformatics tools [183]. Genome engineering by gene editing tools, such as CRISPR-Cas, ZFN, and TALEN, can create functionally improved microorganisms with complex genes that encode catabolic enzymes involved in OPs metabolism [183,184]. However, releasing genetically engineered organisms into the environment requires the approval of various regulatory bodies. So, the direct application of recombinant enzymes, frequently termed cell-free catalytic systems derived from engineered microbes, which are non-replicative, can be used for environmental remediation [185].

The new postgenomic research technologies, called the OMICs approach, may provide tools to investigate microbial interactions with pesticides and construct enzyme-based mechanisms for bioremediation in different environmental settings [185]. The bioinformatics and computational tools in OMICs comprise technologies such as metagenomics, transcriptomics, proteomics, and metabolomics, as well as studies of their interactive pathways, named interactomics [186]. Integration of these technologies creates a multi-omic approach that provides a comprehensive understanding of the processes associated with biodegradation [185]. In this respect, it has been reported that CRISPR-Cas, ZFN, and TALEN as gene editing tools utilizing Pseudomonas, Escherichia coli, and Achromobacter sp. can be employed for remediation of chlorpyrifos, methyl-parathion, carbaryl, triphenyltin, and triazophos by constructing a guide RNA (gRNA) for expressing specific genes for the bioremediation [187]. In addition, computational analysis such as molecular docking, molecular modeling, and simulation analysis can efficiently determine the fate of degraded metabolites, the structural and functional characterization of OPs degrading enzymes, and their binding properties [188]. Developing an artificial microbiome with functionally specific species has also been proposed to facilitate bioremediation processes [183]. There are some concerns that without innovations, such as the New Genomic Techniques

(NGT), the encouragement of gene-editing research, and the revision of the current EU legislation concerning genetically modified organisms (GMO) [189], the targets proposed by the Farm to Fork Strategies will be difficult to reach. Additional research in this area and more in-depth practical implementation of the techniques mentioned in large-scale studies are necessary.

Concerning the use of bioremediation techniques in food processing, it was demonstrated that some fermented foods could be detoxified from pesticides due to the activity of the bacterial microflora [190]. The lactic acid bacteria existing in or added to foods can metabolize a wide range of OPs, using them as a source of carbon and energy [191]. Fermentation by natural microflora or by probiotic strains added to foods is a promising approach for pesticide detoxication. However, since the exact metabolic pathways of degradation are still unknown, further research is necessary [190].

To ensure the development of sustainable agriculture, smart agricultural research employing different artificial intelligence (AI) approaches, such as deep learning (DL), machine learning (ML), agricultural robots, and robotics, has been suggested to resolve the prevailing problems in agriculture and improve productivity [192,193]. Using robotics can increase production and save time on repetitive tasks. It is estimated that low-cost agricultural robots can reduce pesticide usage by 80% if farmers use them for spraying [192,194]. Application of neural networks, DL, and ML techniques can enable early and timely identification of pests and diseases, monitoring of moisture and nitrogen content in the soil, informing irrigation for water saving, detection of herbicide usage, detecting food damage, etc. [193,195,196]. The proposed methods need to be further explored to establish the AI framework that should enable the sustainable development of smart agriculture.

Altogether, to develop a sustainable agriculture and food system, the whole environment in which food is produced should be considered. Besides taking appropriate agricultural management measures, it is necessary to improve and globally harmonize methodologies in the areas of food safety and food quality to protect human health. A holistic approach covering the entire food production chain should be applied to control food contaminants. In addition, the food-producing systems from farm to fork are influenced by several factors, such as climate change, demographics, and the economy, which may create new food safety risks and affect human health [197]. Collaborative efforts among scientists, policymakers, regulators, industry stakeholders, and farmers are crucial to closing the gaps between the EU strategy goals and practical implementations. Embracing innovative technologies, promoting sustainable practices, and fostering knowledge exchange on a global level are required to establish sustainable agriculture and ensure a fair and sustainable food system that does not leave anyone behind.

Author Contributions: Conceptualization, A.L. and S.P.; writing—original draft preparation A.L. and S.P.; writing—review and editing, A.L. and S.P.; supervision, A.L. and S.P. All authors have read and agreed to the published version of the manuscript.

Funding: The research was funded by the Ministry of Science, Technological Development and Innovations of the Republic of Serbia (No. 451-03-47/2023-01/200017).

Data Availability Statement: Data is contained within the article.

Acknowledgments: The authors acknowledge the support provided by the Ministry of Science, Technological Development and Innovations of the Republic of Serbia (No. 451-03-47/2023-01/200017).

Conflicts of Interest: The authors declare no conflict of interest.

References

1. Onyeka Kingsley, N.; Ayibapreye, J. Chemical Pesticides and Food Safety. In *Insecticides*; Ramón Eduardo Rebolledo, R., Ed.; IntechOpen: Rijeka, Croatia, 2022; p. Ch. 3.
2. Antonini, C.; Argilés-Bosch, J.M. Productivity and environmental costs from intensification of farming. A panel data analysis across EU regions. *J. Clean. Prod.* **2017**, *140*, 796–803. [CrossRef]
3. Rajmohan, K.S.; Chandrasekaran, R.; Varjani, S. A Review on Occurrence of Pesticides in Environment and Current Technologies for Their Remediation and Management. *Indian J. Microbiol.* **2020**, *60*, 125–138. [CrossRef]

4. Food and Agriculture Organization of the United Nations (FAO). FAOSTAT Database. Pesticides Use. 2022. Available online: http://www.fao.org/faostat/en/#data/RP (accessed on 29 May 2023).
5. Donley, N. The USA lags behind other agricultural nations in banning harmful pesticides. *Environ. Health* **2019**, *18*, 44. [CrossRef]
6. Ambrus, Á.; Szenczi-Cseh, J.; Doan, V.V.N.; Vásárhelyi, A. Evaluation of Monitoring Data in Foods. *Agrochemicals* **2023**, *2*, 69–95. [CrossRef]
7. European Parliament; Directorate-General for External Policies of the Union; Sarkar, S.; Dias, J.; Gil, B.; Keeley, J.; Möhring, N.; Jansen, K. The Use of Pesticides in Developing Countries and Their Impact on Health and the Right to Food, European Parliament. 2021. Available online: https://data.europa.eu/doi/10.2861/28995 (accessed on 31 May 2023).
8. Food and Agriculture Organization of the United Nations (FAO); World Health Organization (WHO). *Codex Alimentarius, Pesticides*; Online Publication; FAO: Rome, Italy; WHO: Geneva, Switzerland, 2020.
9. Regulation (EC) No 396/2005 of the European Parliament and of the Council of 23 February 2005 on Maximum Residue Levels of Pesticides in or on Food and Feed of Plant and Animal Origin and Amending Council Directive 91/414/EEC Official Journal of the European Union, 70; pp. 1–16. 16 March 2005. Available online: https://eur-lex.europa.eu/legal-content/EN/TXT/?uri=CELEX:02005R0396-20210525 (accessed on 24 March 2023).
10. US Environmental Protection Agency. *Summary of the Federal Food, Drug, and Cosmetic*; US Environmental Protection Agency: District of Columbia, WA, USA, 2002. Available online: https://www.epa.gov/laws-regulations/summary-federal-food-drug-and-cosmetic-act (accessed on 23 March 2023).
11. CIRS China. Available online: https://www.cirs-group.com/en (accessed on 29 May 2023).
12. Agência Nacional de Vigilância Sanitária—Anvisa. Available online: https://www.gov.br/anvisa/pt-br/english/regulation-of-products/pesticides (accessed on 30 May 2023).
13. van den Dries, M.A.; Lamballais, S.; El Marroun, H.; Pronk, A.; Spaan, S.; Ferguson, K.K.; Longnecker, M.P.; Tiemeier, H.; Guxens, M. Prenatal exposure to organophosphate pesticides and brain morphology and white matter microstructure in preadolescents. *Environ. Res.* **2020**, *191*, 110047. [CrossRef] [PubMed]
14. Zikankuba, V.L.; Mwanyika, G.; Ntwenya, J.E.; James, A. Pesticide regulations and their malpractice implications on food and environment safety. *Cogent Food Agric.* **2019**, *5*, 1601544. [CrossRef]
15. Kwong, T.C. Organophosphate Pesticides: Biochemistry and Clinical Toxicology. *Ther. Drug Monit.* **2002**, *24*, 144–149. [CrossRef]
16. Fu, H.; Tan, P.; Wang, R.; Li, S.; Liu, H.; Yang, Y.; Wu, Z. Advances in organophosphorus pesticides pollution: Current status and challenges in ecotoxicological, sustainable agriculture, and degradation strategies. *J. Hazard. Mater.* **2022**, *424*, 127494. [CrossRef]
17. Kim, K.H.; Kabir, E.; Jahan, S.A. Exposure to pesticides and the associated human health effects. *Sci. Total Environ.* **2017**, *575*, 525–535. [CrossRef]
18. Leskovac, A. Organophosphate Pesticides: Cytotoxicity, Genotoxicity and Current Treatment Strategies. In *Organophosphates: Detection, Exposure and Occurrence. Volume 1: Impact on Health and the Natural Environment*; Lazarević-Pašti, T., Ed.; Nova Science Publishers: Hauppauge, NY, USA, 2022; pp. 27–61.
19. Petrovic, S. Organophosphate Pesticides and Human Health: Current Knowledge and Future Prospects. In *Organophosphates: Detection, Exposure and Occurrence. Volume 1: Impact on Health and the Natural Environment*; Lazarević-Pašti, T., Ed.; Nova Science Publishers: Hauppauge, NY, USA, 2022; pp. 1–25.
20. Kubiak-Hardiman, P.; Haughey, S.A.; Meneely, J.; Miller, S.; Banerjee, K.; Elliott, C.T. Identifying Gaps and Challenges in Global Pesticide Legislation that Impact the Protection of Consumer Health: Rice as a Case Study. *Expo. Health* **2022**, *14*, 1–22. [CrossRef]
21. Hejazi, M.; Grant, J.H.; Peterson, E. Trade impact of maximum residue limits in fresh fruits and vegetables. *Food Policy* **2022**, *106*, 102103. [CrossRef]
22. Jors, E.; Neupane, D.; London, L. Pesticide Poisonings in Low- and Middle-Income Countries. *Environ. Health Insights* **2018**, *12*, 1178630217750876. [CrossRef] [PubMed]
23. Pesticides Atlas. *Data and Facts on Toxins in Agriculture*; Kenya, Ed.; Heinrich-Böll-Stiftung: Berlin, Germany, 2022.
24. Wolejko, E.; Lozowicka, B.; Jablonska-Trypuc, A.; Pietruszynska, M.; Wydro, U. Chlorpyrifos Occurrence and Toxicological Risk Assessment: A Review. *Int. J. Environ. Res. Public Health* **2022**, *19*, 12209. [CrossRef]
25. Foong, S.Y.; Ma, N.L.; Lam, S.S.; Peng, W.; Low, F.; Lee, B.H.K.; Alstrup, A.K.O.; Sonne, C. A recent global review of hazardous chlorpyrifos pesticide in fruit and vegetables: Prevalence, remediation and actions needed. *J. Hazard. Mater.* **2020**, *400*, 123006. [CrossRef]
26. Lopez-Carmen, V.A.; Erickson, T.B.; Escobar, Z.; Jensen, A.; Cronin, A.E.; Nolen, L.T.; Moreno, M.; Stewart, A.M. United States and United Nations pesticide policies: Environmental violence against the Yaqui indigenous nation. *Lancet Reg. Health Am.* **2022**, *10*, 100255. [CrossRef] [PubMed]
27. Akomea-Frempong, S.; Ofosu, I.W.; Owusu-Ansah, E.d.-G.J.; Darko, G. Health risks due to consumption of pesticides in ready-to-eat vegetables (salads) in Kumasi, Ghana. *Int. J. Food Contam.* **2017**, *4*, 13. [CrossRef]
28. Ssemugabo, C.; Bradman, A.; Ssempebwa, J.C.; Sille, F.; Guwatudde, D. Pesticide Residues in Fresh Fruit and Vegetables from Farm to Fork in the Kampala Metropolitan Area, Uganda. *Environ. Health Insights* **2022**, *16*, 11786302221111866. [CrossRef] [PubMed]
29. Gill, J.P.S.; Bedi, J.S.; Singh, R.; Fairoze, M.N.; Hazarika, R.A.; Gaurav, A.; Satpathy, S.K.; Chauhan, A.S.; Lindahl, J.; Grace, D.; et al. Pesticide Residues in Peri-Urban Bovine Milk from India and Risk Assessment: A Multicenter Study. *Sci. Rep.* **2020**, *10*, 8054. [CrossRef]

30. El-Sheikh, E.A.; Ramadan, M.M.; El-Sobki, A.E.; Shalaby, A.A.; McCoy, M.R.; Hamed, I.A.; Ashour, M.B.; Hammock, B.D. Pesticide Residues in Vegetables and Fruits from Farmer Markets and Associated Dietary Risks. *Molecules* **2022**, *27*, 8072. [CrossRef]
31. El-Sheikh, E.-S.A.; Li, D.; Hamed, I.; Ashour, M.-B.; Hammock, B.D. Residue Analysis and Risk Exposure Assessment of Multiple Pesticides in Tomato and Strawberry and Their Products from Markets. *Foods* **2023**, *12*, 1936. [CrossRef]
32. Bouhala, A.; Lahmar, H.; Benamira, M.; Moussi, A.; Trari, M. Photodegradation of Organophosphorus Pesticides in Honey Medium by Solar Light Irradiation. *Bull. Environ. Contam. Toxicol.* **2020**, *104*, 792–798. [CrossRef] [PubMed]
33. EFSA (European Food Safety Authority). The 2016 European Union report on pesticide residues in food. *EFSA J.* **2018**, *16*, e05348. [CrossRef]
34. Witczak, A.; Pohoryło, A.; Abdel-Gawad, H.; Cybulski, J. Residues of some organophosphorus pesticides on and in fruits and vegetables available in Poland, an assessment based on the European union regulations and health assessment for human populations. *Phosphorus Sulfur Silicon Relat. Elem.* **2018**, *193*, 711–720. [CrossRef]
35. Mert, A.; Qi, A.; Bygrave, A.; Stotz, H.U. Trends of pesticide residues in foods imported to the United Kingdom from 2000 to 2020. *Food Control.* **2022**, *133*, 108616. [CrossRef]
36. Panseri, S.; Bonerba, E.; Nobile, M.; Di Cesare, F.; Mosconi, G.; Cecati, F.; Arioli, F.; Tantillo, G.; Chiesa, L. Pesticides and Environmental Contaminants in Organic Honeys According to Their Different Productive Areas toward Food Safety Protection. *Foods* **2020**, *9*, 1863. [CrossRef]
37. Nardelli, V.; D'Amico, V.; Ingegno, M.; Della Rovere, I.; Iammarino, M.; Casamassima, F.; Calitri, A.; Nardiello, D.; Li, D.; Quinto, M. Pesticides Contamination of Cereals and Legumes: Monitoring of Samples Marketed in Italy as a Contribution to Risk Assessment. *Appl. Sci.* **2021**, *11*, 7283. [CrossRef]
38. EFSA (European Food Safety Authority); Carrasco Cabrera, L.; Di Piazza, G.; Dujardin, B.; Medina Pastor, P. The 2021 European Union report on pesticide residues in food. *EFSA J.* **2023**, *21*, e07939. [CrossRef]
39. de Jong, E.; van der Voet, H.; Marx-Stoelting, P.; Bennekou, S.H.; Sprong, C.; Bloch, D.; Burchardt, A.; Lasch, A.; Opialla, T.; Rotter, S.; et al. Roadmap for action on Risk Assessment of Combined Exposure to Multiple Chemicals (RACEMiC). *EFSA Support. Publ.* **2022**, *19*, 7555E. [CrossRef]
40. EFSA (European Food Safety Authority). International framework dealing with human risk assessment of combined exposure to multiple chemicals. *EFSA J.* **2013**, *11*, 2125. [CrossRef]
41. EFSA (European Food Safety Authority). Guidance on harmonised methodologies for human health, animal health and ecological risk assessment of combined exposure to multiple chemicals. *EFSA J.* **2019**, *17*, e05634. [CrossRef]
42. EFSA (European Food Safety Authority); Anastassiadou, M.; Choi, J.; Coja, T.; Dujardin, B.; Hart, A.; Hernandez-Jerrez, A.F.; Jarrah, S.; Lostia, A.; Machera, K. Mohimont Cumulative dietary risk assessment of chronic acetylcholinesterase inhibition by residues of pesticides 2021. *EFSA J.* **2021**, *19*, 6392. [CrossRef]
43. Geissen, V.; Silva, V.; Lwanga, E.H.; Beriot, N.; Oostindie, K.; Bin, Z.; Pyne, E.; Busink, S.; Zomer, P.; Mol, H.; et al. Cocktails of pesticide residues in conventional and organic farming systems in EuropeLegacy of the past and turning point for the future. *Environ. Pollut.* **2021**, *278*, 116827. [CrossRef] [PubMed]
44. Cattaneo, I.; Kalian, A.D.; Di Nicola, M.R.; Dujardin, B.; Levorato, S.; Mohimont, L.; Nathanail, A.V.; Carnessechi, E.; Astuto, M.C.; Tarazona, J.V.; et al. Risk Assessment of Combined Exposure to Multiple Chemicals at the European Food Safety Authority: Principles, Guidance Documents, Applications and Future Challenges. *Toxins* **2023**, *15*, 40. [CrossRef] [PubMed]
45. Rempelos, L.; Wang, J.; Baranski, M.; Watson, A.; Volakakis, N.; Hoppe, H.W.; Kuhn-Velten, W.N.; Hadall, C.; Hasanaliyeva, G.; Chatzidimitriou, E.; et al. Diet and food type affect urinary pesticide residue excretion profiles in healthy individuals: Results of a randomized controlled dietary intervention trial. *Am. J. Clin. Nutr.* **2022**, *115*, 364–377. [CrossRef]
46. Vasylieva, N.; Barnych, B.; Wan, D.; El-Sheikh, E.A.; Nguyen, H.M.; Wulff, H.; McMahen, R.; Strynar, M.; Gee, S.J.; Hammock, B.D. Hydroxy-fipronil is a new urinary biomarker of exposure to fipronil. *Environ. Int.* **2017**, *103*, 91–98. [CrossRef]
47. Andersen, H.R.; Rambaud, L.; Riou, M.; Buekers, J.; Remy, S.; Berman, T.; Govarts, E. Exposure Levels of Pyrethroids, Chlorpyrifos and Glyphosate in EU-An Overview of Human Biomonitoring Studies Published since 2000. *Toxics* **2022**, *10*, 789. [CrossRef]
48. EFSA (European Food Safety Authority). Statement on the available outcomes of the human health assessment in the context of the pesticides peer review of the active substance chlorpyrifos. *EFSA J.* **2019**, *17*, e05809. [CrossRef]
49. Qiang, S.; Mohamed, F.; Mackenzie, L.; Roberts, M.S. Rapid determination of polyethoxylated tallow amine surfactants in human plasma by LC-MSMS. *Talanta* **2023**, *254*, 124115. [CrossRef]
50. Mie, A.; Andersen, H.R.; Gunnarsson, S.; Kahl, J.; Kesse-Guyot, E.; Rembiałkowska, E.; Quaglio, G.; Grandjean, P. Human health implications of organic food and organic agriculture: A comprehensive review. *Environ. Health* **2017**, *16*, 111. [CrossRef]
51. Gómez-Ramos, M.d.M.; Nannou, C.; Martínez Bueno, M.J.; Goday, A.; Murcia-Morales, M.; Ferrer, C.; Fernández-Alba, A.R. Pesticide residues evaluation of organic crops. A critical appraisal. *Food Chem. X* **2020**, *5*, 100079. [CrossRef]
52. Rebouillat, P.; Vidal, R.; Cravedi, J.P.; Taupier-Letage, B.; Debrauwer, L.; Gamet-Payrastre, L.; Touvier, M.; Hercberg, S.; Lairon, D.; Baudry, J.; et al. Estimated dietary pesticide exposure from plant-based foods using NMF-derived profiles in a large sample of French adults. *Eur. J. Nutr.* **2021**, *60*, 1475–1488. [CrossRef] [PubMed]
53. Witczak, A.; Abdel-Gawad, H. Comparison of organochlorine pesticides and polychlorinated biphenyls residues in vegetables, grain and soil from organic and conventional farming in Poland. *J. Environ. Sci. Health. Part. B Pestic. Food Contam. Agric. Wastes* **2012**, *47*, 343–354. [CrossRef] [PubMed]

54. EC European Commission Regulation (EC) No 850/2004 of the European Parliament and of the Council of 29 April 2004 on Persistent Organic Pollutants and Amending Directive 79/117/EEC 2009. Available online: http://data.europa.eu/eli/reg/2004/850/2009-05-05 (accessed on 23 June 2023).
55. Sivaperumal, P.; Thasale, R.; Kumar, D.; Mehta, T.G.; Limbachiya, R. Human health risk assessment of pesticide residues in vegetable and fruit samples in Gujarat State, India. *Heliyon* **2022**, *8*, e10876. [CrossRef]
56. Yao, R.; Yao, S.; Ai, T.; Huang, J.; Liu, Y.; Sun, J. Organophosphate Pesticides and Pyrethroids in Farmland of the Pearl River Delta, China: Regional Residue, Distributions and Risks. *Int. J. Environ. Res. Public Health* **2023**, *20*, 1017. [CrossRef]
57. European Parliament and the Council Directive 2009/128/EC of the European Parliament and of the Council of 21 October 2009 Establishing a Framework for Community Action to Achieve the Sustainable Use of Pesticides. Available online: https://eur-lex.europa.eu/legal-content/EN/ALL/?uri=celex%3A32009L0128 (accessed on 6 June 2023).
58. Tataridas, A.; Kanatas, P.; Chatzigeorgiou, A.; Zannopoulos, S.; Travlos, I. Sustainable Crop and Weed Management in the Era of the EU Green Deal: A Survival Guide. *Agronomy* **2022**, *12*, 589. [CrossRef]
59. Helepciuc, F.-E.; Todor, A. Greener European Agriculture? Evaluating EU Member States' Transition Efforts to Integrated Pest Management through Their National Action Plans. *Agronomy* **2022**, *12*, 2438. [CrossRef]
60. Umetsu, N.; Shirai, Y. Development of novel pesticides in the 21st century. *J. Pestic. Sci.* **2020**, *45*, 54–74. [CrossRef]
61. Damalas, C.A.; Koutroubas, S.D. Current Status and Recent Developments in Biopesticide Use. *Agriculture* **2018**, *8*, 13. [CrossRef]
62. Iammarino, M.; Panseri, S.; Unlu, G.; Marchesani, G.; Bevilacqua, A. Editorial: Novel chemical, microbiological and physical approaches in food safety control. *Front. Nutr.* **2022**, *9*, 1060480. [CrossRef]
63. Iammarino, M.; Palermo, C.; Tomasevic, I. Advanced Analysis Techniques of Food Contaminants and Risk Assessment-Editorial. *Appl. Sci.* **2022**, *12*, 4863. [CrossRef]
64. EFSA-SANTE. EFSA-SANTE Action Plan on Cumulative Risk Assessment for Pesticides Residues. Standing Committee for Plants, Animals, Food and Feed, Section Phytopharmaceuticals, Pesticide Residues. SANTE/10178/2021. 2021. Available online: https://ec.europa.eu/food/sites/food/files/plant/docs/pesticides_mrl_cum-risk-ass_sante-10178-2021.pdf (accessed on 24 March 2023).
65. Rawtani, D.; Khatri, N.; Tyagi, S.; Pandey, G. Nanotechnology-based recent approaches for sensing and remediation of pesticides. *J. Environ. Manag.* **2018**, *206*, 749–762. [CrossRef]
66. Zamora-Sequeira, R.; Starbird-Pérez, R.; Rojas-Carillo, O.; Vargas-Villalobos, S. What are the Main Sensor Methods for Quantifying Pesticides in Agricultural Activities? A Review. *Molecules* **2019**, *24*, 2659. [CrossRef]
67. Ghosh, S.; AlKafaas, S.S.; Bornman, C.; Apollon, W.; Hussien, A.M.; Badawy, A.E.; Amer, M.H.; Kamel, M.B.; Mekawy, E.A.; Bedair, H. The application of rapid test paper technology for pesticide detection in horticulture crops: A comprehensive review. *Beni-Suef Univ. J. Basic Appl. Sci.* **2022**, *11*, 73. [CrossRef]
68. Pelton, R. Bioactive paper provides a low-cost platform for diagnostics. *Trends Anal. Chem.* **2009**, *28*, 925–942. [CrossRef] [PubMed]
69. Sicard, C.; Glen, C.; Aubie, B.; Wallace, D.; Jahanshahi-Anbuhi, S.; Pennings, K.; Daigger, G.T.; Pelton, R.; Brennan, J.D.; Filipe, C.D.M. Tools for water quality monitoring and mapping using paper-based sensors and cell phones. *Water Res.* **2015**, *70*, 360–369. [CrossRef] [PubMed]
70. Phongphut, A.; Chayasombat, B.; Cass, A.E.G.; Phisalaphong, M.; Prichanont, S.; Thanachayanont, C.; Chodjarusawad, T. Biosensors Based on Acetylcholinesterase Immobilized on Clay–Gold Nanocomposites for the Discrimination of Chlorpyrifos and Carbaryl. *ACS Omega* **2022**, *7*, 39848–39859. [CrossRef]
71. Jia, M.; Zhombo, E.; Zhai, F.; Bing, X. Rapid Multi-Residue Detection Methods for Pesticides and Veterinary Drugs. *Molecules* **2020**, *25*, 3590. [CrossRef]
72. Garvey, J.; Walsh, T.; Devaney, E.; King, T.; Kilduff, R. Multi-residue analysis of pesticide residues and polychlorinated biphenyls in fruit and vegetables using orbital ion trap high-resolution accurate mass spectrometry. *Anal. Bioanal. Chem.* **2020**, *412*, 7113–7121. [CrossRef]
73. Hajrulai-Musliu, Z.; Uzunov, R.; Jovanov, S.; Jankuloski, D.; Stojkovski, V.; Pendovski, L.; Sasanya, J.J. A new LC-MS/MS method for multiple residues/contaminants in bovine meat. *BMC Chem.* **2021**, *15*, 62. [CrossRef] [PubMed]
74. Pedersen, M.; Hakme, E.; Ninga, E.; Frandsen, H.L. Analysis of veterinary drug- and pesticide residues in pig muscle by LC-QTOF-MS. *Food Control.* **2023**, *148*, 109656. [CrossRef]
75. BS EN 15662:2018; Foods of Plant Origin—Multimethod for the Determination of Pesticide Residues Using GC- and LC-Based Analysis Following Acetonitrile Extraction/Partitioning and Clean-Up by Dispersive SPE. Modular QuEChERS-Method. B S I Standards: Hemel Hempstead, UK, 2018.
76. Ninga, E.; Lehotay, S.J.; Sapozhnikova, Y.; Lightfield, A.R.; Strahan, G.D.; Monteiro, S.H. Analysis of pesticides, veterinary drugs, and environmental contaminants in goat and lamb by the QuEChERSER mega-method. *Anal. Methods* **2022**, *14*, 2761–2770. [CrossRef] [PubMed]
77. Monteiro, S.H.; Lehotay, S.J.; Sapozhnikova, Y.; Ninga, E.; Moura Andrade, G.C.R.; Lightfield, A.R. Validation of the QuEChERSER mega-method for the analysis of pesticides, veterinary drugs, and environmental contaminants in tilapia (Oreochromis Niloticus). *Food Addit. Contam. Part A Chem. Anal. Control. Expo. Risk Assess* **2022**, *39*, 699–709. [CrossRef]
78. Sharma, A.; Kumar, V.; Shahzad, B.; Tanveer, M.; Sidhu, G.P.S.; Handa, N.; Kohli, S.K.; Yadav, P.; Bali, A.S.; Parihar, R.D.; et al. Worldwide pesticide usage and its impacts on ecosystem. *SN Appl. Sci.* **2019**, *1*, 1446. [CrossRef]

79. Nayak, S.B.; Sahoo, A.K.; Kolanthasamy, E.; Rao, K. Role of pesticide application in environmental degradation and its remediation strategies. In *Environmental Degradation: Causes and Remediation Strategies: Volume 1*; Kumar, V., Singh, J., Kumar, P., Eds.; Agro Environ Media: Uttarakhand, India, 2020; pp. 36–46.
80. Hussain, S.; Siddique, T.; Arshad, M.; Saleem, M. Bioremediation and phytoremediation of pesticides: Recent advances. *Crit. Rev. Environ. Sci. Technol.* **2009**, *39*, 843–907. [CrossRef]
81. Ning, J.; Gang, G.; Bai, Z.; Hu, Q.; Qi, H.; Ma, A.; Zhuan, X.; Zhuang, G. In situ enhanced bioremediation of dichlorvos by a phyllosphere *Flavobacterium* strain. *Front. Environ. Sci. Eng.* **2012**, *6*, 231–237. [CrossRef]
82. Ozdal, M.; Ozdal, O.G.; Algur, O.F.; Kurbanoglu, E.B. Biodegradation of α-endosulfan via hydrolysis pathway by Stenotrophomonas maltophilia OG2. *3 Biotech* **2017**, *7*, 113. [CrossRef]
83. Ramu, S.; Seetharaman, B. Biodegradation of acephate and methamidophos by a soil bacterium *Pseudomonas aeruginosa* strain Is-6. *J. Environ. Sci. Health Part B* **2014**, *49*, 23–34. [CrossRef] [PubMed]
84. Huang, Y.; Xiao, L.; Li, F.; Xiao, M.; Lin, D.; Long, X.; Wu, Z. Microbial Degradation of Pesticide Residues and an Emphasis on the Degradation of Cypermethrin and 3-phenoxy Benzoic Acid: A Review. *Molecules* **2018**, *23*, 2313. [CrossRef]
85. Wolicka, D.; Suszek, A.; Borkowski, A.; Bielecka, A. Application of aerobic microorganisms in bioremediation in situ of soil contaminated by petroleum products. *Bioresour. Technol.* **2009**, *100*, 3221–3227. [CrossRef] [PubMed]
86. Tyagi, M.; da Fonseca, M.M.; de Carvalho, C.C. Bioaugmentation and biostimulation strategies to improve the effectiveness of bioremediation processes. *Biodegradation* **2011**, *22*, 231–241. [CrossRef]
87. Iyer, R.; Iken, B.; Damania, A. A comparison of organophosphate degradation genes and bioremediation applications. *Environ. Microbiol. Rep.* **2013**, *5*, 787–798. [CrossRef]
88. Dvořák, P.; Nikel, P.I.; Damborský, J.; de Lorenzo, V. Bioremediation 3.0: Engineering pollutant-removing bacteria in the times of systemic biology. *Biotechnol. Adv.* **2017**, *35*, 845–866. [CrossRef]
89. Briceño, G.; Palma, G.; Durán, N. Influence of Organic Amendment on the Biodegradation and Movement of Pesticides. *Crit. Rev. Environ. Sci. Technol.* **2007**, *37*, 233–271. [CrossRef]
90. Choi, M.K.; Kim, K.D.; Ahn, K.M.; Shin, D.H.; Hwang, J.H.; Seong, C.N.; Ka, J.O. Genetic and phenotypic diversity of parathion-degrading bacteria isolated from rice paddy soils. *J. Microbiol. Biotechnol.* **2009**, *19*, 1679–1687. [CrossRef]
91. Singh, B.K.; Walker, A.; Morgan, J.A.; Wright, D.J. Biodegradation of chlorpyrifos by enterobacter strain B-14 and its use in bioremediation of contaminated soils. *Appl. Environ. Microbiol.* **2004**, *70*, 4855–4863. [CrossRef] [PubMed]
92. Chanika, E.; Georgiadou, D.; Soueref, E.; Karas, P.; Karanasios, E.; Tsiropoulos, N.G.; Tzortzakakis, E.A.; Karpouzas, D.G. Isolation of soil bacteria able to hydrolyze both organophosphate and carbamate pesticides. *Bioresour. Technol.* **2011**, *102*, 3184–3192. [CrossRef] [PubMed]
93. Cycoń, M.; Wójcik, M.; Piotrowska-Seget, Z. Biodegradation of the organophosphorus insecticide diazinon by *Serratia* sp. and *Pseudomonas* sp. and their use in bioremediation of contaminated soil. *Chemosphere* **2009**, *76*, 494–501. [CrossRef]
94. Singh, B.K.; Walker, A. Microbial degradation of organophosphorus compounds. *FEMS Microbiol. Rev.* **2006**, *30*, 428–471. [CrossRef]
95. Caceres, T.P.; Megharaj, M.; Naidu, R. Biodegradation of the pesticide fenamiphos by ten different species of green algae and cyanobacteria. *Curr. Microbiol.* **2008**, *57*, 643–646. [CrossRef]
96. Abdelrazek, M.; Abozeid, A.; Eltholth, M.; Abouelenien, F.; El-Midany, S.; Moustafa, N.; Mohamed, R. Bioremediation of a pesticide and selected heavy metals in wastewater from various sources using a consortium of microalgae and cyanobacteria. *Slov. Vet. Res.* **2019**, *56*, 61–74. [CrossRef]
97. Zhao, R.-B.; Bao, H.-Y.; Liu, Y.-X. Isolation and Characterization of Penicillium oxalicum ZHJ6 for Biodegradation of Methamidophos. *Agric. Sci. China* **2010**, *9*, 695–703. [CrossRef]
98. Zhao, J.; Zhao, D.; Han, J. Isolation and Characterization of Dimethoate Degrading Phytopathogen Fungus from Soil. In Proceedings of the 2009 3rd International Conference on Bioinformatics and Biomedical Engineering, Beijing, China, 11–13 June 2009; pp. 1–4.
99. Tian, J.; Dong, Q.; Yu, C.; Zhao, R.; Wang, J.; Chen, L. Biodegradation of the Organophosphate Trichlorfon and Its Major Degradation Products by a Novel Aspergillus sydowii PA F-2. *J. Agric. Food Chem.* **2016**, *64*, 4280–4287. [CrossRef] [PubMed]
100. Chen, S.; Liu, C.; Peng, C.; Liu, H.; Hu, M.; Zhong, G. Biodegradation of chlorpyrifos and its hydrolysis product 3,5,6-trichloro-2-pyridinol by a new fungal strain Cladosporium cladosporioides Hu-01. *PLoS ONE* **2012**, *7*, e47205. [CrossRef] [PubMed]
101. Jain, R.; Veena, G.; Singh, K.; Sheetal, G. Isolation and characterization of monocrotophos degrading activity of soil fungal isolate Aspergillus Niger MCP1 (ITCC7782.10). *Int. J. Environ. Sci.* **2012**, *3*, 841–850. [CrossRef]
102. Gaber, S.E.; Hussain, M.T.; Jahin, H.S. Bioremediation of diazinon pesticide from aqueous solution by fungal-strains isolated from wastewater. *World J. Chem.* **2020**, *15*, 15–23.
103. Sethunathan, N.; Yoshida, T. A *Flavobacterium* sp. that degrades diazinon and parathion. *Can. J. Microbiol.* **1973**, *19*, 873–875. [CrossRef]
104. Serdar, C.M.; Gibson, D.T.; Munnecke, D.M.; Lancaster, J.H. Plasmid Involvement in Parathion Hydrolysis by *Pseudomonas diminuta*. *Appl. Environ. Microbiol.* **1982**, *44*, 246–249. [CrossRef]
105. Lyagin, I.; Efremenko, E. Enzymes, Reacting with Organophosphorus Compounds as Detoxifiers: Diversity and Functions. *Int. J. Mol. Sci.* **2021**, *22*, 1761. [CrossRef]

106. Theriot, C.M.; Grunden, A.M. Hydrolysis of organophosphorus compounds by microbial enzymes. *Appl. Microbiol. Biotechnol.* **2011**, *89*, 35–43. [CrossRef]
107. Sogorb, M.A.; Vilanova, E. Enzymes involved in the detoxification of organophosphorus, carbamate and pyrethroid insecticides through hydrolysis. *Toxicol. Lett.* **2002**, *128*, 215–228. [CrossRef]
108. Ragnarsdottir, K.V. Environmental fate and toxicology of organophosphate pesticides. *J. Geol. Soc.* **2000**, *157*, 859–876. [CrossRef]
109. Haque, M.A.; Hong, S.Y.; Hwang, C.E.; Kim, S.C.; Cho, K.M. Cloning of an organophosphorus hydrolase (opdD) gene of Lactobacillus sakei WCP904 isolated from chlorpyrifos-impregnated kimchi and hydrolysis activities of its gene product for organophosphorus pesticides. *Appl. Biol. Chem.* **2018**, *61*, 643–651. [CrossRef]
110. Jiang, B.; Zhang, N.; Xing, Y.; Lian, L.; Chen, Y.; Zhang, D.; Li, G.; Sun, G.; Song, Y. Microbial degradation of organophosphorus pesticides: Novel degraders, kinetics, functional genes, and genotoxicity assessment. *Environ. Sci. Pollut. Res.* **2019**, *26*, 21668–21681. [CrossRef] [PubMed]
111. Alejo-Gonzalez, K.; Hanson-Viana, E.; Vazquez-Duhalt, R. Enzymatic detoxification of organophosphorus pesticides and related toxicants. *J. Pestic. Sci.* **2018**, *43*, 1–9. [CrossRef]
112. Caldwell, S.R.; Raushel, F.M. Detoxification of organophosphate pesticides using an immobilized phosphotriesterase from *Pseudomonas diminuta*. *Biotechnol. Bioeng.* **1991**, *37*, 103–109. [CrossRef]
113. Benning, M.M.; Shim, H.; Raushel, F.M.; Holden, H.M. High resolution X-ray structures of different metal-substituted forms of phosphotriesterase from *Pseudomonas diminuta*. *Biochemistry* **2001**, *40*, 2712–2722. [CrossRef]
114. Kang, D.G.; Li, L.; Ha, J.H.; Choi, S.S.; Cha, H.J. Efficient cell surface display of organophosphorous hydrolase using N-terminal domain of ice nucleation protein in Escherichia coli. *Korean J. Chem. Eng.* **2008**, *25*, 804–807. [CrossRef]
115. Horne, I.; Sutherland, T.D.; Harcourt, R.L.; Russell, R.J.; Oakeshott, J.G. Identification of an opd (organophosphate degradation) gene in an Agrobacterium isolate. *Appl. Environ. Microbiol.* **2002**, *68*, 3371–3376. [CrossRef] [PubMed]
116. Dawson, R.M.; Pantelidis, S.; Rose, H.R.; Kotsonis, S.E. Degradation of nerve agents by an organophosphate-degrading agent (OpdA). *J. Hazard. Mater.* **2008**, *157*, 308–314. [CrossRef]
117. Anderson, B.; Phillips, B.; Hunt, J.; Largay, B.; Shihadeh, R.; Tjeerdema, R. Pesticide and toxicity reduction using an integrated vegetated treatment system. *Environ. Toxicol. Chem.* **2011**, *30*, 1036–1043. [CrossRef]
118. Scott, C.; Begley, C.; Taylor, M.J.; Pandey, G.; Momiroski, V.; French, N.; Brearley, C.; Kotsonis, S.E.; Selleck, M.J.; Carino, F.A.; et al. Free-Enzyme Bioremediation of Pesticides. In *Pesticide Mitigation Strategies for Surface Water Quality*; ACS Symposium Series; American Chemical Society: Washington, WA, USA, 2011; Volume 1075, pp. 155–174.
119. DeFrank, J.J.; Cheng, T.C. Purification and properties of an organophosphorus acid anhydrase from a halophilic bacterial isolate. *J. Bacteriol.* **1991**, *173*, 1938–1943. [CrossRef]
120. Jain, M.; Yadav, P.; Joshi, B.; Joshi, A.; Kodgire, P. A novel biosensor for the detection of organophosphorus (OP)-based pesticides using organophosphorus acid anhydrolase (OPAA)-FL variant. *Appl. Microbiol. Biotechnol.* **2021**, *105*, 389–400. [CrossRef] [PubMed]
121. Singh, B.K. Organophosphorus-degrading bacteria: Ecology and industrial applications. *Nat. Rev. Microbiol.* **2009**, *7*, 156–164. [CrossRef] [PubMed]
122. Kulshreshtha, S.; Mathur, N.; Bhatnagar, P. Mushroom as a product and their role in mycoremediation. *AMB Express* **2014**, *4*, 29. [CrossRef]
123. Pandey, C.; Prabha, D.; Negi, Y.K. Mycoremediation of Common Agricultural Pesticides. In *Mycoremediation and Environmental Sustainability: Volume 2*; Prasad, R., Ed.; Springer International Publishing: Cham, Switzerland, 2018; pp. 155–179.
124. Adenipekun, C.O.; Lawal, R. Uses of mushrooms in bioremediation: A Review. *Biotechnol. Mol. Biol. Rev.* **2012**, *7*, 62–68. [CrossRef]
125. Hock, W.; Sisler, H. Metabolism of Chloroneb by Rhizoctonia solani and Other Fungi. *J. Agric. Food Chem.* **1969**, *17*, 123–128. [CrossRef]
126. Singh, H. Fungal Degradation of Pesticides. In *Mycoremediation: Fungal Bioremediation*; Singh, H., Ed.; John Wiley & Sons, Inc.: Hoboken, NJ, USA, 2006; pp. 181–214.
127. Katayama, A.; Matsumura, F. Degradation of organochlorine pesticides, particularly endosulfan, by Trichoderma harzianum. *Environ. Toxicol. Chem.* **1993**, *12*, 1059–1065. [CrossRef]
128. George, N.; Chauhan, P.; Sondhi, S.; Saini, S.; Puri, N.; Gupta, N. Biodegradation and Analytical Methods for Detection of Organophosphorous Pesticide: Chlorpyrifos. *Int. J. Pure Appl. Sci. Technol.* **2014**, *20*, 79–94.
129. Rao, A.V.; Sethunathan, N. Degradation of parathion by Penicillium waksmanii Zaleski isolated from flooded acid sulphate soil. *Arch. Microbiol.* **1974**, *97*, 203–208. [CrossRef]
130. Kim, Y.-H.; Ahn, J.-Y.; Moon, S.-H.; Lee, J. Biodegradation and detoxification of organophosphate insecticide, malathion by *Fusarium oxysporum* f. sp. pisi cutinase. *Chemosphere* **2005**, *60*, 1349–1355. [CrossRef]
131. Mir, Z.A.; Bharose, R.; Lone, A.H.; Malik, Z.A. Review on phytoremediation: An ecofriendly and green technology for removal of heavy metals. *Crop Res.* **2017**, *52*, 74–82.
132. Truua, J.; Truu, J.; Espenberg, M.; Nõlvak, H.; Juhanson, J. Phytoremediation And Plant-Assisted Bioremediation In Soil And Treatment Wetlands: A Review. *Open Biotechnol. J.* **2015**, *9*, 85–92. [CrossRef]
133. Tonelli, F.C.P.; Tonelli, F.M.P.; Lemos, M.S.; Nunes, N.A.d.M. Chapter 3—Mechanisms of phytoremediation. In *Phytoremediation*; Bhat, R.A., Tonelli, F.M.P., Dar, G.H., Hakeem, K., Eds.; Academic Press: Cambridge, MA, USA, 2022; pp. 37–64.

134. Bhalla, G.; Bhalla, B.; Kumar, V.; Sharma, A. Chapter 16—Bioremediation and phytoremediation of pesticides residues from contaminated water: A novel approach. In *Pesticides Remediation Technologies from Water and Wastewater*; Hadi Dehghani, M., Karri, R.R., Anastopoulos, I., Eds.; Elsevier: Amsterdam, The Netherlands, 2022; pp. 339–363.
135. Singh, T.; Singh, D.K. Phytoremediation of organochlorine pesticides: Concept, method, and recent developments. *Int. J. Phytoremediation* **2017**, *19*, 834–843. [CrossRef] [PubMed]
136. Velázquez-Fernández, J.B.; Martínez-Rizo, A.B.; Ramírez-Sandoval, M.; Domínguez-Ojeda, D. Biodegradation and bioremediation of organic pesticides. *Pestic.-Recent Trends Pestic. Residue Assay* **2012**, 253–272. [CrossRef]
137. Takkar, S.; Shandilya, C.; Agrahari, R.; Chaurasia, A.; Vishwakarma, K.; Mohapatra, S.; Varma, A.; Mishra, A. Green technology: Phytoremediation for pesticide pollution. In *Phytoremediation Technology for the Removal of Heavy Metals and Other Contaminants from Soil and Water*; Elsevier: Amsterdam, The Netherlands, 2022; pp. 353–375.
138. Trapp, S.; Karlson, U. Aspects of phytoremediation of organic pollutants. *J. Soils Sediments* **2001**, *1*, 37–43. [CrossRef]
139. Misra, N.N.; Pankaj, S.K.; Walsh, T.; O'Regan, F.; Bourke, P.; Cullen, P.J. In-package nonthermal plasma degradation of pesticides on fresh produce. *J. Hazard. Mater.* **2014**, *271*, 33–40. [CrossRef]
140. Zhang, A.-A.; Sutar, P.P.; Bian, Q.; Fang, X.-M.; Ni, J.-B.; Xiao, H.-W. Pesticide residue elimination for fruits and vegetables: The mechanisms, applications, and future trends of thermal and non-thermal technologies. *J. Future Foods* **2022**, *2*, 223–240. [CrossRef]
141. Hanafi, A.; Elsheshetawy, H.E.; Faied, S.F. Reduction of pesticides residues on okra fruits by different processing treatments. *J. Für Verbraucherschutz Und Leb.* **2016**, *11*, 337–343. [CrossRef]
142. Ranjitha Gracy, T.K.; Sharanyakanth, P.S.; Radhakrishnan, M. Non-thermal technologies: Solution for hazardous pesticides reduction in fruits and vegetables. *Crit. Rev. Food Sci. Nutr.* **2022**, *62*, 1782–1799. [CrossRef]
143. Yang, L.; Hai, C.; Zhang, H.; Feng, C.; Luo, M.; Zhou, P.; Leng, J.; Tian, X.; Zhao, C.; Lai, B. Insights into the role of oxidation and adsorption for degradation of methyl parathion by ferrate (VI). *J. Environ. Chem. Eng.* **2023**, *11*, 110171. [CrossRef]
144. Wang, J.; Yue, W.; Teng, Y.; Zhai, Y.; Zhu, H. Degradation kinetics and transformation pathway of methyl parathion by δ-MnO2/oxalic acid reaction system. *Chemosphere* **2023**, *320*, 138054. [CrossRef] [PubMed]
145. Alhalili, Z. Metal Oxides Nanoparticles: General Structural Description, Chemical, Physical, and Biological Synthesis Methods, Role in Pesticides and Heavy Metal Removal through Wastewater Treatment. *Molecules* **2023**, *28*, 3086. [CrossRef] [PubMed]
146. Kaushal, J.; Khatri, M.; Arya, S.K. A treatise on Organophosphate pesticide pollution: Current strategies and advancements in their environmental degradation and elimination. *Ecotoxicol. Environ. Saf.* **2021**, *207*, 111483. [CrossRef]
147. Xiao, Q.; Xuan, X.; Boczkaj, G.; Yoon, J.Y.; Sun, X. Photolysis for the Removal and Transformation of Pesticide Residues During Food Processing: A State-of-the-Art Minireview. *Front. Nutr.* **2022**, *9*, 888047. [CrossRef] [PubMed]
148. Abedi-Firoozjah, R.; Ghasempour, Z.; Khorram, S.; Khezerlou, A.; Ehsani, A. Non-thermal techniques: A new approach to removing pesticide residues from fresh products and water. *Toxin Rev.* **2021**, *40*, 562–575. [CrossRef]
149. Velioglu, Y.; Fikirdeşici Ergen, S.; Aksu, P.; Altindağ, A. Effects of Ozone Treatment on the Degradation and Toxicity of Several Pesticides in Different Grou. *J. Agric. Sci.* **2018**, *24*, 245–255. [CrossRef]
150. Alsager, O.A.; Alnajrani, M.N.; Alhazzaa, O. Decomposition of antibiotics by gamma irradiation: Kinetics, antimicrobial activity, and real application in food matrices. *Chem. Eng. J.* **2018**, *338*, 548–556. [CrossRef]
151. Khedr, T.; Hammad, A.; Elmarsafy, A.; Halawa, E.; Soliman, M. Degradation of some organophosphorus pesticides in aqueous solution by gamma irradiation. *J. Hazard. Mater.* **2019**, *373*, 23–28. [CrossRef]
152. Yang, L.; Zhou, J.; Feng, Y. Removal of pesticide residues from fresh vegetables by the coupled free chlorine/ultrasound process. *Ultrason. Sonochemistry* **2022**, *82*, 105891. [CrossRef]
153. Zhou, Q.; Bian, Y.; Peng, Q.; Liu, F.; Wang, W.; Chen, F. The effects and mechanism of using ultrasonic dishwasher to remove five pesticides from rape and grape. *Food Chem.* **2019**, *298*, 125007. [CrossRef]
154. Pallares, N.; Sebastia, A.; Martinez-Lucas, V.; Gonzalez-Angulo, M.; Barba, F.J.; Berrada, H.; Ferrer, E. High Pressure Processing Impact on Alternariol and Aflatoxins of Grape Juice and Fruit Juice-Milk Based Beverages. *Molecules* **2021**, *26*, 3769. [CrossRef] [PubMed]
155. Iizuka, T.; Maeda, S.; Shimizu, A. Removal of pesticide residue in cherry tomato by hydrostatic pressure. *J. Food Eng.* **2013**, *116*, 796–800. [CrossRef]
156. Cherif, M.M.; Assadi, I.; Khezami, L.; Ben Hamadi, N.; Assadi, A.A.; Elfalleh, W. Review on Recent Applications of Cold Plasma for Safe and Sustainable Food Production: Principles, Implementation, and Application Limits. *Appl. Sci.* **2023**, *13*, 2381. [CrossRef]
157. Zhou, R.; Zhou, R.; Yu, F.; Xi, D.; Wang, P.; Li, J.; Wang, X.; Zhang, X.; Bazaka, K.; Ostrikov, K. Removal of organophosphorus pesticide residues from Lycium barbarum by gas phase surface discharge plasma. *Chem. Eng. J.* **2018**, *342*, 401–409. [CrossRef]
158. Ranjitha Gracy, T.K.; Gupta, V.; Radhakrishnan, M. Influence of low-pressure non-thermal dielectric barrier discharge (DBD) plasma on chlorpyrifos reduction in tomatoes. *J. Food Process. Eng.* **2019**, *42*, e13242. [CrossRef]
159. Dorraki, N.; Mahdavi, V.; Ghomi, H.; Ghasempour, A. Elimination of diazinon insecticide from cucumber surface by atmospheric pressure air-dielectric barrier discharge plasma. *Biointerphases* **2016**, *11*, 041007. [CrossRef]
160. Zheng, Y.; Wu, S.; Dang, J.; Wang, S.; Liu, Z.; Fang, J.; Han, P.; Zhang, J. Reduction of phoxim pesticide residues from grapes by atmospheric pressure non-thermal air plasma activated water. *J. Hazard. Mater.* **2019**, *377*, 98–105. [CrossRef]

161. Arshad, R.N.; Abdul-Malek, Z.; Roobab, U.; Munir, M.A.; Naderipour, A.; Qureshi, M.I.; El-Din Bekhit, A.; Liu, Z.-W.; Aadil, R.M. Pulsed electric field: A potential alternative towards a sustainable food processing. *Trends Food Sci. Technol.* **2021**, *111*, 43–54. [CrossRef]
162. Lozowicka, B.; Jankowska, M.; Hrynko, I.; Kaczynski, P. Removal of 16 pesticide residues from strawberries by washing with tap and ozone water, ultrasonic cleaning and boiling. *Environ. Monit. Assess.* **2016**, *188*, 51. [CrossRef]
163. Akdemir Evrendilek, G.; Keskin, E.; Golge, O. Interaction and multi-objective effects of multiple non-thermal treatments of sour cherry juice: Pesticide removal, microbial inactivation, and quality preservation. *J. Sci. Food Agric.* **2020**, *100*, 1653–1661. [CrossRef]
164. Tomer, V. Vegetable Processing At Household Level: Effective Tool Against Pesticide Residue Exposure. *IOSR J. Environ. Sci. Toxicol. Food Technol.* **2013**, *6*, 43–53. [CrossRef]
165. Yuan, Z.; Yao, J.; Liu, H.; Han, J.; Trebse, P. Photodegradation of organophosphorus pesticides in honey medium. *Ecotoxicol. Environ. Saf.* **2014**, *108*, 84–88. [CrossRef] [PubMed]
166. Choi, S.W.; Shahbaz, H.M.; Kim, J.U.; Kim, D.-H.; Yoon, S.; Jeong, S.H.; Park, J.; Lee, D.-U. Photolysis and TiO2 Photocatalytic Treatment under UVC/VUV Irradiation for Simultaneous Degradation of Pesticides and Microorganisms. *Appl. Sci.* **2020**, *10*, 4493. [CrossRef]
167. Yang, L.; Li, M.; Li, W.; Jiang, Y.; Qiang, Z. Bench- and pilot-scale studies on the removal of pesticides from water by VUV/UV process. *Chem. Eng. J.* **2018**, *342*, 155–162. [CrossRef]
168. Savic, J.Z.; Petrovic, S.Z.; Leskovac, A.R.; Lazarevic Pasti, T.D.; Nastasijevic, B.J.; Tanovic, B.B.; Gasic, S.M.; Vasic, V.M. UV-C light irradiation enhances toxic effects of chlorpyrifos and its formulations. *Food Chem.* **2019**, *271*, 469–478. [CrossRef]
169. Papagiannaki, D.; Medana, C.; Binetti, R.; Calza, P.; Roslev, P. Effect of UV-A, UV-B and UV-C irradiation of glyphosate on photolysis and mitigation of aquatic toxicity. *Sci. Rep.* **2020**, *10*, 20247. [CrossRef]
170. Baranda, A.B.; Fundazuri, O.; Martínez de Marañón, I. Photodegradation of several triazidic and organophosphorus pesticides in water by pulsed light technology. *J. Photochem. Photobiol. A Chem.* **2014**, *286*, 29–39. [CrossRef]
171. El-Saeid, M.H.; Alotaibi, M.O.; Alshabanat, M.; Alharbi, K.; Altowyan, A.S.; Al-Anazy, M. Photo-Catalytic Remediation of Pesticides in Wastewater Using UV/TiO2. *Water* **2021**, *13*, 3080. [CrossRef]
172. Kaur, R.; Singh, D.; Kumari, A.; Sharma, G.; Rajput, S.; Arora, S.; Kaur, R. Pesticide residues degradation strategies in soil and water: A review. *Int. J. Environ. Sci. Technol.* **2023**, *20*, 3537–3560. [CrossRef]
173. Jafari, S.J.; Moussavi, G.; Hossaini, H. Degradation and mineralization of diazinon pesticide in UVC and UVC/TiO2 process. *Desalination Water Treat.* **2016**, *57*, 3782–3790. [CrossRef]
174. Gupta, V.K.; Eren, T.; Atar, N.; Yola, M.L.; Parlak, C.; Karimi-Maleh, H. CoFe2O4@TiO2 decorated reduced graphene oxide nanocomposite for photocatalytic degradation of chlorpyrifos. *J. Mol. Liq.* **2015**, *208*, 122–129. [CrossRef]
175. Li, W.; Zhao, Y.; Yan, X.; Duan, J.; Saint, C.P.; Beecham, S. Transformation pathway and toxicity assessment of malathion in aqueous solution during UV photolysis and photocatalysis. *Chemosphere* **2019**, *234*, 204–214. [CrossRef] [PubMed]
176. Di Vaio, A.; Boccia, F.; Landriani, L.; Palladino, R. Artificial Intelligence in the Agri-Food System: Rethinking Sustainable Business Models in the COVID-19 Scenario. *Sustainability* **2020**, *12*, 4851. [CrossRef]
177. Tian, Z.; Wang, J.W.; Li, J.; Han, B. Designing future crops: Challenges and strategies for sustainable agriculture. *Plant J.* **2021**, *105*, 1165–1178. [CrossRef] [PubMed]
178. European Union. *Farm to Fork Strategy. For a Fair, Healthy and Environmentally-Friendly Food System*; European Union: Brussels, Belgium, 2020.
179. McGinley, J.; Healy, M.G.; Ryan, P.C.; Harmon O'Driscoll, J.; Mellander, P.E.; Morrison, L.; Siggins, A. Impact of historical legacy pesticides on achieving legislative goals in Europe. *Sci. Total Environ.* **2023**, *873*, 162312. [CrossRef] [PubMed]
180. Kaur, N.; Khunger, A.; Wallen, S.L.; Kaushik, A.; Chaudhary, G.R.; Varma, R.S. Advanced green analytical chemistry for environmental pesticide detection. *Curr. Opin. Green Sustain. Chem.* **2021**, *30*, 100488. [CrossRef]
181. Chaudhary, V.; Rustagi, S.; Kaushik, A. Bio-derived smart nanostructures for efficient biosensors. *Curr. Opin. Green Sustain. Chem.* **2023**, *42*, 100817. [CrossRef]
182. Rani, M.; Yadav, J.; Chaudhary, S.; Shanker, U. An updated review on synthetic approaches of green nanomaterials and their application for removal of water pollutants: Current challenges, assessment and future perspectives. *J. Environ. Chem. Eng.* **2021**, *9*, 106763. [CrossRef]
183. Bala, S.; Garg, D.; Thirumalesh, B.V.; Sharma, M.; Sridhar, K.; Inbaraj, B.S.; Tripathi, M. Recent Strategies for Bioremediation of Emerging Pollutants: A Review for a Green and Sustainable Environment. *Toxics* **2022**, *10*, 484. [CrossRef]
184. Dangi, A.K.; Sharma, B.; Hill, R.T.; Shukla, P. Bioremediation through microbes: Systems biology and metabolic engineering approach. *Crit. Rev. Biotechnol.* **2019**, *39*, 79–98. [CrossRef]
185. Dash, D.M.; Osborne, W.J. A systematic review on the implementation of advanced and evolutionary biotechnological tools for efficient bioremediation of organophosphorus pesticides. *Chemosphere* **2023**, *313*, 137506. [CrossRef] [PubMed]
186. Patil, A.; Yesankar, P.; Bhanse, P.; Maitreya, A.; Kapley, A.; Qureshi, A. Omics Perspective: Molecular Blueprint for Agrochemical Bioremediation Process in the Environment. In *Agrochemicals in Soil and Environment: Impacts and Remediation*; Naeem, M., Bremont, J.F.J., Ansari, A.A., Gill, S.S., Eds.; Springer Nature: Singapore, 2022; pp. 585–608.
187. Hassan, S.; Ganai, B.A. Deciphering the recent trends in pesticide bioremediation using genome editing and multi-omics approaches: A review. *World J. Microbiol. Biotechnol.* **2023**, *39*, 151. [CrossRef] [PubMed]

188. Sharma, B.; Shukla, P. Designing synthetic microbial communities for effectual bioremediation: A review. *Biocatal. Biotransformation* **2020**, *38*, 405–414. [CrossRef]
189. Zimny, T. New genomic techniques and their European Union reform. Potential policy changes and their implications. *Front. Bioeng. Biotechnol.* **2022**, *10*, 1019081. [CrossRef] [PubMed]
190. Armenova, N.; Tsigoriyna, L.; Arsov, A.; Petrov, K.; Petrova, P. Microbial Detoxification of Residual Pesticides in Fermented Foods: Current Status and Prospects. *Foods* **2023**, *12*, 1163. [CrossRef] [PubMed]
191. Petrova, P.; Arsov, A.; Tsvetanova, F.; Parvanova-Mancheva, T.; Vasileva, E.; Tsigoriyna, L.; Petrov, K. The Complex Role of Lactic Acid Bacteria in Food Detoxification. *Nutrients* **2022**, *14*, 2038. [CrossRef]
192. Sachithra, V.; Subhashini, L.D.C.S. How artificial intelligence uses to achieve the agriculture sustainability: Systematic review. *Artif. Intell. Agric.* **2023**, *8*, 46–59. [CrossRef]
193. Ghatrehsamani, S.; Jha, G.; Dutta, W.; Molaei, F.; Nazrul, F.; Fortin, M.; Bansal, S.; Debangshi, U.; Neupane, J. Artificial Intelligence Tools and Techniques to Combat Herbicide Resistant Weeds—A Review. *Sustainability* **2023**, *15*, 1843. [CrossRef]
194. Azmi, H.N.; Hajjaj, S.S.H.; Gsangaya, K.R.; Sultan, M.T.H.; Mail, M.F.; Hua, L.S. Design and fabrication of an agricultural robot for crop seeding. *Mater. Today Proc.* **2023**, *81*, 283–289. [CrossRef]
195. Yang, J.; Ma, S.; Li, Y.; Zhang, Z. Efficient Data-Driven Crop Pest Identification Based on Edge Distance-Entropy for Sustainable Agriculture. *Sustainability* **2022**, *14*, 7825. [CrossRef]
196. Zhou, Z.; Majeed, Y.; Diverres Naranjo, G.; Gambacorta, E.M.T. Assessment for crop water stress with infrared thermal imagery in precision agriculture: A review and future prospects for deep learning applications. *Comput. Electron. Agric.* **2021**, *182*, 106019. [CrossRef]
197. Marvin, H.J.P.; Bouzembrak, Y.; Janssen, E.M.; van der Fels-Klerx, H.J.; van Asselt, E.D.; Kleter, G.A. A holistic approach to food safety risks: Food fraud as an example. *Food Res. Int.* **2016**, *89*, 463–470. [CrossRef] [PubMed]

Disclaimer/Publisher's Note: The statements, opinions and data contained in all publications are solely those of the individual author(s) and contributor(s) and not of MDPI and/or the editor(s). MDPI and/or the editor(s) disclaim responsibility for any injury to people or property resulting from any ideas, methods, instructions or products referred to in the content.

Article

The Effect of Washing, Blanching and Frozen Storage on Pesticide Residue in Spinach

Federica Flamminii [1,*], Silvia Minetti [2,3], Adriano Mollica [3], Angelo Cichelli [1] and Lorenzo Cerretani [2]

[1] Department of Innovative Technologies in Medicine and Dentistry, University "G. d'Annunzio" of Chieti-Pescara, Via dei Vestini, 66100 Chieti, Italy; angelo.cichelli@unich.it

[2] Società Abruzzese Lavorazione Prodotti Agricoli (S.A.L.P.A.) S.A.C.arl, Via Nazionale, 64026 Roseto degli Abruzzi, Italy; silvia.minetti@salparoseto.it (S.M.); lorenzo.cerretani@salparoseto.it (L.C.)

[3] Department of Pharmacy, University "G. d'Annunzio" of Chieti-Pescara, Via dei Vestini, 66100 Chieti, Italy; a.mollica@unich.it

* Correspondence: federica.flamminii@unich.it

Abstract: Spinach (*Spinacia oleracea* L.) is a representative green leafy vegetable commonly consumed fresh or as a ready-to-cook frozen product, with increasing consumption because of its many health-related properties. Among leafy vegetables, spinach poses a major concern in terms of pesticide residue detection due to common phytotechnical practices. In this study, spinach leaves were treated in the open field with three commercial pesticide formulations containing propamocarb, lambda-cyhalothrin, fluopicolide and chlorantraniliprole at the highest concentration. The effects of the successive processing steps of washing, blanching, freezing and frozen storage were evaluated on the levels of the four pesticide residues and the degradation product (propamocarb n-desmethyl). The washing step caused a reduction of fluopicolide and chlorantraniliprole of −47% and −43%, respectively, while having a mild effect on lambda-cyhalothrin content (+5%). A two-minute blanching step allowed for the reduction of pesticides content ranging from −41% to −4% with respect to the washed sample. Different behaviors were depicted for longer blanching times, mainly for propamocarb, reaching −56% after 10 min of treatment. Processing factors higher than 1 were reported mainly for lambda-cyhalothrin and fluopicolide. Frozen storage led to a slight increase in the pesticide content in samples treated for 6 and 10 min. The optimal blanching treatment for spinach, submitted to freezing and frozen storage, seems to be 2 min at 80 °C.

Keywords: spinach; household processing; washing; blanching; pesticide residues; processing factor; frozen storage; propamocarb; lambda-cyhalothrin; propamocarb n-desmethyl

1. Introduction

Global demand for food is ever increasing and accompanied by changes in lifestyles and choices of food. To feed the ever-growing global population, it is estimated that global food production will need to increase 70% by the year 2050, and the wide use of pesticides in the agricultural sector in some way is encouraged to boost crop production [1].

Pesticides, referred to also as plant protection products (PPPs), are globally used in agriculture systems to control plant diseases, insect pests and weeds, providing high yields and product quality [2]. Despite the positive effects on agricultural production, there are many ecological and health risks associated with pesticides and their residues in foodstuffs. Indeed, toxic effects on humans, ranging from short-term effects such as headaches and nausea to chronic effects like cancer, reproductive damage, endocrine disruption or autism, learning disabilities and neurological disorders, including Parkinson's and Alzheimer's diseases, have been frequently reported [3–5]. The issue of food contamination caused by pesticide residues is becoming increasingly important due to the growing global demand for food. The European Union introduced several policies regulating the use of PPPs, namely, Regulation (EC) n. 1107/2009 and Regulation (EC) n. 396/2005, concerning the placing of plant protection products on the market and the maximum residue levels (MRLs) of pesticides in/on food and feed of plant and animal origin, respectively [6,7]. Along with the promotion of organic farming, the adoption of integrated pest management (IPM) represents one of the tools for low-pesticide-input pest management and one of the major pathways for progress in reducing pesticide risks. It offers a framework for managing economic, health and environmental risks while minimizing undue outcomes for crop production [8].

The persistence of pesticides in plants varies depending on environmental conditions that lead to their hydrolysis, oxidation, reduction, photolysis [9] and the alteration of their properties such as solubility, volatilization, dissociation constant and formulation; moreover, their persistence induces changes in vegetable characteristics such as plant morphology and metabolic activity. Propamocarb (propyl [3-(dimethylamino)propyl]carbamate), for example, is a systemic fungicide with protective action against phycomycetous diseases. It is known to degrade into a variety of metabolite as propamocarb hydrochloride, n-oxide propamocarb, oxazoline-2-one propamocarb, 2-hydroxypropamocarb and n-desmethyl propamocarb [10]. Despite the fact that metabolites can be more toxic than the parent compound, little is known about the dissipation of both the parent and metabolite compounds, and the toxicity of metabolites has not been studied yet [11]. Lambda-cyhalothrin, a nonsystemic and lipophilic active substance which belongs to the group of pyrethroid compounds, is a broad-spectrum insecticide that acts on the nervous system, resulting in a product that is highly stable to light and high temperature and has a half-life of seven days [12].

Agricultural products are generally consumed after being processed. Washing represents the first and most common domestic or industrial unit of operation, useful to remove surface pesticide residues in fruits and vegetables [13], and major portions of polar compounds [14]. Wu and co-authors investigated the effect on different agricultural products of different home/commercial washing strategies for pesticide removal, showing that sodium bicarbonate solution, ozone water and active oxygen solution were more effective in decontamination of spinach than in kumquat and cucumber [13]. Washing with diluted salt or/and chemical solution can be a more effective decontamination strategy than the use of pure water [15].

Further processing steps such as peeling, blanching, baking, pasteurization, frying and different operation-unit combinations are identified as critical processing operations and significantly reduce pesticide residues by up to 100% removal. Peeling or trimming the outer skin of some fruits and fruiting vegetables represents the most efficient approach to reducing pesticide residues [13]. Hot-water blanching increases pesticide removal and may hydrolyze substantial fractions of nonpersistent compounds [14]. Numerous factors impact the level and degradation of residues after the cooking process; these include time, temperature, pH, a decrease in moisture content and the type of cooking system (open or closed). Blanching, as a single operation, was recognized to be the most effective treatment to reduce boscalid (45%) and propamocarb (72%), enhancing the elimination of the residues in spinach leaves [4].

Although broadly studied and well-established as decontamination methods, the aforementioned procedures are deleterious to the physical and phytochemical (flavonoid, carotenoid, etc.) and micronutrient content of fruits and vegetables [9]. The recent approaches in sanitation of fruits and vegetables, based on the oxidation processes (generation of free radicals such as hydroxyl radicals), include modern nonthermal techniques such as ozonation, ultrasonication, high hydrostatic pressure (HHP), electrolyzed water (EW), gamma radiation, and nonthermal plasma technology, which avoid the undesirable effects of heat treatments [16,17].

Despite the positive effects of technological treatments, some toxic compounds or metabolites may be formed during processing, and in some cases, processes can induce an increase in pesticide residues [2]. Different studies have been performed to measure the concentrations of pesticide residues after home or industrial processing. Processing factors

(PFs) indices for pesticides in foods, related to the ratio between residue concentrations in the processed commodity and the same in the raw material, represent the effect of cumulative processes on the residues levels [18]. These PFs indices highlight the higher reduction in pesticide concentrations as a result of washing, blanching and sterilization than in washing, blanching and microwave cooking [4]

Human consumption of green leafy vegetables, which are cheap and easily accessible, has increased worldwide because they offer multiple health benefits due to the presence of vitamins, minerals, fiber, essential amino acids, polyphenols and flavonoids [2].

Among leaf vegetables, spinach (*Spinacia oleracea* L.) is commonly consumed fresh in salads, soups or as a ready-to-cook frozen product. This leafy vegetable poses a major concern in terms of pesticide residue detection, in relation to common phytotechnical practices [4,19], and its morphology—indeed, the morphology of all leafy vegetables including spinach–means it is not peeled and is difficult to clean and, only attached dust, insects and outer substances can be removed [20]. In addition, because of the large surface area of the leaves, pesticide residues are likely to remain on them [20] and in some cases, such as in the drying process, the residue content can increase as a result of the water loss during processing [13] or lead to the production of more toxic products or metabolites [21].

The abundance of agricultural commodities at a reasonably low cost is related to the use of pesticides, which are able to increase the production yield; this abundance, along with, consumer demand for a wide variety of products and the desire for them to be available the whole year long, has stimulated the development of frozen products that can generally be found worldwide in retail [22]. Despite numerous studies concerning the effects of home/commercial processes on the pesticide residue in foods, few research studies have focused on the effect of frozen storage on the pesticide residue content in foods, in particular zucchini, tomato, apple, mango and fish, where controversial results have been observed [23–28].

Therefore, based on this scenario, the aim of this research is to study the effect of the washing and blanching processes on the residual content of chlorantraniliprole, lambda-cyhalothrin, fluopicolide, propamocarb and propamocarb n-desmethyl in spinach. Furthermore, the effect of freezing and frozen storage ($-20\,°C$) is also evaluated during a period of ten months.

2. Materials and Methods

2.1. Materials

HPLC-grade methanol (MeOH) and acetonitrile (ACN) were purchased from Honeywell (Charlotte, NC, USA); and formic acid (purity 99.7%) and ammonium formate (purity 99%) were purchased from Sigma-Aldrich (St. Louis, MO, USA). The QuEChERS extraction kit, which contains 4 g of magnesium sulfate ($MgSO_4$), 1 g of sodium chloride (NaCl) and dispersive SPE (dSPE), and the cleanup kit tubes were obtained from Interchim (Montlucon, France). Analytical reference standards of fluopicolide, propamocarb hydrochloride, propamocarb n-desmethyl, chlorantraniliprole and lambda-cyhalothrin compounds were purchased from Labinstruments (Castellana Grotte, Bari, Italy).

2.2. Application of Pesticides and Harvesting

Spinach (*Spinacia oleracea*) var. Monterey was cultivated in a commercial open field located in Corropoli (Teramo, Italy) (42°49′29.4″ north latitude, 13°53′07.9″ east longitude). Different commercial pesticide formulations—such as VOLARE® (Bayer AG, Leverkusen, Germany), composed of 5.53% (*w/w*) fluopicolide and 55.3% (*w/w*) propamocarb hydrochloride; ALTACOR® (FMC Agro Italia Srl, Bergamo, Italy) composed of 35% (*w/w*) chlorantraniliprole; and KARATE ZEON® (Syngenta Crop Protection AG, Basil, Switzerland) with 9.48% (*w/w*) lambda-cyhalothrin—were nebulized at the highest concentration defined by the manufacturers on 4/5 spinach leaves. Physicochemical characteristics of the main active compounds contained in each commercial preparation are reported in Table 1. Fresh samples of spinach leaves were manually harvested two days (48 h) after the treatment, in order to have the highest concentration of pesticides, and were carried to Società Abruzzese Lavorazione Prodotti Agricoli (S.A.L.P.A.) S.A.C.arl industry (Roseto degli Abruzzi, Teramo, Italy), which was involved in this research, for the processing treatments, storage and analysis steps.

2.3. Methods of Washing, Blanching and Freezing

Fresh spinach samples (F) were harvested, divided into different batches (a, b, c) and subjected to successive processing steps, as shown in the process flowchart (Figure 1). The washing step (W) was conducted with running tap water (15 °C for 1 min); subsequently, each sample batch was independently blanched with hot water (80 ± 1 °C) in a 1:2 ratio. Different heat treatment timings were defined: 2 min (B2), 6 min (B6) and 10 min (B10). At the end of the treatments, each sample was lightly drained, placed into square shapes, frozen with dry ice (CO_2) and stored at -20 °C for further analysis during storage time at T1 (day after the treatment), T2 (after 4 months) and T3 (after 10 months).

2.4. Extraction and Purification of Pesticides

The extraction of pesticides was performed on fresh and frozen samples. In the first step, spinach samples were comminuted using the lab knife mill Retsch GM 200 (Haan, Germany). All equipment was carefully cleaned and rinsed with water between the processing of each sample. The extraction was performed using the QuEChERS (quick, easy, cheap, effective, rugged and safe) method. An amount of homogenized sample (10 ± 0.1 g) was weighed into a 50 mL extraction centrifuge tube and filled with 1 g of sodium chloride and 4 g of anhydrous magnesium sulfate; then, an addition was made of 10 mL of a solution of 0.02% (v/v) formic acid in acetonitrile with bromo dimethylphenyl carbamate (BDC) and triphenyl phosphate (TPP), as internal standards. The mixture was shaken for 20 min and then centrifuged for 5 min at 5000 rpm. An aliquot of 3 mL of supernatant was transferred into a 15 mL centrifuge tube for the cleanup step, which contained 25 mg primary secondary amine (PSA), 2.5 mg graphitized carbon black (GCB) and 150 mg magnesium sulfate ($MgSO_4$). PSA sorbent removes polar inferences including organic acids and sugars; GCB removes pigments such as chlorophyll and carotenoids; and magnesium sulfate adsorbs water. The tubes were shaken for 5 min and centrifuged for 5 min at 5000 rpm. An amount of 200 microliters of supernatant was diluted with 800 mL of ultrapure water or acetone directly in the vial for LC or GC analysis, respectively.

2.5. LC-MS/MS Analysis

Determinations of pesticide residues were performed in a Waters XEVO TQ-S micro HPLC (Waters, Milford, MA, USA) equipped with a reversed-phase column (KINETEX C18, 2.6 μm, 100 mm × 2.1 mm, Phenomenex, Torrance, CA, USA) at 40 °C. The mobile phase used was water/methanol (80:20) (containing 0.1% formic acid and ammoniac 0.02%) as phase A and methanol (containing 0.1% formic acid and ammoniac 0.02%) as phase B, with a flow rate of 0.4 mL min^{-1}. The injection volume was 10 μL. For the mass spectrometric analysis, an XEVO triple quadrupole LC/MS system was applied. The ESI source was operated in positive ionization mode, and its parameters were as follows: source temperature, 150 °C; gas desolvation temperature, 600 °C; cone gas flow, 17 L h^{-1}; desolvation gas flow, 1000 L h^{-1}; nitrogen gas used as the nebulizer and collision gas. MassLynks MS (Waters, Milford, MA, USA) software v4.1 was used for method development and data acquisition, qualitative analysis and quantitative analysis. The multiple reaction monitoring (MRM) mode was selected to monitor the precursor-to-product ion transitions. The retention times and chromatographic parameters of pesticides are shown in Table 2.

Table 1. The main properties of pesticide contents in commercial formulations.

Commercial Product	Pesticide	Class and Mode of Action	MRLs mg/kg	Log p at pH 7, 20 °C	Water Solubility at 20 °C mg/L	Melting Point	Henry's Constant at 25 °C Pa m^3 mol^{-1}	Stability *
VOLARE®	Fluopicolide	Benzamide and pyridine fungicides; systemic; absorbed by the roots and leaves and transported acropetally; modifies the distribution of fungal spectrinlike proteins.	6	2.9	2.80	150	4.15×10^{-5}	Stable in tightly closed and dry conditions; slow degradation to photolysis and hydrolysis; degradation occurs through hydroxylation in the bridge carbon.
	Propamocarb hydrochloride	Carbamate fungicide; systemic; reduces mycelial growth and development of sporangia, influences the biochemical synthesis of membranes; absorbed by the roots and leaves and transported acropetally.	40	−1.3	1,005,000	64.2	8.50×10^{-9}	Stable to hydrolysis, temperature up to 400 °C and photolysis.
ALTACOR®	Chlorantraniliprole	Systemic anthranilic diamide insecticide, it binds to ryanodine receptor, causing impaired muscle regulation, paralysis and insect death.	20	2.86	0.88	209	3.2×10^{-9}	Fast aqueous photolysis; slow aqueous hydrolysis.
KARATE ZEON®	Lambda-cyhalothrin	Pyrethroid insecticide with contact and stomach action and repellent properties; nonsystemic.	0.6	5.5	0.005	49.2	2.00×10^{-2}	Stable to light, stable to storage >6 months at 15–25 °C and stable to decomp and cis–trans isomerization for at least 4 years in the dark at 50 °C.

* Information about stability was obtained from the 13th edition of the *Pesticide Manual: A World Compendium of Pesticides* [29].

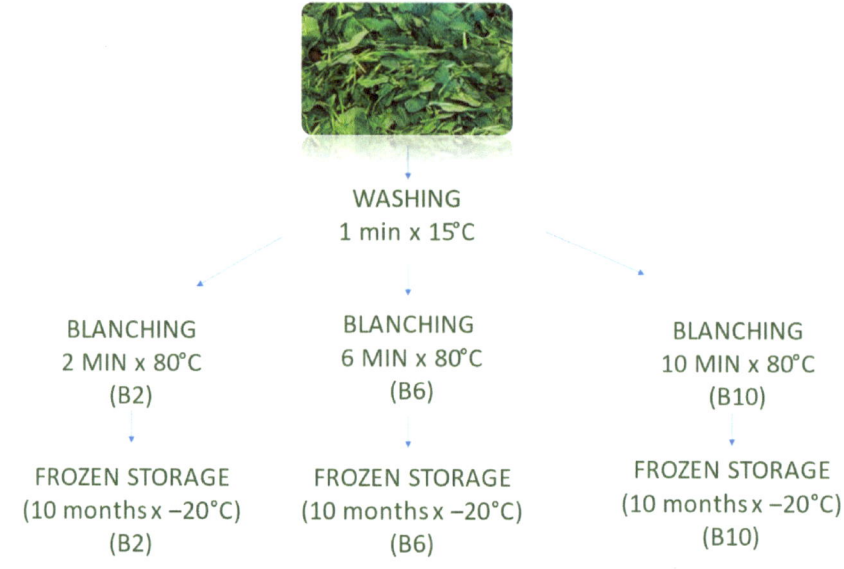

Figure 1. Flow diagram of spinach processing steps and samplings.

Table 2. LC-MS and GC-MS parameters of identified pesticides.

Pesticide	Retention Time (min)	Precursor Ions (m/z)	Product Ions (m/z)	Cone Voltage	Collision Energy (V)	Quantifying Ions
LC-MS						
Lambda-cyhalothrin	7.95	467.2	225	16	23	x
		467.2	141.2	16	57	
Fluopicolide	5.68	383	173	40	20	x
		383	109	40	66	
Propamocarb hydrochloride	0.91	189.1	102	15	15	x
		189.1	144	15	10	
Propamocarb n-desmethyl	0.90	175.2	102.1	30	14	x
		175.2	144.2	30	8	
Chlorantraniliprole	5.12	484	453	20	14	x
		482	451	20	14	
GC-MS						
Lambda-cyhalothrin	33.55	181.1	152.1		30	
		181.1	127.1		35	

"x" identifies the quantifying ion.

2.6. GC–MS/MS Analysis

Pesticide residues were also analyzed by a gas chromatograph (Agilent GC 7890A, Agilent Technologies, Santa Clara, CA, USA) MS 7000 (Agilent) triple quadrupole equipped with two capillary columns (15 m length × 0.25 mm i.d. × 0.25 µm film thickness) and connected with a back flush system; helium was used as the gas carrier (1.4 mL min^{-1} column flow). Injection was set in a programmable temperature vaporizer (PTV) mode (initial injection 70 to final injection 280 °C), the source temperature was 230 °C. The column temperature was initially set at 65 °C for 1 min, raised at 30 °C min^{-1} to 100 °C × 0 min, 5 °C min^{-1} to 280 °C and maintained for 6 min. The injection volume was 3 µL in

PTV solvent vent. Under the conditions described above, the retention time and further chromatographic parameters are reported in Table 2.

2.7. Statistical Analysis

Results were expressed as mean ± standard deviation of three replicates ($n = 3$) for each sample. Analysis of variance (ANOVA) was used to assess the effect of different processes on the pesticide, and Tukey's test was used to establish the statistical significance (0.05) among samples. The data analysis was performed with XLSTAT software package v2016 (Addinsoft, New York, NY, USA).

3. Results and Discussion

3.1. Preliminary Evaluation of Pesticide Residues in Spinach

With the aim of evaluating, on an exploratory basis, the pesticide residues that can be frequently found in fresh spinach and frozen derived products, a preliminary evaluation was conducted on the S.A.L.P.A. S.A.C.a.r.l's database on the latest three years of crop production (2020–2022). Spinach samples, derived from standard (ST), integrated pest management (IPM) and organic (O) cultivation methods, were considered. The first cultivation method (ST) is not bound by any type of legislation except that of food safety defined by the European Union and corresponds to conventional agricultural production. The integrated pest management (IPM) has the purpose of ensuring the defense of productions by guaranteeing the lowest environmental impact in the framework of eco-compatible sustainable agriculture and refers to regional regulations. Finally, organic agriculture (O) is included in the cultivation methods governed by the reference legislation [30]. The preliminary results, mainly concerning ST products, highlighted that in the considered period (2020–2022), the propamocarb residue and its metabolite (propamocarb n-desmethyl) were identified on about 33% of the samples, with a mean concentration of ≈0.3 mg kg^{-1}; while 23% and 24% of the samples contained lambda-cyhalothrin and chlorantraniliprole with mean values of 0.09 and 0.10 mg kg^{-1}, respectively. In smaller percentages (5% and 6%) pyraclostrobin, fluopicolide and boscalid were also identified. Similar trends, in terms of detected substances and quantities, were found by analyzing the database as it referred to the two different crop seasons, i.e., summer (15/05–14/09) and winter (15/09–14/05). The results of this preliminary screening highlighted that all the identified residues were found to be up to 50 times lower than the maximum residual levels (MRLs) established for the spinach category and for each specific compound according to European regulations [31–34]. Based on the above discussed outcomes, four pesticides were selected and used for the experimental plan of this research work. Propamocarb, chlorantraniliprole, lambda-cyhalothrin and fluopicolide were selected; moreover, the presence of the propamocarb metabolite, propamocarb n-desmethyl, was also assessed in spinach samples.

3.2. Effects of Washing and Blanching Treatments for Pesticide Removal

The concentration of the pesticides and the degradation products measured in spinach samples are shown in Figure 2a,b; chromatograms of residues of some replicates of fresh samples are reported in Figure 3, while those related to the treatments are shown in Figure S1. All the pesticide compounds were detected above the maximum residue levels (MRLs) in the raw samples (F), and therefore, it was useful to perform a study on PFs as also indicated by Bonnechère and co-authors [4]. The initial concentration of residues in fresh samples was 5.9, 0.5, 21.4, 117.5 and 7.4 mg kg^{-1} on the dry weight of chlorantraniliprole, lambda-cyhalothrin, fluopicolide, propamocarb and propamocarb n-desmethyl, respectively. Comparing the fresh samples with the treated ones, regardless of the type of operation used, a general reduction in the concentration of all the residues was observed for all the samples without any significant differences ($p > 0.05$) (Figure 2).

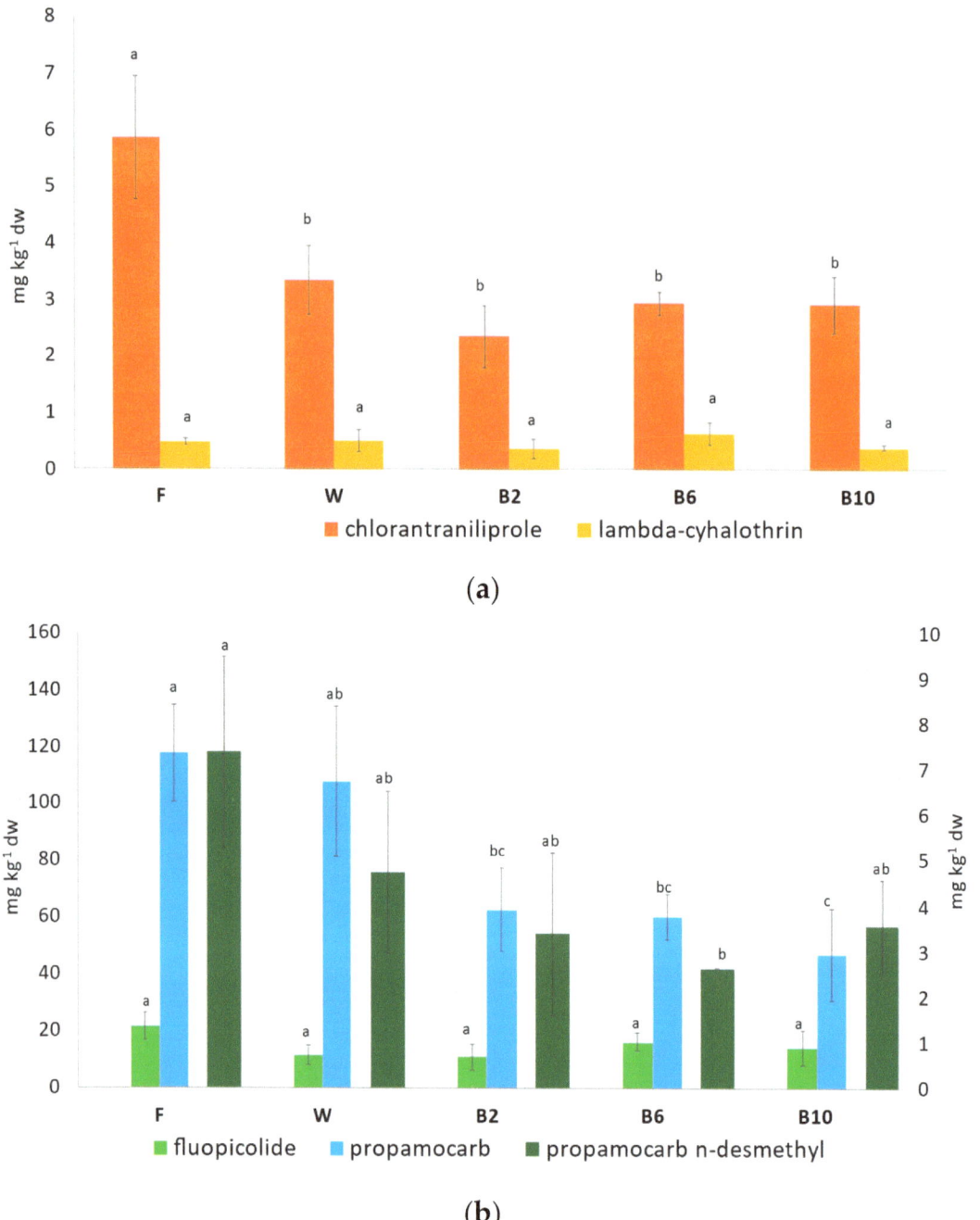

Figure 2. (**a**,**b**) Content of pesticide residues in treated spinach samples (mg kg^{-1} on dry weight). In (**b**), fluopicolide and propamocarb are reported on the primary axis, while propamocarb n-desmethyl is on the secondary axis. Different lowercase letters indicate significant differences between mean values of each residue concentration at different processing points ($p < 0.05$).

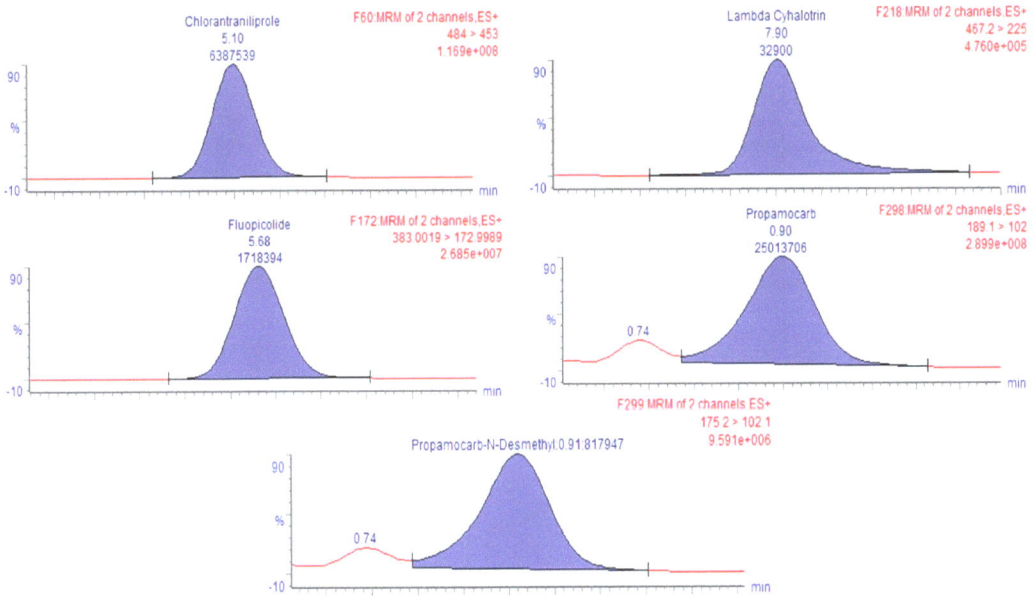

Figure 3. LC-MS/MS chromatograms of residues in fresh samples.

Washing is considered the most common and straightforward form of processing. It is generally the first step in various types of treatments (household and commercial preparation) applied to food commodities [14]. In this study, the variation (Δ%) of pesticide residues at different operation steps is depicted in Figure 4. The washing step, operated with running water, led to a significant reduction of chlorantraniliprole (−43%) and fluopicolide (−47%), with respect to the fresh sample data; a slight reduction was observed for propamocarb (−8%) and propamocarb n-desmethyl (−8%), while conversely, a slight increase in lambda-cyhalothrin was depicted (+5%), although without significant difference ($p > 0.05$). As reported by Yang et al., washing with running water led to the highest removal efficiency among all methods used for different leafy vegetables; in spinach, this method accounted for about an 87% reduction in the pesticide content [35]. Different factors influence the persistence and the removal of pesticides; for example, polar, water-soluble pesticides, such as propamocarb, are more readily removed than low-polarity molecules [4,36]; furthermore, the systemic action mode, for which the active substance is absorbed into the system of a plant, renders its parts (the roots, stems and leaves) poisonous to plant pests and pathogens [4]. In this study, the small reduction, caused by the washing step, of propamocarb and the slight increase in lambda-cyhalothrin could be related to the systemicity and the hydrophobicity characteristics of the first and the second molecules, respectively. Furthermore, as reported by Wu et al., it was difficult to remove pesticides from spinach by tap water, showing a small reduction of lambda-cyhalothrin of 5%, when compared with other washing methods such as ozone water, detergents and alkaline solution [13]. The leaf characteristics (surface area, wax amount on cuticle, thickness) [35] and the age of chemicals could also be considered factors affecting the removal of pesticides; indeed, it was observed that pesticides were easily remove 1 day after spraying than 1 week after [37]. It is important to point out that the commercial pesticide formulations used in this study contained different coformulants that may have altered the kinetic degradation of each pesticide.

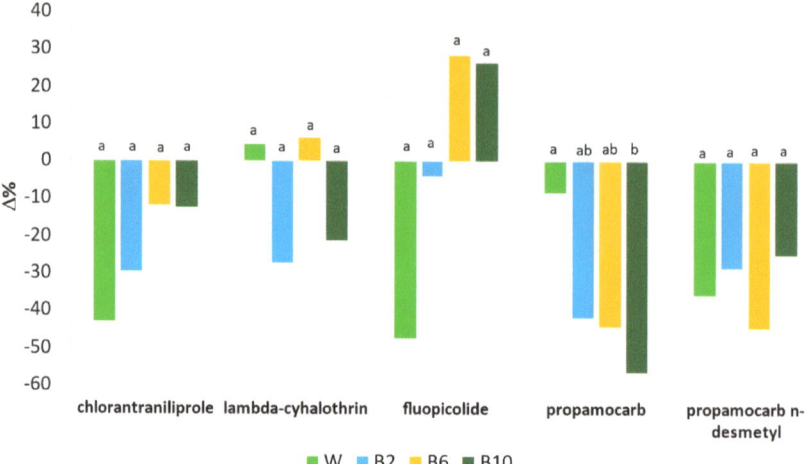

Figure 4. Variation (Δ%) of pesticide residues in spinach samples affected by different treatments. Different lowercase letters indicate significant differences between mean values of each residue for different processing ($p < 0.05$). Blanched samples are referred to the washed one.

Concerning the blanching step, a decrease in residues was observed when 2 min treatment was applied (B2), compared to the washed samples. The highest reduction was depicted for propamocarb (−41%); about −30%, −29% and −27% reductions were observed for chlorantraniliprole, propamocarb n-desmetyl and lambda-cyhalothrin, respectively, while a small reduction was identified for fluopicolide (−4%). The reduction of propamocarb could be associated with the polarity of the molecules with the weakest log–octanol–water partition coefficient (−1.3 Log p), as observed also by Bonnechère et al. in blanched spinach [4].

Blanching for 6 min (B6) led to a decrease in propamocarb and chlorantraniliprole of −44% and −11%, respectively, while an increase in fluopicolide (+28%) and lambda-cyhalothrin (+6%) was observed. The extension of the treatment time to 10 min (B10) caused a decrease, with respect to the washed sample (W), of chlorantraniliprole (−13%), lambda-cyhalothrin (−21%) and propamocarb and its metabolite of −56% and −25%, respectively; conversely, an increase in fluopicolide (+26%) was observed. The controversial behavior of fluopicolide and lambda-cyhalothrin could be related to the log p value and the system of action of each compound, besides the operation conditions used, as the prolonged time of blanching (10 min) in an open pan. Concerning propamocarb content, the decrease is negatively correlated with the increase in the treatment time (−0.93). As also described and observed by Yang and co-authors, the prevalent physicochemical parameter affecting the rate of pesticide residue removal during heat treatment (blanching and boiling) was the partition coefficient [35]. In this study, the low partition coefficient (log p) of the propamocarb denotes its hydrophilic behavior, confirming the results reported by Nagayama that showed a negative correlation between the log p value and a reduction of residue content after blanching or making jam. Conversely, the higher the partition coefficient value is, the lower the pesticide reduction [38].

Concerning propamocarb n-desmetyl, a slight reduction was observed after 10 min of treatment with respect to 6 min, without any correlation. Despite the higher toxicity of the metabolite propamocarb n-desmetyl than the parent compound, to date, the toxicity values and physicochemical properties have not been studied, except in preharvest kinetic dissipation studies regarding cucumber, zucchini and tomato [39], both in water and soil [11]; therefore, it is quite difficult to infer the behavior of this compound.

Considering the residual concentration in the respective blanching waters, the highest content was observed for propamocarb and its metabolite, while lambda-cyhalothrin was not detected, regardless of the time of blanching applied, probably associated with the concentration effect in the spinach (B6) or because of the degradation or volatilization caused from the prolonged blanching (B10).

Furthermore, the different behaviors and fates of the pesticide compounds observed in this study could also be associated with different factors, linked to the process, such as the reduction of the intracellular water content, caused by the prolonged effect of blanching, resulting in a concentration effect of the compound [36] or due to the accumulation of pesticides as a result of moisture evaporation via heating in an open environment [40], as observed for chlorfenapyr in crow daisy for lufenuron in perilla leaves and ssamchoo (*Brassica lee* ssp. *namai*) after boiling (5 min 100 °C) [35], or for acetamiprid in green chilis after boiling [41].

Based on the obtained results, despite the fact that no significant differences were observed among the three blanching treatments, the optimal blanching condition seems to be 2 min treatment (B2) at 80 °C, which also reduces the negative effect of the blanching process over the vegetable structure.

The processing factor (PF), derived by the ratio of the residue concentration in the processed commodity to that in the RACs (raw agricultural commodities), was determined for all the residues with respect to the fresh sample to the washed one, and to the washed sample to the respectively treated ones (blanched). A factor < 1 (=reduction factor) indicates a reduction of the residue in the processed commodity, whereas a factor > 1 (=concentration factor) indicates a concentration effect of the processing procedures.

Based on the results reported in Figure 5, chlorantraniliprole showed values lower than 1 at all the operation levels; lambda-cyhalothrin and fluopicolide showed variable accumulation behavior (W, B2, B6, B10) with values ranging from 0.7 to 1.3 and from 0.5 to 1.4 for the first and second compound, respectively, without significant differences ($p > 0.05$). Concerning propamocarb hydrochloride, PF values lower than 1 were observed, with significant differences only between the washed (0.92) and 10 min treated sample (B10) with a value of 0.4 ($p < 0.05$). Our results are consistent with previous findings from studies in spinach that showed for propamocarb a mean washing and blanching processing factor of 0.89 and 0.30, respectively, while a mean processing factor of 1.6 was shown for deltamethrin [4], similar to that detected for lambda-cyhalothrin in our study. Furthermore, the median values of processing factors reported in the EFSA database are similar to the processing factors for propamocarb (0.88) and lambda-cyhalothrin (1.7), despite that fact that no harmonized list of processing factors is available within Europe and worldwide [18].

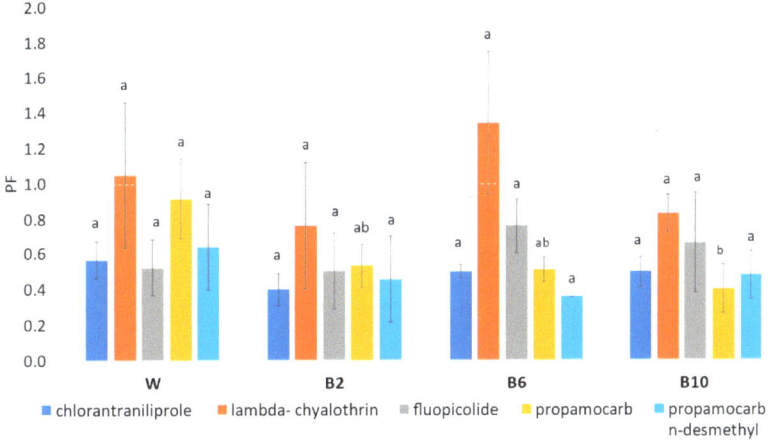

Figure 5. Pesticide residues' processing factors (PFs). Different lowercase letters indicate significant differences between mean values of residue for each processing ($p < 0.05$). TF: <1: reduction of residue; >1: accumulation of residue.

Different PF values observed in this study could be associated with physicochemical properties of the compounds; indeed, as reported by Timme and Walz-Tylla, the persistence of pesticides in processed food depends on their octanol–water partition coefficient, which is an indicator of a pesticide's hydrophilic or lipophilic properties and also to its higher tendency to accumulate inside the matrix [36]. As described by Bozena et al. for strawberry jam, the adsorption of the pesticides onto plant tissues and their solubility in water strongly influenced the removal of pesticide residues during the heat treatment process [42]. Therefore, as well-discussed in the literature, the characteristics of the pesticides, such as the solubility as well as the evaporation of water during the thermal process, may have influenced the residual amounts [35].

3.3. Effect of Frozen Storage on Pesticide Residue

The results of pesticide residue contents in frozen spinach samples, treated with different times of blanching (B2, B6, B10) and stored for 10 months at $-20\ °C$, are presented in Table 3. Fresh and washed samples were not considered because the aim of the study was to assess the effect of the process and the content of residue in the final frozen product, ready for the consumer. Generally, no significant variations ($p > 0.05$) were depicted after the first day of frozen storage at $-20\ °C$ (t1) for all the pesticide residues, regardless of the time of blanching used, except for propamocarb n-desmethyl treated for 10 min ($p < 0.05$). It can be concluded that the freezing process, after blanching, did not influence the content of pesticides in spinach samples.

During the frozen storage period (0–300 days), important variations were depicted in all the treated samples. Chlorantraniliprole showed a significant increase ($p < 0.05$) in B2, B6 and B10 samples, reaching at the end of the storage (t300) values of 5.83 ± 1.11, 12.15 ± 2.00 and 11.94 ± 2.20 mg kg^{-1}, respectively. Concerning lambda-cyhalothrin, an increasing trend was depicted, with a significant variation for B2, B6 and B10 samples during storage, starting mainly at t105 and ending at t300 with 4.17 ± 1.60, 6.38 ± 2.90 and 8.63 ± 0.30 mg kg^{-1}, respectively, Fluopicolide content remained almost stable during frozen storage in B2 samples ($p > 0.05$), while a significant increase was observed at the end of the storage period for B6 and B10 samples. Propamocarb and propamocarb n-desmethyl showed similar trends. No significant variations ($p > 0.05$) were observed in B2 samples, and both the residue concentrations remained almost constant during the refrigeration period (0 to 300 days); B10 samples showed the same behavior observed in B2, while propamocarb n-desmethyl showed a significant reduction at t1, with a value of 1.23 ± 0.16 mg kg^{-1} compared to t0 and t300. Concerning B6 samples, a significant increase in propamocarb n-desmethyl was depicted in the sample stored for 300 days with respect to the initial content reaching a value of 3.87 ± 1.10 mg kg^{-1}. Considering the influence of blanching time on the residue pesticide contents (Table 3), significant variations were highlighted at the end of storage period (t300) and mainly for chlorantraniliprole, lambda-cyhalothrin and fluopicolide, probably associated with the mechanical disruption of cells' structure during processing, leading to a loss of cell vegetable water along with a dehydration effect of freezing [43] resulting in a concentration of pesticide molecules; conversely, this was trend not associated with propamocarb and its metabolite, for which the time of blanching did not significantly influence their content ($p > 0.05$).

Table 3. Content of pesticides residues (mg kg^{-1} dw) in spinach samples treated with different times of blanching.

Blanching Methods	Storage (Days)	Chlorantraniliprole	Lambda-Cyhalothrin	Fluopicolide	Propamocarb	Propamocarb n-Desmethyl
B2	t0	2.34 ± 0.55 [a,B]	0.36 ± 0.17 [b,D]	10.81 ± 4.68 [a,C]	62.61 ± 10.45 [a,A]	3.38 ± 1.79 [a,AB]
	t1	2.57 ± 0.27 [a,B]	0.31 ± 0.06 [b,D]	12.61 ± 2.81 [a,C]	77.93 ± 6.09 [a,A]	3.83 ± 1.03 [a,A]
	t105	2.93 ± 1.12 [a,B]	2.93 ± 0.88 [a,CD]	16.53 ± 10.44 [a,C]	63.77 ± 21.80 [a,A]	1.99 ± 0.11 [a,A]
	t300	5.83 ± 1.11 [a,B]	4.17 ± 1.60 [a,BC]	24.85 ± 2.30 [a,BC]	66.93 ± 10.30 [a,A]	3.45 ± 0.40 [a,AB]
B6	t0	2.94 ± 0.20 [b,B]	0.64 ± 0.19 [b,D]	16.23 ± 3.31 [b,C]	60.09 ± 7.93 [ab,A]	2.63 ± 0.10 [b,AB]
	t1	2.30 ± 0.46 [b,B]	0.24 ± 0.04 [b,D]	11.40 ± 2.01 [b,C]	43.86 ± 5.32 [b,A]	1.62 ± 0.40 [b,AB]
	t105	4.67 ± 1.51 [b,B]	4.77 ± 1.37 [ab,BD]	37.58 ± 1.04 [ab,AB]	69.76 ± 10.66 [ab,A]	1.67 ± 0.11 [b,A]
	t300	12.15 ± 2.00 [a,A]	6.38 ± 2.90 [a,AB]	44.12 ± 1.00 [a,A]	73.04 ± 11.90 [a,A]	3.87 ± 1.10 [a,A]
B10	t0	2.92 ± 0.50 [b,B]	0.39 ± 0.05 [c,D]	14.22 ± 6.04 [b,C]	46.94 ± 16.09 [a,A]	3.56 ± 1.10 [a,AB]
	t1	2.37 ± 0.54 [b,B]	0.33 ± 0.11 [c,D]	8.53 ± 0.10 [b,C]	36.04 ± 5.92 [a,A]	1.23 ± 0.16 [b,B]
	t105	3.88 ± 1.24 [b,B]	4.11 ± 1.30 [b,BC]	18.53 ± 8.28 [b,C]	58.10 ± 14.59 [a,A]	1.65 ± 0.45 [ab,A]
	t300	11.94 ± 2.20 [a,A]	8.63 ± 0.30 [a,A]	45.58 ± 0.70 [a,A]	78.61 ± 24.70 [a,A]	4.05 ± 0.60 [a,A]

Different lowercase letters in a column indicate significant differences among each blanching method during storage time ($p < 0.05$). Different uppercase letters in a column indicate significant differences among all the blanching methods during storage time ($p < 0.05$).

Our results generally agree with the few presented in the literature, such as those observed in fish stored at −70 °C for four weeks for organochlorines, organophosphorus, pyrethroid and carbamates compounds [23] and for insecticides and fungicides in zucchini at −30 °C [24], where no significant reduction of pesticides was observed during frozen storage. On the contrary, reductions of difenoconazole residues were observed in mango at low storage temperature (−20 °C) [25]; in fresh, raw mackerel fillets at −20 °C for different pyrethroid pesticides [27]; in apples at −25 °C for fungicides [28]; and in tomatoes stored at −10 °C for organochlorines and organophosphorus compounds [26]. It is well-known that low temperatures reduce the chemical reactivity and decay of organic compounds. The storage at very low temperature of food contaminated with pesticides has been reported to have no significant effects on the pesticides' levels [23]. Further, a sample's matrix has abundant enzymes and microbes that can cause the degradation of the pesticide residue. Low storage temperature limits both the enzymatic and microbic activity that otherwise result in improved concentrations when higher storage temperature is used (−4 °C, +4 °C) [23,25]. The blanching treatments applied in this study could have inhibited the enzymatic activity, and therefore this degradation aspect cannot be considered. Furthermore, we should not exclude the effect of the freezing process on the texture modification of the spinach vegetable structure. Indeed, ice formation causes important damage in the plant tissue, influencing the dehydration, partitioning, concentration or degradation [44] that could affect the pesticide residue contents in frozen foods. Due to the limited number and controversial nature of existing studies, the main factors that affect the stability of the residues in frozen spinach remain unclear; even though the reactivity of pesticides by their molecular structure can be predicted, it is still difficult to know their degradation process, which depends upon the structure and transport behavior of pesticide, as well as the environmental conditions [9]. Furthermore, as previously discussed, the presence of coformulants may have modified the physicochemical interaction of the molecules with the vegetable matrix.

4. Conclusions

This study examined the effects of sequential processing operations such as washing, blanching, freezing and frozen storage on chlorantraniliprole, lambda-cyhalothrin, fluopicolide, propamocarb and propamocarb n-desmethyl content in spinach, considered as the main residues in spinach based on a preliminarily evaluation. The washing step was mainly effective for chlorantraniliprole and fluopicolide reduction and less on propamocarb and lambda-cyhalothrin content. Blanching was effective for the reduction of all the pesticides, regardless of the time used (2, 6 and 10 min). The freezing step did not influence the residue content, while during ten months of frozen storage, an increase in chlorantraniliprole, lambda-cyhalothrin, fluopicolide and propamocarb n-desmethyl was observed for the samples blanched for 6 and 10 min. No variation of pesticide contents was highlighted in frozen spinach samples blanched for 2 min. These findings highlighted the positive effect of washing and short-time blanching treatment (2 min) in pesticide residue content with positive effect also during spinach frozen storage. The knowledge of the physicochemical features, specific for each pesticide compound, represents an important point in order to optimize good agricultural practices (GAP) and the correct household or industrial food production treatments. Furthermore, research outcomes emphasize both the effects of frozen storage and the frozen products as possible health hazards due to pesticide content, opening this question to further studies because of the increasing consumption of frozen products, from a "field to fork" food safety perspective.

Supplementary Materials: The following supporting information can be downloaded at: https://www.mdpi.com/article/10.3390/foods12142806/s1, Figure S1: Example of LC-MS/MS chromatograms of chlorantraniliprole, fluopicolide, lambda-cyhalothrin (S1a) and propamocarb and propamocarb n-desmethyl (S1b) in spinach samples before (F) and after each treatment: washing (W), blanching 2 min (B2), blanching 6 min (B6) and blanching 10 min (B10).

Author Contributions: Conceptualization, A.C. and L.C.; methodology, F.F., L.C. and S.M.; formal analysis, F.F.; investigation, F.F.; resources, S.M. and L.C.; data curation, F.F. and S.M.; writing—original draft preparation, F.F.; writing—review and editing, F.F., A.C., S.M., L.C. and A.M.; visualization, F.F.; supervision, A.C. and L.C.; project administration, A.C., A.M. and L.C. All authors have read and agreed to the published version of the manuscript.

Funding: This research received no external funding.

Data Availability Statement: Data available on request due to restrictions. The data presented in this study are available on request from the corresponding author.

Acknowledgments: The authors are grateful to all the analysts of the S.A.L.P.A. S.A.C.arl industry for their help during the analysis of the samples.

Conflicts of Interest: The authors declare no conflict of interest. The authors S.M. and L.C. disclose that they have neither financial nor nonfinancial interests and declare that their participation has not affected the authenticity and objectivity of the experimental results.

References

1. Farina, Y.; Abdullah, M.P.; Bibi, N.; Khalik, W.M.A.W.M. Determination of pesticide residues in leafy vegetables at parts per billion levels by a chemometric study using GC-ECD in Cameron Highlands, Malaysia. *Food Chem.* **2017**, *224*, 186–192. [CrossRef]
2. Yigit, N.; Velioglu, Y.S. Effects of processing and storage on pesticide residues in foods. *Crit. Rev. Food Sci. Nutr.* **2019**, *60*, 3622–3641. [CrossRef]
3. Santarelli, G.A.; Migliorati, G.; Pomilio, F.; Marfoglia, C.; Centorame, P.; D'Agostino, A.; D'Aurelio, R.; Scarpone, R.; Battistelli, N.; Di Simone, F.; et al. Assessment of pesticide residues and microbial contamination in raw leafy green vegetables marketed in Italy. *Food Control* **2018**, *85*, 350–358. [CrossRef]
4. Bonnechère, A.; Hanot, V.; Jolie, R.; Hendrickx, M.; Bragard, C.; Bedoret, T.; Van Loco, J. Effect of household and industrial processing on levels of five pesticide residues and two degradation products in spinach. *Food Control* **2012**, *25*, 397–406. [CrossRef]
5. Islam, M.S.; Azim, F.; Saju, H.; Zargaran, A.; Shirzad, M.; Kamal, M.; Fatema, K.; Rehman, S.; Azad, M.A.M.; Ebrahimi-Barough, S. Pesticides and Parkinson's disease: Current and future perspective. *J. Chem. Neuroanat.* **2021**, *115*, 101966. [CrossRef]
6. European Commission. *Regulation (EC) No 396/2005 of the European Parliament and of the Council of 23 February 2005 on Maximum Residue Levels of Pesticides in or on Food and Feed of Plant and Animal Origin and Amending Council Directive 91/414/EEC Text with EEA Relevance*; European Commission: Brussels, Belgium, 2005.
7. European Commission. *Regulation (EC) No 1107/2009 of the European Parliament and of the Council of 21 October 2009 Concerning the Placing of Plant Protection Products on the Market and Repealing Council Directives 79/117/EEC and 91/414/EEC*; European Commission: Brussels, Belgium, 2009.
8. Jepson, P.C.; Murray, K.; Bach, O.; Bonilla, M.A.; Neumeister, L. Selection of pesticides to reduce human and environmental health risks: A global guideline and minimum pesticides list. *Lancet Planet Health* **2020**, *4*, e56–e63. [CrossRef]
9. Bhilwadikar, T.; Pounraj, S.; Manivannan, S.; Rastogi, N.K.; Negi, P.S. Decontamination of Microorganisms and Pesticides from Fresh Fruits and Vegetables: A Comprehensive Review from Common Household Processes to Modern Techniques. *Compr. Rev. Food Sci. Food Saf.* **2019**, *18*, 1003–1038. [CrossRef]
10. Brancato, A.; Brocca, D.; De Lentdecker, C.; Erdos, Z.; Ferreira, L.; Greco, L.; Jarrah, S.; Kardassi, D.; Leuschner, R.; Lythgo, C.; et al. Modification of the existing maximum residue level for propamocarb in chards/beet leaves. *EFSA J.* **2017**, *15*, e05055. [CrossRef]
11. López-Ruiz, R.; Romero-González, R.; Garrido Frenich, A. Dissipation kinetics of fenamidone, propamocarb and their metabolites in ambient soil and water samples and unknown screening of metabolites. *J. Environ. Manag.* **2020**, *254*, 109818. [CrossRef]
12. Henry, L.N.; Ngowo, J.N. Degradation of Lambda-Cyhalothrin in Spinach (Spinacia Oleracea) & Collard Green (Brassica Oleracea) Under Tropical Conditions. *Int. J. Sci. Res.* **2016**, *5*, 2013–2016.
13. Wu, Y.; An, Q.; Li, D.; Wu, J.; Pan, C. Comparison of different home/commercial washing strategies for ten typical pesticide residue removal effects in kumquat, spinach and cucumber. *Int. J. Environ. Res. Public Health* **2019**, *16*, 472. [CrossRef] [PubMed]
14. Kaushik, G.; Satya, S.; Naik, S.N. Food processing a tool to pesticide residue dissipation—A review. *Food Res. Int.* **2009**, *42*, 26–40. [CrossRef]
15. Djordjevic, T.; Djurovic-Pejcev, R. Food processing as a means for pesticide residue dissipation. *Pestic. i Fitomedicina* **2016**, *31*, 89–105. [CrossRef]
16. Misra, N.N. The contribution of non-thermal and advanced oxidation technologies towards dissipation of pesticide residues. *Trends Food Sci. Technol.* **2015**, *45*, 229–244. [CrossRef]
17. Gavahian, M.; Sarangapani, C.; Misra, N.N. Cold plasma for mitigating agrochemical and pesticide residue in food and water: Similarities with ozone and ultraviolet technologies. *Food Res. Int.* **2021**, *141*, 110138. [CrossRef] [PubMed]
18. Scholz, R.; van Donkersgoed, G.; Herrmann, M.; Kittelmann, A.; von Schledorn, M.; Graven, C.; Mahieu, K.; van der Velde-Koerts, T.; Anagnostopoulos, C.; Bempelou, E.; et al. Database of processing techniques and processing factors compatible with the EFSA food classification and description system FoodEx 2 Objective 3: European database of processing factors for pesticides in food. *EFSA Support Publ.* **2018**, *15*. [CrossRef]

19. DGISAN. *Controllo Ufficiale sui Residui dei Prodotti Fitosanitari Negli Alimenti–Rapporto 2020*; DGISAN: Daegu, Republic of Korea, 2023.
20. Farha, W.; Abd El-Aty, A.M.; Rahman, M.M.; Jeong, J.H.; Shin, H.C.; Wang, J.; Shin, S.S.; Shim, J.H. Analytical approach, dissipation pattern and risk assessment of pesticide residue in green leafy vegetables: A comprehensive review. *Biomed. Chromatogr.* **2018**, *32*, e4134. [CrossRef]
21. Holland, P.T.; Hamilton, D.; Ohlin, B.; Skidmore, M.W. Effects of storage and processing on pesticide residue in plant products. *Pure Appl. Chem.* **1994**, *66*, 335–356. [CrossRef]
22. Concha-meyer, A.; Grandon, S.; Sepúlveda, G.; Diaz, R.; Yuri, J.A.; Torres, C. Pesticide residues quantification in frozen fruit and vegetables in Chilean domestic market using QuEChERS extraction with ultra-high-performance liquid chromatography electrospray ionization Orbitrap mass spectrometry. *Food Chem.* **2019**, *295*, 64–71. [CrossRef]
23. Alaboudi, A.R.; Almashhadany, D.A.; Jarrah, B.S. Effect of Cooking and Freezing on Levels of pesticides residues in Local Fresh Fish. *Bull. Univ. Agric. Sci. Vet. Med. Cluj-Napoca Anim. Sci. Biotechnol.* **2021**, *78*, 28. [CrossRef]
24. Oliva, J.; Cermeño, S.; Cámara, M.A.; Martínez, G.; Barba, A. Disappearance of six pesticides in fresh and processed zucchini, bioavailability and health risk assessment. *Food Chem.* **2017**, *229*, 172–177. [CrossRef]
25. Zhao, F.; Liu, J.; Han, B.; Luo, J. Investigation and validation of detection of storage stability of difenoconazole residue in Mango. *J. Food Qual.* **2019**, *2019*, 5641643. [CrossRef]
26. Abou-Arab, A.A.K. Behavior of pesticides in tomatoes during commercial and home preparation. *Food Chem.* **1999**, *65*, 509–514. [CrossRef]
27. Wongmaneepratip, W.; Gao, X.; Yang, H. Effect of food processing on reduction and degradation pathway of pyrethroid pesticides in mackerel fillet (*Scomberomorus commerson*). *Food Chem.* **2022**, *384*, 132523. [CrossRef]
28. Słowik-Borowiec, M.; Szpyrka, E. Selected food processing techniques as a factor for pesticide residue removal in apple fruit. *Environ. Sci. Pollut. Res.* **2020**, *27*, 2361–2373. [CrossRef]
29. Tomlin, C.D.S. *Pesticide Manual: A World Compendium of Pesticides*; British Crop Protection Council: Hampshire, UK, 2003.
30. European Commission. *Regulation (EU) 2018/848 of the European Parliament and of the Council of 30 May 2018 on Organic Production and Labelling of Organic Products and Repealing Council Regulation (EC) No 834/2007*; European Commission: Brussels, Belgium, 2018.
31. European Commission. *Commission Regulation (EU) 2021/590 of 12 April 2021 Amending Annexes II and IV to Regulation (EC) No 396/2005 of the European Parliament and of the Council as Regards Maximum Residue Levels for Aclonifen, Boscalid, Cow Milk, Etofenprox, Ferric Pyrophosph*; European Commission: Brussels, Belgium, 2021.
32. European Commission. *Commission Regulation (EU) 2021/616 of 13 April 2021 Amending Annexes II, III and V to Regulation (EC) No 396/2005 of the European Parliament and of the Council as Regards Maximum Residue Levels for Benalaxyl, Benalaxyl-M, Dichlobenil, Fluopicolide, Proqu*; European Commission: Brussels, Belgium, 2021.
33. European Commission. *Commission Regulation (EU) 2022/1324 of 28 July 2022 Amending Annexes II and III to Regulation (EC) No 396/2005 of the European Parliament and of the Council as Regards Maximum Residue Levels for Benzovindiflupyr, Boscalid, Fenazaquin, Fluazifop-P, Flupyr*; European Commission: Brussels, Belgium, 2022.
34. European Commission. *Commission Regulation (EU) 2020/856 of 9 June 2020 Amending Annexes II and III to Regulation (EC) No 396/2005 of the European Parliament and of the Council as Regards Maximum Residue Levels for Cyantraniliprole, Cyazofamid, Cyprodinil, Fenpyroximate, Flud*; European Commission: Brussels, Belgium, 2020.
35. Yang, S.-J.; Mun, S.; Kim, H.J.; Han, S.J.; Kim, D.W.; Cho, B.S.; Kim, A.G.; Park, D.W. Effectiveness of Different Washing Strategies on Pesticide Residue Removal: The First Comparative Study on Leafy Vegetables. *Foods* **2022**, *11*, 2916. [CrossRef]
36. Timme, G.; Walz-Tylla, B. Effects of Food Preparation and Processing on Pesticide Residues in Commodities of Plant Origin. In *Pesticide Residues in Food and Drinking Water: Human Expsure and Risks*; John Wiley & Sons, Ltd.: Hoboken, NJ, USA, 2004; pp. 121–148.
37. Guardia-Rubio, M.; Ayora-Cañada, M.J.; Ruiz-Medina, A. Effect of Washing on Pesticide Residues in Olives. *J. Food Sci.* **2007**, *72*, C139–C143. [CrossRef]
38. Nagayama, T. Behavior of Residual Organophosphorus Pesticides in Foodstuffs during Leaching or Cooking. *J. Agric. Food Chem.* **1996**, *44*, 2388–2393. [CrossRef]
39. López-Ruiz, R.; Romero-González, R.; Serra, B.; Garrido Frenich, A. Dissipation kinetic studies of fenamidone and propamocarb in vegetables under greenhouse conditions using liquid and gas chromatography coupled to high-resolution mass spectrometry. *Chemosphere* **2019**, *226*, 36–46. [CrossRef]
40. Keikotlhaile, B.M.; Spanoghe, P.; Steurbaut, W. Effects of food processing on pesticide residues in fruits and vegetables: A meta-analysis approach. *Food Chem. Toxicol.* **2010**, *48*, 1–6. [CrossRef]
41. Yang, A.; Park, J.H.; Abd El-Aty, A.M.; Choi, J.H.; Oh, J.H.; Do, J.A.; Kwon, K.; Shim, K.H.; Choi, O.J.; Shim, J.H. Synergistic effect of washing and cooking on the removal of multi-classes of pesticides from various food samples. *Food Control* **2012**, *28*, 99–105. [CrossRef]
42. Bozena, L.; Magdalena, J.; Izabela, H.; Piotr, K. Removal of 16 pesticide residues from strawberries by washing with tap and ozone water, ultrasonic cleaning and boiling. *Environ. Monit. Assess.* **2016**, *188*, 1–19. [CrossRef]

43. Kidmose, U.; Martens, H.J. Changes in texture, microstructure and nutritional quality of carrot slices during blanching and freezing. *J. Sci. Food Agric.* **1999**, *79*, 1747–1753. [CrossRef]
44. van der Sman, R.G.M. Impact of Processing Factors on Quality of Frozen Vegetables and Fruits. *Food Eng. Rev.* **2020**, *12*, 399–420. [CrossRef]

Disclaimer/Publisher's Note: The statements, opinions and data contained in all publications are solely those of the individual author(s) and contributor(s) and not of MDPI and/or the editor(s). MDPI and/or the editor(s) disclaim responsibility for any injury to people or property resulting from any ideas, methods, instructions or products referred to in the content.

Article

Green Pre-Treatment Strategy Using Ionic Liquid-Based Aqueous Two-Phase Systems for Pesticide Determination in Strawberry Samples

Ana Jocić [1], Slađana Marić [1,*], Danijela Tekić [1], Jasmina Mušović [1], Jelena Milićević [2], Sanja Živković [1] and Aleksandra Dimitrijević [1]

[1] Department of Physical Chemistry, VINČA Institute of Nuclear Sciences—National Institute of the Republic of Serbia, University of Belgrade, Mike Petrovića Alasa 12-14, 11351 Belgrade, Serbia; ana.jocic@vin.bg.ac.rs (A.J.); danijela.tekic@vin.bg.ac.rs (D.T.); jasmina.musovic@vin.bg.ac.rs (J.M.); sanjaz@vin.bg.ac.rs (S.Ž.); daleksandra@vin.bg.ac.rs (A.D.)

[2] Department for Bioinformatics and Computational Chemistry, VINČA Institute of Nuclear Sciences—National Institute of the Republic of Serbia, University of Belgrade, Mike Petrovića Alasa 12-14, 11351 Belgrade, Serbia; jdjordjevic@vin.bg.ac.rs

* Correspondence: sladjana.maric@vin.bg.ac.rs

Abstract: Pesticides used in agriculture can contaminate foods like fruits and vegetables, posing health risks to consumers and highlighting the need for effective residue monitoring. This study explores aqueous two-phase systems (ATPSs) comprising phosphonium or ammonium ionic liquids (ILs) combined with ammonium sulfate as an alternative pre-treatment method for extracting and concentrating the pesticides clomazone, pyraclostrobin, and deltamethrin from strawberry samples. Liquid–liquid equilibrium measurements for each ATPS were conducted, followed by extraction experiments to determine the most efficient systems for pesticide extraction. Results showed that all three pesticides migrated effectively to the IL-rich phase across the tested ATPSs. For the most promising system, tetrabutylphosphonium salicylate ([TBP][Sal]) with ammonium sulfate, extraction efficiencies for each pesticide exceeded 98% under optimized conditions for parameters such as pH, temperature, and ATPS composition. Application of this ATPS to strawberries resulted in significant pesticide preconcentration, reaching mg/L levels suitable for detection by liquid chromatography. The method's sustainability was supported by green chemistry metrics, with AGREEprep and AGREE scores of 0.68 and 0.55, respectively, underscoring its alignment with eco-friendly practices.

Keywords: aqueous two-phase system; ionic liquid; pesticide in food; preconcentration; strawberry; clomazone; pyraclostrobin; deltamethrin

1. Introduction

As the global population continues to rise, the demand for increased food production has grown substantially [1]. To meet these demands and protect agricultural products, the widespread use of pesticides has become essential. Without such protective measures, it is estimated that crop yields would only achieve about 30% of their full potential [2]. Despite their essential role, the prolonged use of pesticides has led to environmental pollution and raised significant food safety concerns due to the accumulation of pesticide residues. Even pesticides considered low in toxicity can persist in the environment, leading to pollution of soil and water and threatening non-target organisms and ecosystems. Of particular concern are pesticides, which are suspected of acting as endocrine disruptors, potentially leading to long-term health risks such as hormonal imbalances and reproductive disorders in humans [3–5]. These risks underscore the importance of routinely monitoring pesticide residues in food products, especially in fruits and vegetables, which are common sources of exposure. Regulatory bodies, such as the European Commission, have established maximum residue limits (MRLs) for numerous pesticides to ensure safe consumption levels [6].

These regulations help ensure that pesticide levels remain within safe limits, safeguarding public health. However, studies consistently detect residues from pesticides in fruits like strawberries, apples, cherries, and grapes. Strawberries consistently top the 'dirty dozen' list, which identifies the 12 fruits and vegetables with the highest pesticide residue levels, alongside apples and grapes [7–9]. For instance, a study from Egyptian markets found pesticide residues in all tested fresh strawberries and strawberry-based products, including yogurt, juice, jam, and dried strawberries, with average concentrations ranging from 0.006 to 0.568 mg/kg, depending on the product [10]. Similarly, research from Shanghai on strawberry samples from the 2018 and 2019 harvest seasons revealed 51 pesticide residues, with 97.91% of samples containing at least one pesticide and 2.39% exceeding the maximum residue limits set by Chinese regulations [11]. This underscores the urgent need for reliable, sensitive, and environmentally friendly detection methods for food samples. Detecting pesticide residues accurately is not only essential for adhering to regulatory standards like MRLs but also for minimizing long-term exposure to potentially harmful chemicals. The development of new pesticides and the growing number of pesticide formulations further increase the necessity for ongoing advancements in detection techniques.

Pesticide detection in complex matrices often requires advanced analytical techniques like high-performance liquid chromatography (HPLC) [9,12,13] and gas chromatography (GC), known for their high sensitivity and precision in quantifying trace pesticide levels [4,14,15]. Due to the complexity of matrices like fruits, pre-treatment steps are typically essential to extract analytes effectively and achieve low detection limits. Conventional methods for pesticide preconcentration and extraction, such as solid-phase extraction (SPE) and liquid–liquid extraction (LLE), are effective but come with certain limitations: LLE requires large amounts of organic solvents, while SPE can be labor-intensive. In addition to these techniques, the widely used QuEChERS (Quick, Easy, Cheap, Rugged, Effective, and Safe) method offers a simpler and more efficient approach for pesticide preconcentration, although it predominantly relies on acetonitrile (ACN) as the organic solvent. To mitigate these drawbacks, microextraction techniques like solid-phase microextraction (SPME), liquid-phase microextraction (LPME), and dispersive liquid–liquid microextraction (DLLME) have been developed. These methods are more efficient in minimizing solvent use while still offering effective preconcentration [16,17]. In recent years, ionic liquids (ILs) have emerged as a promising alternative to traditional organic solvents in these extraction processes. Ionic liquids are salts with depressed melting points and possess tunable properties such as low volatility and reduced environmental impact compared to organic solvents. Their tunable nature makes them ideal for optimizing extraction conditions [18,19]. Extensive research has been conducted on the use of ILs in DLLME for pesticide preconcentration, usually from water samples, with acetonitrile as the most common organic solvent. Typically, HPLC is favored over GC in systems involving ionic liquids due to the low volatility of ILs, which complicates GC analysis [20]. As part of the shift toward sustainable solvents, researchers have proposed ionic liquid-based aqueous two-phase systems (IL-ATPSs). These IL-ATPSs, formed by mixing an IL aqueous solution with a salt or polymer solution under controlled conditions (pH, temperature), offer unique advantages, including operational simplicity, low cost, and high biocompatibility. Unlike other methods, which often rely on hazardous organic solvents, ATPSs provide a gentler, predominantly water-based extraction strategy, further enhancing its appeal for environmentally friendly extraction processes. Compared to QuEChERS and DLLME, IL-ATPSs offer a more sustainable and efficient approach to preconcentration. QuEChERS is faster and simpler but has limited preconcentration capacity, while DLLME achieves high enrichment with minimal solvent but suffers from scalability challenges. IL-ATPSs combine the advantages of both, providing high preconcentration efficiency with a greener solvent profile [16,18,21]. Despite the promising benefits of IL-ATPSs, surprisingly few studies focus on their application in the preconcentration of pesticides from food matrices prior to analytical methods. A study by Tian et al., demonstrated the effectiveness of a cholinium acetate ([Ch][Ac]) and EOPO-2500 copolymer IL-ATPS for herbicide extraction from honey, significantly improving sample preparation and preconcentration for accurate HPLC–MS/MS quantification [22]. A similar system, employing EOPO-2500 and cholinium L-lysinate ([Ch][L-Lys]), has been used to extract target fungicides from commercial fruit juice and homemade apple juice. Although this setup showed a modest preconcentration factor, it effectively isolated analytes into the EOPO phase. After phase separation, the enriched phase was analyzed by HPLC-MS/MS, demonstrating reliable quantification of fungicides in complex food matrices [23]. Conversely, ionic liquid and salt-based systems have demonstrated high efficiency and high preconcentration factors in IL-ATPS applications, especially those utilizing ammonium, phosphonium, imidazolium, and cholinium ILs, for the extraction and preconcentration of various contaminants, including pharmaceuticals, bisphenols, and parabens, typically sourced from wastewater samples [24–26]. Furthermore, a substantial number of studies have demonstrated the high, often near-complete extraction efficiency of IL-ATPSs for pesticide removal from wastewater. These systems have proven particularly effective as an initial step in sample preparation, concentrating pesticides for subsequent analytical detection. For example, a study using an ATPS composed of choline chloride ([Ch]Cl) and polypropylene glycol 400 (PPG400) demonstrated high extraction efficiencies for pesticides like dicamba, clomazone, pyraclostrobin, and deltamethrin. Nearly complete extraction of these pesticides from agricultural wastewater in one step into the polymer-rich phase highlighted the potential of IL-ATPSs for high-efficiency extraction, with significant preconcentration factors achieved by optimizing the system composition [27]. Additionally, research by Jocic et al. explored the use of tetrabutylphosphonium salicylate ([TBP][Sal]) in combination with a citrate buffer, highlighting its efficacy in the extraction of organophosphate pesticides from wastewater. The study also found that the trace amounts of ionic liquid remaining in the aqueous phase after extraction posed no cytotoxic effects, underscoring the potential of IL-based ATPS as a greener alternative to traditional extraction methods [28]. Given the promising results, this study aimed to evaluate and compare the effectiveness of phosphonium and ammonium ILs in developing an aqueous two-phase system specifically for the preconcentration of pesticides, enabling their detection from complex fruit matrices, such as strawberries. Notably, this research addresses a gap in the literature, as no previous studies have used ionic liquid and salt-based ATPS for this purpose.

In this study, both phosphonium and ammonium ionic liquids with anions derived from natural acids (salicylate, acetate, and lactate) were tested as an alternative pretreatment strategy for pesticide determination in food samples. Three pesticides from distinct agrochemical groups were selected: clomazone (CLO), a herbicide used for weed control in vegetable crops; pyraclostrobin (PYR), a fungicide targeting fungal infections in plants; and deltamethrin (DLM), a pyrethroid insecticide for pest management [29–31]. Initially, phase diagrams for ammonium ionic liquids (TBA-ILs) with ammonium sulfate (as the salting-out agent) were established and compared to those of phosphonium ILs (TBP-ILs). Extraction experiments in the two-phase medium were conducted to identify the most favorable system for the extraction of pesticides. The effects of temperature, pH, and ATPS compositions on the extraction performance of a [TBP][Sal]-based ATPS were systematically investigated. After identifying optimal conditions, the [TBP][Sal]-based ATPS was applied to the preconcentration of pesticides from the strawberry sample, with quantification carried out using the ultra-performance liquid chromatography (UPLC) method. To assess the environmental impact of the proposed method, green analytical metrics AGREEprep and AGREE were utilized. These metrics enabled a comprehensive evaluation of the method's environmental footprint, from sample preparation through to final analysis, ensuring alignment with the principles of Green Analytical Chemistry [32,33].

2. Materials and Methods

2.1. Materials

Three pesticides, namely clomazone, pyraclostrobin, and deltamethrin, were obtained from Sigma-Aldrich (USA), and their chemical structures are shown in Figure 1. HPLC-grade acetonitrile (C_2H_3N), ethanol (CH_3CH_2OH, ≥99.8 wt%), and acetone (≥99.5 wt%) were procured from Honeywell (Germany). Ammonium sulfate ((NH_4)$_2SO_4$, ≥99 wt%), tetrabutylammonium hydroxide ($C_{16}H_{37}NO$, 40 wt% in H_2O), and salicylic acid ($C_7H_6O_3$, ≥99 wt%) were supplied from Sigma-Aldrich (USA), while lactic acid ($C_3H_6O_3$, ≥90 wt%) was acquired from Fluka Chemie (Switzerland) and acetic acid (CH_3COOH, ≥99 wt%) was provided by MSK Kikinda (Serbia). Phosphonium ILs, including tetrabutylphosphonium acetate ([TBP][Ac]), tetrabutylphosphonium lactate ([TBP][Lac]), and tetrabutylphosphonium salicylate ([TBP][Sal]) were synthesized in previous study [34]. Ammonium ILs—tetrabutylammonium acetate ([TBA][Ac]), tetrabutylammonium lactate ([TBA][Lac]), and tetrabutylammonium salicylate ([TBA][Sal])—were synthesized through the neutralization of tetrabutylammonium hydroxide with salicylic, acetic, or lactic acid, respectively [35]. The chemical structures of ILs used in this study are provided in Figure 1. The ILs were prepared by mixing equimolar amounts of hydroxide and the corresponding acid, with the mixture being stirred for 24 h at room temperature. Subsequently, water was removed from the obtained ILs under vacuum using a R-210 Rotavapor System BÜCHI Labortechnik AG (Switzerland). Additionally, all ILs underwent a drying process, where they were heated at 50 °C under vacuum for approximately 48 h. Water content was determined via Karl Fischer titration, using an 831 Karl Fischer coulometer from Metrohm, and found to be less than 2 wt% in all ILs. The chemical structures of the ILs were confirmed using a Thermo Fisher Scientific Nicolet iS5 spectrometer (USA), and the corresponding spectra, along with peak assignments, are provided in the Supplementary Materials (Figures S1–S3).

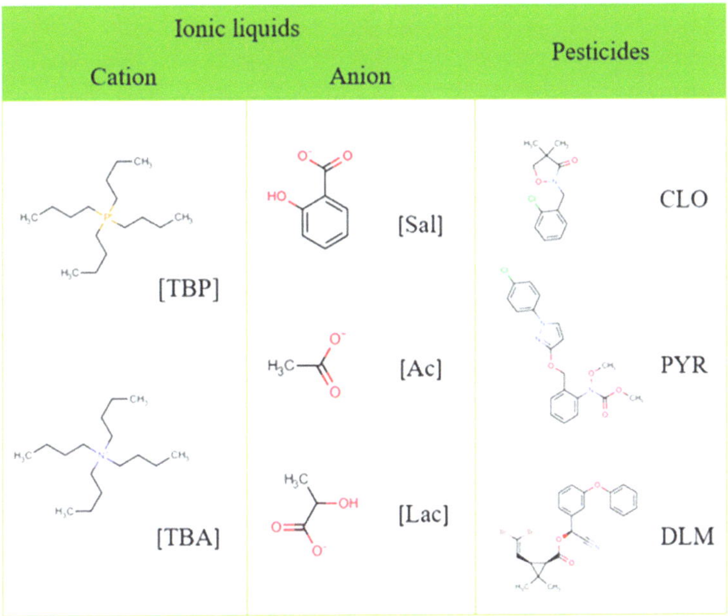

Figure 1. Chemical structures and abbreviations of studied ILs and pesticides.

2.2. Determination of the Phase Diagrams and Tie-Lines

The liquid–liquid ternary phase diagrams of phosphonium ILs, ammonium sulfate, and water were obtained in previous work using the cloud point titration method at (25 ± 1) °C and atmospheric pressure (0.10 ± 0.01 MPa) [34]. In this study, phase diagrams of ammonium ILs with ammonium sulfate and water were determined using the same method. Briefly, this method involves the dropwise addition of a salt solution (~40 wt%) to a known mass of an ammonium IL aqueous solution (~70 wt%) until the mixture becomes cloudy, followed by the addition of water until the solution becomes clear. After each addition of salt or water, the solution was shaken using a vortex agitator at 2500 rpm (Reax Top, Heidolph), and masses were recorded using an analytical balance (KERN ADJ 210-4) with an uncertainty of 10^{-4} g. This procedure was repeated to obtain enough experimental data points to construct the binodal curve. The experimental obtained binodal curves were fitted by the Merchuk equation [36]:

$$Y = A \exp\left[\left(CX^{0.5}\right) - \left(BX^3\right)\right] \quad (1)$$

where Y and X are, respectively, the IL and salt weight percentages; A, B, and C are regression parameters.

Except for the [TBP][Sal]-based ATPS, each tie-line (TL) and tie-line length (TLL) were determined using the gravimetric method described by Merchuk et al. [36]. Mixtures within the biphasic region (corresponding to the mixture compositions used in the extraction experiments) were prepared by combining a specific amount of IL, salt, and water to reach a total system mass of 0.5 g. After vigorous vortex stirring and allowing the systems to equilibrate for a minimum of 12 h at (25 ± 1) °C, ATPSs were centrifuged for 5 min at 10,000 rpm (LLG-uniCFUGE5). The phases were then carefully separated and weighed, with all calculations performed as suggested by the literature [36]. For the [TBP][Sal]-based ATPS, tie-lines were determined analytically, as the equations and mass balance approaches proposed by Merchuk et al. were inadequate for describing the solubility data. Mixtures within the biphasic region were prepared by adding known amounts of [TBP][Sal], salt, and water to microtubes (total mass of 1 g), followed by vigorous vortex mixing and equilibrating for 12 h at (25 ± 1) °C. After centrifugation and phase separation, approximately 100 mg of each phase was weighed and diluted with water as needed to ensure proper detection. The content of [TBP][Sal] in the top and bottom phases was determined by UV spectroscopy using an LLG-uniSPEC2 spectrophotometer at wavelengths of 296 nm. Quantification was based on calibration curves ($R^2 > 0.9996$) within the concentration range of 1–50 mg L^{-1}. Water content was determined by drying each phase at 70 °C under constant stirring until a constant weight was achieved. The amount of $(NH_4)_2SO_4$ was determined by difference based on mass balance.

2.3. Screening of IL-Based ATPSs for the Extraction of Pesticides

Considering the previously established liquid–liquid ternary phase diagrams, compositions of (a) 3 wt% of $(NH_4)_2SO_4$ + 20 wt% of IL + 77 wt% of H_2O + 5 μL of pesticide solution for [TBP][Sal] and [TBA][Sal], and (b) 23 wt% of $(NH_4)_2SO_4$ + 20 wt% of IL + 57 wt% of H_2O + 5 μL of pesticide solution for [TBP][Ac], [TBA][Ac], [TBP][Lac], and [TBA][Lac]-based ATPSs, were chosen for pesticide extraction. These compositions correspond to a TLL of approximately 50–60. The pesticide solution in methanol (~2.5 g/L of each compound) was used for all extraction experiments. Pre-weighted ATPS constituents (total mass of 0.5 g) were stirred and left to equilibrate for 2 h at 25 °C. After centrifugation at 6000 rpm for 5 min, the phases were separated, followed by measurement of their weight and volume. Fifty microliters of each phase were diluted with 40 vol% ethanol to a final volume of 500 μL and filtered, and the pesticide concentration was determined by ultra-performance liquid chromatography using the Waters ACQUITY UPLC system, with a PDA detector and BEH C18 column (1.7 μm, 100 mm × 2.1 mm, Waters). The mobile phase used was a mixture of water with 10 vol% acetonitrile (A) and 100 vol% acetonitrile (B). The elution gradient started at 60% (v/v) acetonitrile and was held constant for 6 min. Then, the proportion of acetonitrile was raised from 60% to 80% (v/v) over 3 min and maintained for 4 min. Subsequently, the acetonitrile proportion decreased to 60% (v/v) within 1 min and kept at this composition for an additional 1 min before the next injection. It was subsequently decreased back to 60% (v/v) within 1 min and held constant for 1 additional minute before the next injection. Detection and quantification were carried out at 220 nm for CLO and DLM and 280 nm for PYR. The elution was performed at a flow rate of 0.2 mL/min with a sample injection volume of 5 μL. The detailed experimental parameters for the UPLC method can be found in Table S1 of the Supplementary Materials. Extraction efficiencies (EE, %) of the targeted pesticides toward the IL-rich phase are calculated as

$$\text{EE (\%)} = \frac{[\text{Pest}]_{\text{IL-rich phase}} \cdot V_{\text{IL-rich phase}}}{[\text{Pest}]_{\text{IL-rich phase}} \cdot V_{\text{IL-rich phase}} + [\text{Pest}]_{\text{salt-rich phase}} \cdot V_{\text{salt-rich phase}}} \cdot 100 \quad (2)$$

where $[\text{Pest}]_{\text{IL-rich phases}}$ is the concentration of the pesticide in the IL-rich phase, $[\text{Pest}]_{\text{salt-rich phase}}$ is the concentration of pesticide in the salt-rich phase, and $V_{\text{IL-rich phase}}$ and $V_{\text{salt-rich phase}}$ represent the volumes of the IL- and salt-rich phases, respectively.

2.4. Influence of Experimental Parameters on the Extraction of Pesticides

Following the initial extraction of pesticides, the [TBP][Sal] system was selected for further in-depth investigations based on the quantities of ATPS components and the observed extraction efficiencies. These further studies focused on the influence of key parameters, such as tie-line length, mixture composition along the same tie line, temperature, and pH. To explore the effect of temperature and pH, the system consisting of 3 wt% salt and 20 wt% IL was analyzed at three different temperatures (25, 35, and 45 °C) and three pH values (3, 5, and 6). The pH values were adjusted using 1M sulfuric acid and measured with an Orion-Star A111 pH Benchtop Meter Kit at 25 ± 1 °C, with an accuracy of ±0.01 pH units. To investigate the effect of TLL on pesticide extraction efficiency, ATPS mixtures were prepared with varying concentrations of $(NH_4)_2SO_4$ while maintaining a constant amount of [TBP][Sal]. The extraction efficiency was evaluated by varying the initial composition of the ATPS mixtures along the characterized tie line. Once the ATPS constituents were prepared (as outlined in Table 1), all systems were subjected to the procedures described in Chapter 2.3. The preconcentration factor (PF) was calculated (Equation (3)) by dividing the concentration of each pesticide in the IL-rich phase after preconcentration and $[\text{Pest}]_{\text{IL-rich phase}}$ by its initial concentration in the system, $[\text{Pest}]_0$.

$$PF = \frac{[\text{Pest}]_{\text{IL-rich phase}}}{[\text{Pest}]_0} \quad (3)$$

Table 1. Selected compositions (wt%), temperature (°C), and pH for evaluating pesticide extraction in ATPSs: $(NH_4)_2SO_4$ (X) + [TBP][Sal] (Y) + H_2O (Z) + 5 μL pesticide solution.

	100·X	100·Y	100·Z	T	pH
		Temperature influence			
T25	3	20	77	25	6
T35	3	20	77	35	6
T40	3	20	77	45	6
		pH influence			
pH3	3	20	77	25	3
pH5	3	20	77	25	5
pH6	3	20	77	25	6
		Tie-line length influence			
TL1	3	20	77	25	6
TL2	10	20	70	25	6
TL3	15	20	65	25	6
		Phase ratio influence			
TL3	15	20	65	25	6
TL3b	17	10	73	25	6
TL3c	18	5	77	25	6

2.5. Preconcentration of Pesticides Using IL-Based ATPS

After conducting a comprehensive screening of IL-based ATPSs and evaluating various factors influencing the extraction of pesticides (CLO, PYR, and DLM), the system comprising 8 wt% [TBP][Sal] + 27 wt% $(NH_4)_2SO_4$ + 65 wt% H_2O at 25 °C was selected for the preconcentration of pesticides in food samples. Fresh strawberries, sourced from a local market in Belgrade, were homogenized into a uniform puree using a blender. One gram of the blended strawberry sample was spiked with 5 µL of a standard pesticide solution, with each pesticide (CLO, PYR, and DLM) at a concentration of approximately 2.5 g/L. The spiked sample was left for 2 h to ensure thorough absorption of the pesticides into the strawberry matrix, after which a 12% aqueous solution of [TBP][Sal] was added to achieve a total volume of 5 mL. This mixture was agitated on a rotary disk for 10 min and then filtered. In order to minimize salt usage, a smaller biphasic system was prepared by mixing approximately 1.6 g of the filtered sample solution (supernatant) with 0.6 g of ammonium sulfate to achieve a selected biphasic system composition. The mixture was thoroughly mixed and allowed to equilibrate for 30 min at 25 °C to promote effective phase separation. Once equilibrium was achieved, the mixture was centrifuged at 5000 rpm for 3 min to separate the phases. The salt-rich phase was carefully removed, and 50 µL was diluted with 40 vol% of ethanol to obtain a final volume of 500 µL. The IL-rich phase was further centrifuged at 15,000 rpm for an additional 2 min to ensure complete separation from any remaining salt-rich phase and interphase (third phase). Finally, 50 µL of the IL-rich phase was carefully extracted with a Hamilton syringe, followed by a dilution with 40% ethanol. After dilution, the phases were filtered through syringe filters (0.45 µm), and the samples were quantified using UPLC. Figure 2 shows the schematic steps in pesticide preconcentration.

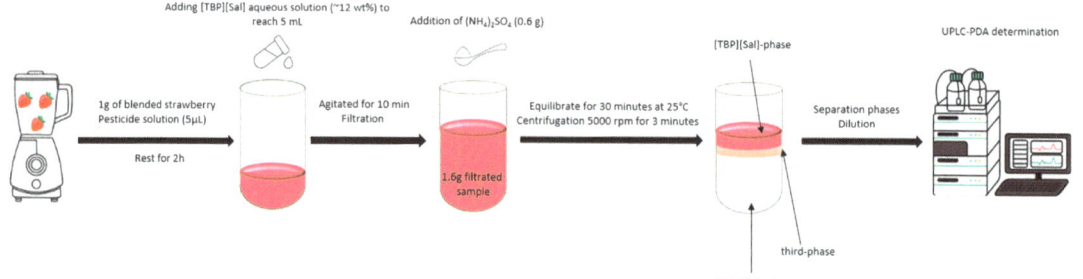

Figure 2. Preconcentration procedure based on IL-salt ATPS for pesticide determination in strawberry samples.

2.6. Environmental and Analytical Assessment of ATPS-Based Preconcentration Using Green Metrics

The greenness assessment of the proposed ATPS-UPLC-PDA method was conducted using the AGREEprep and AGREE metrics. AGREEprep is based on 10 principles of Green Sample Preparation (GSP), while AGREE follows the 12 principles of Green Analytical Chemistry (GAC). The software for applying these metrics is freely available and cited in studies [32,33].

3. Results and Discussion

3.1. Liquid–Liquid Phase Diagrams of Phosphonium and Ammonium-Based ATPSs

The phase diagram analysis provided insight into the required quantities of each ATPS component needed to achieve two-phase separation, represented by the binodal curve, which outlines the boundaries between monophasic and biphasic regions. By comparing the binodal curves for different ionic liquids combined with the same salt, the effects of specific IL cations and anions on phase formation can be better understood. In this study, ATPSs were designed using phosphonium- and ammonium-based ILs with ammonium sulfate, a salt chosen for its strong salting-out capability and mild acidity, which aligns well with the pH of most fruit samples, promoting analyte stability and efficient partitioning.

The ternary phase diagrams of phosphonium ILs (tetrabutylphosphonium acetate, tetrabutylphosphonium lactate, and tetrabutylphosphonium salicylate) with ammonium sulfate were previously determined [34], while the phase diagrams of ammonium-based ILs (tetrabutylammonium acetate, tetrabutylammonium lactate, and tetrabutylammonium salicylate) combined with the same salt were established in this work. Experimental solubility data for each ATPS, obtained through cloud-point titration, are available in the Supplementary Materials (Table S2), alongside the parameters of the Merchuk equation (A, B, and C), their standard errors (σ), and coefficients of determination (R^2) (Table S3). Typically, these systems show an IL-rich top phase and a salt-enriched bottom phase. However, for the [TBP][Sal]-based system, when the mixture contains 20 wt% [TBP][Sal] and 3 wt% ammonium sulfate, a phase inversion occurs, producing an IL-rich bottom phase and a salt-rich top phase. In other compositions, the system behaves conventionally, with the IL forming the upper layer and the salt-rich phase at the bottom. Due to the inability to fit experimental data from the [TBP][Sal]-based ATPS with the Merchuk equation, tie lines for these systems were determined analytically. The phase compositions at equilibrium for each extraction experiment were assessed through gravimetric or analytical methods, with results presented in Tables S4 and S5 of the Supplementary Materials.

All tested phosphonium- and ammonium-based ILs underwent liquid–liquid demixing in the presence of ammonium sulfate. Despite the overlapping same binodal curves (Figure 3), the efficacy of these ILs in forming ATPSs at 20 wt% (the concentration used in extraction studies) follows the order: [TBP][Sal] > [TBA][Sal] > [TBP][Ac] > [TBP][Lac] > [TBA][Lac] > [TBA][Ac]. Among these, [TBP][Sal] shows the highest capacity to promote two-phase formation with ammonium sulfate, while [TBA][Ac] demonstrates the lowest. Systems with [TBP][Sal] and [TBA][Sal] displayed substantial biphasic regions with binodal curves close to the origin; however, [TBP][Sal] required significantly lower concentrations to induce ATPS formation than [TBA][Sal]. The same trend was observed when lactate or acetate anions were used instead of salicylate, with ILs of the same anion following the order: [TBP] > [TBA]. This behavior is due to the higher hydrophobicity of the phosphonium cation, which promotes phase separation more effectively than the ammonium cation [37]. The same behavior of [TBP][Sal] and [TBA][Sal] was observed when using citrate salt as the salting-out agent [28].

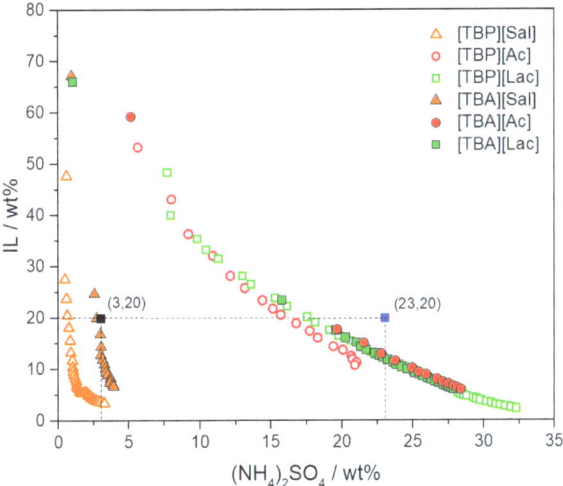

Figure 3. Ternary phase diagrams composed of [TBP]-IL + $(NH_4)_2SO_4$ + H_2O at 25 °C and p = 0.1 MPa (open symbols) [34] and [TBA]-IL + $(NH_4)_2SO_4$ + H_2O at 25 °C and p = 0.1 MPa (full symbols). Compositions of ATPS mixture selected for TLs and extraction study (■, [TBP][Sal] or [TBA][Sal]-based ATPS, and ■, [TBP][Ac] or [TBA][Ac] or [TBP][Lac] or [TBA][Lac]-based ATPS).

The effect of anions on ATPS formation was assessed using ILs with the same cation but different anions. For ILs with the [TBP] cation, the ability to form ATPSs decreased in the following order: [Sal] > [Ac] > [Lac], aligning with their octanol-water partition coefficients (logK_{ow}) of 1.977, −0.223, and −0.472, respectively [38]. For ammonium ILs, the trend in efficiency followed [Sal] > [Lac] > [Ac]. This discrepancy can be attributed to the stronger cation–anion interactions in TBA-based ILs, as confirmed by the solid state of [TBA][Lac] at room temperature, in contrast to the liquid state of [TBP][Lac]. Previously, it has been shown that ion-pair binding energy between the cation and anion in ILs correlates well with melting point [39]. The higher degree of molecular packing and binding energy in [TBA][Lac] reduces interactions with water and enhances phase-separation tendencies in ATPSs. Notably, [TBP][Sal] formed an ATPS more easily than other investigated ILs, requiring exceptionally low sulfate concentrations (<3%) to induce phase separation. This underscores the importance of the salicylate anion hydrophobicity, as its high logK_{ow} value facilitates the formation of two-phase systems.

3.2. Screening of IL-Based ATPSs for the Extraction of Pesticides

The extraction capabilities of phosphonium and ammonium-based ATPSs for three pesticides (CLO, PYR, and DLM) from aqueous solutions were evaluated. Optimal compositions of ternary ATPS mixtures (illustrated in Figure 3) were selected as follows: (a) 3 wt% of $(NH_4)_2SO_4$ + 20 wt% of IL + 77 wt% of H_2O + 5 µL of pesticide solution for [TBP][Sal] and [TBA][Sal], and (b) 23 wt% of $(NH_4)_2SO_4$ + 20 wt% of IL + 57 wt% of H_2O + 5 µL of pesticide solution for [TBP][Ac], [TBA][Ac], [TBP][Lac], and [TBA][Lac]-based ATPSs. To maintain a consistent comparison, a tie line length of 50–60 with a constant wt% of IL was targeted for each mixture. The extraction efficiencies for CLO, PYR, and DLM with respective standard deviations are shown in Figure 4. At the pH values of the ATPS mixtures (6–7), all pesticides are present in neutral forms, meaning that their hydrophobicity largely determines their partitioning behavior, as reflected in the extraction efficiency trend (EE_{CLO} < EE_{PYR} < EE_{DLM}), aligning with the pesticides logK_{ow} values (2.93 < 4.70 < 5.74) [38].

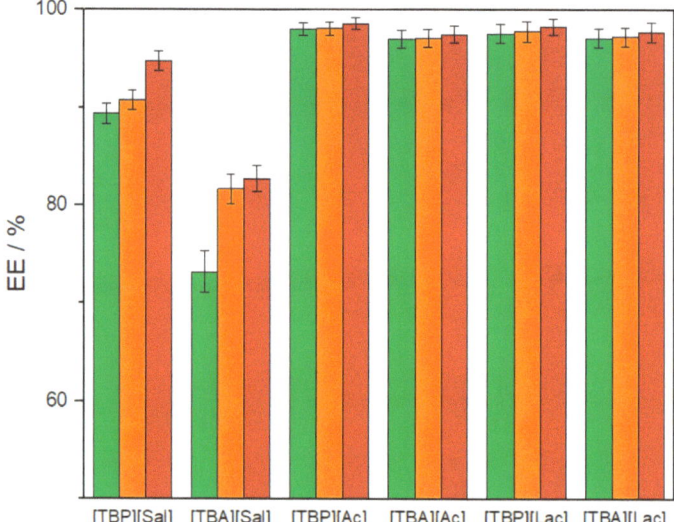

Figure 4. Extraction efficiencies of CLO (green bar), PYR (orange bar) and DLM (red bar) in ATPSs composed of: 3 wt% of $(NH_4)_2SO_4$ + 20 wt% of IL + 77 wt% of H_2O for [TBP][Sal] and [TBA][Sal], and 23 wt% of $(NH_4)_2SO_4$ + 20 wt% of IL + 57 wt% of H_2O for [TBP][Ac], [TBA][Ac], [TBP][Lac], and [TBA][Lac] at 25 °C and p = 0.1 MPa.

The highest EE for each pesticide was observed in the $(NH_4)_2SO_4/[TBP][Ac]$ ATPS: $98.01 \pm 0.66\%$ for CLO, $98.07 \pm 0.63\%$ for PYR, and $98.59 \pm 0.59\%$ for DLM. In contrast, the $(NH_4)_2SO_4/[TBA][Sal]$ ATPS had slightly lower efficiency ($EE_{CLO} = 73.23 \pm 2.09\%$, $EE_{PYR} = 81.7 \pm 1.5\%$, and $EE_{DLM} = 82.75 \pm 1.33\%$). However, the trend in EE for each pesticide across ATPS compositions—[TBP][Ac] > [TBP][Lac] > [TBA][Lac] > [TBA][Ac] > [TBP][Sal] > [TBA][Sal]—did not fully follow the order of ATPS formation. This discrepancy can be attributed to differing tie-line lengths; systems with longer TLs tend to have higher EEs and preconcentration factors, along with lower cross-contamination between phases. For instance, in systems with long TLs, the IL-rich phase contains minimal sulfates, while the salt-rich phase primarily consists of water and salt. The [TBA][Sal]-ATPS system exhibits lower extraction efficiency, with a TLL of 32, while others range from 50 to 60. Table S4 illustrates that the $[TBP][Ac]/(NH_4)_2SO_4$ system contains less than 1% IL in the salt-rich phase, resulting in the highest extraction efficiency across all pesticides. Conversely, the [TBA][Sal]-based ATPS has the lowest extraction efficiency, with over 12% IL in the salt-rich phase. A high concentration of IL in the salt-rich phase can reduce the salting-out effect, weakening the ionic strength and decreasing the phase's ability to exclude analytes. Considering the component quantities, phase volume ratios, and extraction efficiencies, the [TBP][Sal]/ammonium sulfate ATPS presents itself as a promising extraction medium for CLO, PYR, and DLM from food matrices. Furthermore, [TBP][Sal] exhibits low cytotoxicity [28] with minimal loss or cross-contamination into the salt-rich phase.

3.3. Optimization of Experimental Parameters for Extraction of Pesticides

After conducting the initial screening of ATPSs for extracting clomazone, pyraclostrobin, and deltamethrin, the results revealed that the most effective candidate for extracting all three pesticides is the $[TBP][Ac] + (NH_4)_2SO_4 + H_2O$ system, exhibiting a high extraction efficiency exceeding 98%. This efficiency was achieved using a system comprising 20 wt% IL and 23 wt% salt. Conversely, utilizing 20 wt% [TBP][Sal] with significantly lower salt content (3 wt%) yielded slightly lower efficiencies ranging from 90 to 94%. Given the minimal differences in extraction efficiencies among these systems and the substantial variation in required constituent amounts, the $[TBP][Sal]/(NH_4)_2SO_4$ system was chosen for further investigation. Several operational parameters, including temperature, pH, TLL, and mixture composition along the same TL (phase ratio), were investigated for their effect on the extraction of pesticides. This approach enables the identification of optimal conditions for the extraction and preconcentration process while utilizing fewer components, thereby reducing costs, energy consumption, and environmental impact.

Temperature and pH influence. The influence of the temperature (25, 35, and 45 °C) on EE values of CLO, PYR, and DLM are shown in Figure 5a. At 35 °C, extraction efficiencies for all three pesticides increase by approximately 7% compared to those at 25 °C, while a smaller increase of 5% is observed at 45 °C. This suggests a slightly positive effect of temperature on extraction efficiency. Given the slight dependence of EEs on temperature changes ($89.39 \pm 0.80\% > EE_{CLO} > 96.22 \pm 0.82\%$, $90.79 \pm 0.95\% > EE_{PYR} > 96.62 \pm 0.95\%$ and $94.77 \pm 0.79\% > EE_{DLM} > 98.14 \pm 0.79\%$) and aiming to minimize energy and the operational costs of the process, the room temperature, i.e., 25 °C, can be taken as optimal.

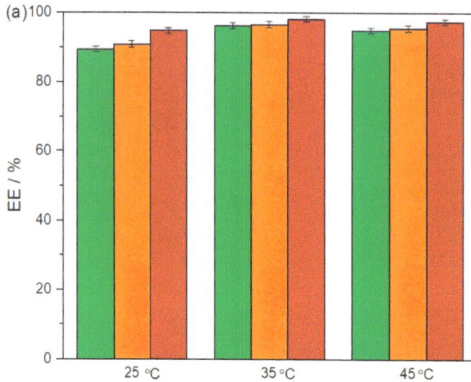
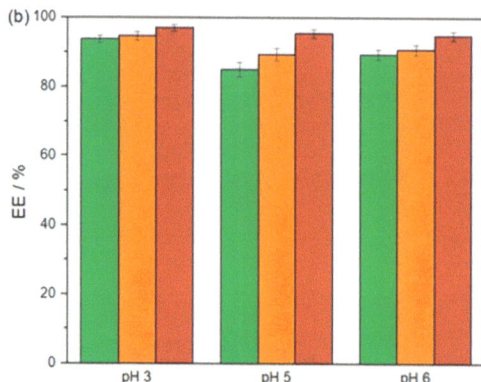

Figure 5. Extraction efficiencies of CLO (green bar), PYR (orange bar) and DLM (red bar) in ATPSs at $p = 0.1$ MPa composed of 3 wt% of $(NH_4)_2SO_4$ + 20 wt% of [TBP][Sal] + 77 wt% of H_2O (**a**) on 25, 35 and 45 °C and (**b**) on 25 °C at pH 3, pH 5, and pH 6.

The pH effect on extraction pesticides was examined for systems without pH adjustment (pH ≈ 6) and for systems adjusted to pH ≈ 5 and ≈3 using 1M sulfuric acid. This pH range aligns with the typical acidity levels found in most fruit juices, such as strawberries, which have a pH of 3.54 ± 0.01. PYR and DLM are in neutral form in the whole pH working range, while clomazone is neutral at pH > 4 and at pH = 3 is 50% negatively charged [38]. However, the results (85.01 ± 2.13% > EE_{CLO} > 93.69 ± 1.06%, 89.33 ± 1.78% > EE_{PYR} > 94.66 ± 1.25%, and 95.58 ± 1.20% > EE_{DLM} > 97.00 ± 0.96%) indicate that pH has a small effect on EE parameter of each compound (Figure 5b). Based on the minimal influence of pH on extraction efficiency observed in this study, future experiments will use an unadjusted ATPS without acid addition. This approach reduces the number of components involved and eliminates the need for acidic modifiers, making the process more environmentally friendly and aligned with green chemistry principles. Eliminating pH adjustments in the extraction process enhances sustainability and simplifies the method, making it more feasible for practical applications.

TLL and phase ratio influence. The system formed by [TBP][Sal] and $(NH_4)_2SO_4$, as mentioned before, exhibits a wide biphasic region, resulting in a long tie-line length. A long TLL reduces cross-contamination between phases enriched with opposite components (the IL phase contains very little ammonium salt, while the salt phase contains only salt and water). Adjusting the mixture compositions along the same tie line allows for the tailoring of the volumes of the coexisting phases without altering their composition and enables higher preconcentration factors [40].

The impact of TLL on pesticide extraction was examined by varying the concentrations of $(NH_4)_2SO_4$ (3.28, 10.07, and 14.88 wt%) while maintaining [TBP][Sal] at ~20 wt%. Figure 6 demonstrates how the extraction of CLO, PYR, and DLM is influenced by increasing TLL values: 57.17, 65.94, and 66.96. By increasing the TLL under selected conditions (pH ≈ 6 and T = 25 °C), the extraction efficiencies increase from 89.39 ± 1.45% to 98.94 ± 1.26% for CLO, from 90.79 ± 1.33% to 99.12 ± 1.04% for PYR, and from 94.77 ± 1.55% to 99.48 ± 0.98% for DLM. Given that the longest TLL (TL3, ~67) resulted in over 98% extraction efficiency for all tested pesticides and only 0.01% ionic liquid remained in the salt-rich phase (Table S5), this condition was selected for further investigation.

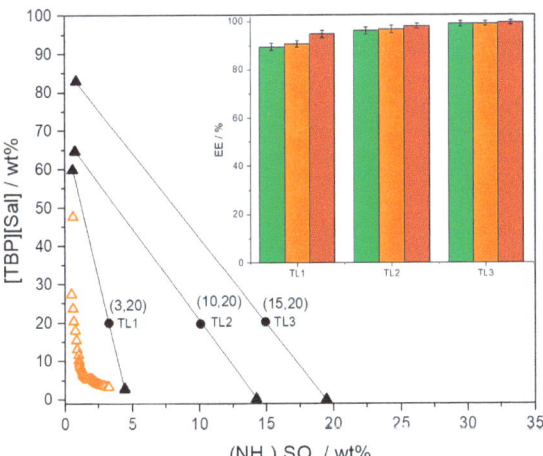

Figure 6. Effect of different TLs on extraction efficiencies of CLO (green bar), PYR (orange bar), and DLM (red bar) in ATPSs composed of $(NH_4)_2SO_4$ + [TBP][Sal] + H_2O on 25 °C at 25 °C and p = 0.1 MPa. Binodal curve data (△), TL (▲), and mixture compositions (●).

Several systems with varying $V_{IL\text{-phase}}/V_{salt\text{-phase}}$ ratios were investigated along TL3 (~67 in length), and the obtained extraction efficiencies for pesticides are shown in Figure 7. The EEs decreased from 98.94 ± 1.26% to 94.91 ± 1.66% for CLO, from 99.12 ± 1.04% to 95.1.60 ± 1.40% for PYR and from 99.48 ± 0.98% to 97.8 ± 1.4 for DLM, as ratios of phases decline from 0.36 to 0.13. Preconcentration factors of 3.3- to 11.5-fold for CLO, 3.7- to 12.3-fold for PYR, and 3.0- to 14.6-fold for DLM were achieved, demonstrating the potential to concentrate pesticides in the IL-rich phase by up to 14-fold with only a 4% decrease in extraction efficiency. These initial findings suggest that the [TBP][Sal]-based ATPS offers a promising method for pesticide preconcentration, facilitating their detection and quantification by UPLC.

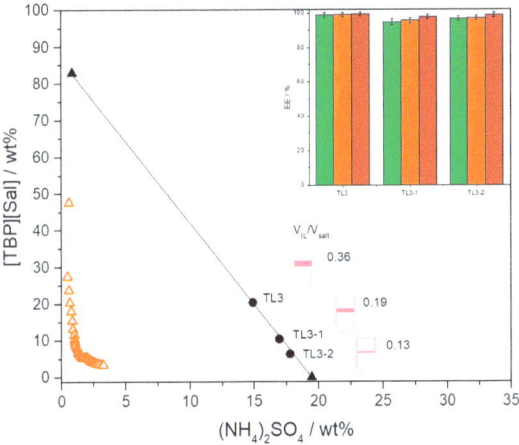

Figure 7. Effect of different phase ratio (VIL/Vsalt) on extraction efficiencies of CLO (green bar), PYR (orange bar), and DLM (red bar) in ATPSs composed of (NH4)$_2$SO$_4$ + [TBP][Sal] + H$_2$O on 25 °C at 25 °C and p = 0.1 MPa. Binodal curve data (△), TL (▲), and mixture compositions (●).

3.4. Preconcentration of Pesticides Using IL-Based ATPSs

Pesticide monitoring, particularly in food samples where concentrations often fall below the MRLs, places greater importance on preconcentration than on extraction efficiency. Achieving adequate preconcentration is essential for detecting even trace levels of pesticides, as analytical devices may struggle to identify them at such low concentrations without a significant preconcentration step [16]. By increasing the preconcentration factor, achieved through maximizing the ratio of pesticide concentration in the IL-rich phase to its initial concentration, the method significantly enhances the preconcentration of pesticides. This is largely due to the substantial volume difference between the IL-rich phase and the original sample volume, allowing trace amounts of pesticides to be more readily detected and quantified. Higher volume ratios are attained by selecting a biphasic system composition located deep within the biphasic region [24,26]. This composition requires minimal ionic liquid for phase formation but results in a higher salt concentration. As a result, a composition of 8% IL, 27% salt, and 65% water, with a volume ratio of 0.07 (0.125 mL/1.800 mL) at 25 °C, was chosen for the preconcentration of pesticides from the real sample. Initially, 12 wt% aqueous solution of [TBP][Sal] was introduced to the spiked strawberry, followed by agitation for 10 min to promote interaction with the target compounds. The mixture was then filtered, and ammonium sulfate was added to the supernatant to induce ATPS formation at 25 °C. Once equilibrium was reached, the mixture was centrifuged, resulting in three distinct phases: a top IL-rich phase, a bottom salt-rich phase, and a third phase (TP) (Figure 8).

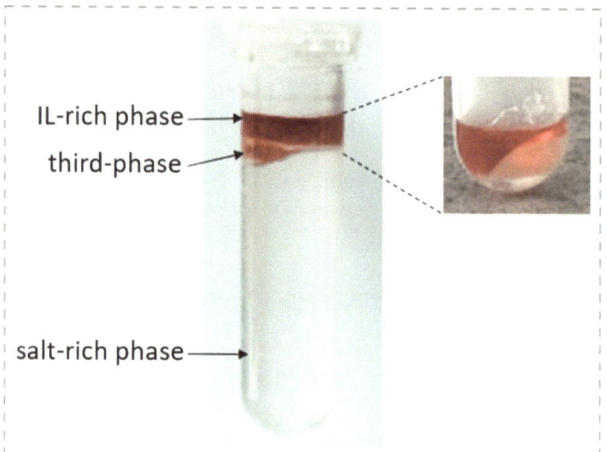

Figure 8. A photograph of the [TBP][Sal]/$(NH_4)_2SO_4$ ATPS used for pesticide preconcentration in a strawberry sample, showing three distinct phases: IL-rich phase, third-phase, and salt-rich phase. The inset highlights the IL-rich phase and the third phase after the salt-rich phase has been removed.

The third phase only appeared after the addition of salt, whereas a supernatant without salt did not form an interphase. This suggests that components from the strawberry matrix, such as fibers and carbohydrates, were salted out, leading to the formation of the third phase. Despite the presence of the third phase, it did not interfere with the experimental sampling, as only 50 µL of phases were required for analysis. Initially, the entire salt-rich bottom phase was withdrawn using a syringe, followed by further centrifugation of the top IL-rich phase along with the interphase layer. After this additional step, 50 µL of the top IL-rich phase was collected, diluted, and analyzed using UPLC to quantify pesticide residue. The UPLC analysis showed that pesticides were extracted into the IL-rich phase, with extraction efficiencies of 72.20 ± 1.98% for CLO, 74.24 ± 1.89% for PYR, and 82.40 ± 1.56% for DLM. The slightly lower extraction efficiencies can be attributed to the complexity of the real sample, as well as the existence of the third phase, which likely retained some pesticide residues.

Figure 9 presents chromatograms at 220 nm and 280 nm of the spiked strawberry sample, both without ATPS pre-treatment and after preconcentration with ATPS. In the post-preconcentration chromatogram, the pesticide peaks are clearly visible and quantifiable, demonstrating a 16-fold preconcentration in a single-step process. Given that the LOD values for UPLC are 0.1 mg/L for CLO and PYR and 0.05 mg/L for DLM (Table S1), this method enables the quantification of pesticide-contaminated samples at concentrations around the µg/L range. Considering the established MRLs for strawberries (CLO 0.01 mg/L, PYR 1.5 mg/L, and DLM 0.2 mg/L) [41], the proposed aqueous two-phase extraction method, using tetrabutylphosphonium salicylate, proves effective for monitoring pesticide residues in strawberries. The analysis of the chromatograms reveals distinct peaks for each pesticide without any interference from other compounds found in strawberries, indicating that both cleaning and preconcentration were effectively accomplished in a single step. This clarity in the chromatographic data facilitates accurate quantification of clomazone, pyraclostrobin, and deltamethrin individually. The findings underscore the efficacy of the developed one-step clean-up and preconcentration method utilizing the [TBP][Sal]-based aqueous two-phase system. Additionally, the IL-ATPS approach could be expanded to target a broader range of pesticide classes, including those with varying polarities or solubilities. By adjusting ionic liquid or salt combinations, the system can be tailored to selectively extract diverse pesticides, enhancing its application for comprehensive food safety monitoring. In addition to phosphonium- and ammonium-based ionic liquids, cholinium-based ionic liquids present a compelling alternative due to their low toxicity and biodegradability. Their use could further improve the method's green chemistry profile and broaden its validation for other pesticides and matrices. The flexibility and adaptability of this method, combined with the exploration of different ionic liquid types, establish it as a promising solution for addressing food safety challenges across diverse and complex matrices. Future research should prioritize testing various types of ionic liquids and optimizing the system for more challenging matrices, such as grains, high-fiber foods, and root vegetables, to fully validate its robustness and expand its applicability.

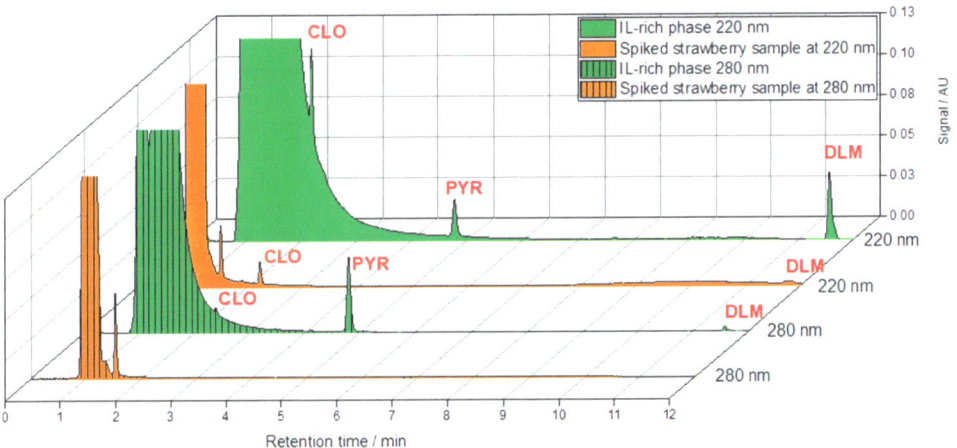

Figure 9. Chromatograms at 220 nm and 280 nm of the spiked strawberry sample with clomazone (CLO), pyraclostrobin (PYR), and deltamethrin (DLM) without ATPS pre-treatment and chromatograms at 220 nm and 280 nm of the IL-rich phase with clomazone (CLO), pyraclostrobin (PYR) and deltamethrin (DLM) after preconcentration with ATPSs.

3.5. Environmental and Analytical Assessment of IL-Based ATPSs for Preconcentration of Pesticides Using Green Metrics

The proposed method for preconcentrating pesticides using aqueous two-phase systems with ionic liquids aligns with several principles of green chemistry, such as eliminating the need for derivatization, minimizing analytical waste, and utilizing low-toxicity reagents like ionic liquids [42]. Notably, Jocić et al. [28] demonstrated that [TBP][Sal] concentrations as low as 0.01% in the bottom aqueous phase exhibited minimal cytotoxicity, showing no significant adverse effects on MRC-5 cells. This reinforces the environmental and safety advantages of using [TBP][Sal] in pesticide preconcentration methods.

To assess the environmental sustainability of the proposed ATPS-UPLC-PDA method, both the AGREEprep and AGREE metrics were applied. AGREEprep focuses specifically on sample preparation, evaluating compliance with the 10 Green Sample Preparation principles, while AGREE examines the entire analytical workflow through the 12 Green Analytical Chemistry principles. The evaluations use a color-coded system (green for high compliance to red for low) and provide an overall score between 0 and 1, where a higher score indicates better environmental performance [32,33].

In this study, as shown in Figure 10, AGREEprep scored 0.68, with most parameters aligning well with Green Analytical Chemistry principles. According to AGREEprep, the greenest features of the proposed IL-ATPS method are the use of non-hazardous materials (criterium 2), as this method involves only a low-cytotoxicity IL and a safe inorganic salt as reagents, contributing to a green rating for operator safety (criterium 10). Additionally, low energy consumption (criterium 8) is achieved using a disk rotator, vortex, and centrifugation, requiring just 10 Wh of energy input. Less eco-friendly aspects include the 1 g sample size (criterium 5) and a throughput of 25 samples per hour (criterium 6). However, a few areas—criterium 1 (ex situ preparation), criterium 4 (5.6 g waste), criterium 7 (manual preparation with six steps or fewer), and criterium 9 (UPLC use)—were flagged as less eco-friendly. While the AGREEprep evaluates only the sample preparation process, AGREE extends its analysis to include the entire workflow, covering both sample preparation and detection. With an AGREE score of 0.55, the method demonstrates notable green chemistry features, such as 1 g of sample (criterium 2), the absence of a derivatization step (criterium 6), and analysis of three analytes in a single run with a throughput of five samples per hour (criterium 8). The UPLC, consuming only 0.217 kWh (criterium 9), contributes to its eco-friendly profile, alongside operator safety (criterium 12), with highly flammable acetonitrile being the only hazardous solvent used. However, some aspects of the ATPS-UPLC-PDA method are less green: manual preparation of miniaturized samples (criterium 5), the use of bio-based reagents (criterium 10), and the small quantities of acetonitrile (1.2 mL) in the mobile phase (criterium 11). Additionally, the method's less green features include six or more distinct steps in sample preparation (criterium 4) and the generation of 8.6 g of waste (criterium 7). Similar to the AGREEprep method, the least green steps are the multi-step sample pre-treatment (criterium 1) and off-line measurement (criterium 3). The UPLC instrumentation, selected for its low energy requirements and minimized use of ACN, contributes significantly to the method's eco-friendliness. Furthermore, UPLC generally requires smaller volumes of mobile phase compared to HPLC, making it a greener choice in this context. This reduced solvent usage aligns well with sustainable practices, even though the method includes off-line measurements. Additionally, as previously noted, ionic liquid-based samples are particularly suited to liquid chromatography due to their low melting points, which prevent volatility issues common in gas chromatography. These factors together strengthen the environmental profile of the method by balancing resource efficiency with practical analytical performance.

Figure 10. The results of AGREEprep and AGREE assessment of procedure for preconcentration of pesticides using [TBP][Sal]/(NH$_4$)$_2$SO$_4$ ATPS without UPLC (**left**) and with UPLC (**right**).

For comparison, Figure S4 in Supplementary Materials displays the AGREEprep and AGREE assessment diagrams applied to a study reported in the literature [9]. A QuEChERS-based sample preparation method combined with ultra-performance liquid chromatography–tandem mass spectrometry (UPLC-MS/MS) was established for the determination of 43 pesticide residues in strawberries. Optimized clean-up sorbents of 50 mg PSA, 50 mg C18, and 150 mg MgSO$_4$ demonstrated the best performance in dispersive solid-phase extraction (d-SPE). The proposed ATPS-UPLC-PDA method in this study achieved higher green scores in both AGREE and AGREEprep compared to the QuEChERS-UPLC-MS/MS method, scoring 0.68 and 0.55, respectively, while the method from the literature scored only 0.35 and 0.3. These differences are primarily due to the QuEChERS method's reliance on hazardous chemicals, such as 10 mL of highly flammable organic solvents (ACN) as extractant, a larger sample size, and consequently more waste generated. The findings indicate that IL-ATPS provides a greener alternative for pesticide preconcentration, with minimized solvent use, reduced waste, and alignment with sustainable analytical practices.

4. Conclusions

This study introduces a greener method for monitoring pesticides in food samples, focusing on strawberries, by addressing the challenge of low pesticide concentrations through an ionic liquid-based aqueous two-phase system (IL-ATPS) for preconcentration and UPLC-PDA detection. To achieve this, phase diagrams and initial extraction studies were conducted for clomazone, pyraclostrobin, and deltamethrin using phosphonium- and ammonium-ionic-based IL-ATPS with ammonium sulfate. Extraction efficiencies exceeded 90%, except for tetrabutylammonium salicylate-based ATPS, which yielded 73%. Among the tested ILs, the most hydrophobic—tetrabutylphosphonium salicylate—was the most efficiently salted out, achieving over 98% extraction efficiency for each pesticide after optimizing parameters such as temperature, pH, tie-line length, and system composition. The developed ATPS technology was successfully applied to strawberry samples, achieving preconcentration factors of up to 16-fold, reaching mg/L levels that allow for detection using UPLC-PDA.

The environmental assessments conducted with AGREEprep and AGREE tools confirmed the alignment of the ATPS-UPLC-PDA method for pesticide determination in strawberries with green chemistry principles. The AGREEprep score of 0.68 and AGREE score of 0.55 underscore the method's sustainable aspects, including low solvent use, minimized waste, the use of a low-cytotoxicity ionic liquid, and minimal use of hazardous reagents like acetonitrile. The inclusion of UPLC, selected for its low energy consumption and minimal ACN use, enhances the overall environmental profile in the AGREE evaluation, emphasizing the sustainability of both the sample preparation and detection stages. Overall, this IL-ATPS platform not only achieves efficient pesticide preconcentration from strawberries but also holds promise for broader application with reduced environmental impact. Future work will explore expanding this method to other pesticide types and matrices while further optimizing its green profile.

Supplementary Materials: The following supporting information can be downloaded at https://www.mdpi.com/article/10.3390/foods13244106/s1, Figures S1–S3: FTIR spectra of the synthesized ionic liquids; Figure S4: AGREEprep and AGREE assessment for QuEChERS-UPLC-MS/MS; Table S1: Parameters for the UPLC-PDA analytical method; Table S2: Experimental binodal mass fraction data for the salt (X) + IL (Y) + H$_2$O ATPS at 25 °C and at p = 0.1 MPa.; Table S3: Correlation parameters, standard deviations and determination coefficients (R2) of the salt (X) + IL (Y) + H$_2$O ATPS obtained by the Merchuk equation at 25 °C and at p = 0.1 MPa); Table S4: Experimental tie-lines data in percentage weight fraction for the ATPS composed of (salt (X) + IL (Y) + H$_2$O (Z)) at 25 °C, and volume ratios (Vr).; Table S5: Experimental tie-lines data in percentage weight fraction for the ATPS composed of (salt (X) + [TBP][Sal] (Y) + H$_2$O (Z)) at 25 °C, and volume ratios (Vr).

Author Contributions: Conceptualization, A.J., S.M. and A.D.; methodology, A.J., S.M., D.T., J.M. (Jasmina Mušović), S.Ž., J.M. (Jelena Milićević), and A.D.; validation, A.J. and S.M.; formal analysis, D.T. and J.M. (Jasmina Mušović); investigation, D.T., J.M. (Jasmina Mušović), J.M. (Jelena Milićević), and S.Ž.; writing—original draft preparation, A.J. and S.M.; writing—review and editing, A.J. and S.M.; visualization, A.J. and S.M.; supervision, A.D. All authors have read and agreed to the published version of the manuscript.

Funding: This work was supported by the Ministry of Science, Technological Development and Innovation of the Republic of Serbia (Contract number: 451-03-66/2024-03/200017).

Institutional Review Board Statement: Not applicable.

Informed Consent Statement: Not applicable.

Data Availability Statement: The original contributions presented in this study are included in the article/Supplementary Materials. Further inquiries can be directed to the corresponding author.

Conflicts of Interest: The authors declare no conflicts of interest.

References

1. Fróna, D.; Szenderák, J.; Harangi-Rákos, M. The challenge of feeding the world. *Sustainability* **2019**, *11*, 5816. [CrossRef]
2. Tudi, M.; Ruan, H.D.; Wang, L.; Lyu, J.; Sadler, R.; Connell, D.; Chu, C.; Development, A. Pesticide Application and Its Impact on the Environment. *Environ. Rsearch Public Health* **2021**, *18*, 1112. [CrossRef] [PubMed]
3. Ogaly, H.A.; Khalaf, A.A.; Ibrahim, M.A.; Galal, M.K.; Abd-Elsalam, R.M. Influence of green tea extract on oxidative damage and apoptosis induced by deltamethrin in rat brain, Neurotoxicol. *Teratol.* **2015**, *50*, 23. [CrossRef]
4. Liang, Y.; Wei, L.; Hu, J. Residues and dietary intake risk assessments of clomazone, fomesafen, haloxyfop-methyl and its metabolite haloxyfop in spring soybean field ecosystem. *Food Chem.* **2021**, *360*, 129921. [CrossRef]
5. Zhang, C.; Wang, J.; Zhang, S.; Zhu, L.; Du, Z.; Wang, J. Acute and subchronic toxicity of pyraclostrobin in zebrafish (*Danio rerio*). *Chemosphere* **2017**, *188*, 510. [CrossRef]
6. European Parliament and Council. Regulation (EC) No 396/2005 on Maximum Residue Levels of Pesticides in or on Food and Ffeed of Plant and Animal Origin and Aamending Council Directive 91/414/EEC. *Off. J. Eur. Union* **2005**, *L70*, 1–16.
7. Ambrus, Á.; Vásárhelyi, A.; Ripka, G.; Szemánné-Dobrik, H.; Szenczi-Cseh, J. Evaluation of the Results of Pesticide Residue Analysis in Food Sampled between 2017 and 2021. *Agrochemicals* **2023**, *2*, 409–435. [CrossRef]
8. Nguyen, T.T.; Rosello, C.; Bélanger, R.; Ratti, C. Fate of residual pesticides in Fruit and Vegetable Waste (FVW) processing. *Foods* **2020**, *9*, 1468. [CrossRef] [PubMed]
9. Wang, H.; Ping, H.; Liu, Q.; Han, P.; Guo, X. Determination of Pesticide Residues in Strawberries by Ultra-performance Liquid Chromatography-Tandem Mass Spectrometry. *Food Anal. Methods* **2022**, *15*, 85. [CrossRef]
10. Li, D.; Hamed, I.; Ashour, M.; Hammock, B.D. Pesticides in Tomato and Strawberry and Their Products from Markets. *Foods* **2023**, *12*, 1936. [CrossRef]
11. Shao, W.C.; Zang, Y.Y.; Ma, H.Y.; Ling, Y.; Kai, Z.P. Concentrations and Related Health Risk Assessment of Pesticides, Phthalates, and Heavy Metals in Strawberries from Shanghai, China. *J. Food Prot.* **2021**, *84*, 2116. [CrossRef]
12. Jafari, A.; Shoeibi, S.; Amini, M.; Amirahmadi, M.; Rastegar, H.; Ghaffarian, A.; Ghazi-Khansari, M. Monitoring dithiocarbamate fungicide residues in greenhouse and non-greenhouse tomatoes in Iran by HPLC-UV. *Food Addit. Contam. Part. B Surveill.* **2012**, *5*, 87. [CrossRef] [PubMed]

13. Chen, Z.; Li, Q.; Yang, T.; Zhang, Y.; He, M.; Zeng, H.; Mai, X.; Liu, Y.; Fan, H. Sequential extraction and enrichment of pesticide residues in Longan fruit by ultrasonic-assisted aqueous two-phase extraction linked to vortex-assisted dispersive liquid-liquid microextraction prior to high performance liquid chromatography analysis. *J. Chromatogr. A* **2020**, *1619*, 460929. [CrossRef]
14. Fernandes, V.C.; Domingues, V.F.; Mateus, N.; Delerue-Matos, C. Pesticide residues in Portuguese strawberries grown in 2009–2010 using integrated pest management and organic farming. *Environ. Sci. Pollut. Res* **2012**, *19*, 4184. [CrossRef] [PubMed]
15. da Silva Sousa, J.; Nascimento, H.O.D.; de Oliveira Gomes, H.; Nascimento, R.F.D. Pesticide residues in groundwater and surface water: Recent advances in solid-phase extraction and solid-phase microextraction sample preparation methods for multiclass analysis by gas chromatography-mass spectrometry. *Microchem. J.* **2021**, *168*, 106359. [CrossRef]
16. Samsidar, A.; Siddiquee, S.; Shaarani, S.M. A review of extraction, analytical and advanced methods for determination of pesticides in environment and foodstuffs. Trends Food Sci. Technol. **2018**, *71*, 188.
17. Narenderan, S.T.; Meyyanathan, S.N.; Babu, B. Review of pesticide residue analysis in fruits and vegetables Pre-treatment, extraction and detection techniques. *Food Res. Int.* **2020**, *133*, 109141. [CrossRef]
18. Freire, M.G. *Ionic-Liquid-Based Aqueous Biphasic Systems: Fundamentals and Applications*; Springer: Berlin/Heidelberg, Germany, 2016.
19. Tobiszewski, M.; Mechlinska, A.; Namie, J. Green analytical chemistry—Theory and practice. *Chem. Soc. Rev.* **2010**, *39*, 2869. [CrossRef] [PubMed]
20. Musarurwa, H.; Tavengwa, N.T. Emerging green solvents and their applications during pesticide analysis in food and environmental samples. *Talanta* **2021**, *223 Pt 1*, 121507. [CrossRef] [PubMed]
21. Ventura, S.P.M.; Silva, F.A.E.; Quental, M.V.; Mondal, D.; Freire, M.G.; Coutinho, J.A.P. Ionic-Liquid-Mediated Extraction and Separation Processes for Bioactive Compounds: Past, Present, and Future Trends. *Chem. Rev.* **2017**, *117*, 6984. [CrossRef]
22. Tian, H.; Fu, H.; Xu, C.; Xu, C. Simultaneous Determination of Three Herbicides in Honey Samples Using an Aqueous Biphasic System Coupled with HPLC–MS/MS. *Chromatographia* **2019**, *82*, 1571. [CrossRef]
23. Tian, H.; Xu, C.; Shi, Z.; Fu, H.; Li, X. Enantioseparation and determination of triazole fungicides in fruit juice by aqueous biphasic system coupled with HPLC-MS/MS. *J. Sep. Sci.* **2021**, *44*, 3407. [CrossRef]
24. Almeida, H.F.D.; Freire, M.G.; Marrucho, I.M. Improved monitoring of aqueous samples by the preconcentration of active pharmaceutical ingredients using ionic-liquid-based systems †. *Green. Chem.* **2017**, *19*, 4651–4659. [CrossRef] [PubMed]
25. González-Martín, R.; Silva, F.A.E.; Trujillo-Rodríguez, M.J.; Díaz, D.D.; Lorenzo-Morales, J.; Freire, M.G.; Pino, V. Ionic liquid-based aqueous biphasic systems as one-step clean-up, microextraction and preconcentration platforms for the improved determination of salivary biomarkers. *Green Chem.* **2023**, *25*, 8544. [CrossRef]
26. Šulc, J.; Pacheco-Fernández, I.; Ayala, J.H.; Bajerová, P.; Pino, V. A green miniaturized aqueous biphasic system prepared with cholinium chloride and a phosphate salt to extract and preconcentrate personal care products in wastewater samples. *J. Chromatogr. A* **2021**, *1648*, 1. [CrossRef] [PubMed]
27. Marić, S.; Jocić, A.; Tekić, D.; Mušović, J.; Milićević, J.; Dimitrijević, A. Customizable cholinium-based aqueous biphasic systems as ecofriendly extraction platform for removal of pesticide from wastewaters. *Sep. Purif. Technol.* **2024**, *340*, 126609. [CrossRef]
28. Jocić, A.; Marić, S.; Tekić, D.; Lazarević-Pašti, T.; Mušović, J.; Tričković, J.F.; Dimitrijević, A. Complete removal of organophosphate pesticides from wastewaters with sustainable ionic liquid-based aqueous two-phase strategy. *J. Water Process Eng.* **2024**, *64*, 105621. [CrossRef]
29. Zanella, R.; Primel, E.G.; Machado, S.L.O.; Gonçalves, F.F.; Marchezan, E. Monitoring of the herbicide clomazone in environmental water samples by solid-phase extraction and high-performance liquid chromatography with ultraviolet detection. *Chromatographia* **2002**, *55*, 573. [CrossRef]
30. Skandalis, N.; Dimopoulou, A.; Beri, D.; Tzima, A.; Malandraki, I.; Theologidis, I.; Bitivanos, S.; Varveri, C.; Klitsinaris, T.; Vassilakos, N. Effect of pyraclostrobin application on viral and bacterial diseases of tomato. *Plant Dis.* **2016**, *100*, 1321. [CrossRef]
31. Rehman, H.; Aziz, A.T.; Saggu, S.; Abbas, Z.K.; Mohan, A.; Ansari, A.A. Systematic review on pyrethroid toxicity with special reference to deltamethrin Medicinal Plants View project Toxicity of Pesticides and Herbicides View project Systematic review on pyrethroid toxicity with special reference to deltamethrin. *J. Entomol. Zool. Stud.* **2014**, *2*, 60.
32. Wojnowski, W.; Tobiszewski, M.; Pena-Pereira, F.; Psillakis, E. AGREEprep—Analytical greenness metric for sample preparation. *TrAC—Trends Anal. Chem.* **2022**, *149*, 116553. [CrossRef]
33. Pena-Pereira, F.; Wojnowski, W.; Tobiszewski, M. AGREE—Analytical GREEnness Metric Approach and Software. *Anal. Chem.* **2020**, *92*, 10076. [CrossRef] [PubMed]
34. Mušović, J.; Tekić, D.; Marić, S.; Jocić, A.; Stanković, D.; Dimitrijević, A. Sustainable recovery of cobalt and lithium from lithium-ion battery cathode material by combining sulfate leachates and aqueous biphasic systems based on tetrabutylphosphonium-ionic liquids. *Sep. Purif. Technol.* **2024**, *348*, 127707. [CrossRef]
35. Quental, M.V.; Pereira, M.M.; Silva, F.A.E.; Coutinho, J.A.P.; Freire, M.G. Aqueous Biphasic Systems Comprising Natural Organic Acid-Derived Ionic Liquids. *Separations* **2022**, *9*, 46. [CrossRef]
36. Merchuk, J.C.; Andrews, B.A.; Asenjo, J.A. Aqueous two-phase systems for protein separation studies on phase inversion. *J. Chromatogr. B Biomed. Appl.* **1998**, *711*, 285. [CrossRef]
37. Bridges, N.J.; Gutowski, K.E.; Rogers, R.D. Investigation of aqueous biphasic systems formed from solutions of chaotropic salts with kosmotropic salts (salt-salt ABS). *Green Chem.* **2007**, *9*, 177. [CrossRef]
38. Available online: https://chemicalize.com/app/calculation (accessed on 18 June 2024).

39. Bernard, U.L.; Izgorodina, E.I.; MacFarlane, D.R. New insights into the relationship between Ion-Pair binding energy and thermodynamic and transport properties of ionic liquids. *J. Phys. Chem. C* **2010**, *114*, 20472. [CrossRef]
40. Freire, M.G.; Cláudio, A.F.M.; Araújo, J.M.M.; Coutinho, J.A.P.; Marrucho, I.M.; Lopes, J.N.C.; Rebelo, L.P.N. Aqueous biphasic systems: A boost brought about by using ionic liquids. *Chem. Soc. Rev.* **2012**, *41*, 4966. [CrossRef]
41. Available online: https://ec.europa.eu/food/plant/pesticides/eu-pesticides-database/start/screen/products/details/39 (accessed on 15 September 2024).
42. Gałuszka, A.; Migaszewski, Z.; Namieśnik, J. The 12 principles of green analytical chemistry and the SIGNIFICANCE mnemonic of green analytical practices. *TrAC—Trends Anal. Chem.* **2013**, *50*, 78. [CrossRef]

Disclaimer/Publisher's Note: The statements, opinions and data contained in all publications are solely those of the individual author(s) and contributor(s) and not of MDPI and/or the editor(s). MDPI and/or the editor(s) disclaim responsibility for any injury to people or property resulting from any ideas, methods, instructions or products referred to in the content.

Article

Evaluating the Human Risks of Consumption of Foods of Bovine Origin with Ivermectin Residues in Ecuador

Valeria Paucar-Quishpe [1,2,3], Darío Cepeda-Bastidas [4], Richar Rodríguez-Hidalgo [3,5], Ximena Pérez-Otáñez [3,6], Cecilia Perez [5], Sandra Enríquez [3], Erika Guzman [4], Fernanda Ulcuango [4], Jorge Grijalva [5], Sophie O. Vanwambeke [6], Lenin Ron-Garrido [3,5,*,†] and Claude Saegerman [1,2,*,†]

[1] Research Unit of Epidemiology and Risk Analysis Applied to Veterinary Science (UREAR-ULiège), Faculty of Veterinary Medicine, University of Liege, 4000 Liège, Belgium; avpaucar@uce.edu.ec

[2] Fundamental and Applied Research for Animals & Health (FARAH) Center, Faculty of Veterinary Medicine, University of Liege, 4000 Liège, Belgium

[3] Instituto de Investigación en Zoonosis (CIZ), Universidad Central del Ecuador, Quito 170521, Ecuador; rrodriguez@uce.edu.ec (R.R.-H.); ximena.perezotanez@uclouvain.be (X.P.-O.); senriquez@uce.edu.ec (S.E.)

[4] Facultad de Ciencias Agrícolas, Universidad Central del Ecuador, Quito 170521, Ecuador; dacepedab@uce.edu.ec (D.C.-B.); ezguzman@uce.edu.ec (E.G.); fvulcuango@uce.edu.ec (F.U.)

[5] Facultad de Medicina Veterinaria y Zootecnia, Universidad Central del Ecuador, Quito 170521, Ecuador; bcperez@uce.edu.ec (C.P.); jgrijalva@uce.edu.ec (J.G.)

[6] Georges Lemaitre Centre for Earth and Climate Research, Earth & Life Institute, UCLouvain, 1348 Louvain-la-Neuve, Belgium; sophie.vanwambeke@uclouvain.be

* Correspondence: ljron@uce.edu.ec (L.R.-G.); claude.saegerman@uliege.be (C.S.)

† These authors contributed equally to this work.

Abstract: Ivermectin is a widely used antiparasitic in livestock, but its use can result in residues in bovine products and excretions. The objective of the present study was to determine the presence of ivermectin residues in cattle meat, liver, milk, faeces, and urine and assess consumer risk from chronic exposure through contaminated bovine products using a deterministic approach. To determine the presence of ivermectin residues, 124 samples were analysed by liquid chromatography. Residues were found in 68% of faeces samples and small percentages (3%) in liver, milk, and urine, with no residues detected in meat. The mean ivermectin residue in the liver (16.46 µg/kg) remained below the maximum residue limit (MRL); however, in milk (12.46 µg/kg), the residues exceeded the permitted MRL. The results obtained from chronic dietary exposure show that the consumption of ivermectin residues was low, and the risk was assessed as being rare to very rare. Additionally, this study reveals concerning levels of ivermectin residues in milk that may far exceed established safety limits. This situation emphasises the urgent need for stricter regulations and monitoring in milk production, particularly from small farms, to protect vulnerable populations. However, from a one health perspective, the presence of residues in faeces poses potential environmental hazards, warranting further research. Moreover, the detection of residues in milk, despite the ban on ivermectin use in dairy cattle, underscores the importance of compliance with food safety regulations and the need for continued vigilance in this area.

Keywords: ivermectin residues; milk; meat; liver; urine; faeces; risk assessment; consumption

1. Introduction

The livestock sector plays a critical role in global food supply and security, providing essential nutrition through products like meat, milk, and eggs. These products account for 18% of global calorie intake and 34% of protein consumption, significantly enhancing human diets [1]. Meat and meat-derived products are an energy-dense source of high-quality protein enriched in micronutrients such as vitamin B12, iron, zinc, selenium, and phosphorus [2]. Milk and dairy products contribute significantly to calcium, phosphorus,

Citation: Paucar-Quishpe, V.; Cepeda-Bastidas, D.; Rodríguez-Hidalgo, R.; Pérez-Otáñez, X.; Perez, C.; Enríquez, S.; Guzman, E.; Ulcuango, F.; Grijalva, J.; Vanwambeke, S.O.; et al. Evaluating the Human Risks of Consumption of Foods of Bovine Origin with Ivermectin Residues in Ecuador. *Foods* **2024**, *13*, 3470. https://doi.org/10.3390/foods13213470

Academic Editor: Dapeng Peng

Received: 1 September 2024
Revised: 24 October 2024
Accepted: 24 October 2024
Published: 29 October 2024

Copyright: © 2024 by the authors. Licensee MDPI, Basel, Switzerland. This article is an open access article distributed under the terms and conditions of the Creative Commons Attribution (CC BY) license (https://creativecommons.org/licenses/by/4.0/).

iodine, riboflavin, and vitamins A and B12 [3]. In 2023, global milk production reached 965.5 million tons and meat production 76.6 million tons [4,5].

In Ecuador, the livestock sector is a vital component of the agricultural economy. The country has a cattle population of about 3.7 million heads [6], with 81% of this population raised by small producers who manage 20 or fewer cattle heads [7]. Cattle husbandry in Ecuador primarily relies on grazing, with about 80% located in tropical or subtropical areas. These areas create proper conditions conducive to diseases caused by endo- and ectoparasites, which are the leading causes of illness and production losses [8].

Various drugs are used to manage cattle internal and external parasites; one of the most widely used is avermectins. Avermectins are a class of macrocyclic lactones produced by the soil actinomycete *Streptomyces avermitilis* [9]. This drug was discovered in 1973 and introduced to massive commercial success in the animal health market in 1981 [10]. The most widely utilised derivative of avermectin is ivermectin. Five years after its introduction, it was sold in 46 countries and administered to 320 million cattle heads [11]. Its success in the livestock market is due to its strong activity against nematode and arthropod parasites. Ivermectin is used to treat billions of cattle heads, helping to boost the production of food and leather products, as well as keeping cattle healthy around the world [10].

Although the health benefits of ivermectin are particularly important in ectoparasiticide control in livestock, its high level of faecal and urine excretion represents a potential environmental risk [12]. Additionally, the indiscriminate use of these drugs can result in trace amounts of residues and their metabolites persisting in edible tissues and animal products, such as meat, liver, and milk (Figure 1), which may pose potential health risks to people who consume these products [13].

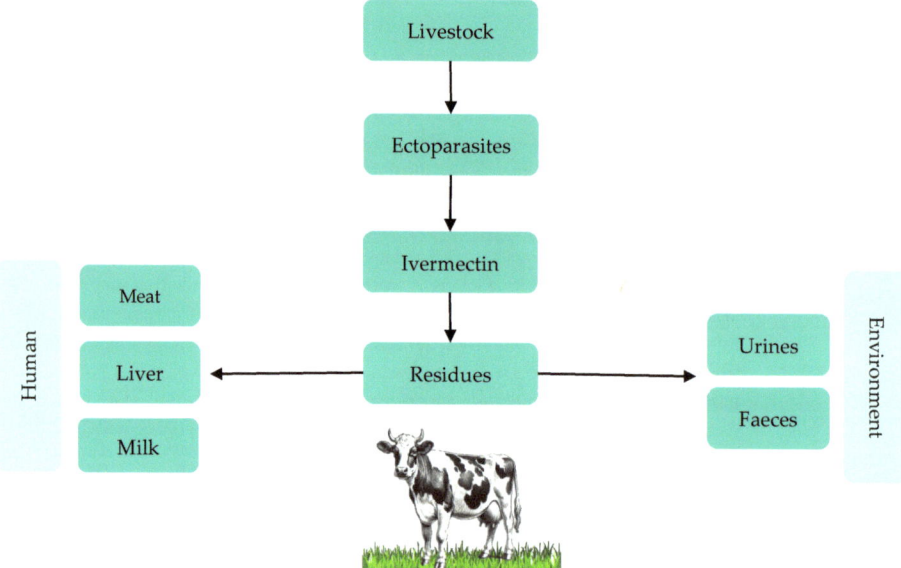

Figure 1. Interactions between ivermectin treatment, foods of bovine origin, the environment, and humans.

To ensure that levels of acaricides and/or their metabolites in food of animal origin remain below thresholds which are hazardous to consumers, the livestock industry must strictly comply with mandated withdrawal periods to reduce residue levels. The Food and Agriculture Organisation of the United Nations (FAO) and the World Health Organisation (WHO) jointly administer Expert Committees on Food Additives (JECFA) that evaluate the safety of veterinary drug residues. These evaluations serve as the basis for national and international food safety standards set by the Codex Alimentarius Commission [14].

JECFA has introduced several measures for food safety assessment, including the maximum residue limit (MRL), which represents the maximum allowable level of veterinary drug residues in food of animal origin, the acceptable daily intake (ADI), indicating the maximum amount of a veterinary drug which can be consumed daily over a lifetime without appreciable health risks, and the global estimate of chronic dietary exposure (GECDE), which assumes that, in the longer term, an individual would be a high-level consumer of only one category of food and that their consumption of other foods containing the residue would remain at the population average [14]. The MRL for ivermectin allowed in foods of bovine origin to ensure that the ADI limit (10 µg/kg body weight per day) is not exceeded varies according to the type of tissue. The limits are 800 µg/kg in the liver, 400 µg/kg in fat, 100 µg/kg in the kidney, 30 µg/kg in muscles, and 10 µg/kg in milk [15]. However, in some countries, the MRL for ivermectin in milk is set to zero. Ivermectin is banned in milk production due to its highly lipophilic nature, which causes its residues to persist in milk and dairy products [16]. Therefore, in order to respect the withdrawal period, farmers must dispose of the milk and not slaughter the animal during the resting period, which represents an additional cost to tick control [17]. Dairy farmers continue to rely on the use of pesticides to control pests and increase productivity, looking only at the immediate advantages of pest control without considering the potential short- and long-term risks of residue accumulation.

This challenge is particularly acute for small dairy farmers, who must compete with larger producers that enjoy better market access and lower production costs. Compared to large industries, small farmers often grapple with daily issues such as inadequate infrastructure, limited resources, and insufficient access to veterinary care, making it difficult for them to produce high-quality milk at competitive costs [18]. In Ecuador, inconsistent supervision leads many small producers to operate in informal markets where quality controls are minimal [19]. Economic pressures and competition further exacerbate the situation, often compromising milk quality and posing potential health risks to consumers.

In this context, the objectives of this study were the following: (1) to determine the prevalence of ivermectin residues in small-scale dairy farms located in two subtropical areas of Ecuador in foods of bovine origin such as milk, meat, liver, and excretions like urine, and faeces; and (2) to assess the risk of consuming these foods of bovine origin through the measurement of ivermectin concentrations in these food items.

2. Materials and Methods

2.1. Study Area

This research was conducted in two livestock areas of Ecuador. Area 1, known as the "Northwest of Pichincha", is situated in the province of Pichincha and includes the localities of Nanegal, Nanegalito, Pacto, Gualea, San Miguel de los Bancos, and Pedro Vicente Maldonado. This area is part of the Chocó Andino Biosphere Reserve, characterised by forests, rivers, waterfalls, and a diverse range of flora and fauna [20]. Area 2, known as the "Quijos River Valley", is located in the province of Napo and comprises the localities of San Francisco de Borja, Sumaco, Linares, Sardinas, El Chaco, and Baeza. This area, situated in the foothills of the Andes Mountains and the high jungle of the Amazon region, is part of the protected areas of the Antisana Ecological Reserve, Cayambe Coca National Park, and Sumaco Napo Galeras National Park [21]. The residents in both areas are mainly involved in ecotourism, agriculture (tropical fruits, sugar cane, cacao, coffee, and palm hearts), fish farming (tilapia and trout), and cattle breeding [20,22,23].

Livestock in these areas primarily consists of small- and medium-sized cattle herds dedicated to dairy or dual-purpose production. The farmers in these areas frequently use dairy breeds such as Brown Swiss, Holstein, and Jersey or their crosses [22,24]. Feeding is mainly through grazing, but there is also the use of supplemental feeding based on concentrates or agro-industrial byproducts [18]. The cattle population in the study areas is around 100,000 heads, distributed across 4087 herds [25].

2.2. Sampling and Chemical Analysis

From 2021 to 2023, samples of milk, beef, and liver were collected from the two study areas. Raw cow milk samples (N = 70) were obtained from small milk tanks designated for collection. Each milk collector tank represented one farm and held approximately 40 litres from 5 to 7 cows. The selection of milk-tank samples was at the convenience of the researcher. It was based mainly on the farmer's acceptance of participation in this study and the accessibility of the samples. Samples were collected following the primary route taken by the local milk collection trucks. Additionally, samples of urine (N = 39) and faeces (N = 40) were collected. Each of these samples consisted of a pool of 6 randomly selected cows from the farm. Meat (N = 46) and liver (N = 30) samples were collected at local slaughterhouses in each area. As one local slaughterhouse in Area 2 was closed, samples were acquired from a nearby slaughterhouse where animals from the study areas were relocated (Figure 2). The sampling process was carried out randomly, and official animal movement guides were reviewed to ensure that the animals were from the study areas.

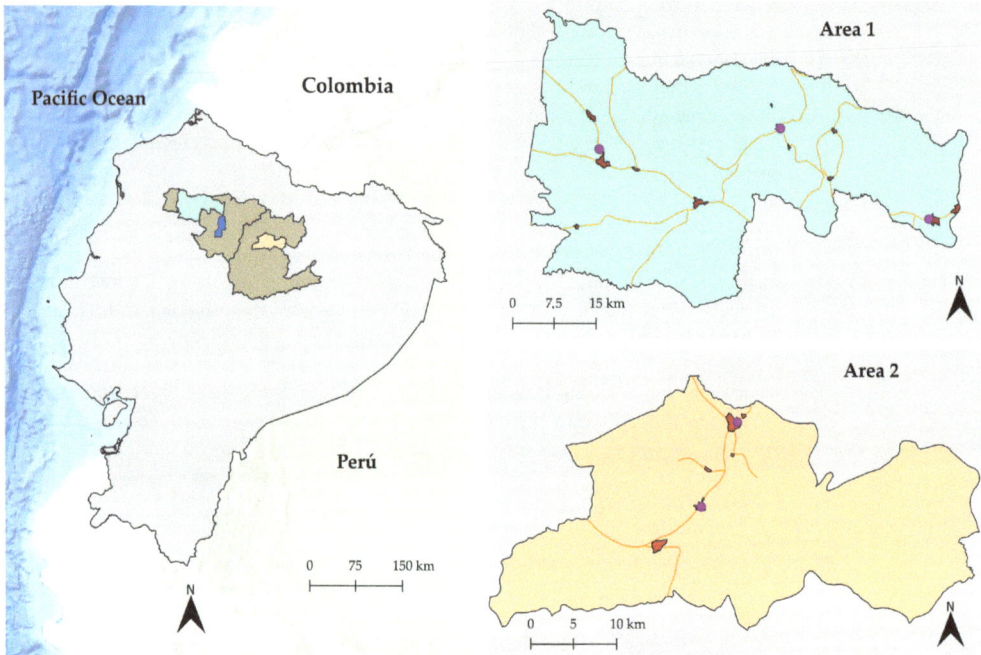

Figure 2. Location of study areas and sampling sites. Legend: ▨ provinces of Pichincha (right) and Napo (left); ▨ Quito (country capital); ▨ Area 1 (northwest of Pichincha); ▨ Area 2 (Quijos River Valley); ▨ populated areas where surveys were conducted; ● location of slaughterhouses; and — main routes of the milk collection trucks.

Raw milk samples (100 mL) were collected in polyethene plastic vials, and the meat and liver samples (100 g) were stored individually in zip-lock plastic bags. All samples were kept in a cooler with ice blocks until they were transported to the laboratory. Analyses were conducted in the EcuaChemLab Chemical and Microbiological Laboratory of Ecuador, which is accredited according to NTE INEN ISO/IEC 17025 [26]. The concentration of the B1a component of ivermectin was analysed using a reverse-phase high-performance liquid chromatographic method. The analysis was conducted on a Perkin Elmer Series 200 HPLC system (Shelton, CT, USA) equipped with a JASCO UV-975 detector (Hachioji, Tokyo, Japan) and a ZORBAX Eclipse Plus C18 column. The limit of detection (LOD) was established at <10 µg/kg (Supplementary File S1).

2.3. Food Consumption Survey

A questionnaire was used to estimate milk, meat, and liver consumption. The questionnaire was validated by national and international experts in the field. It was pilot-tested with a small group of volunteers who commented on the clarity of the questions. The participants interviewed were male and female inhabitants of the populated parts of the study areas (Figure 2), heads of household, and over 18 years of age in 2024. The data were collected in paper-and-pencil format and contained questions on socio-demographic information (gender, age, and number of persons living in the household) and consumption habits of foods of bovine origin at the household level. The sample size was estimated using household data from the "Servicio Ecuatoriano de Normalización" (INEC) (2010) census [27], with a total of 17,194 households as the population size reference, a confidence level of 95%, and a margin of error of 5%. Consequently, the study included a sample size of 631 households. Information on beef, liver, and milk consumption was expressed in grammes, with the conversion factor of 1 mL of milk corresponding to 1.03 g.

2.4. Risk to Consumer Health

For human health, the Joint FAO/WHO Expert Committee on Food Additives (JECFA) has established for ivermectin an acceptable daily intake (ADI) for consumers of 0–10 µg/kg of their body weight [15].

Concentrations of ivermectin residues measured in milk, meat, and liver were compared to the maximum residue level (MRL) for human consumption. The MRLs are 10 µg/kg in milk, 30 µg/kg in muscle, and 800 µg/kg in liver [15]. There are several approaches for chronic exposure assessment and risk assessment. The World Health Organization (WHO) and the European Food Safety Authority (EFSA) recommend three scenarios for dealing with contamination data below the quantification limits. These approaches are named as follows: lower bound that induced underestimation (LB), middle bound that induced overestimation (MB), and upper bound that induced most overestimation (UB) [28,29]. In this study, we used the MB, whereby results below the limit of detection were replaced by LOD/2.

There is no international consensus on the age groups of consumers [30]. In this study, the risk was assessed for two groups—(1) individuals younger than 10 years and (2) individuals older than 10 years—in such a way that the latter group included adolescents and adults, following the recommendations of the American Academy of Pediatrics [31].

Two scenarios were analysed: (A) the overall study population and (B) specifically people who consumed foods of bovine origin.

The estimated daily intake (EDI) of ivermectin residues by the consumer was calculated as follows:

$$EDI = \frac{contamination\left(\frac{mg}{g}\right) * consumption\ (g)}{bw\ (kg)} \quad (1)$$

where *contamination* is the mean concentration of ivermectin in the meat, liver, and milk; consumption stands for the daily mean *consumption* of these products in the study region; and bw represents the body weight. The mean *bw* in the study area for a person (man or woman) under 10 years of age was 13.50 kg (standard error (SE): 8.19 kg) and 68.39 kg (SE: 15.70 kg) for a person (man or woman) over 10 years of age. These data were obtained from a database provided by a local health clinic (N = 15,223).

The JECFA uses the global estimate of chronic dietary exposure (GECDE) for chronic dietary exposure assessment to veterinary drug residues [14]. According to the JECFA, ivermectin's GECDE recommendation level for adults and the elderly is lower than 0.72 µg/kg bw per day, which represents 7.2% of the upper bound of the ADI of 10 µg/kg bw. The GECDE recommendation level for children is lower than 0.93 µg/kg bw per day, which represents 9.3% of the upper bound of the ADI of 10 µg/kg bw [15].

The Global Estimate Chronic Dose Exposure (GECDE) to ivermectin residues for the population in the study areas was the highest exposure calculated using the 97.5th percentile consumption figure for a single food selected from all the foods plus the mean dietary exposure from all the other relevant foods [32]:

$$\text{Highest exposure from each animal product} = 97.5^{th} \text{percentile consumption} * \text{Median residue} \quad (2)$$

$$\begin{aligned}\text{Global Estimate Chronic Dose Exposure (GECDE)}\\ = \text{Highest exposure from one animal product}\\ + \text{Total mean exposure from all other products}\end{aligned} \quad (3)$$

3. Results and Discussion

3.1. Ivermectin Residues

Of the total samples analysed (N = 225), the presence of ivermectin residues was determined in 68% of the faeces samples (27/40; values between 30 and 420 µg/kg) and around 3% in the liver (1/30; value of 340 µg/kg), milk (2/70; values of 90 and 440 µg/kg), and urine (1/39; value of 60 µg/kg) samples. No residue over the LOD was detected in the 46 meat samples (Figures 3 and 4). Taking into account that the middle bound (MB) was used in this study, samples under the LOD (10 µg/kg) were assigned a value of 5 µg/kg. The mean value of ivermectin in faeces reached 118.13 µg/kg (95% CI: 85.61–155.00 µg/kg), 6.41 µg/kg (95% CI: 5.00–9.23 µg/kg) in urines, 12.46 µg/kg (95% CI: 5.00–26.16 µg/kg) in milk, 16.17 µg/kg (95% CI: 5.00–38.50 µg/kg) in liver, and 5.00 µg/kg (95% CI: 5.00–5.00 µg/kg) in meat.

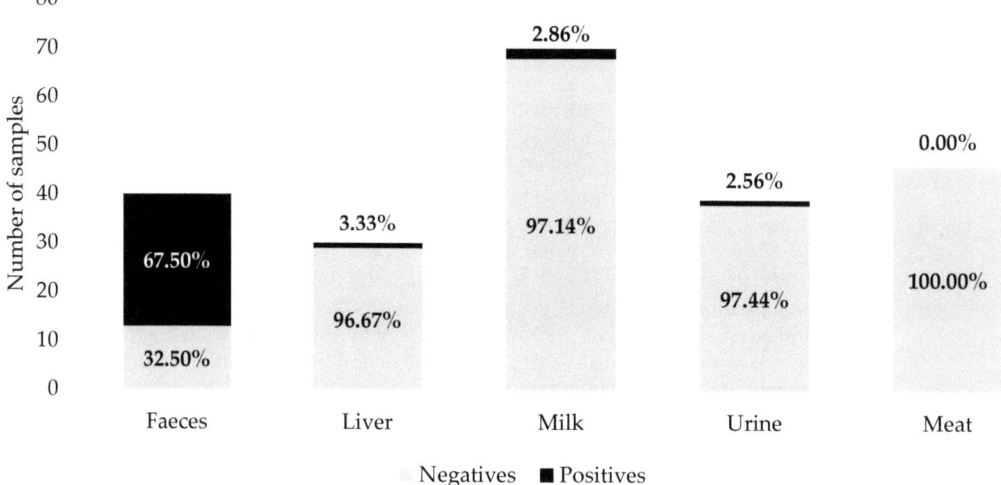

Figure 3. Prevalence of ivermectin residues in the analysed samples (decreasing order).

This study was the first to detect ivermectin in foods of bovine origin such as milk, meat, and liver and excretions like urine and faeces from small-scale dairy farms. While HPLC did not detect residues above the maximum residue limit (MRL) in the meat and liver samples, the positive milk samples did exceed the MRL established by FAO [15]. Regionally, a study conducted in the Brazilian retail dairy market reported that ivermectin was substantially used in dairy cows; the authors reported that 46% of milk samples had some level of residues detected/quantified. Although these residues did not exceed the maximum residue limit (MRL), their presence in nearly half of the samples is concerning [33].

Although the Codex Alimentarius sets an MRL of 10 µg/kg for ivermectin, other regulatory bodies, such as the European Medicines Agency [34] and Health Canada (2024) [35], do not approve the use of ivermectin in dairy cattle, resulting in no established legal MRL for milk in these regions. Moreover, another study conducted in the areas of the present study indicates that ivermectin was used in dairy cattle at a rate of 50% [36]. The discovery that 68% of faeces samples contain ivermectin residues strongly confirms the widespread use of this drug among local farmers. Despite the fact that, in this study, the number of samples over 90 and 440 µg/kg, the MRL was small, and there are no additional studies conducted in the field where this has been studied, although there are several studies conducted on raw milk and meat, identifying the presence of antibiotics and heavy metals [37–40]. Additionally, some undergraduate research projects have conducted experimental studies investigating the occurrence and elimination of antiparasitics, such as eprinomectin, ivermectin, and fipronil, in meat and milk [41,42]. Freire [41], in 2017, evaluated the presence of fipronil and ivermectin residues in meat and found that the ivermectin levels (38 µg/kg) exceeded the permitted limit (30 µg/kg) [15] on day 15 post application (pour-on). Similarly, Balseca [42], in 2017, reported eprinomectin residues (19.06 µg/L) in milk close to the permitted limit (20 µg/L) [15] on day 19 after application (pour-on). Considering that two-thirds of the milk in Ecuador is marketed informally, where quality controls are minimal and represent the most accessible market for small producers [19], there is a pressing need for larger-scale studies to assess risk and ensure food safety and quality accurately. These comprehensive investigations are especially important given that milk is a fundamental food for children's development, providing essential nutrients like calcium, protein, and vitamins [3]. A better understanding of the prevalence of ivermectin residues in milk is needed for informing regulatory decisions aimed at safeguarding public health.

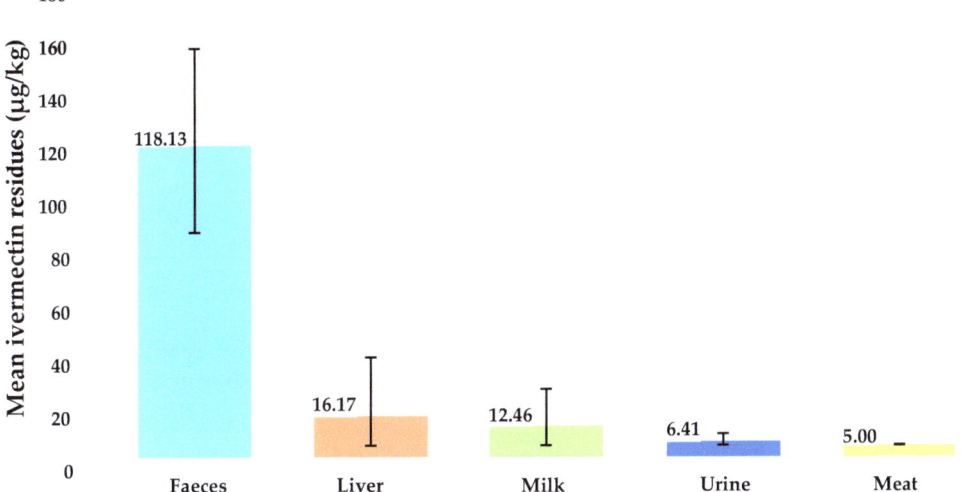

Figure 4. Mean of ivermectin residues in the analysed samples (decreasing order).

While this study primarily focuses on the presence of ivermectin residues in faeces and urines, it is important to recognise the broader environmental implications. Low doses of ivermectin residues have been shown to cause significant short- and long-term ecological effects, including alterations in decomposer insect communities, the disruption of manure degradation processes, and changes in soil properties and functions [43,44]. Research has extensively examined the impact of ivermectin on decomposer insect communities, particularly dung beetles. It has been demonstrated that ivermectin can remain at toxic concentrations for insects for 28 days in pour-on treatments and 35 days in subcutaneous injections [45] and can persist in dung for as long as 180 days [46]. Studies indicate that adult beetles are attracted to dung with residues, leading to a 90% mortality rate when consuming fallen dung 2–3 days after injection [47]. Moreover, adult beetles that colonise dung with residues do not reduce the number of eggs laid, but the resulting larvae die early in their development [44]. The impact of residues on dung beetles is of particular interest due to their role in maintaining healthy pasture growth by facilitating dung removal, promoting aeration through tunnel formation, and facilitating nutrient recycling [43]. As a result, the effects of endectocide residues on soil degradation are an important consideration for the agricultural economy [44]. In addition, ivermectin residues may affect aquatic invertebrates, which are especially sensitive to its effects. In the case of ivermectin, the direct deposition of faeces from treated livestock into small watercourses, ponds, or lakes has been considered to be a major threat to aquatic ecosystems. Contamination through leaching has been considered unlikely, given that ivermectin is strongly adsorbed into soil and organic matter [48,49].

It is important to note that the absence of ivermectin residues in meat does not automatically ensure food safety. In our study, sampling was conducted in formal slaughterhouses with some level of sanitary control. However, in Ecuador, approximately 36% of foods of bovine origin come from informal slaughtering (clandestine or homemade) [50]. Furthermore, only larger farms have better access to official slaughterhouses, while smaller producers depend on intermediate dealers, who collect animals from different farms and transport them to livestock markets or directly to slaughterhouses [51]. Sampling local butcher shops could help determine the safety of the food reaching consumers.

3.2. Consumption of Foods of Bovine Origin

It was determined that most of the households surveyed consumed meat and milk (91% and 97%, respectively), but only 30% consumed liver. The daily consumption (DC) mean values of meat, liver, and milk, considering the two scenarios, are shown in Table 1.

Table 1. Daily consumption of meat, liver, and milk in g/person/day.

Product	Households (%)	Less or Equal to 10 Years Old			Higher than 10 Years Old		
		Consumers (%)	DC ^{A}	DC ^{B}	Consumers (%)	DC ^{A}	DC ^{B}
Milk	97	98	94.76	98.24	97	95.85	102.15
Meat	91	91	5.88	6.45	92	30.95	35.62
Liver	30	32	0.14	0.35	30	0.75	2.27

Legend. (DC) Daily consumption (grammes): (A) scenario with the mean consumption of the inhabitants of the study area; and (B) scenario with the mean consumption only of the inhabitants that consume each food of bovine origin.

Based on the survey data, the annual consumption in the study area was 10 kg of meat and 0.26 kg of liver for the whole population. There are no data available regarding the amount of liver consumption at the local or national level. However, meat consumption in our study was close to the national average of 13 kg per capita [52]. The national per capita consumption is similar to that of neighbouring countries like Colombia (14 kg) but lower than the reported consumption in South American countries such as Argentina (48 kg) or Brazil (35 kg), both of which are substantial consumers of meat globally [52].

Furthermore, our findings reveal an annual milk consumption per person of 32 L, which is significantly lower than the national per capita average of 110 L [53] and well below the recommended 180 L by the FAO and OMS [54]. Previous studies in the country indicate that the highest per capita milk consumption is in the Highlands region of Ecuador. In contrast, in the Amazonian and coastal regions, where the study areas are situated, consumption is much lower, reaching a quarter of the consumption in the Highlands areas [55].

3.3. Risk Assessment for Consumers of Ivermectin-Contaminated Foods of Bovine Origin

The risk in this study was estimated based on the amount of ivermectin present in milk, liver, and meat, the consumption of these foods in the study areas, and the data evaluated by the JECFA [15].

For milk, 2 samples (90 µg/kg and 440 µg/kg) had an ivermectin concentration above the MRL (10 µg/kg), and 68 samples below the LOD (10 µg/kg). For the liver, 1 sample (340 µg/kg) had an ivermectin concentration between the LOD (10 µg/kg) and the MRL (800 μg/kg) and 45 samples below the LOD (10 μg/kg). For meat, all samples had ivermectin concentrations under the MRL (30 μg/kg) and LOD (10 μg/kg).

Considering (1) the body weight (bw) of a person (man or woman) under 10 years of age (Avg. 13.49) and of a person (man or woman) over 10 years of age (Avg. 68.39), (2) the individual consumption data, (3) and the results of the estimated amount of ivermectin, the estimate daily intake (EDI) of ivermectin residues through milk, meat, and liver was close to zero (between 0.02 and 0.0935 μg/kg bw/day), i.e., the lower limit of the ADI (0 μg/kg bw), and also largely lower than the upper limit of the ADI (10 μg/kg bw) [15]. Furthermore, the GECDE for ivermectin residues was between 0.0029% (Scenario A) and 0.2959% (Scenario B) of the ADI for a person older than 10 years, i.e., lower than the 7.2% recommended. In addition, the GECDE for ivermectin residues was between 0.9131% (Scenario A) and 1.0246% (Scenario B) of the ADI for a person younger than or equal to 10 years, i.e., also largely lower than the recommended level of 9.3% [15] (Table 2). Complementary stochastic modelling confirmed the same picture, with only 1 EDI simulation among the 10,000 simulations with values higher than the ADI for a person younger than or equal to 10 years of age. Indeed, the expression of the risk should be qualified as rare to very rare. Additionally, despite the maximum residue limit (MRL) being set to 10 μg/kg for ivermectin, the results show common use of the drug, with reported residue levels in positive samples between 5.00 and 38.50 μg/kg, at least in some milk samples. While population-level exposure may remain under the MRL, the situation becomes alarming for children. Given their lower body weight (below 10 kg) and, for instance, the recommended daily milk consumption of 0.42 kg, their exposure greatly exceeds the MRL, raising concerns about milk safety, particularly from small-scale farms. This finding suggests a need for stricter regulations and enforcement in milk production to protect vulnerable populations like children [56].

Table 2. Daily intake of ivermectin and chronic dietary exposure.

		Food of Bovine Origin	Median Residue Concentration (μg/kg)	Mean Residue Concentration (μg/kg)	Consumption Percentile 97.5th (kg/day)	Consumption Means (kg/day)	bw (kg)	EDI (μg/bw/day)	EDI (μg/kg bw/day)	Exposure (μg/kg bw/day) 97.5th	Exposure Mean	GECDE μg/kg bw/day	GECDE %ADI
A	Less or equal to 10 years old	Milk	5.0000	12.4571	0.2404	0.0948	13.4935	1.1804	0.0875	0.0891	0.0351	0.0891	0.8908
		Meat	5.0000	5.0000	0.0118	0.0059	13.4935	0.0294	0.0022	0.0044	0.0022	0.0022	0.0218
		Liver	5.0000	16.1667	0.0008	0.0001	13.4935	0.0022	0.0002	0.0003	0.0001	0.0001	0.0005
		TOTAL						1.2120	0.0899	0.0937	0.0373	0.0913	0.9131
A	Higher than 10 years old	Milk	5.0000	12.4571	0.3668	0.0958	68.3891	1.1940	0.0175	0.0268	0.0070	0.0268	0.0027
		Meat	5.0000	5.0000	0.0724	0.0309	68.3891	0.1547	0.0023	0.0053	0.0023	0.0023	0.0002
		Liver	5.0000	16.1667	0.0056	0.0008	68.3891	0.0122	0.0002	0.0004	0.0001	0.0001	0.0000
		TOTAL						1.2609	0.0200	0.0325	0.0093	0.0291	0.0029
B	Less or equal to 10 years old	Milk	5.0000	12.4571	0.2697	0.0982	13.4935	1.2238	0.0907	0.0999	0.0364	0.0999	0.9993
		Meat	5.0000	5.0000	0.0133	0.0065	13.4935	0.0323	0.0024	0.0049	0.0024	0.0024	0.0239
		Liver	5.0000	16.1667	0.0011	0.0004	13.4935	0.0057	0.0004	0.0004	0.0001	0.0001	0.0013
		TOTAL						1.2618	0.0935	0.1053	0.0389	0.1025	1.0246
B	Higher than 10 years old	Milk	5.0000	12.4571	0.3668	0.1022	68.3891	1.2725	0.0186	0.0268	0.0075	0.0268	0.2682
		Meat	5.0000	5.0000	0.0733	0.0356	68.3891	0.1781	0.0026	0.0054	0.0026	0.0026	0.0260
		Liver	5.0000	16.1667	0.0072	0.0023	68.3891	0.0367	0.0005	0.0005	0.0002	0.0002	0.0017
		TOTAL						1.4873	0.0217	0.0327	0.0102	0.0296	0.2959

Legend. (A) Scenario with the mean consumption of the inhabitants of the study area; (B) scenario with the mean consumption only of the inhabitants that consume foods of bovine origin; estimated daily intake (EDI); acceptable daily intake (ADI); global estimated chronic dietary exposure (GECDE); and body weight (bw).

Although no long-term toxicity studies have been conducted with repeated doses in humans or other laboratory animals, studies with abamectin in mice (94 weeks) have found carcinogenic effects [57]. In addition, short-term studies (4 weeks) in young rats have shown increased sensitivity to ivermectin due to an underdeveloped blood–brain barrier [58]. Given its recent use during the COVID-19 pandemic, clinical effects, including neurotoxicity, gastrointestinal symptoms, and musculoskeletal complaints, have been reported. Patients taking high doses of veterinary ivermectin reported neurotoxicity, with altered mental status. On the other hand, patients taking lower doses of ivermectin over a prolonged period reported milder toxicity, with no cases of severe altered mental status [59].

It is important to note that milk consumption in the study areas was about a quarter of the national average. This suggests that risk assessments could yield different results in regions with higher milk consumption. While no immediate risk was identified, the potential danger remains, and measures should be implemented to ensure food safety. Given that milk is a basic product in the basic food basket and provides essential micro- and macronutrients, particularly crucial during infancy and childhood, when bone mass growth is critical, ensuring its safety is paramount [60].

There may be other potentially more dangerous scenarios, such as the consumption of milk from cows recently treated with ivermectin, particularly if consumed during the short-term withdrawal period. While this study focused on the mean consumption of foods of bovine origin in the study areas, the scenario mentioned is certainly plausible among small farmers, where the accumulation of milk consumption over several days could lead to hazardous or toxic situations.

Given that, currently, only 56% of the 823 formal slaughterhouses in operation meet the Slaughterhouse Under Official Inspection (MABIO) certification standards, which guarantee that the protein is safe and processed in authorised facilities [61] and that around 36% of the meat consumed in households and restaurants comes from informal slaughterhouses [50], this study does not guarantee food safety for consumers of beef protein in the area studied. It therefore recommends that future studies employ a more comprehensive sampling strategy, including samples from local butchers and a larger number of samples. On the other hand, not all provinces in Ecuador have official livestock slaughterhouses, forcing small and medium livestock farmers to incur higher costs to transport their animals to neighbouring provinces, sell their livestock at lower prices to intermediaries, or resort to home or clandestine slaughter, making it even more difficult to have a traceability system in place. All this underlines the need for governments or industry stakeholders to be able to offer financial incentives to small-scale producers who demonstrate compliance with withdrawal periods and good agricultural practices. In addition, the creation and improvement of active slaughterhouses and the updating of inspection programmes to include residue testing for pesticides, antibiotics, and heavy metals will help ensure safe and high-quality food for Ecuadorian consumers. Equally important are training programmes for farmers and informal market vendors, which can raise awareness about the importance of withdrawal periods and food safety, thereby improving compliance across the sector.

4. Conclusions

This study provides critical insights into ivermectin residue levels across various bovine products and excretions, with the most significant findings observed in faeces. The mean ivermectin residue in the liver (16.46 µg/kg) and meat (5 µg/kg) remained within the acceptable limits set by the JECFA (LMR = 800 µg/kg and 30 µg/kg, respectively). However, in milk, residues (12.46 µg/kg) exceeded the permitted MRL (10 µg/kg), raising concerns about food safety compliance. Despite this, the risk from chronic dietary exposure was deemed low, with the likelihood of adverse health effects considered rare to very rare. While the health risks from consuming products such as meat, liver, and milk appear minimal under normal conditions, this study highlights the potential dangers of scenarios such as consuming milk from cows recently treated with ivermectin, especially during the short-term withdrawal period. This raises important questions about the regulation of antiparasitics in livestock and the safety of dairy products particularly from small-scale farms. This finding suggests a need for stricter regulations and enforcement in milk production to protect vulnerable populations like children. Additionally, given the high presence of ivermectin in faecal samples, from one health perspective, the presence of residues in faeces poses potential environmental hazards, warranting further research. This study emphasises the importance of collaborative and intersectoral efforts. Veterinary professionals, public health experts, biologists, and ecologists must work together to address this issue, ensuring good animal health, food safety, and human health through sustainable and environmentally friendly livestock practices.

Supplementary Materials: The following supporting information can be downloaded at: https://www.mdpi.com/article/10.3390/foods13213470/s1, Supplementary File S1: HPLC Protocol. References [62,63] are cited in the Supplementary Materials.

Author Contributions: Conceptualization, V.P.-Q., X.P.-O., R.R.-H., S.O.V., L.R.-G., and C.S.; methodology, L.R.-G. and C.S.; software, V.P.-Q., L.R.-G., and C.S.; validation, L.R.-G. and C.S.; formal analysis, V.P.-Q.; investigation, V.P.-Q. and X.P.-O.; resources, R.R.-H., L.R.-G., S.O.V., and C.S.; data curation, V.P.-Q., F.U., E.G., L.R.-G., and C.S.; writing—original draft preparation, V.P.-Q.; writing—review and editing, all authors; visualisation, L.R.-G. and C.S.; supervision, L.R.-G. and C.S.; project administration, R.R.-H., L.R.-G., and C.S.; and funding acquisition, R.R.-H., C.P., D.C.-B., J.G., S.E., S.O.V., L.R.-G., and C.S. All authors have read and agreed to the published version of the manuscript.

Funding: This research was funded by the Academy of Research and Higher Education (ARES) through the Research for Development Project (PRD) entitled "Socio-eco-epidemiology of ticks, tickborne parasites, acaricide resistance and residual effects of acaricides in tropical Ecuadorian livestock: environmental, animal and public health impacts (Project number: 03E-2020), which involves universities from Ecuador (CIZ, Universidad Central del Ecuador) and Belgium (UCLouvain and ULiège).

Institutional Review Board Statement: This study was conducted as part of the project "Socio-eco-epidemiology of ticks, tick-borne para-sites, acaricide resistance and residual effects of acaricides in Ecuadorian tropical livestock: environmental, animal and public health impacts". The study protocol was reviewed and approved by the Human Research Ethics Committee (COIF-FMVZ) of the Universidad Central del Ecuador (Code: 017-DOC-FMVZ-2023), ensuring that all ethical guidelines were followed.

Informed Consent Statement: ICS Informed consent was obtained from all subjects involved in the study.

Data Availability Statement: The original contributions presented in the study are included in the article/Supplementary Materials, further inquiries can be directed to the corresponding authors.

Acknowledgments: Our thanks go to the Académie de Recherche et d'Enseignement Supérieur (ARES) for funding this research and to the Universidad Central del Ecuador (UCE), UCLouvain, and the University of Liège for hosting this project. In addition, our thanks go to all the farmers who participated, the community leaders, the field technicians, and the students of the Faculty of Veterinary Medicine and Agronomy of the Universidad Central del Ecuador, who participated in the field and laboratory work and thanks to whom this research was possible.

Conflicts of Interest: The authors declare no conflicts of interest.

References

1. FAO; CIRAD; ILRI. *Livestock Sector Investment and Policy Toolkit (LSIPT)—Making Responsible Decisions*; FAO: Rome, Italy, 2020.
2. Boateng, E.F.; Nasiru, M.M.; Agyemang, M. Meat: Valuable Animal-Derived Nutritional Food. A Review. *Asian Food Sci. J.* **2020**, *15*, 9–19. [CrossRef]
3. Woźniak, D.; Cichy, W.; Dobrzyńska, M.; Przysławski, J.; Drzymała-Czyż, S. Reasonableness of Enriching Cow's Milk with Vitamins and Minerals. *Foods* **2022**, *11*, 1079. [CrossRef] [PubMed]
4. FAO. *Dairy Market Review: Overview of Global Market Developments in 2023*; FAO: Rome, Italy, 2024. Available online: http://www.nmpf.org/files/DMReport_March31.pdf (accessed on 5 June 2024).
5. FAO. *Meat Market Review: Overview of Global Market Developments in 2023*; FAO: Rome, Italy, 2024.
6. INEC. *Encuesta de Superficie y Producción Agropecuaria Continua*; Instituto Nacional de Estadística y Censos: Quito, Ecuador, 2024.
7. Agrocalidad. Categorías de Población de Ganado Bovino de Ecuador. 2018. Available online: https://www.agricultura.gob.ec/wp-content/uploads/2019/09/ANEXO-1.pdf (accessed on 28 March 2024).
8. Rodríguez-Hidalgo, R.; Pérez-Otáñez, X.; Garcés-Carrera, S.; Vanwambeke, S.O.; Madder, M.; Benítez-Ortiz, W. The current status of resistance to alpha-cypermethrin, ivermectin, and amitraz of the cattle tick (*Rhipicephalus microplus*) in Ecuador. *PLoS ONE* **2017**, *12*, e0174652. [CrossRef]
9. D'Auria, M.; Guarnaccio, A.; Racioppi, R.; Stoia, S.; Emanuele, L. Photodegradation of Drugs and Crop Protection Products. In *Photochemistry of Heterocycles*; Elsevier: Amsterdam, The Netherlands, 2023. [CrossRef]
10. Pandit, M.A.; Tarkeshwar. Ivermectin: An Anthelminthic and Insecticide. In *Chemistry and Biological Activities of Ivermectin*; Wiley: Hoboken, NJ, USA, 2023; pp. 163–197. [CrossRef]

11. Pecenka, J.R.; Lundgren, J.G. Effects of herd management and the use of ivermectin on dung arthropod communities in grasslands. *Basic Appl. Ecol.* **2019**, *40*, 19–29. [CrossRef]
12. Mancini, L.; Lacchetti, I.; Chiudioni, F.; Cristiano, W.; Kevin, D.D.; Marcheggiani, S.; Carere, M.; Bindi, L.; Borrello, S. Need for a sustainable use of medicinal products: Environmental impacts of ivermectin. *Ann. Ist. Super. Sanita* **2020**, *56*, 492–496. [CrossRef]
13. Soares, V.M.; Pereira, J.G.; Barreto, F.; Jank, L.; Rau, R.B.; Ribeiro, C.B.D.; dos Santos Castilhos, T.; Tomaszewski, C.A.; Hilleshiem, D.R.; Mondadori, R.G.; et al. Residues of Veterinary Drugs in Animal Products Commercialized in the Border Region of Brazil, Argentina, and Uruguay. *J. Food Prot.* **2022**, *85*, 980–986. [CrossRef]
14. FAO/WHO. *Evaluation of Certain Veterinary Drug Residues in Food: Ninety-Fourth Report of the Joint FAO/WHO Expert Committee on Food Additives*; WHO Technical Report Series No. 1041; WHO: Geneva, Switzerland, 2022; pp. 301–302. [CrossRef]
15. FAO/WHO. *Maximum Residue Limits (MRLs) and Risk Management Recommendations (RMRs) for Residues of Veterinary Drugs in Foods*; World Health Organization—Technical Report Series; WHO: Geneva, Switzerland, 2023.
16. Dedavid e Silva, L.A.; Ali, A.; Termignoni, C.; Vaz Júnior, I.d.S. Vaccination against Rhipicephalus microplus: An alternative to chemical control? *Cienc. Rural* **2024**, *54*, e20230161. [CrossRef]
17. Nyokabi, S.; Luning, P.A.; de Boer, I.J.; Korir, L.; Muunda, E.; Bebe, B.O.; Lindahl, J.; Bett, B.; Oosting, S.J. Milk quality and hygiene: Knowledge, attitudes and practices of smallholder dairy farmers in central Kenya. *Food Control* **2021**, *130*, 108303. [CrossRef]
18. Paucar-Quishpe, V.; Pérez-Otáñez, X.; Rodríguez-Hidalgo, R.; Cepeda-Bastidas, D.; Pérez-Escalante, C.; Grijalva-Olmedo, J.; Enríquez, S.; Arciniegas-Ortega, S.; Sandoval-Trávez, L. An economic evaluation of cattle tick acaricide-resistances and the financial losses in subtropical dairy farms of Ecuador: A farm system approach. *PLoS ONE* **2023**, *18*, e0287104. [CrossRef]
19. CIL. *En Ecuador, dos de cada tres litros de leche comercializan de modo informal*; Centro De La Industria Láctea Del Ecuador: Quito, Ecuador, 2023. Available online: https://www.cil-ecuador.org/post/en-ecuador-dos-de-cada-tres-litros-de-leche-comercializan-de-modo-informal#:~:text=2min.-,EnEcuador,dosdecadatreslitrosdelechecomercializan,5′500.000litrospord%C3%ADa (accessed on 14 August 2024).
20. Choco Andino Pichincha. Reserva de biósfera del Chocó Andino De Pichincha. Available online: https://www.chocoandinopichincha.com/ (accessed on 24 October 2022).
21. Ministerio de Ambiente y Agua. *Plan de Manejo Parque Nacional Cayambe Coca*; Ministerio de Ambiente y Agua: Quito, Ecuador, 2022. [CrossRef]
22. Benavides, B. *Análisis de los Sistemas Productivos Agropecuarios Ganaderos en el Noroccidente de Pichincha*; Repositorio Digital-Universidad Central del Ecuador: Quito, Ecuador, 2022.
23. PDOT Quijos. *Plan de Desarrollo y Ordenamiento Territorial del Cantón Quijos*; Dirección de Planificación y Ordenamiento Territorial, GAD Municipal Quijos: Quito, Ecuador, 2024; pp. 1–394.
24. Paucar, V.; Ron-Román, J.; Benítez-Ortiz, W.; Celi, M.; Berkvens, D.; Saegerman, C.; Ron-Garrido, L. Bayesian Estimation of the Prevalence and Test Characteristics (Sensitivity and Specificity) of Two Serological Tests (RB and SAT-EDTA) for the Diagnosis of Bovine Brucellosis in Small and Medium Cattle Holders in Ecuador. *Microorganisms* **2021**, *9*, 1815. [CrossRef]
25. Agrocalidad. *Reporte Vacunacion Fiebre Aftosa*; Agencia de Regulación y Control Fito y Zoosanitario: Quito, Ecuador, 2023.
26. INEN Acreditación de laboratorios de ensayo y calibración según. *NTE INEN- ISO/IEC 17025:2018*; INEN: Quito, Ecuador, 2018.
27. INEC Censo De Población Y Vivienda (PV 2010). 2010. Available online: https://www.ecuadorencifras.gob.ec/censo-de-poblacion-y-vivienda/ (accessed on 31 May 2024).
28. Carrasco Cabrera, L.; Medina Pastor, P.; EFSA. The 2020 European Union report on pesticide residues in food. *EFSA J.* **2022**, *20*, e07215. [CrossRef] [PubMed]
29. FAO/WHO. *Instructions for Electronic Submission of Data on Chemical Contaminants in Food and the Diet. Global Environment Monitoring System—Food Contamination Monitoring and Assessment Programme (GEMS/Food)*; Food and Agriculture Organization of the United Nations (FAO): Rome, Italy, 2023.
30. Hubal, E.A.C.; de Wet, T.; Du Toit, L.; Firestone, M.P.; Ruchirawat, M.; van Engelen, J.; Vickers, C. Identifying important life stages for monitoring and assessing risks from exposures to environmental contaminants: Results of a World Health Organization review. *Regul. Toxicol. Pharmacol.* **2014**, *69*, 113–124. [CrossRef] [PubMed]
31. Hagan, J.; Shaw, J.; Duncan, P. *Bright Futures: Guidelines for Health Supervision of Infants, Children, and Adolescents*, 4th ed.; American Academy of Pediatrics: Itasca, Illinois, USA, 2017. [CrossRef]
32. FAO/WHO. *Joint FAO/WHO Expert Meeting on Dietary Exposure Assessment Methodologies for Residues of Veterinary Drugs: Final Report Including Report of Stakeholder Meeting*; World Health Organization: Geneva, Switzerland, 2012. Available online: https://www.fao.org/fileadmin/user_upload/agns/pdf/jecfa/Dietary_Exposure_Assessment_Methodologies_for_Residues_of_Veterinary_Drugs.pdf (accessed on 22 July 2024).
33. Novaes, S.F.D.; Schreiner, L.L.; Silva, I.P.E.; Franco, R.M. Residues of veterinary drugs in milk in Brazil. *Ciência Rural* **2017**, *47*, e20170215. [CrossRef]
34. EMA. Opinion of the Committee for Medicinal Products for Veterinary Use on the establishment of maximum residue limits; procedure no: EU/09/170/SCM; name of the substance: Octenidine dihydrochloride (INN). *Vet. Med. Prod. Data Manag. Opin.* **2013**, *44*, 1–3.
35. Health Canada. List of Maximum Residue Limits (MRLs) for Veterinary Drugs in Foods. In *Food and Drugs Act*; Health Canada: Ottawa, ON, Canada, 2024.

36. Paucar, V.; Pérez-Otáñez, X.; Rodríguez-Hidalgo, R.; Perez, C.; Cepeda-Bastidas, D.; Grijalva, J.; Enríquez, S.; Arciniegas-Ortega, S.; Vanwambeke, S.O.; Ron-Garrido, L.; et al. The Associated Decision and Management Factors on Cattle Tick Level of Infestation in Two Tropical Areas of Ecuador. *Pathogens* **2022**, *11*, 403. [CrossRef]
37. Brito, S.N. *Determinación de residuos de antibióticos en carne de ganado bovino por el método de ELISA en el Centro de Faenamiento de la Empresa Pública Metropolitana de rastro Quito- La Ecuatoriana*; Repositorio Digital-Universidad Central del Ecuador: Quito, Ecuador, 2017; p. 168.
38. Puga-Torres, B.; Aragón Vásquez, E.; Ron, L.; Álvarez, V.; Bonilla, S.; Guzmán, A.; Lara, D.; De la Torre, D. Milk Quality Parameters of Raw Milk in Ecuador between 2010 and 2020: A Systematic Literature Review and Meta-Analysis. *Foods* **2022**, *11*, 3351. [CrossRef]
39. Puga-Torres, B.; Aragón, E.; Contreras, A.; Escobar, D.; Guevara, K.; Herrera, L.; López, N.; Luje, D.; Martínez, M.; Sánchez, L.; et al. Analysis of quality and antibiotic residues in raw milk marketed informally in the Province of Pichincha–Ecuador. *Food Agric. Immunol.* **2024**, *35*, 2291321. [CrossRef]
40. de la Cueva, F.; Naranjo, A.; Torres, B.P.; Aragón, E. Presence of heavy metals in raw bovine milk from Machachi, Ecuador. *Granja* **2021**, *33*, 21–30. [CrossRef]
41. Freire, N. Evaluacion cuantitativa de residuos en carne de ganado de engorde, post aplicacion pour-on, del producto fipronil mas ivermectina. *Univ. Cent. Ecuad.* **2017**, *14*, 1–49.
42. Balseca, P. *Determinacion cuantitativa de residuos en leche de epronomectina usado como mosquicida en vacas lecheras*; Universidad Central del Ecuador: Quito, Ecuador, 2017; pp. 1–44.
43. Junco, M.; Iglesias, L.E.; Sagués, M.F.; Guerrero, I.; Zegbi, S.; Saumell, C.A. Effect of macrocyclic lactones on nontarget coprophilic organisms: A review. *Parasitol. Res.* **2021**, *120*, 773–783. [CrossRef]
44. Finch, D.; Schofield, H.; Floate, K.D.; Kubasiewicz, L.M.; Mathews, F. Implications of Endectocide Residues on the Survival of Aphodiine Dung Beetles: A Meta-Analysis. *Environ. Toxicol. Chem.* **2020**, *39*, 863–872. [CrossRef] [PubMed]
45. Herd, R.P.; Sams, R.A.; Ashcraft, S.M. Persistence of ivermectin in plasma and faeces following treatment of cows with ivermectin sustained-release, pour-on or injectable formulations. *Int. J. Parasitol.* **1996**, *26*, 1087–1093. [CrossRef] [PubMed]
46. Suárez, V.H.; Lifschitz, A.L.; Sallovitz, J.M.; Lanusse, C.E. Effects of faecal residues of moxidectin and doramectin on the activity of arthropods in cattle dung. *Ecotoxicol. Environ. Saf.* **2009**, *72*, 1551–1558. [CrossRef] [PubMed]
47. Lumaret, J.P.; Galante, E.; Lumbreras, C.; Mena, J.; Bertrand, M.; Bernal, J.L.; Cooper, J.F.; Kadiri, N.; Crowe, D. Field Effects of Ivermectin Residues on Dung Beetles. *J. Appl. Ecol.* **1993**, *30*, 428–436. [CrossRef]
48. Kövecses, J.; Marcogliese, D.J. *Avermectins: Potential Environmental Risks and Impacts on Freshwater Ecosystems in Quebec*; Scientific and Technical Report ST-233E; Environment Canada, Quebec Region, Environmental Conservation, St. Lawrence Centre: Montréal, QC, Canada, 2005.
49. Wohde, M.; Blanckenhorn, W.U.; Floate, K.D.; Lahr, J.; Lumaret, J.P.; Römbke, J.; Scheffczyk, A.; Tixier, T.; Düring, R.A. Analysis and dissipation of the antiparasitic agent ivermectin in cattle dung under different field conditions. *Environ. Toxicol. Chem.* **2016**, *35*, 1924–1933. [CrossRef]
50. Mendoza, M.; El Comercio. El 36% de la carne sale de camales clandestinos. 2017. Available online: https://www.elcomercio.com/actualidad/quito/carne-camales-clandestinos-quito-normativa.html (accessed on 7 June 2024).
51. Castillo, M.J.; Carpio, C.E. Demand for High-Quality Beef Attributes in Developing Countries: The Case of Ecuador. *J. Agric. Appl. Econ.* **2019**, *51*, 568–590. [CrossRef]
52. Ritchie, H.; Roser, M.; Meat and Dairy Production. OurWorldInData.org. Available online: https://ourworldindata.org/meat-production (accessed on 7 June 2024).
53. Baquerizo, V.; Córdova, V. Impacto económico del sector lácteo: Un estudio de los gastos publicitarios y las ventas en tiempos de pandemia. *593 Digit. Publ. CEIT* **2022**, *7*, 310–321. [CrossRef]
54. CIL. *El sector lácteo ecuatoriano se reactiva con miras positivas para el 2022*; Centro de Industria Láctea del Ecuador: Quito, Ecuador, 2022. Available online: https://www.cil-ecuador.org/post/el-sector-l%C3%A1cteo-ecuatoriano-se-reactiva-con-miras-positivas-para-el-2022 (accessed on 21 October 2023).
55. Bermeo, F. Seguridad alimentaria; Responsabilidad de los Gobiernos Autónomos Descentralizados Provinciales. *Flacsoandes* **2015**, *13*, 1–4.
56. CIL. *Entre 2022 y 2023 el consumo de lácteos en Ecuador cayó un 12%*; Centro De La Industria Láctea Del Ecuador: Quito, Ecuador, 2024.
57. JECFA. *Toxicological Evaluation of Certain Veterinary Drug Residues in Food- Eighty-first meeting of the Joint FAO/WHO Expert Committee on Food Additives*; WHO: Geneva, Switzerland, 2016; Volume 18.
58. Lankas, G.R.; Minsker, D.H.; Robertson, R.T. Effects of ivermectin on reproduction and neonatal toxicity in rats. *Food Chem. Toxicol.* **1989**, *27*, 523–529. [CrossRef]
59. Hoang, R.; Temple, C.; Correia, M.S.; Clemons, J.; Hendrickson, R.G. Characteristics of ivermectin toxicity in patients taking veterinary and human formulations for the prevention and treatment of COVID-19. *Clin. Toxicol.* **2022**, *60*, 1350–1355. [CrossRef]
60. Vragović, N.; Bažulić, D.; Njari, B. Risk assessment of streptomycin and tetracycline residues in meat and milk on Croatian market. *Food Chem. Toxicol.* **2011**, *49*, 352–355. [CrossRef] [PubMed]
61. Agrocalidad. *Centros de faenamiento habilitados como MABIO utilizarán logo para diferenciar sus productos*; Agencia de Regulación y Control Fito y Zoosanitario: Quito, Ecuador, 2020; Available online: https://actoresproductivos.com/agrocalidad-presenta-oficialmente-el-logotipo-mabio// (accessed on 9 June 2024).

62. *USP 42, Official Monographs, Ivermectin*; United Stated Pharmacopeial Convention: Rockville, MD, USA, 2020; pp. 2555–2556.
63. Nuñez, M.; Palma, M.; Araneda, M.; Pérez, R. Validation of an analytical method and determination of ivermectin residues in sheep tissues. *Rev. Cient.* **2007**, *17*, 6.

Disclaimer/Publisher's Note: The statements, opinions and data contained in all publications are solely those of the individual author(s) and contributor(s) and not of MDPI and/or the editor(s). MDPI and/or the editor(s) disclaim responsibility for any injury to people or property resulting from any ideas, methods, instructions or products referred to in the content.

Article

Dietary Risk Assessment of Cadmium Exposure Through Commonly Consumed Foodstuffs in Mexico

Alejandra Cantoral [1], Sonia Collado-López [2,*,†], Larissa Betanzos-Robledo [3,*,†], Héctor Lamadrid-Figueroa [4], Betzabeth A. García-Martínez [5], Camilo Ríos [6], Araceli Díaz-Ruiz [7], Rosa María Mariscal-Moreno [1] and Martha María Téllez-Rojo [2]

1. Health Department, Iberoamericana University, Mexico City 01219, Mexico; alejandra.cantoral@ibero.mx (A.C.); rosa.mariscal@ibero.mx (R.M.M.-M.)
2. Center for Nutrition and Health Research, National Institute of Public Health, Cuernavaca 62100, Mexico; mmtellez@insp.mx
3. Doctoral Program in Epidemiology, Department of Public Health, National Autonomous University of Mexico, Mexico City 04510, Mexico
4. Department of Perinatal Health, Center for Population Health Research, National Institute of Public Health, Cuernavaca 62100, Mexico; hlamadrid@insp.mx
5. Service of Basic Neuroscience, National Institute of Rehabilitation, Mexico City 14389, Mexico; bgarcia@correo.xoc.uam.mx
6. Research Direction, National Institute of Rehabilitation, Mexico City 14389, Mexico; camrios@yahoo.com.mx
7. Department of Neurochemistry, National Institute of Neurology and Neurosurgery Manuel Velasco Suárez, Mexico City 14269, Mexico; adiaz@innn.edu.mx
* Correspondence: sonia.collado@insp.edu.mx (S.C.-L.); larissa.betanzos@comunidad.unam.mx (L.B.-R.)
† These authors contributed equally to this work.

Abstract: Cadmium (Cd) is a toxic heavy metal widely distributed in foodstuffs. In Mexico, few studies have evaluated Cd content in foods. This study aimed to determine Cd concentrations in foodstuffs that are highly consumed and bought in Mexico City to identify foods exceeding the Maximum Level (ML) and to assess the health risks of theoretical Cd intake from a diet following the Mexican Dietary Guidelines. A total of 143 foodstuffs were analyzed by atomic absorption spectrophotometry. Theoretical Cd intake was estimated in portions per week and compared with the Cd Tolerable Weekly Intake (TWI = 2.5 µg/kg per body weight). A total of 68.5% of the foodstuffs had detectable Cd concentrations. Higher concentrations were found in oyster mushrooms (0.575 mg/kg), romaine lettuce (0.335 mg/kg), and cocoa powder (0.289 mg/kg). Food groups with higher mean concentrations were vegetables (0.084 mg/kg) and snacks, sweets, and desserts (0.049 mg/kg). Ancho chili and romaine lettuce exceed the ML. The theoretical Cd intake estimation was 1.80, 2.05, and 3.82 µg/kg per body weight for adults, adolescents, and school-age children, respectively. This theoretical Cd intake represents a health risk only for school children exceeding the TWI by 53.2%. Our study confirms the presence and risk of Cd in Mexican foodstuffs and highlights the importance of monitoring programs.

Keywords: cadmium; health risk assessment; tolerable weekly intake; monitoring study; food contamination

1. Introduction

Cadmium (Cd) is an environmental pollutant of increasing worldwide concern with numerous adverse health effects [1]. Cd occurs naturally in the environment in its organic form from sources such as volcanic emissions and rock erosion [2]. Additionally, the Cd level has increased as a consequence of human activities, such as production of batteries, electronic instruments, insecticides, fertilizers, and synthetic chemicals [3]. Consequently, it enters into the soil and water, and it can then be absorbed and accumulated in plants and animals, resulting in its consumption and absorption by the human body through the food chain [4,5].

The food supply is the primary source of Cd exposure for the non-smoking general population [6,7]. After dietary exposure, absorption of Cd is estimated to be low (3% to 5%); however, for children, it has been suggested that absorption can be as high as 44% [8]. Within the body, Cd is accumulated in different tissues and organs, particularly in the liver and kidneys [9]. Cd has a long biological half-life in the human body, estimated to be from 16 to 30 years, and chronic exposure to low concentrations of the metal is associated with kidney damage [10], some lung diseases, and high blood pressure [11] and is also a risk factor for osteoporosis [12]. Additionally, Cd has been classified as a human carcinogen by the International Agency for Research on Cancer [13].

Worldwide evidence documented that cereals, vegetables, nuts and pulses, starchy roots and potatoes, as well as meat and derivate products, contribute the most to human Cd exposure [14]. Other foods, such as chocolate and spices, also contain high concentrations of the metal [15,16]. The importance of the presence of Cd in food and the health risk repercussions has led agencies such as the Food and Agriculture Organization/World Health Organization (FAO/WHO) and the European Food Safety Authority (EFSA) Panel to establish the Maximum Levels (MLs) for foods [17] and the Tolerable Weekly Intake (TWI) of 2.5 µg/kg of Cd per body weight (BW) [18].

Additionally, another concern emerges when the foodstuffs recommended by Dietary Guidelines as a part of a healthy diet are also commonly reported with high Cd concentrations [19]. As an example, leafy vegetables, potatoes and grains, peanuts, soybeans, and sunflower seeds are foods with high nutritional properties; however, high concentrations of Cd have been reported in these products [7,18], which could negatively impact human health. In this sense, Cd food monitoring studies and the health risk assessments of its intake are important to ensure sufficient protection for consumers and human health by keeping the concentration of Cd intake as low as possible. However, in Mexico, there is no evidence of these types of studies.

Therefore, this study aimed to determine the Cd concentrations in commonly consumed foodstuffs by the Mexican population and purchased in Mexico City. Secondly, it aimed to identify foods and food groups that exceed the ML for Cd established by the FAO/WHO. Finally, it assessed the health risks of theoretical Cd intake according to a diet that follows the most recent Mexican Dietary Guidelines and compared the results with the Cd TWI.

2. Materials and Methods

A list of 100 foodstuffs most consumed by the Mexican population was identified for Cd concentration analysis using data from the National Health and Nutrition Survey 2018 (ENSANUT) and complemented with additional food items identified in the literature as sources of Cd [20]; therefore, a final sample of 143 was analyzed. Sample collection was performed in retail outlets from Mexico City from 12 April 2022 to 30 January 2023. The selected markets represent the most popular food purchase places (more details on the sample collection have already been published) [21].

Foodstuffs were analyzed at the Neurochemistry Laboratory of the National Institute of Neurology and Neurosurgery in Mexico City; more information on sample management before analysis has already been published [21]. The Cd concentrations were determined using an Atomic Absorption Spectrophotometer (AAS) (Perkin Elmer AAnalyst-600) equipped with a graphite furnace HGA-600 and coupled to an AS800 autosampler. The temperature programming for the method is shown in Table S1. The calibration curve solutions (0.5–2 µg/L) were prepared each day of the analysis by diluting a standard solution with 0.2% ultrapure HNO_3 (Merck, Darmstadt, Germany), and the coefficient of determination was at least 0.99. Extrapolation was used to calculate the cadmium concentration in the sample. A volume of 20 µg/L of the acid digestion was injected directly into the graphite furnace. Cd concentrations were determined in duplicate for each sample. System validation tests were carried out following the recommendations established by the Commission for Analytical Control and Expansion of Coverage (CCAYAC-CR-03/0), and a repeatability test was performed. This test involves evaluating the lower limit of quantification (LoQ) by five-fold (0.5 µg/L, the lower value of the calibration curve) and three concentration values located within the calibration curve: low level 1.5 µg/L, medium level 3 µg/L, and high level 5 µg/L. For compliance with this test, the values obtained should have a maximum coefficient of variation of 20% for the LoQ and 15% for the other determinations. The results obtained were 13.9%, 3.6%, 7.8%, and 1.8%, respectively. Therefore, the proposed method complied with the repeatability test. We considered compliance through validation of the method used to perform duplicate testing. The standard addition technique was used to measure Cd levels for non-perishable foods, while direct acid digestion quantification (no standard added) was performed for perishable foods. The products with high concentrations were re-assessed, and if the values were above the maximum concentration of the calibration curve, the samples were diluted and re-assayed. As an internal control, on every day of analysis, a solution of acid digestion of bovine liver standard (similar digestion as the samples) equivalent to 5 µg/L of Cd was analyzed every 30 samples. A percentage recovery of 100.9 ± 13.6% was obtained from the controls analyzed.

2.1. Chemicals and Reagents Standard

Solutions for Cd AAS (1000 µg/mL) were used as certified calibration standards (Perkin Elmer, Norwalk, CT, USA). Bovine Liver Standard NIST 1577c (Sigma-Aldrich, St. Louis, MO, USA) was used as the internal control. Nitric acid (HNO3) 65% Suprapur® (Merck, Darmstadt, Germany) was used to prepare acid digestions and calibration curves. Dibasic ammonium phosphate (Sigma-Aldrich, St. Louis, MO, USA) and Triton X-100 (Sigma-Aldrich, St. Louis, MO, USA) were used to prepare the matrix modifier. All solutions were prepared with deionized water obtained from a Direct-Q 3 UV purification system (Millipore, Bedford, MA, USA).

2.2. Sample Treatment

All samples were unpacked and inspected, and those that could decompose were stored at −20 °C until further experimentation to avoid decomposition. Foods packaged in paper or cardboard, plastic, or other containers were only cleaned with deionized water twice. The fruits were washed with soap and deionized water (twice). Those with inedible peel (such as bananas, mangoes, and oranges) were peeled, and only the pulp was used for the next steps. In the case of fruits such as apples (commonly consumed with the peel), the whole fruit was used. Meats (chicken, beef, pork) and eggs (without shell) were not cooked and were sampled raw for the next steps. The cereals and legumes were analyzed without processing (raw samples). The solid items underwent dehydration at 80 °C for 72 h, and the liquid samples were processed on a wet-weight basis. All solid items were ground using a house grinder and finally stored in polypropylene tubes until analysis. The digestion was carried out considering the recommendations outlined in Mexican Official NOM-117-SSA1-1994. The liquid and solid samples and the reference material were weighed in duplicate (0.1–0.2 g for solids) in test tubes, and 2 mL of 65% Suprapur® HNO_3 was added. The tubes were covered, mixed, and left at room temperature for 12 h. Subsequently, they were placed in a water bath (Labline instrument: shaking water bath) at 60 °C until a clear solution was obtained. Samples of 100 µL of acid digestion were placed directly into the autosampler for analysis. Those samples that presented concentrations higher than the calibration curve were diluted with 0.2% ultrapure HNO_3.

2.3. Statistical Analysis

First, to perform the data characterization and summary, we averaged the Cd concentrations (mg/kg) obtained by duplicate per foodstuff analyzed and expressed as means and standard deviations (SD). All food items were encompassed in 14 food groups according to the Mexican Dietary Guidelines or to their nutritional characteristics [22,23]: (1) baby foods; (2) beverages; (3) cereals; (4) condiments and spices; (5) dairy; (6) fats and oils; (7) fruits; (8) legumes; (9) meats, sausages, and eggs; (10) nuts and seeds; (11) seafood; (12) snacks, sweets, and desserts; (13) soups and creams; and (14) vegetables. For each food group, we calculated the percentage of foodstuffs with non-detected Cd concentrations (<LoQ) and the percentage of foodstuffs with detected Cd concentrations (>LoQ). The mean, SD, and range (minimum and maximum) of Cd concentrations were estimated for each of the 14 food groups, considering only detected Cd values >LoQ.

Second, in order to identify the foodstuffs that exceeded the ML, the obtained Cd concentrations (mg/kg) were compared with the Cd ML in foods established by the Codex Alimentarius Commission of the FAO/WHO [17].

Finally, to perform the health risk assessment of Cd intake from Mexican foodstuffs, we determined the theoretical Cd intake in a diet that complies with the Mexican Healthy and Sustainable Guidelines recommendations [23]. Mexican guidelines established portion recommendations of intake according to age and sex population groups for the different food groups. On average, these guidelines recommend a daily intake of 4 to 5 portions of vegetables, 2 to 3 portions of fruits, 1 to 2 portions of legumes, 5 to 12 portions of cereals, and 1 to 2 portions of dairy, as along with weekly recommendations for the intake of food of animal origin, as part of a healthy diet. We obtained the average recommended number of portions per week for adults, adolescents, and school children. As a first step, we assigned at least one portion of each foodstuff analyzed per food group (considering a varied diet); as a second step, we considered those foods that are most consumed (i.e., maize) as a portion repeated throughout the week [24]. Then, we estimated the theoretical Cd intake (μg/per portion) for each portion of each food item and food group, considering moisture and fat changes that could occur during cooking [25].

Finally, for the health risk assessment analysis, we calculated the theoretical Cd intake for each age group, considering the average BW for adults, adolescents, and school children of 70 kg, 56 kg, and 25 kg, respectively [26]. Finally, we compared these results to the Cd TWI established by the EFSA of 2.5 μg/kg BW to ensure sufficient protection against Cd intake [18].

As an additional analysis, to ensure the robustness of our analysis, we employed Chauvenet's criterion [27] to identify extreme values. To accomplish this, we compared the Cd concentrations obtained in our study with those reported from the U.S. [28] and the U.K. [29] (Supplementary Figure S1), and we excluded foodstuffs identified with extreme Cd concentrations from the theoretical Cd calculation.

All the analyses were conducted using STATA 17 statistical software (StataCorp LLC, College Station, TX, USA).

3. Results

3.1. Cadmium Concentrations in Mexican Foodstuffs

A total of 143 foodstuffs were analyzed, and the detected Cd (>LoQ) concentrations were identified in 68.5% ($n = 98$), ranging from 0.004 mg/kg to 0.575 mg/kg. We identified 45 foodstuffs below the LoQ. The top ten foodstuffs with the highest Cd concentrations were oyster mushrooms "*Pleorotus ostreatus*" (0.575 mg/kg), romaine lettuce "*Lactuca sativa* L. var. *longifolia*" (0.335 mg/kg), cocoa powder (0.289 mg/kg), chocolate powder (0.117 mg/kg), saladette tomatoes "*Solanum lycopersicum roma*" (0.095 mg/kg), bread-crumbs (0.069 mg/kg), chocolate bars (0.060 mg/kg), ancho chilies "*Capsicum annuum*" (0.059 mg/kg), chard "*Beta vulgaris* var. *cicla*" (0.058 mg/kg), and mushrooms "*Agaricus bisporus*" (0.055 mg/kg). All these items came from vegetable sources. The complete list of the 143 foodstuff samples evaluated is presented in Table S2.

Table 1 shows the average estimate of Cd concentration by food group. Almost all food groups contain foodstuffs with detected Cd concentrations, from 22 to 100% of the analyzed samples per group. The food groups with the highest percentage of detected Cd concentrations were seafood, the food group with 100% detected Cd concentrations, followed by cereals (96%, n = 23), vegetables (89%, n = 17), legumes (87%, n = 7), condiments and spices (82%, n = 9), snacks, sweets, and deserts (78%, n = 14), and baby foods (75%, n = 9).

Table 1. Cadmium concentrations (mg/kg) by food group analyzed (n = 143).

Food Group	Foodstuffs Analyzed	Number of Foodstuffs Analyzed (n)	Foodstuffs with Non-Detectable Cd Concentrations n (%)	Foodstuffs with Detectable Cd Concentrations n (%)	Mean (SD) of Detectable Cd Concentrations (mg/kg)	Range of Detectable Cd Concentrations (min–max)
Baby foods	Apple Juice; Carrot Porridge, Chicken, Vegetables and Rice Porridge; Infant Formula Soy Milk (Brand 1); Infant Formula Whole Milk (Brand 2); Infant Formula Whole Milk (Brand 1); Infant Formula Whole Milk (Brand 2); Infant Formula Whole Milk (Brand 3); Infant Grow and Gain Strawberry Shake; Infant Rice Cereal (Brand 1); Infant Rice Cereal (Brand 2); Strawberry and Apple Cereal Snack.	12	3 (25%)	9 (75%)	0.010 (0.002)	0.004–0.023
Beverages	Bottled Soft Drink; Soluble Coffee.	2	1 (50%)	1 (50%)	na	na
Cereals	Amaranth "*Amaranthus* spp."; Breadcrumbs; Breakfast Cereal; Maize "*Zea mays*"; Maize grain "*Zea mays*"; Maize Flour; Crackers; French Fries; Oat "*Avena sativa*"; Potato "*Solanum tuberosum*"; Pre-Cooked Rice; Rice "*Oryza sativa*" (Brand 1); Rice "*Oryza sativa*" (Brand 2); Rice "*Oryza sativa*" (Brand 3); Rice Cake; Rice Flour; Wheat Cookies; Wheat Flour (Brand 1); Wheat Flour (Brand 2); Wheat Tortillas; White Bread (Bakery); White Bread (Brand 1); Whole Wheat Bread (Bakery); Whole Wheat Bread (Brand 1).	24	1 (4%)	23 (96%)	0.026 (0.002)	0.005–0.069
Condiments and spices	Ancho Chilies "*Capsicum annuum*"; Black Pepper "*Piper nigrum L*"; Canned Green Chilies; Chicken Broth Cubes; Chili Powder; Chilies "*Capsicum annuum* 'Guajillo'"; Guajillo Industrialized Sauce; Mole; Paprika "*Capsicum annuum*"; Saffron "*Crocus sativus*"; Turmeric "*Curcuma longa*".	11	2 (18%)	9 (82%)	0.028 (0.004)	0.009–0.059

Table 1. Cont.

Food Group	Foodstuffs Analyzed	Number of Foodstuffs Analyzed (n)	Foodstuffs with Non-Detectable Cd Concentrations n (%)	Foodstuffs with Detectable Cd Concentrations n (%)	Mean (SD) of Detectable Cd Concentrations (mg/kg)	Range of Detectable Cd Concentrations (min–max)
Dairy	Asadero Cheese; Manchego Cheese; Natural Yogurt; Petit Suisse; Whole Liquid Milk (Brand 1); Whole Liquid Milk (Brand 2).	6	6 (100%)	0 (0%)	na	na
Fats and oils	Hass Avocado "*Persea americana*"; Butter; Lard; Margarine; Mayonnaise; Sour Cream (Brand 1); Sour Cream (Brand 2); Vegetable Oil (Brand 1); Vegetable Oil (Brand 2).	9	7 (78%)	2 (22%)	0.008 (0.001)	0.008–0.009
Fruits	Golden Yellow Apple "*Malus domestica* 'golden delicious'"; Grape "*Vitis vinifera*"; Grapefruit "*Citrus paradisi*"; Guava "*Psidium guajava*"; Lime "*Citrus autantifolia*"; Melon "*Cucumis melo*"; Orange "*Citrus sinensis* L."; Strawberry "*Fragaria x anassa*"; Ataulfo Mango "*Mangifera indica* 'Ataulfo'"; Manila Mango "*Mangifera indica* 'Manila'"; Banana Tabasco "*Musa × paradisiaca*"; Papaya "*Carica papaya*"; Watermelon "*Citrullus lanatus*".	13	10 (77%)	3 (23%)	0.037 (0.006)	0.016–0.050
Legumes	Black Beans "*Phaseolus vulgaris*"; Black Canned Beans; Chickpeas "*Cicer arietinum*"; Fava Beans "*Vicia faba*"; Lentils "*Lens culinaris*"; Lentils Instant Soup; Soybean "*Glycine max* L. Merr."; White Beans "*Phaseolus vulgaris*".	8	1 (13%)	7 (87%)	0.009 (0.002)	0.006–0.017
Meats, sausages and eggs	Beef (Brand 1); Beef (Brand 2); Chicken; Chicken Liver; Eggs (Brand 1); Eggs (Brand 2); Pork (Brand 1); Pork (Brand 2); Pork Ham (Brand 1); Pork Ham (Brand 2); Sausage "Longaniza"; Turkey Sausages (Brand 1); Turkey Sausages (Brand 2).	13	6 (46%)	7 (54%)	0.011 (0.002)	0.006–0.015

Table 1. Cont.

Food Group	Foodstuffs Analyzed	Number of Foodstuffs Analyzed (n)	Foodstuffs with Non-Detectable Cd Concentrations n (%)	Foodstuffs with Detectable Cd Concentrations n (%)	Mean (SD) of Detectable Cd Concentrations (mg/kg)	Range of Detectable Cd Concentrations (min–max)
Nuts and seeds	Peanuts; Sunflower Seeds.	2	1 (50%)	1 (50%)	na	na
Seafood	Canned Tuna (Brand 1); Canned Tuna (Brand 2); Fresh Tuna.	3	0 (0%)	3 (100%)	0.016 (0.001)	0.010–0.019
Snacks, sweets and desserts	Chamoy Candy; Chewing Gum; Chocolate Bar; Chocolate Powder; Cocoa Powder; Dark Chocolate Bars; Honey; Jelly; Marzipan Candy; Pastries; Popsicle; Potato Chips (Brand 1); Potato Chips (Brand 2); Sweet Cookie; Tamarind Candy; Tamarind Poblano Candy; Tamarind Popsicle; Wheat Chips.	18	4 (22%)	14 (78%)	0.049 (0.005)	0.004–0.289
Soups	Canned Vegetable Soup; Instant Pasta Soup; Pasta Soup to Prepare.	3	1 (33%)	2 (67%)	0.032 (0.003)	0.030–0.033
Vegetables	Broccoli "*Brassica oleracea* var. *italica*"; Cabbage "*Brassica oleracea* var. *capitata* L."; Carrot "*Daucus carota* subsp. *sativus*"; Cauliflower "*Brassica oleracea* var. *botrytis*"; Chard "*Beta vulgaris* var. *cicla*"; Chayote "*Sechium edule*"; Coriander "*Coriandrum sativum*"; Cucumbers "*Cucumis sativus*"; Fresh Green Chilies "*Capsicum annuum* 'Serrano'"; Green Beans "*Phaseolus vulgaris*"; Jicama "*Pachyrhizus erosus*"; Mushrooms "*Agaricus bisporus*"; Nopal Cactus "*Opuntia ficus-indica* L."; Romaine Lettuce "*Lactuca sativa l.* var. *longifolia*"; Saladette Tomatoes "*Solanum lycopersicum*"; Oyster Mushrooms "*Pleurotus ostreatus*"; Spinach "*Spinacia oleracea*"; White Onions "*Allium cepa*"; Zucchini "*Cucurbita pepo* L.".	19	2 (11%)	17 (89%)	0.084 (0.011)	0.006–0.575

Abbreviations: Cd: cadmium; na: not applicable because the value corresponds to a concentration <LoQ. Non-detectable concentrations were defined as concentrations below the limit of quantification (<LoQ). Detectable concentrations were defined as concentrations above the limit of quantification (>LoQ).

From these food groups with detectable Cd concentrations, higher means were found in vegetables [0.084 mg/kg (range: 0.006–0.575 mg/kg)], followed by snacks, sweets, and desserts [0.049 mg/kg (range: 0.004–0.289 mg/kg)], fruits [0.037 mg/kg (range: 0.016–0.050 mg/kg)], condiments and spices (0.028 mg/kg [range: 0.009–0.059 mg/kg]), and cereals [0.026 mg/kg (range: 0.005–0.069 mg/kg)]. The lowest detected mean Cd concentrations were found in the food groups of fats and oils [0.008 mg/kg (range: 0.008–0.009 mg/kg)], legumes [0.009 mg/kg (range: 0.006–0.017)], and baby foods [0.010 mg/kg (range: 0.004–0.023)]. The only food group with no detected concentrations was dairy, for which all items were below the LoQ.

3.2. Mexican Foodstuffs That Exceed the Food and Agriculture Organization/World Health Organization Maximum Level (ML) Established for Cadmium

In Table 2, we observed that, of the 98 foodstuffs with detectable Cd concentrations, 61 had a reference value of ML established by the FAO/OMS. Of these 61 foodstuffs, two exceeded the ML: ancho Chili "*Capsicum annuum*" (0.059 mg/kg) and romaine lettuce "*Lactuca sativa* L. var. *longifolia*" (0.335 mg/kg) Broccoli "*Brassica oleracea* var. *italica*" (0.046 mg/kg) was within the limited concentration. In addition, it is important to highlight that foodstuffs such as saladette tomatoes "*Solanum lycopersicum*", mushrooms "*Agaricus bisporus*" and "*Pleorotus ostreatus*" have no reference ML values to compare, and both foodstuffs presented the highest Cd concentrations in this study.

Table 2. Comparison between cadmium concentrations in Mexican foodstuffs and the Food and Agriculture Organization/World Health Organization Maximum Level established for cadmium ($n = 61$).

Mexican Foodstuffs Analyzed per Food Group	Cd Mean Concentration (mg/kg)	FAO/WHO Product Name Classification	FAO/WHO ML for Cd (mg/kg)
Baby foods			
Carrot Porridge	0.004	Root and tuber vegetables ˆ	0.1
Infant Rice Cereal (Brand 1)	0.016	Rice, Polished	0.4
Infant Rice Cereal (Brand 2)	0.023	Rice, polish	0.4
Strawberry and Apple Cereal Snack	0.013	Cereal grains	0.1
Cereals			
Amaranth "*Amaranthus* spp."	0.019	Cereal grains	0.1
Breadcrumbs	0.069	Wheat ˆ	0.2
Breakfast Cereals	0.008	Cereal grains	0.1
Crackers	0.011	Wheat ˆ	0.2
French Fries (Street Stand)	0.042	Root and tuber vegetables ˆ	0.1
Maize Flour	0.005	Cereal grains	0.1
Maize "*Zea mays*"	0.007	Cereal grains	0.1
Oat "*Avena sativa*"	0.005	Cereal grains	0.1
Pre-Cooked Rice	0.017	Rice, polish	0.4
Potato "*Solanum tuberosum*"	0.009	Root and tuber vegetables	0.1
Rice Cake	0.024	Rice, polish ˆ	0.4
Rice Flour	0.037	Rice, polish ˆ	0.4
Rice "*Oryza sativa*" (Brand 1)	0.030	Rice, polish	0.4
Rice "*Oryza sativa*" (Brand 2)	0.027	Rice, polish	0.4
Rice "*Oryza sativa*" (Brand 3)	0.031	Rice, polish	0.4
Wheat Flour (Brand 1)	0.032	Wheat	0.2
Wheat Flour (Brand 2)	0.040	Wheat	0.2
Wheat Tortillas	0.028	Wheat	0.2
Wheat Cookies	0.033	Wheat ˆ	0.2
White Bread (Brand 1)	0.033	Wheat ˆ	0.2
White Bread (Bakery)	0.037	Wheat ˆ	0.2
Whole Wheat Bread (Bakery)	0.037	Wheat ˆ	0.2
Whole Wheat Bread (Brand 1)	0.033	Wheat ˆ	0.2
Condiments and spices			

Table 2. Cont.

Mexican Foodstuffs Analyzed per Food Group	Cd Mean Concentration (mg/kg)	FAO/WHO Product Name Classification	FAO/WHO ML for Cd (mg/kg)
Ancho Chili "*Capsicum annuum*"	**0.059**	Fruiting vegetable	**0.05**
Guajillo Chili "*Capsicum annuum 'Guajillo'*"	0.018	Fruiting vegetable	0.05
Legumes			
Black Beans "*Phaseolus vulgaris*"	0.017	Pulses	0.1
Black Canned Beans	0.007	Pulses	0.1
Chick Beans "*Cicer arietinum*"	0.008	Pulses	0.1
Fava Bean "*Vicia faba*"	0.009	Pulses	0.1
Lentils "*Lens culinaris*"	0.006	Pulses	0.1
Lentils Instant Soup	0.007	Pulses	0.1
Soybean "*Glycine max L. merr.*"	0.009	Pulses	0.1
Snacks, sweets, and desserts			
Cocoa Powder	0.289	Chocolate ≥70% total cocoa solids	0.9
Chocolate Bars	0.060	Chocolate containing or declaring ≥ 50% to < 70% total cocoa solids	0.8
Chocolate Powder	0.117	Chocolate containing or declaring ≥ 50% to < 70% total cocoa solids	0.8
Dark Chocolate Bars	0.029	Chocolate containing or declaring ≥ 50% to < 70% total cocoa solids	0.8
Pastries	0.028	Wheat ^	0.2
Potato Chips (Brand 1)	0.029	Root and tuber vegetables ^	0.1
Potato Chips (Brand 2)	0.041	Root and tuber vegetables ^	0.1
Sweet Cookies	0.018	Wheat ^	0.2
Wheat Chips	0.011	Wheat ^	0.2
Soups			
Instant Pasta Soup	0.030	Wheat	0.2
Pasta Soup to Prepare	0.033	Wheat	0.2
Vegetables			
Broccoli "*Brassica oleracea* var. *italica*"	0.046	Brassica vegetables	0.05
Cabbage "*Brassica oleracea* var. *capitata* L."	0.007	Brassica vegetables	0.05
Cauliflower "*Brassica oleracea* var. *botrytis*"	0.036	Brassica vegetables	0.05
Carrot "*Daucus carota* subsp. *sativus*"	0.016	Root and tuber vegetables	0.1
Chard "*Beta vulgaris* var. *cicla*"	0.058	Leafy vegetables	0.2
Coriander "*Coriandrum sativum*"	0.023	Leafy vegetables	0.2
Cucumber "*Cucumis sativus*"	0.041	Fruiting vegetable	0.05
Fresh Green Chilies "*Capsicum annuum 'Serrano'*"	0.014	Fruiting vegetable	0.05
Green Beans "*Phaseolus vulgaris*"	0.006	Legume vegetables	0.1
Nopal Cactus "*Opuntia ficus-indica* L."	0.016	Leafy vegetables ^	0.2
Romaine Lettuce "*Lactuca sativa* L. var. *longifolia*"	**0.335**	Leafy vegetables	**0.2**
Spinach "*Spinacia oleracea*"	0.059	Leafy vegetables	0.2
White Onions "*Allium cepa*"	0.022	Bulb vegetable	0.05
Zucchini "*Cucurbita pepo* L."	0.025	Fruiting vegetables	0.05

Abbreviations: ML: Maximum Level; Cd: cadmium; FAO: Food and Drug Administration; WHO: World Health Organization. ^ Foods were grouped into this category based on their nature, even though these specific foods are not detailed by the FAO/WHO product name classification. Source: FAO and WHO General Standard for Contaminants and Toxins in Food and Feed. Note: FAO/WHO product name classification of fruiting vegetables does not apply to tomatoes and edible fungi. Cd concentrations in bold exceed ML.

3.3. Health Risk Assessment of Theoretical Cadmium Intake Based on the Adherence to the Healthy and Sustainable Dietary Guidelines for the Mexican Population

Table 3 and Figure 1 present the results of the estimated theoretical Cd intake per kg of BW for adults, adolescents, and school children, following the recommended weekly food portions per food group in the Mexican Healthy and Sustainable Dietary Guidelines.

Table 3. Calculation of the theoretical cadmium intake per age group according to average body weight, following the recommended weekly portions in the Healthy and Sustainable Guidelines for the Mexican Population.

	Theoretical Cd Intake for Adults			Theoretical Cd Intake for Adolescents			Theoretical Cd Intake for School Children		
Food Group	Recommended Portions per Week	Cd Intake (µg/per Portion)	Theoretical Cd Intake per 70 kg of BW	Recommended Portions per Week	Cd Intake (µg/per Portion)	Theoretical Cd Intake per 56 kg of BW	Recommended Portions per Week	Cd Intake (µg/per Portion)	Theoretical Cd Intake per 25 kg of BW
Vegetables	31.5	85.415	1.220	31.5	73.627	1.315	28.0	60.267	2.411
Fruits	14.0	10.560	0.151	21	10.560	0.189	21.0	10.560	0.422
Legumes	10.5	4.418	0.063	10.5	4.068	0.073	10.5	4.068	0.163
Cereals	59.5	21.996	0.314	66.5	23.317	0.416	49.0	17.427	0.697
Dairy	10.5	0.000	0.000	10.5	0.000	0.000	7.0	0.000	0.000
Food of animal origin									
Beef	3.5	0.525	0.008	3.5	0.525	0.009	3.0	0.450	0.018
Other red meat	3.0	0.000	0.000	3.0	0.000	0.000	2.5	0.000	0.000
Chicken	9.0	0.000	0.000	8.0	0.000	0.000	6.5	0.000	0.000
Seafood	3.5	2.073	0.030	3.5	2.073	0.037	3.5	2.073	0.083
Eggs	7.0	1.078	0.015	7.0	1.078	0.019	5.0	0.770	0.031
Total per week		126.07 µg	1.80 µg/kg per BW		115.25 µg	2.06 µg/kg per BW		95.62 µg	3.83 µg/kg per BW

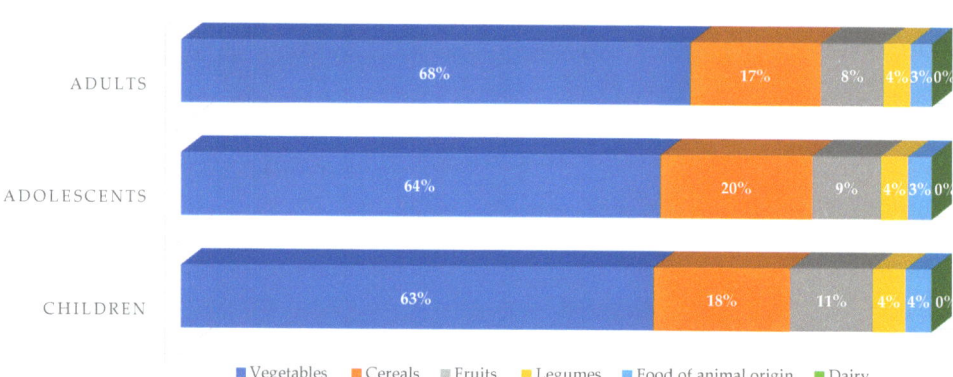

Figure 1. Contribution per food group to the total theoretical dietary weekly cadmium intake by age group, following the recommendations of the Healthy and Sustainable Guidelines for the Mexican population.

Following this guideline would result in a weekly theoretical Cd intake of 1.80 µg/kg of BW, 2.06 µg/kg per BW, and 3.83 µg/kg per BW for adults, adolescents, and school children, respectively. Compared to the TWI (2.5 µg/kg per BW), this theoretical Cd intake does not represent a health risk for adults and adolescents, but for school children, a theoretical Cd intake of 3.83 µg/kg of BW exceeds the TWI by 53.2% (Table 3).

The food groups with the highest contribution to the theoretical Cd intake in all age groups were vegetables and cereals, with more than 80% of the combined contribution (Figure 1).

It is important to mention that these guidelines do not include unhealthy foods with detectable Cd concentrations, such as chocolate powder (0.117 mg/kg), chocolate bars (0.060 mg/kg), dark chocolate bars (0.029 mg/kg), potato chips (0.035 mg/kg), pastries (0.028 mg/kg), and different candies (average 0.011 mg/kg). Therefore, including these foodstuffs in the theoretical Cd intake would imply a greater risk of exceeding Cd TWI, with prolonged consumption potentially adversely affecting health.

3.4. Additional Analysis: Comparison Between Cd Concentrations Detected in Mexican Foodstuffs with Total Diet Studies from the U.S. and the U.K.

Among the 143 Mexican foodstuffs that were analyzed, we found 79 that were comparable to those reported by TDSs in the U.S. or U.K. (between 77 and 54 were equivalent to the U.S. and U.K. data, respectively). Overall, Mexico had a high Cd concentration in many foodstuffs (Figure S1). In the U.S., the highest Cd concentrations were identified in sunflower seeds (0.333 mg/kg), spinach (0.222 mg/kg), lettuce (0.061 mg/kg), and French fries (0.058 mg/kg). In Mexico, sunflower seeds had Cd concentrations <LoQ, while spinach (0.050 mg/kg) and French fries (0.042 mg/kg) showed lower concentrations compared to the U.S., but lettuce presented a higher concentration (0.335 mg/kg). For the U.K., the highest Cd concentrations were reported in potato chips (0.078 mg/kg), chicken liver (0.032 mg/kg), French fries (0.031 mg/kg), and flour (0.030 mg/kg). In Mexico, lower concentrations were identified in potato chips (0.035 mg/kg) and chicken liver (0.012 mg/kg), but higher concentrations were found in French fries (0.042 mg/kg) and wheat flour (0.036 mg/kg). Consistently, romaine lettuce, French fries, potato chips, potatoes, wheat flour, whole wheat bread, breakfast cereals, white bread, canned tuna, dark chocolate bars, sweet cookies, carrots, oatmeal, and pastries were identified as relevant sources of Cd concentrations across all three countries.

4. Discussion

This study is the first effort to evaluate Cd concentrations in commonly consumed Mexican foodstuffs and the possible health risks associated with their consumption. We found that 68.5% (n = 98) of the foodstuff samples had detectable concentrations of Cd, with concentrations ranging between 0.004 mg/kg and 0.575 mg/kg. As expected, crop-sourced food items such as mushrooms, lettuce, tomato, and cocoa presented the highest Cd concentrations. Concerning mushrooms, these results are not surprising. It is documented that they have a very effective mechanism to accumulate metals such as Cd from the environment [30]. In the case of lettuce, Cd possesses a high capacity of being transferred from the soil to the leaves [31]. With respect to cocoa, a product derived from cacao, consistent literature reports that cacao beans tend to bioaccumulate Cd [32], which is related to environmental factors and the chemical composition of cacao, which allows strong binding of Cd in cacao tissues and cocoa products. This has become a global concern; a review on ML of Cd in cocoa reported high Cd levels in the main Latin American producing countries. In addition, a study that assessed the concentration of different elements, including Cd, in 155 chocolate samples from the U.S. market identified high mean concentrations of Cd in samples from Central America countries (which included five samples from Mexico), where cocoa Cd concentrations were found to be as high as 689 μg/kg [33].

In terms of food groups, we identified that vegetables (0.084 mg/kg), snacks, sweets and desserts (0.049 mg/kg), and cereals (0.025 mg/kg) were the food groups with the highest Cd concentrations; additionally, cereals were the food group with the most items with detectable Cd concentrations (96%). Consistently cereal crops such as rice, wheat, and maize were identified as major sources of Cd exposure for humans. Leading to extensive efforts to understand the mechanisms of Cd accumulation in these foods, specific genes and transporters involved in Cd uptake and transport have been identified in rice, with similar pathways observed in wheat [34].

Similar to our results, diverse studies have shown that Cd concentrations in vegetable and cereal groups are the main contributors to Cd intake. Using data from the National Health and Nutrition Examination Survey (NHANES), a group of researchers in the United States found that the food groups that contributed the most to Cd intake were cereals and bread (34%), leafy vegetables (20%), potatoes (11%), legumes and nuts (7%), and stem/root vegetables (6%) [7]. EFSA's reports state that cereals and cereal products, vegetables, nuts and pulses, starchy roots and potatoes, as well as meat and meat products, are the food groups that contribute the most to human Cd exposure [18]. In China, cereals (46.2%), vegetables (19.2%), and aquatic food (18.4%) contributed the most to the dietary Cd exposure of the Chinese population [35]. In addition, in a study of the Shenzhen population, vegetables were identified as the principal food group contributor to dietary Cd exposure at 32.6%, followed by rice and its products (19.2%), fish, seafood, and shellfish (18.5%), and legumes, nuts, and their products (14.5%) [36]. Similarly, in a contaminated area of Thailand, the major food groups identified as the principal contributors to Cd exposure were rice and grains, shellfish and seafood, meat, and vegetables [37].

In the present analysis the dairy group was the only food group with no detected Cd concentrations. Similarly, a report from Egypt estimated that raw milk contributed only 1.59% of the estimated weekly intake [38]. Moreover, in Peru, 19 milk samples were evaluated, and the Cd concentrations were low (0.007 ± 0.006 mg/kg), with the intake at all ages being below the estimated Cd TWI [39]. A review conducted in China identified 13 studies that evaluated Cd concentrations in milk and milk products, showing contrasting results, highlighting that Cd concentrations in milk in China were lower than those reported in milk from other developing countries but were higher than those in developed countries [40].

We identified one previous analysis in Mexico concerning dietary Cd sources. The authors identified potential dietary contributors of Cd, such as chorizo (mean 0.0302 mg/kg), sausages (mean 0.0032 mg/k), ham 0.0056 mg/kg), and chicken breast (0.0007 mg/kg). Similarly, in this analysis, we also identified Cd in chorizo (sausage "longaniza") (mean 0.012 mg/kg), in turkey sausages (Brand 2) (0.015 mg/kg), and in two brands of pork ham (mean 0.011 mg/kg), but not in chicken breast (<LoQ) [41].

The vegetable food group was the food group with the highest Cd values, as many vegetables had very high Cd concentrations. This highlights the need to maintain a varied diet within the food groups to reduce the daily intake of the foods identified with the highest concentration. A previous cohort study in Mexico City identified that refined grains and vegetables contributed the most to dietary Cd intake for pregnant women and children, with over 60% of the vegetable contribution coming from leafy greens [42]. This result is similar to our estimation, where over 60% of the Cd contribution came from the vegetable group.

Concerning the Cd intake results, a diet that follows the adherence to the Mexican Healthy and Sustainable Guidelines would result in a weekly theoretical Cd intake below the TWI for both adults and adolescents (1.80 µg/kg BW and 2.05 µg/kg BW, respectively), and above the TWI for school-age children (3.82 µg/kg of BW). According to another study, with information from individuals aged 2 years and older from the National Health and Nutrition Examination Survey (NHANES) 2007–2012, the average dietary Cd consumption in the U.S. general population was 0.54 µg/kg BW/week (22% of the TWI) [7], which was lower than the levels found in our study [7]. Additionally, data from the sixth Chinese TDS reported a mean Cd intake for male adults of 8.26 µg/kg BW per month (2.60–30.02 µg/kg BW per month), which would represent a Cd intake of 2.07 µg/kg BW per week, which is higher than our estimated Cd intake for adults (1.80 µg/kg BW/week) [43].

Regarding the intake results for children, another study in the U.S. found that the Cd intake in infants and young children who regularly consumed rice, spinach, oats, barley, potatoes, and wheat did not exceed the daily intake set by the EFSA but exceeded the Cd exposure set by the Agency for Toxic Substances and Disease Registry (ATSDR), which is considered a stricter level [44].

Similar to our results, a cohort study based in Mexico City evaluated children's dietary Cd intake using data from various countries on Cd concentrations in foods. The study reported Cd intakes of 4.43 ± 2.53 µg/d, 4.65 ± 2.45 µg/d; 6.00 ± 3.45 µg/d; 6.83 ± 3.15 µg/d and 8.09 ± 4.33 µg/d at 1, 2, 4, 6 and 9 years of age, respectively. It also identified leafy greens as top contributors at ages 1 and 2 (1 year: 16.0%, 2 years: 9.0%); at age 4, sweets (6.8%); at age 6, lettuce (6.8%) and sweets (5.5%). Lastly, at age 9, the major contributors were lettuce (6.0%), pasta soup (5.7%), and sweets (5.5%) [45]. Moreover, a previous study in Mexico indicated that 16–64% of children exceed the TWI at ages 1 to 9 years [45].

In our study, vegetables, and cereals were the food groups with the highest contribution to the theoretical Cd intake in a diet adhering to the Mexican Healthy and Sustainable Diet Guidelines across all age groups, with more than 80% of the combined contribution. This is consistent with the major sources of Cd reported in the U.S., where cereals and bread represent 34% and leafy vegetables 20% [7]. Vegetable consumption is recommended because they are a source of vitamins, fiber, and bioactive components; however, the increasing evidence of the accumulation of toxic metals such as Cd requires routine monitoring and highlights the recommendation of consuming a varied diet to avoid excess exposure to a particular foodstuff [46].

Our study has important limitations. First, we did not have a representative foodstuff sample, as all samples were bought from a specific area of Mexico City. In addition, a random selection of stores and foodstuffs was not carried out, so the variability in Cd content in foods from different regions of the country, as well as variations due to seasonality, could not be ascertained. Second, we had a small sample size of the analyzed foodstuffs. Third, our Cd concentrations were measured by AAS, and these results differ from other methods, such as ICP-MS. Third, the dietary theoretical Cd estimation was based on a diet that follows the Healthy and Sustainable Mexican Guidelines recommendations and differed from those obtained using other methods reported in the literature that based Cd intake estimations on Food Frequency Questionnaires. It is important to mention that these guidelines do not include unhealthy foods; therefore, our theoretical Cd intake does not include foodstuffs "not recommended or unhealthy", such as the snacks, sweets, and desserts group, which was the second food group with the highest mean Cd concentrations (0.049 mg/kg). In our results, this group included items with detectable Cd concentrations, such as chocolate powder (0.117 mg/kg), chocolate bars (0.060 mg/kg), dark chocolate bars (0.029 mg/kg), potato chips (0.035 mg/kg), pastries (0.028 mg/kg), and different candies (average 0.011 mg/kg). Therefore, including these foodstuffs in the estimation of theoretical Cd intake would imply a greater risk of exceeding Cd TWI, whose prolonged consumption could adversely affect health, especially for young children, because of their higher absorption and lower detoxification capacities [47].

5. Conclusions

Our study confirms that foodstuffs bought from retail stores in Mexico City are an important source of Cd intake and could represent a health risk for school children. This highlights the importance of monitoring the presence of contaminants in the foods consumed by the population, especially in those widely recommended by the Healthy and Sustainable Dietary Guidelines. The frequent consumption of those foods with detectable Cd concentrations could be considered chronic exposure, which is particularly important for children. Therefore, food quality and safety are among the most important public health concerns, and the food available on the market should be free of chemical contaminants that pose a risk to consumer health. Food safety is not only the responsibility of food producers but also of state governments and agencies, which should systematically monitor their safety as an essential strategy for public health actions.

Supplementary Materials: The following supporting information can be downloaded at: https://www.mdpi.com/article/10.3390/foods13223649/s1, Table S1: Instrumental condition used in the analysis of Cd, Table S2: Complete information for each of the Mexican Foodstuffs analyzed for the determination of Cd concentrations (n = 143); Figure S1: Comparison of cadmium concentrations (mg/kg) between Mexico and reported Cadmium concentrations in U.S and U.K total diet studies (n = 79).

Author Contributions: Conceptualization, A.C. and M.M.T.-R.; methodology, A.C., B.A.G.-M., A.D.-R., M.M.T.-R., S.C.-L., L.B.-R., R.M.M.-M., C.R., and H.L.-F.; software, S.C.-L. and H.L.-F.; validation, C.R., B.A.G.-M., and R.M.M.-M.; formal analysis, S.C.-L., L.B.-R., and H.L.-F.; investigation, A.C., S.C.-L., and L.B.-R.; data curation, S.C.-L.; writing—original draft preparation, A.C, S.C.-L., and L.B.-R.; writing—review and editing, M.M.T.-R., H.L.-F., B.A.G.-M., C.R., and R.M.M.-M.; visualization, S.C.-L. and L.B.-R.; supervision, M.M.T.-R.; resources A.D.-R.; project administration B.A.G.-M.; funding acquisition, A.C. All authors have read and agreed to the published version of the manuscript.

Funding: This research was funded by Pure Earth Mexico/Blacksmith Institute. Collado-López (#556419) and Betanzos-Robledo (#854860) are supported by Consejo Nacional de Humanidades Ciencias y Tecnologías (CONAHCYT) as a research assistant and a doctoral student, respectively.

Institutional Review Board Statement: Not applicable.

Informed Consent Statement: Not applicable.

Data Availability Statement: The data presented in this study are available on request from the corresponding author. The data are not publicly available due to privacy restrictions.

Acknowledgments: The authors acknowledge the Pure Earth Mexico/Blacksmith Institute for funding the analysis of the samples. We thank Gabriel Velázquez Elizalde, Fernanda Reyna Cantoral, and Luis Javier Villa Cervantes for their help with the food purchase and laboratory support.

Conflicts of Interest: The authors declare no conflicts of interest.

Abbreviations

BW	Body Weight
Cd	Cadmium
EFSA	European Food Safety Authority
EU	European Union
FDA	Food and Drug Administration
FAO	Food and Agriculture Organization of the United Nations
JECFA	Joint FAO/WHO Expert Committee on Food Additives
MLs	Maximum Levels of contaminants and toxins in food
NHANES	National Health and Nutrition Examination Survey
PTMI	Provisional Tolerable Monthly Intake
TDS	Total Diet Study
TWI	Tolerable Weekly Intake
WHO	World Health Organization

References

1. WHO. *Exposure to Cadmium: A Major Public Health Concern*; World Health Organization: Geneva, Switzerland, 2019.
2. Nordberg, G.F.; Bernard, A.; Diamond, G.L.; Duffus, J.H.; Illing, P.; Nordberg, M.; Bergdahl, I.A.; Jin, T.; Skerfving, S. Risk assessment of effects of cadmium on human health (IUPAC Technical Report). *Pure Appl. Chem.* **2018**, *90*, 755–808. [CrossRef]
3. Rakib, M.R.J.; Rahman, M.A.; Onyena, A.P.; Kumar, R.; Sarker, A.; Hossain, M.B.; Islam, A.; Islam, M.S.; Rahman, M.M.; Jolly, Y.N.; et al. A comprehensive review of heavy metal pollution in the coastal areas of Bangladesh: Abundance, bioaccumulation, health implications, and challenges. *Environ. Sci. Pollut. Res. Int.* **2022**, *29*, 67532–67558. [CrossRef]
4. Yang, Q.; Li, Z.; Lu, X.; Duan, Q.; Huang, L.; Bi, J. A review of soil heavy metal pollution from industrial and agricultural regions in China: Pollution and risk assessment. *Sci. Total Environ.* **2018**, *642*, 690–700. [CrossRef]
5. Islam, M.M.; Karim, M.R.; Zheng, X.; Li, X. Heavy Metal and Metalloid Pollution of Soil, Water and Foods in Bangladesh: A Critical Review. *Int. J. Environ. Res. Public Health* **2018**, *15*, 2825. [CrossRef]
6. Bernhoft, R.A. Cadmium toxicity and treatment. *Sci. World J.* **2013**, *2013*, 394652. [CrossRef]

7. Kim, K.; Melough, M.M.; Vance, T.M.; Noh, H.; Koo, S.I.; Chun, O.K. Dietary Cadmium Intake and Sources in the US. *Nutrients* **2018**, *11*, 2. [CrossRef] [PubMed]
8. Vesey, D.A. Transport pathways for cadmium in the intestine and kidney proximal tubule: Focus on the interaction with essential metals. *Toxicol. Lett.* **2010**, *198*, 13–19. [CrossRef]
9. Koons, A.L.; Rajasurya, V. *Cadmium Toxicity*; National Library of Medicine: Treasure Island, FL, USA, 2023.
10. Smereczanski, N.M.; Brzoska, M.M. Current Levels of Environmental Exposure to Cadmium in Industrialized Countries as a Risk Factor for Kidney Damage in the General Population: A Comprehensive Review of Available Data. *Int. J. Mol. Sci.* **2023**, *24*, 8413. [CrossRef]
11. Charkiewicz, A.E.; Omeljaniuk, W.J.; Nowak, K.; Garley, M.; Niklinski, J. Cadmium Toxicity and Health Effects-A Brief Summary. *Molecules* **2023**, *28*, 6620. [CrossRef] [PubMed]
12. Genchi, G.; Sinicropi, M.S.; Lauria, G.; Carocci, A.; Catalano, A. The Effects of Cadmium Toxicity. *Int. J. Environ. Res. Public Health* **2020**, *17*, 3782. [CrossRef] [PubMed]
13. IARC. *Monographs*; IARC: Lyon, France, 2024.
14. Puente De La Cruz, L.N.; Giorgione, R.; Marini, F.; Astolfi, M.L. Rice sample preparation method for ICP-MS and CV-AFS analysis: Elemental levels and estimated intakes. *Food Chem.* **2024**, *461*, 140831. [CrossRef] [PubMed]
15. Arguello, D.; Chavez, E.; Lauryssen, F.; Vanderschueren, R.; Smolders, E.; Montalvo, D. Soil properties and agronomic factors affecting cadmium concentrations in cacao beans: A nationwide survey in Ecuador. *Sci. Total Environ.* **2019**, *649*, 120 127. [CrossRef] [PubMed]
16. Kowalska, G. The Safety Assessment of Toxic Metals in Commonly Used Herbs, Spices, Tea, and Coffee in Poland. *Int. J. Environ. Res. Public Health* **2021**, *18*, 5779. [CrossRef] [PubMed]
17. WHO. *General Standard for Contaminants and Toxins in Food and Feed*; WHO: Geneva, Switzerland, 1999.
18. EFSA. European Food Safety Authority; Cadmium dietary exposure in the European population. *EFSA J.* **2012**, *10*, 2551. [CrossRef]
19. Camara, M.; Giner, R.M.; Gonzalez-Fandos, E.; Lopez-Garcia, E.; Manes, J.; Portillo, M.P.; Rafecas, M.; Dominguez, L.; Martinez, J.A. Food-Based Dietary Guidelines around the World: A Comparative Analysis to Update AESAN Scientific Committee Dietary Recommendations. *Nutrients* **2021**, *13*, 3131. [CrossRef]
20. Collado-Lopez, S.; Betanzos-Robledo, L.; Tellez-Rojo, M.M.; Lamadrid-Figueroa, H.; Reyes, M.; Rios, C.; Cantoral, A. Heavy Metals in Unprocessed or Minimally Processed Foods Consumed by Humans Worldwide: A Scoping Review. *Int. J. Environ. Res. Public Health* **2022**, *19*, 8651. [CrossRef]
21. Cantoral, A.; Betanzos-Robledo, L.; Collado-Lopez, S.; Garcia-Martinez, B.A.; Lamadrid-Figueroa, H.; Mariscal-Moreno, R.M.; Diaz-Ruiz, A.; Rios, C.; Tellez-Rojo, M.M. Lead Levels in the Most Consumed Mexican Foods: First Monitoring Effort. *Toxics* **2024**, *12*, 318. [CrossRef]
22. Gaona-Pineda, E.B.; Rodriguez-Ramirez, S.; Medina-Zacarias, M.C.; Valenzuela-Bravo, D.G.; Martinez-Tapia, B.; Arango-Angarita, A. Consumidores de grupos de alimentos en población mexicana. Ensanut Continua 2020–2022. *Salud Publica Mex* **2023**, *65*, s248–s258. [CrossRef]
23. SSA. *Guías Alimentarias Saludables y Sostenibles para la Población Mexicana 2023*; Secretaría de Salud: Mexico, 2023.
24. FAO. *Food Balance Sheets 2010–2022. Global, Regional and Country Trends*; FAO: Roma, Italy, 2024.
25. Bognár, A. *Tables on Weight Yield of Food and Retention Factors of Food Constituents for the Calculation of Nutrient Composition of Cooked Foods (Dishes)*; Bundesforschungsanstalt für Ernährung: Karlsruhe, Germany, 2002.
26. Cantoral, A.; Batis, C.; Basu, N. National estimation of seafood consumption in Mexico: Implications for exposure to methylmercury and polyunsaturated fatty acids. *Chemosphere* **2017**, *174*, 289–296. [CrossRef]
27. Taylor, J.R. *An Introduction to Error Analysis: The Study of Uncertainties in Physical Measurements*, 3rd ed.; University Science Books: Sausalito, CA, USA, 2022.
28. FDA. U.S. Food and Drug Administration. *Total Diet Study Report*; FDA: Silver Spring, MD, USA, 2022.
29. Food Standards Agency. *Total Diet Study of Metals and Other Elements in Food*; Report for the UK Food Standards Agency (FS102081); Food and Environment Research Agency: London, UK, 2014.
30. Melgar, M.J.; Alonso, J.; Garcia, M.A. Cadmium in edible mushrooms from NW Spain: Bioconcentration factors and consumer health implications. *Food Chem. Toxicol.* **2016**, *88*, 13–20. [CrossRef]
31. Shen, T.; Kong, W.; Liu, F.; Chen, Z.; Yao, J.; Wang, W.; Peng, J.; Chen, H.; He, Y. Rapid Determination of Cadmium Contamination in Lettuce Using Laser-Induced Breakdown Spectroscopy. *Molecules* **2018**, *23*, 2930. [CrossRef] [PubMed]
32. Maddela, N.R.; Kakarla, D.; Garcia, L.C.; Chakraborty, S.; Venkateswarlu, K.; Megharaj, M. Cocoa-laden cadmium threatens human health and cacao economy: A critical view. *Sci. Total Environ.* **2020**, *720*, 137645. [CrossRef] [PubMed]
33. Godebo, T.R.; Stoner, H.; Kodsup, P.; Bases, B.; Marzoni, S.; Weil, J.; Frey, M.; Daley, P.; Earnhart, A.; Ellias, G.; et al. Occurrence of heavy metals coupled with elevated levels of essential elements in chocolates: Health risk assessment. *Food Res. Int.* **2024**, *187*, 114360. [CrossRef]
34. Hu, J.; Chen, G.; Xu, K.; Wang, J. Cadmium in Cereal Crops: Uptake and Transport Mechanisms and Minimizing Strategies. *J. Agric. Food Chem.* **2022**, *70*, 5961–5974. [CrossRef]
35. Qing, Y.; Li, Y.; Cai, X.; He, W.; Liu, S.; Ji, Y.; Jiang, M.; Yang, L.; Wang, J.; Ping, S.; et al. Assessment of Cadmium Concentrations in Foodstuffs and Dietary Exposure Risk Across China: A Metadata Analysis. *Expo. Health* **2023**, *15*, 10. [CrossRef]

36. Wang, Z.; Pan, L.; Liu, G.; Zhang, H.; Zhang, J.; Jiang, J.; Xiao, Y.; Bai, W.; Jiao, R.; Huang, W. Dietary exposure to cadmium of Shenzhen adult residents from a total diet study. *Food Addit. Contam. Part A Chem. Anal. Control Expo. Risk Assess.* **2018**, *35*, 706–714. [CrossRef]
37. Chunhabundit, R. Cadmium Exposure and Potential Health Risk from Foods in Contaminated Area, Thailand. *Toxicol. Res.* **2016**, *32*, 65–72. [CrossRef]
38. Amer, A.A.E.; El-Makarem, H.S.A.; El-Maghraby, M.A.; Abou-Alella, S.A. Lead, cadmium, and aluminum in raw bovine milk: Residue level, estimated intake, and fate during artisanal dairy manufacture. *J. Adv. Vet. Anim. Res.* **2021**, *8*, 454–464. [CrossRef] [PubMed]
39. Chirinos-Peinado, D.; Castro-Bedrinana, J.; Rios-Rios, E.; Castro-Chirinos, G.; Quispe-Poma, Y. Lead, Cadmium, and Arsenic in Raw Milk Produced in the Vicinity of a Mini Mineral Concentrator in the Central Andes and Health Risk. *Biol. Trace Elem. Res.* **2024**, *202*, 2376–2390. [CrossRef]
40. Yan, M.; Niu, C.; Li, X.; Wang, F.; Jiang, S.; Li, K.; Yao, Z. Heavy metal levels in milk and dairy products and health risk assessment: A systematic review of studies in China. *Sci. Total Environ.* **2022**, *851*, 158161. [CrossRef]
41. Torres-Sanchez, L.; Vazquez-Salas, R.A.; Vite, A.; Galvan-Portillo, M.; Cebrian, M.E.; Macias-Jimenez, A.P.; Rios, C.; Montes, S. Blood cadmium determinants among males over forty living in Mexico City. *Sci. Total Environ.* **2018**, *637–638*, 686–694. [CrossRef] [PubMed]
42. Moynihan, M.; Peterson, K.E.; Cantoral, A.; Song, P.X.K.; Jones, A.; Solano-Gonzalez, M.; Meeker, J.D.; Basu, N.; Tellez-Rojo, M.M. Dietary predictors of urinary cadmium among pregnant women and children. *Sci. Total Environ.* **2017**, *575*, 1255–1262. [CrossRef] [PubMed]
43. Zhao, X.; Shao, Y.; Ma, L.; Shang, X.; Zhao, Y.; Wu, Y. Exposure to Lead and Cadmium in the Sixth Total Diet Study—China, 2016–2019. *China CDC Wkly.* **2022**, *4*, 176–179. [CrossRef] [PubMed]
44. Pokharel, A.; Wu, F. Dietary exposure to cadmium from six common foods in the United States. *Food Chem. Toxicol.* **2023**, *178*, 113873. [CrossRef]
45. Rodriguez-Lopez, E.; Tamayo-Ortiz, M.; Ariza, A.C.; Ortiz-Panozo, E.; Deierlein, A.L.; Pantic, I.; Tolentino, M.C.; Estrada-Gutierrez, G.; Parra-Hernandez, S.; Espejel-Nunez, A.; et al. Early-Life Dietary Cadmium Exposure and Kidney Function in 9-Year-Old Children from the PROGRESS Cohort. *Toxics* **2020**, *8*, 83. [CrossRef]
46. Moskwa, J.; Bronikowska, M.; Socha, K.; Markiewicz-Zukowska, R. Vegetable as a Source of Bioactive Compounds with Photoprotective Properties: Implication in the Aging Process. *Nutrients* **2023**, *15*, 3594. [CrossRef]
47. Kordas, K.; Cantoral, A.; Desai, G.; Halabicky, O.; Signes-Pastor, A.J.; Tellez-Rojo, M.M.; Peterson, K.E.; Karagas, M.R. Dietary Exposure to Toxic Elements and the Health of Young Children: Methodological Considerations and Data Needs. *J. Nutr.* **2022**, *152*, 2572–2581. [CrossRef]

Disclaimer/Publisher's Note: The statements, opinions and data contained in all publications are solely those of the individual author(s) and contributor(s) and not of MDPI and/or the editor(s). MDPI and/or the editor(s) disclaim responsibility for any injury to people or property resulting from any ideas, methods, instructions or products referred to in the content.

 *foods*

Review

Mycotoxins in Food: Cancer Risks and Strategies for Control

Alice N. Mafe [1] and Dietrich Büsselberg [2,*]

[1] Department of Biological Sciences, Faculty of Sciences, Taraba State University, Main Campus, Jalingo 660101, Taraba State, Nigeria; mafealice1991@gmail.com
[2] Weill Cornell Medicine-Qatar, Education City, Qatar Foundation, Doha Metropolitan Area, Doha P.O. Box 22104, Qatar
* Correspondence: dib2015@qatar-med.cornell.edu

Abstract: Mycotoxins are toxic compounds produced by fungi such as *Aspergillus*, *Penicillium*, and *Fusarium*, contaminating various food crops and posing severe risks to food safety and human health. This review discusses mycotoxins' origins, significance, and impact, particularly in relation to cancer risk. Major mycotoxins like aflatoxins, ochratoxins, fumonisins, zearalenone, and patulin are examined, along with their sources and affected foods. The carcinogenic mechanisms of these toxins, including their biochemical and molecular interactions, are explored, as well as epidemiological evidence linking mycotoxin exposure to cancer in high-risk populations. The review also highlights critical methodologies for mycotoxin detection, including HPLC, GC-MS, MS, and ELISA, and the sample preparation techniques critical for accurate analysis. Strategies for controlling mycotoxin contamination, both pre- and post-harvest, are discussed, along with regulations from organizations like the FAO and WHO. Current challenges in detection sensitivity, cost, and control effectiveness are noted. Future research is needed to develop innovative analytical techniques, improve control strategies, and address the influence of climate change on mycotoxin production. Finally, global collaboration and emerging technologies are essential for advancing mycotoxin control and enhancing food safety.

Keywords: mycotoxins; food; cancer risk; control strategies; analysis

1. Introduction

Mycotoxins are toxic secondary metabolites produced by various species of fungi, primarily molds, that pose significant risks to food safety and public health. These fungi, which include notable genera such as *Aspergillus*, *Penicillium*, and *Fusarium*, thrive on a wide range of food crops, especially under warm and humid conditions [1]. They can grow in the field, during harvest, and even during food storage, contaminating essential commodities like cereals, nuts, dried fruits, coffee, and spices. As a result, mycotoxins are a significant food safety concern, especially in regions where environmental conditions favor fungal growth and food preservation systems may be inadequate [2].

The production of mycotoxins by fungi is a natural defense mechanism, typically triggered under stressful conditions such as drought, insect damage, or improper food storage [3]. These toxic metabolites can contaminate food at any stage in the supply chain, from pre-harvest to processing and storage. Several mycotoxins are of particular concern due to their prevalence and toxicity. Aflatoxins, produced by *Aspergillus* species, are commonly found in peanuts, maize, and other grains and are well known for their carcinogenic properties [4]. Ochratoxins, produced by *Aspergillus* and *Penicillium* species, are often detected in cereals, coffee, and dried fruits [5]. Fumonisins, predominantly made by *Fusarium* species, are found in maize, while zearalenone and deoxynivalenol (DON), also produced by *Fusarium*, are common contaminants in wheat, barley, and corn [6].

The health risks posed by mycotoxins are significant. Some mycotoxins, such as aflatoxins, are highly carcinogenic and are directly linked to liver cancer, while others cause immunosuppression, kidney damage, and reproductive disorders [7]. Chronic exposure to low levels of mycotoxins can be detrimental to health. This challenge is enhanced in developing countries where food contamination is more prevalent and diets heavily depend on susceptible crops [8]. The health impacts are not limited to humans as livestock consuming contaminated feed may suffer, leading to economic losses in agriculture and food-producing sectors [9].

From an economic standpoint, mycotoxin contamination has far-reaching consequences. Contaminated crops may be rejected for sale, reducing yields and causing substantial financial losses for farmers and food producers [10]. Furthermore, the costs associated with mycotoxin detection, management, and control measures increase the economic burden, affecting local food security and international trade [11]. Countries with stricter food safety standards may reject imported contaminated food products, resulting in trade barriers that affect global food markets [12].

Due to mycotoxins' health risks and economic impacts, regulatory bodies established stringent guidelines and limits for mycotoxin levels in food and animal feed [13]. International organizations like the Codex Alimentarius, in collaboration with the World Health Organization (WHO) and the Food and Agriculture Organization (FAO), set global standards aimed at minimizing mycotoxin contamination and protecting public health [14]. These regulatory efforts are critical in ensuring food safety, yet challenges remain in achieving comprehensive control across all stages of the food production process [15].

In summary, mycotoxins produced by molds that grow on various crops represent a significant concern in food safety due to their toxic effects on health, including the potential to cause cancer [16]. The economic implications of mycotoxin contamination further complicate food security and trade. Effective monitoring, control measures, and international regulatory standards are crucial in mitigating the impact of mycotoxins on public health and the global food supply chain [17].

Mycotoxins are crucial due to their significant impact on public health and food safety, particularly their association with cancer risk. Mycotoxins, such as aflatoxins, are highly potent carcinogens and are directly linked to liver cancer, among other health issues [18]. The International Agency for Research on Cancer (IARC) has classified aflatoxins as Group 1 carcinogens, indicating clear evidence of their cancer-causing potential in humans [19]. Such toxins, produced by fungi that contaminate staple food crops like maize, peanuts, and grains, can accumulate in the food chain, posing chronic health risks when consumed over time [20]. Therefore, regular food analysis and monitoring of mycotoxins are essential to prevent long-term exposure that could elevate cancer risks, particularly in vulnerable populations with limited dietary diversity [21].

The synergistic interactions between various mycotoxins, along with other environmental and dietary factors, significantly amplify their toxicity and complicate public health risks [22]. Multiple mycotoxins often co-occur in contaminated food or feed, such as aflatoxins and fumonisins in maize, where their combined presence enhances the carcinogenic potential beyond the effects of each toxin alone [23]. This synergism not only exacerbates liver cancer risk but also increases the likelihood of other adverse health outcomes, including immunosuppression and impaired growth [24]. Factors like dose and exposure levels further influence this synergism, where low doses of multiple mycotoxins can have enhanced effects than higher doses of a single one [25].

Metabolic interactions also play a role, with the metabolites of one mycotoxin potentially enhancing the toxic effects of another [26]. The immune system, often suppressed by certain mycotoxins, becomes more vulnerable to further toxic impact, while disruptions in gut microbiota and impaired detoxification processes can increase susceptibility to multiple mycotoxins. Nutritional deficiencies, particularly in essential nutrients like proteins, vitamins, and minerals, worsen these effects by reducing the body's ability to detoxify mycotoxins, increasing vulnerability to chronic diseases, including cancer [27]. Additionally, mycotoxins can interfere with liver enzymes responsible for detoxification and induce oxidative stress, weakening the body's defenses against toxins. Understanding these complex interactions is crucial for effective risk assessment, food safety interventions, and developing strategies to mitigate the health risks associated with multiple mycotoxin exposures [28].

In addition to cancer risks, mycotoxins present broader food safety concerns that necessitate thorough analysis. Contaminated foods can lead to a range of health issues beyond carcinogenicity, including immune suppression, gastrointestinal disorders, and reproductive problems [29]. These effects are particularly concerning in regions with poor food safety infrastructure, where contaminated foods may be widely consumed due to limited regulation or insufficient post-harvest management [30]. Ensuring food safety through analyzing mycotoxin levels helps mitigate these health risks and safeguard consumers and the global food supply chain [31]. Therefore, effective mycotoxin management is pivotal in food safety and cancer prevention and sustaining economic viability in the food industry [32].

1.1. Types of Mycotoxins

Different species of fungi produce common mycotoxins, which are a significant concern due to their toxic effects on human and animal health [33]. The most prevalent mycotoxins include aflatoxins, ochratoxins, fumonisins, zearalenone, and patulin, each with specific characteristics, sources, and health implications (Table 1) [34]. These mycotoxins contaminate a wide range of food products, making their presence a critical issue in food safety management.

Table 1. Types of mycotoxins in foods.

Mycotoxins	Description
Aflatoxins	Aflatoxins are among the most studied and dangerous mycotoxins, primarily produced by *Aspergillus flavus* and *Aspergillus parasiticus* species. They commonly contaminate crops such as maize, peanuts, and tree nuts, especially in warm, humid climates. Aflatoxins are highly toxic and carcinogenic, with aflatoxin B1 being the most potent. Chronic exposure to aflatoxins has been linked to liver cancer, particularly in regions with high consumption of contaminated grains. Aflatoxins can also cause acute poisoning (aflatoxicosis), liver damage, immune suppression, and stunted growth in children. Due to their severe health effects, aflatoxins are highly regulated globally in food and feed products [35].
Ochratoxins	Ochratoxins, particularly ochratoxin A, are produced by species of *Aspergillus* and *Penicillium* and are commonly found in cereals, dried fruits, coffee, and wine. Ochratoxin A is nephrotoxic, meaning it can cause damage to the kidneys, and is also considered a potential carcinogen. Long-term exposure to ochratoxins has been associated with kidney disorders, such as Balkan Endemic Nephropathy, and may also have immunosuppressive effects. Ochratoxin contamination often occurs during improper food storage, especially in humid conditions, making post-harvest management crucial in preventing its occurrence [36].
Fumonisins	Fumonisins are primarily produced by *Fusarium* species, with *Fusarium verticillioides* being a common contaminant of maize. These mycotoxins are particularly prevalent in regions where maize is a dietary staple. Fumonisin B1 is the most toxic form, and it is associated with a range of health effects, including esophageal cancer, neural tube defects, and liver and kidney toxicity. In animals, fumonisins have been linked to diseases such as equine leukoencephalomalacia (ELEM) in horses and pulmonary edema in pigs. Controlling fumonisin contamination is vital to reduce both human and animal health risks [37].
Zearalenone	Zearalenone is another mycotoxin produced by *Fusarium* species, often found in maize, wheat, and barley. Zearalenone mimics estrogen in animals and humans, making it a significant concern for reproductive health. Exposure to zearalenone can cause reproductive disorders, including infertility, in livestock and may also disrupt hormonal balance in humans. It is particularly problematic in livestock feed, leading to economic losses in animal husbandry. While not classified as a potent carcinogen, zearalenone's endocrine-disrupting effects highlight the importance of controlling its levels in food and feed [38].
Patulin	Patulin is a mycotoxin primarily produced by *Penicillium* and *Aspergillus* species and is most commonly associated with moldy fruits, particularly apples. Contamination by patulin can occur during the production of fruit juices, mainly when damaged or decayed fruits are processed. Although patulin is not a potent carcinogen, it can cause gastrointestinal distress and is mutagenic in certain studies. Regulations limit patulin levels in fruit products, particularly apple-based foods, to protect consumers from its toxic effects [39].

Aflatoxins, ochratoxins, fumonisins, zearalenone, and patulin are the major mycotoxins of concern due to their prevalence in food products and significant health impacts. These toxins pose various risks, from carcinogenicity to reproductive and kidney disorders, emphasizing the need for rigorous monitoring and control in food production to ensure food safety.

1.2. Sources and Affected Foods of Mycotoxins

Mycotoxins contaminate various food products at multiple stages of production, from pre-harvest to post-harvest, due to fungal growth. Different fungi are responsible for producing these toxic compounds, with certain foods being more susceptible to contamination based on environmental conditions and storage practices [40]. Foods are commonly contaminated by major mycotoxins (Table 2) and the fungi responsible for their production.

Table 2. Major mycotoxins in several food items.

Mycotoxin	Responsible Fungi	Affected Foods
Aflatoxins	Aflatoxins are mainly produced by *Aspergillus flavus* and *Aspergillus parasiticus*. These fungi thrive in warm and humid environments, making aflatoxin contamination more common in tropical and subtropical regions.	Aflatoxins commonly contaminate maize (corn), peanuts, tree nuts (almonds, pistachios, and walnuts), cottonseed, and some spices. Improper storage conditions, particularly in humid environments, increase the likelihood of aflatoxin contamination in these crops. Additionally, dairy products can be affected by aflatoxin M1 when livestock consume contaminated feed [41].
Ochratoxins	Ochratoxins, particularly ochratoxin A, are produced by species of *Aspergillus* (notably *Aspergillus ochraceus*) and *Penicillium* (*Penicillium verrucosum*).	Ochratoxins are commonly found in cereals such as wheat, barley, and oats, as well as in coffee, dried fruits (like raisins and figs), wine, beer, and grape juice. Contamination often occurs in stored grains, particularly under poor storage conditions with high moisture levels. Additionally, ochratoxins have been found in spices and cured meats [42].
Fumonisins	Fumonisins are produced primarily by *Fusarium verticillioides* (formerly *Fusarium moniliforme*) and *Fusarium proliferatum*, widespread in maize-growing regions.	Fumonisins are most commonly associated with maize (corn) and its products, including cornmeal, popcorn, cornflakes, and animal feed made from corn. In regions where maize is a dietary staple, fumonisin contamination is particularly concerning. The toxin may also be found in other cereal grains like sorghum [43].
Zearalenone	Zearalenone is produced by *Fusarium graminearum* and *Fusarium culmorum and is* commonly found in temperate climates.	This mycotoxin frequently contaminates maize, wheat, barley, oats, and rye. It is also commonly found in animal feed, disrupting reproductive functions in livestock, particularly pigs. Zearalenone contamination often occurs during harvest or storage when grains are exposed to moist conditions [44].
Patulin	Patulin is mainly produced by *Penicillium expansum*, and some species of *Aspergillus* and *Byssochlamys*. It grows primarily on decaying or damaged fruits.	Patulin is most found in apples and apple-derived products, such as apple juice and cider, mainly when damaged or moldy apples are processed. Other fruits that may be contaminated with patulin include pears, peaches, grapes, and apricots. Contamination can also occur in fruit juices and jams if compromised fruit is used during processing [39].

Aflatoxins, ochratoxins, fumonisins, zearalenone, and patulin are the most concerning mycotoxins due to their prevalence in various food products and harmful health effects. The fungi producing these toxins thrive in specific environmental conditions, contaminating crops like maize, peanuts, cereals, dried fruits, and apples. Proper storage, handling, and monitoring of these foods are essential to reduce the risk of mycotoxin contamination and ensure food safety [45]. Table 3 provides an overview of the regulatory limits and health risk levels for various mycotoxins in food products, emphasizing the need for continuous monitoring and control to ensure food safety. Figure 1 represents the percentages for the estimated contribution of each mycotoxin to the total global mycotoxin burden based on toxicity and occurrence.

Figure 1. Molds generate mycotoxins in foods. (**A**) *Aspergilus fumigatus*, which releases aflatoxin, (**B**) molds on red corn, (**C**) molds on maize flour, (**D**) molds on red corn, (**E**) molds on soybean flour, (**F**) molds in rice grains, (**G**) mold on white maize grains, and (**H**) molds on white corn. All pictures by Alice N. Mafe.

Table 3. Toxic levels of various mycotoxins in food.

Mycotoxin	Toxic Level in Food	Food Sources
Aflatoxins	Maximum allowable limit: 4 µg/kg (EU regulation). Toxic level associated with health risks: 0.5 µg/kg [46].	Nuts, grains, corn, and spices. Peanuts, tree nuts, and maize.
Zearalenone	Maximum allowable limit: 100 µg/kg (EU regulation). Toxic level associated with health risks: 50 µg/kg [47].	Cereals, grains, maize, and animal feed. Maize products, wheat, barley.
Patulin	Maximum allowable limit: 50 µg/kg (EU regulation). Toxic level associated with health risks: 25 µg/kg.	Apples, apple juice, and apple-based products. Processed fruit products.
Fumonisins	Maximum allowable limit: 4000 µg/kg (EU regulation) [48]. Toxic level associated with health risks: 2000 µg/kg.	Corn and corn-based products. Maize and maize flour.
Ochratoxin A	Maximum allowable limit: 3 µg/kg (EU regulation). Toxic level associated with health risks: 1 µg/kg [49].	Coffee, cereals, and dried fruit. Wine, grains, and legumes.
Deoxynivalenol (DON)	Maximum allowable limit: 1750 µg/kg (EU regulation). Toxic level associated with health risks: 1000 µg/kg [50].	Wheat, barley, and oats. Cereal products and animal feed.
T-2 Toxin	Maximum allowable limit: 1000 µg/kg (EU regulation). Toxic level associated with health risks: 500 µg/kg [51].	Cereal grains and animal feed. Wheat, barley, and oats.

1.3. Global Toxic Levels of Mycotoxins in Foods

The global distribution of mycotoxins presents a significant concern for food safety due to their toxic effects on human and animal health. Based on global toxicity levels, aflatoxins are the most prevalent, accounting for 35% of the global toxic load. These highly toxic compounds are commonly found in crops like maize and peanuts, particularly in warm and humid climates, and are known for their carcinogenic properties [52].

Following aflatoxins, ochratoxin A and deoxynivalenol (DON) each contributes 20% to the total mycotoxin burden. Ochratoxin A is often found in cereals, coffee, and dried fruits and is associated with nephrotoxicity, while DON, commonly referred to as "vomitoxin", occurs in grains and can cause acute gastrointestinal distress [53].

Fumonisins, which make up 15% of the global mycotoxin contamination, are prevalent in maize and are linked to esophageal cancer and neural tube defects. Lastly, zearalenone, responsible for 10% of the global burden, is an estrogenic mycotoxin found in grains that disrupt hormonal balance, particularly in livestock [34].

2. Cancer Risk Associated with Mycotoxins

Mycotoxins are widely recognized for their carcinogenic potential, with certain types posing significant cancer risks to humans. The primary mycotoxin associated with cancer risk is aflatoxin, which has been classified as a Group 1 carcinogen by the International Agency for Research on Cancer (IARC). Other mycotoxins, such as ochratoxin A and fumonisin B1, are also considered potential carcinogens. These toxins contribute to cancer development through various biochemical and molecular mechanisms that lead to genetic damage, cell cycle disruption, and immune suppression [54].

2.1. Mechanisms of Carcinogenicity

Mycotoxins, particularly aflatoxins, contribute to cancer development through multiple biochemical and molecular mechanisms. These include direct DNA damage and mutagenesis, oxidative stress, cell cycle regulation, disruption, apoptosis inhibition, immune suppression, and epigenetic modifications [55]. The cumulative effects of these mechanisms (Figure 2 and Table 4) can lead to the initiation, promotion, and progression of cancer, making mycotoxins a significant concern for public health, particularly in regions with high exposure to contaminated foods. Understanding these mechanisms is crucial for developing effective strategies to reduce the cancer risk associated with mycotoxin exposure [56].

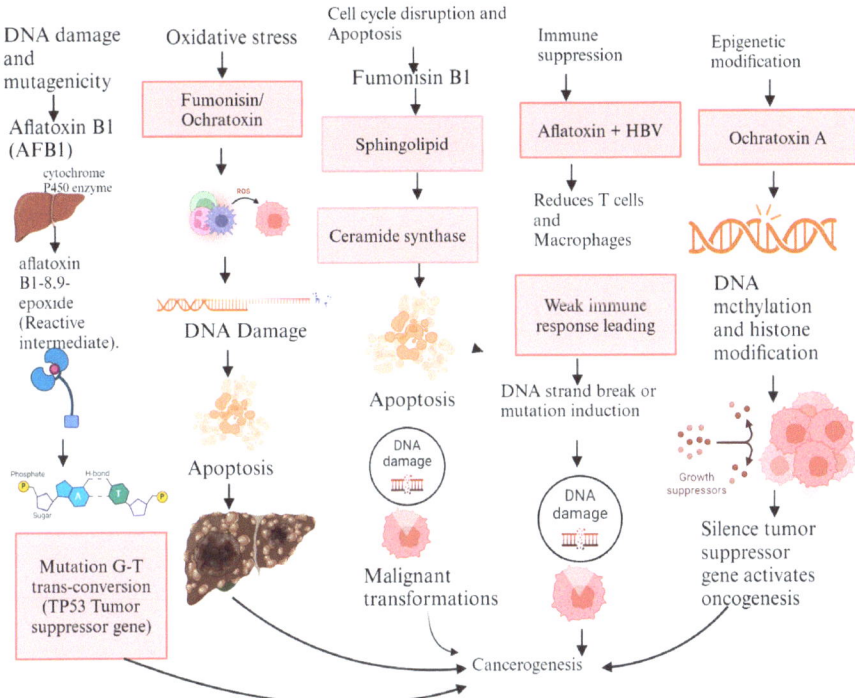

Figure 2. Mechanisms of carcinogenicity for aflatoxin B1, fumonisins, and ochratoxin A. Legend: AFB1: aflatoxin B1, HBV: hepatitis B virus, ROS: reactive oxygen species, TP53: tumor suppressor gene. The figure was generated using BioRender.

Table 4. Mechanisms of carcinogenicity caused by mycotoxins.

Mechanism/Toxin	Description
DNA Damage and Mutagenicity	Aflatoxins, particularly aflatoxin B1 (AFB1), are potent carcinogens that exert their effects by directly damaging DNA. Once ingested, AFB1 is metabolized in the liver by cytochrome P450 enzymes into a reactive intermediate, aflatoxin B1-8,9-epoxide. This metabolite can bind covalently to DNA, forming DNA adducts, particularly at the guanine base, leading to mutations. One of the most common mutations caused by AFB1 is the G-to-T transversion in the TP53 tumor suppressor gene, which is crucial in regulating cell growth and apoptosis. Mutations in TP53 result in uncontrolled cell proliferation and are strongly associated with hepatocellular carcinoma (liver cancer). Aflatoxin-induced DNA damage is thus a key mechanism driving the initiation of cancer [57].
Oxidative Stress	Mycotoxins can also induce oxidative stress, a condition where there is an imbalance between the production of reactive oxygen species (ROS) and the body's ability to detoxify them. Aflatoxins and other mycotoxins, such as fumonisins and ochratoxins, can generate ROS during their metabolism, leading to oxidative damage to cellular components like DNA, proteins, and lipids. This oxidative damage can cause mutations, promote inflammation, and contribute to the initiation and progression of cancer. In addition, chronic oxidative stress can disrupt cellular signaling pathways that control cell growth and apoptosis, further promoting carcinogenesis [58].

Table 4. Cont.

Mechanism/Toxin	Description
Cell Cycle Disruption and Apoptosis Inhibition	Mycotoxins can interfere with normal cell cycle regulation, contributing to the development of cancer. For instance, fumonisin B1, commonly found in maize, disrupts sphingolipid metabolism by inhibiting ceramide synthase. Sphingolipids are essential in regulating cell growth, differentiation, and apoptosis. The disruption of sphingolipid pathways can impair apoptosis (programmed cell death), allowing damaged cells to survive and proliferate uncontrollably, a hallmark of cancer development. Additionally, by blocking apoptosis, mycotoxins facilitate the survival of cells with DNA damage, increasing the likelihood of malignant transformation [59].
Immune Suppression	Chronic exposure to certain mycotoxins can lead to immune suppression, which further increases cancer risk. Aflatoxins, for example, are known to impair the immune system by reducing the production and function of immune cells like T-cells and macrophages. This weakened immune response hampers the body's ability to recognize and eliminate cancerous or pre-cancerous cells. Furthermore, immune suppression can promote the persistence of viral infections, such as hepatitis B virus (HBV), which is a significant cofactor in aflatoxin-induced liver cancer. Individuals who are exposed to both aflatoxins and HBV are at a much higher risk of developing liver cancer due to the combined effects of viral infection and toxin-induced DNA damage [60].
Epigenetic Modifications	In addition to directly damaging DNA, mycotoxins can cause epigenetic changes that alter gene expression without affecting the underlying DNA sequence. For instance, mycotoxins like ochratoxin A have been shown to induce changes in DNA methylation and histone modifications, which can silence tumor suppressor genes or activate oncogenes. These epigenetic alterations can promote carcinogenesis by disrupting normal cellular functions and facilitating uncontrolled cell growth [61].
Zearalenone (ZEA)	Zearalenone is a nonsteroidal estrogenic mycotoxin primarily produced by Fusarium species, commonly found in cereals and grains. The carcinogenicity of ZEA is primarily linked to its estrogenic properties, as it mimics the action of natural estrogens by binding to estrogen receptors (ERs) in target tissues. This interaction leads to hormonal disruption, which promotes the proliferation of estrogen-sensitive cells, particularly in reproductive tissues. Over time, the hyperproliferation of these cells increases the risk of hormone-dependent cancers, such as breast, ovarian, and endometrial cancers. ZEA's ability to activate ER signaling can also induce DNA damage and oxidative stress, further contributing to its carcinogenic potential. Oxidative stress generates reactive oxygen species (ROS), which can cause mutations, impair DNA repair mechanisms, and lead to genomic instability. Additionally, ZEA may disrupt normal cell cycle regulation, promoting abnormal cell division and enhancing the risk of cancer development [62].
Patulin	Patulin, produced by Penicillium and Aspergillus species, is a mycotoxin primarily found in apples and apple products. Its carcinogenicity is associated with its ability to induce oxidative stress and DNA damage. Patulin promotes the generation of ROS, which can damage cellular components, including lipids, proteins, and nucleic acids. This oxidative damage leads to mutations and chromosomal aberrations, increasing the risk of malignant transformations. Furthermore, patulin interferes with key cellular pathways involved in apoptosis (programmed cell death) and cell cycle regulation. By inhibiting apoptosis, patulin allows damaged cells to survive and proliferate, which may contribute to cancer initiation and progression. Patulin also impairs the function of tumor suppressor proteins, such as p53, which generally help to maintain genomic integrity by halting the cell cycle in response to DNA damage. When p53 function is disrupted, cells with damaged DNA can continue to divide uncontrollably, further contributing to the carcinogenic process [63].

2.2. Epidemiological Evidence Linking Mycotoxins to Cancer Risk

Epidemiological studies provide strong evidence linking exposure to mycotoxins, particularly aflatoxins, to an increased risk of cancer (Table 5). Most notably, aflatoxin exposure is extensively associated with liver cancer or hepatocellular carcinoma (HCC). Ochratoxin A and fumonisins have also been implicated in developing kidney and esophageal cancers, respectively. Below is an overview of critical studies and data illustrating the correlation between mycotoxin exposure and cancer risk (Table 5), focusing on high-risk populations and regions.

Table 5. Key studies and data illustrating the correlation between mycotoxin exposure and cancer risk for zearalenone and patulin.

Mycotoxin	Findings	Cancer Type
Zearalenone	Reported increased estrogenic activity of zearalenone in human breast cancer cells, leading to cell proliferation [64].	Breast Cancer
	Found that dietary exposure to zearalenone in rats led to a significant increase in uterine weight and hyperplasia [65].	Uterine Cancer
	Identified a link between zearalenone exposure and increased risk of reproductive cancers through hormonal disruption [66].	Reproductive Cancers
	Showed that zearalenone exposure caused oxidative stress and DNA damage in liver cells, potentially increasing liver cancer risk [67].	Liver Cancer
Patulin	Documented the DNA-damaging effects of patulin in human liver cells, leading to mutagenic changes [68].	Liver Cancer
	Investigated the carcinogenic potential of patulin in mouse models, noting an increase in tumor incidence [69].	Multiple Cancer Types
	Found that patulin exposure induced oxidative stress and apoptosis in colon cancer cells, highlighting its potential role in colorectal cancer [70].	Colorectal Cancer
	Reported that dietary exposure to patulin in rats resulted in liver toxicity and increased cancer risk [71].	Liver Cancer

2.3. Cancer Risk Associated with Mycotoxins

Epidemiological evidence strongly supports the link between mycotoxin exposure and increased cancer risk, particularly liver cancer due to aflatoxins and esophageal and kidney cancers from fumonisins [72]. Populations in regions with high contamination and limited food safety measures are at risk. This underscores the urgent need for effective monitoring, regulatory measures, and food safety interventions to reduce exposure to mycotoxins and mitigate their associated cancer risks.

2.4. Non-Cancer Risks Associated with Aflatoxins

Aflatoxins, a group of mycotoxins produced by *Aspergillus* species, primarily *Aspergillus flavus* and *Aspergillus parasiticus*, are known for their carcinogenic properties. However, the non-cancer risks, particularly acute toxicities and their potential to cause stunting in infants, are significant public health concerns that merit further examination [41].

2.4.1. Acute Toxicities of Aflatoxins

Aflatoxins can induce a range of acute toxic effects upon ingestion, particularly at high exposure levels. Acute aflatoxicosis is characterized by rapid-onset symptoms, which can vary depending on the dose and route of exposure. The main acute toxic effects include the following:

- Hepatotoxicity: The liver is the primary target organ for aflatoxins. Acute exposure can lead to liver damage, manifesting as jaundice, abdominal pain, and elevated liver enzymes. Severe cases can progress to liver failure, which may be fatal. The hepatotoxic effects are often attributed to the bioactivation of aflatoxins to reactive epoxide intermediates, leading to cellular damage and necrosis [73].
- Gastrointestinal Symptoms: Ingestion of contaminated food can cause gastrointestinal disturbances such as nausea, vomiting, abdominal cramps, and diarrhea. These symptoms result from direct irritation of the gastrointestinal tract and liver dysfunction.
- Immune System Suppression: Aflatoxins can impair immune function, making individuals more susceptible to infections. This is particularly concerning in infants, who already have immature immune systems. Immune suppression can lead to higher rates of morbidity and mortality from infectious diseases.

- Neurological Effects: In some cases, aflatoxin exposure has been linked to neurological symptoms, including headaches, confusion, and altered mental status. These effects may be due to hepatic encephalopathy resulting from liver dysfunction or direct neurotoxicity.

2.4.2. Stunting in Infants

The relationship between aflatoxin exposure and stunting in infants is an emerging area of research, with significant implications for child health and development. Stunting refers to impaired growth and development in children, characterized by low height-for-age. It is a critical public health issue, as it can lead to long-term consequences for physical and cognitive development [74]. The mechanisms through which aflatoxin exposure may contribute to stunting include the following:

- Nutritional Deficiencies: Aflatoxins can interfere with nutrient absorption and metabolism. They can cause malabsorption syndromes by damaging the intestinal lining, leading to nutrient deficiencies, particularly of proteins, vitamins, and minerals essential for growth. Infants exposed to aflatoxins may not receive adequate nutrition, exacerbating the risk of stunting [75].
- Chronic Inflammation: Aflatoxin exposure can provoke an inflammatory response, resulting in chronic inflammation that impairs growth. Prolonged inflammation can alter metabolic processes and hinder the body's ability to utilize nutrients effectively, which is critical for growth and development during infancy [76].
- Impaired Immune Function: As mentioned earlier, aflatoxins can suppress the immune system. Infants who experience repeated infections due to immune compromise may have increased metabolic demands and reduced nutrient absorption, contributing to stunting. Frequent illness can also lead to increased energy expenditure, diverting resources away from growth and development [77].
- Hormonal Disruption: Aflatoxins have been shown to affect the endocrine system, potentially disrupting growth hormone pathways. Any disruption in growth hormone signaling can have significant effects on growth and development, leading to stunted growth in infants [66].
- Maternal Exposure: The effects of aflatoxins are not limited to direct exposure in infants. Pregnant and lactating women exposed to aflatoxins can transfer these toxins to their infants through placental transfer and breast milk. This transference can adversely affect the growth and development of infants, compounding the risk of stunting [78].

2.4.3. Public Health Implications

The non-cancer risks associated with aflatoxins, particularly acute toxicities and stunting in infants, highlight the urgent need for public health interventions. Effective strategies to mitigate aflatoxin exposure include the following:

- Food Safety Regulations: implementing strict regulations and monitoring systems to limit aflatoxin levels in food supplies, particularly in high-risk regions where staple crops are often contaminated [79].
- Education and Awareness: raising awareness among farmers, food processors, and consumers about the risks of aflatoxins, safe storage practices, and proper food handling techniques [80].
- Nutritional Interventions: providing nutritional support and supplementation for vulnerable populations, particularly in areas with high aflatoxin exposure, to mitigate the adverse effects of malnutrition and improve overall health outcomes.
- Research and Monitoring: continued research into the health effects of aflatoxins, particularly in children, and ongoing monitoring of aflatoxin levels in food sources will help to inform public health policies and interventions [81].

While aflatoxins are widely recognized for their carcinogenic properties, the acute toxicities and potential for stunting in infants represent significant non-cancer risks. Addressing these concerns is essential for improving child health and preventing long-term developmental consequences associated with aflatoxin exposure [82].

3. Methods of Analyzing Mycotoxins

Accurate detection and quantification of mycotoxins in food products are critical for ensuring food safety and preventing mycotoxin-related health risks, including cancer. Various analytical techniques are employed to identify and measure mycotoxins in agricultural products, processed foods, and animal feed. These methods differ in sensitivity, specificity, and complexity, allowing for qualitative and quantitative analysis across a range of mycotoxin types [83]. Table 6 describes the primary techniques used for mycotoxin analysis while Table 7 focuses on the recent publications on the immuno-detection of mycotoxins in food.

Table 6. Primary techniques used for mycotoxin analysis.

Method	Description		
Chromatography	Chromatography is one of the most widely used techniques for analyzing mycotoxins due to its high sensitivity, accuracy, and ability to separate and identify multiple mycotoxins simultaneously. Two common forms of chromatography used in mycotoxin analysis are high-performance liquid chromatography (HPLC) and gas chromatography–mass spectrometry (GC-MS) [84].		
	Principle	Application	Advantage
High-Performance Liquid Chromatography (HPLC)	HPLC involves the separation of mycotoxins based on their interaction with a stationary phase (usually a column) and a mobile phase (usually a solvent). The different affinities of mycotoxins for the stationary phase allow them to be separated, detected, and quantified [85].	HPLC is commonly used to analyze aflatoxins, fumonisms in various foods, including cereals, nuts, and dairy products. It is highly effective when coupled with fluorescence or UV detection methods, which enhance sensitivity for specific mycotoxins.	HPLC offers high resolution, accuracy, and the ability to detect low levels of mycotoxins. It is widely accepted in regulatory testing and can be used for routine food safety monitoring.
Gas Chromatography–Mass Spectrometry (GC-MS)	GC-MS vaporizes mycotoxin samples, separates them via gas chromatography, and identifies them by mass spectrometry. Mycotoxins are derivatized to ensure volatility [86].	GC-MS is beneficial for the detection of volatile mycotoxins like patulin. It is susceptible and specific, making it suitable for detecting trace levels of mycotoxins in complex food matrices.	GC-MS provides high specificity and sensitivity, making it the gold standard for detecting mycotoxins like patulin in fruit juices.
Spectrometry	Mass spectrometry (MS) is often combined with chromatography to improve the sensitivity and specificity of mycotoxin detection. MS measures the mass-to-charge ratio of ionized mycotoxin molecules, providing precise molecular identification and quantification.		
Mass Spectrometry (MS)	MS works by ionizing chemical compounds and measuring the mass-to-charge ratio of the resulting ions. Coupled with HPLC or GC, it allows for separating and identifying mycotoxins based on their mass [87].	HPLC-MS and GC-MS are widely used to analyze various mycotoxins, including aflatoxins, ochratoxins, and fumonisins. These techniques are valuable in multi-mycotoxin analysis, where several toxins may exist in a single sample.	MS provides high accuracy and detects multiple mycotoxins at low concentrations, which is crucial for regulatory testing and detailed mycotoxin profiling in food products.
Immunoassays	Immunoassays are rapid, sensitive, and cost-effective techniques for detecting mycotoxins in food. They rely on antibodies' specific binding to mycotoxins and are suitable for quickly screening large numbers of samples [88].		

Table 6. Cont.

Method	Description		
Enzyme-Linked Immunosorbent Assay (ELISA)	ELISA is based on antibodies binding to mycotoxins, followed by an enzyme–substrate reaction that produces a detectable signal, usually colorimetric or fluorescent. The intensity of the signal corresponds to the concentration of mycotoxins in the sample [89].	ELISA is commonly used to detect aflatoxins, ochratoxns, zearalenone, and fumonisms in food products such as grains, nuts, and milk. It is often employed for routine screening in food industries and regulatory bodies.	ELISA is a quick, affordable method for mycotoxin detection but may lack the specificity of chromatographic techniques due to cross-reactivity.
Lateral Flow Immunoassay (LFIA)	LFIA is similar to ELISA but uses a test strip format. Mycotoxin–antibody interactions produce a visible line or signal on the test strip, indicating the presence of mycotoxins [90].	LFIA is used for rapid, on-site testing of mycotoxins in agricultural products. It is commonly applied to detect aflatoxins, fumonisins, and zearalenone in grains, nuts, and animal feed.	LFIA is a portable, quick method for mycotoxin detection, ideal for field testing, though less precise than lab methods.
Hyperspectral Analysis	Utilizes the spectral signature of materials across a wide range of wavelengths to identify and quantify mycotoxin contamination [91].	Sorting and detecting mycotoxin presence in grains, nuts, and other food products.	Non-destructive, rapid analysis, can be applied in real-time sorting, and high-throughput screening.
Immuno-detection	Employs specific antibodies that bind to mycotoxins, allowing their detection through various methods (e.g., ELISA, lateral flow assays) [90].	Food safety testing, monitoring mycotoxin levels in processed and raw food products.	High specificity and sensitivity, can detect low concentrations of mycotoxins, and suitable for various matrices.

Table 7. Recent publications on immuno-detection of mycotoxins.

Technique	Findings
ELISA	Developed a novel ELISA method for detecting aflatoxins in peanuts with high sensitivity [92].
Lateral Flow Immunoassay	Created a lateral flow immunoassay for rapid detection of zearalenone in cereal products [93].
Immunoaffinity Columns	Utilized immunoaffinity columns for the extraction and detection of patulin in fruit juices, achieving high recovery rates [94].
Magnetic Nanoparticles	Developed magnetic nanoparticles coupled with immunoassays for the detection of multiple mycotoxins in grains [95].

The analysis of mycotoxins in food relies on a combination of advanced techniques to ensure accurate detection and quantification. Chromatographic methods such as high-performance liquid chromatography (HPLC) and gas chromatography–mass spectrometry (GC-MS) are susceptible and precise, especially when combined with mass spectrometry [96]. Immunoassays, including enzyme-linked immunosorbent assay (ELISA) and lateral flow immunoassay (LFIA), offer rapid, cost-effective screening, making them suitable for routine testing [97]. Each method has its advantages depending on the type of mycotoxin, the complexity of the food matrix, and the desired level of accuracy, enabling comprehensive monitoring and control of mycotoxin contamination.

Sample Preparation for Mycotoxin Analysis

Sample preparation is critical in accurately detecting and quantifying mycotoxins in food. The preparation process involves several stages aimed at isolating mycotoxins from complex food matrices while maintaining the integrity and concentration of the target mycotoxins [98]. Proper sample preparation reduces interference from food components, improves extraction efficiency, and enhances the sensitivity of subsequent analytical techniques [99]. The main steps in preparing food samples for mycotoxin analysis (Table 8) are sampling, homogenization, extraction, cleanup, and concentration.

Table 8. Main steps in preparing food samples for mycotoxin analysis.

Sample Preparation	Importance	Procedure	Challenges
Sampling	Accurate mycotoxin analysis relies on proper sampling, as uneven distribution in food can cause incorrect results.	The process begins with collecting food samples from different parts of a batch to account for variability in mycotoxin contamination. To form a composite sample, random sub-samples are collected and combined for solid foods like grains, nuts, and cereals. Mixing ensures even distribution before sampling for liquids, milk, or fruit juices [100].	Heterogeneous mycotoxin contamination complicates sampling, requiring larger samples for bulk goods like grains to minimize errors.
Homogenization	After sampling, food must be homogenized to ensure even distribution of mycotoxins, particularly in solid or semi-solid foods [101].	Homogenization involves grinding or blending the sample into a fine, uniform consistency. Equipment such as mills or blenders reduces the particle size of solid foods like grains or nuts. Mixing ensures consistency in liquids.	Prevent contamination during homogenization, as excessive grinding heat can degrade sensitive mycotoxins.
Extraction	The extraction aims to separate mycotoxins from the food matrix into a solvent, isolating them from interfering compounds like proteins, fats, and carbohydrates for easier analysis.	The solvent choice depends on the food type and mycotoxin analyzed. Common organic solvents like methanol and acetonitrile effectively dissolve mycotoxins. The homogenized food is mixed with the solvent and agitated for solid samples, while liquid samples require filtration or centrifugation to remove debris before analysis [102].	The selection of an appropriate solvent system is crucial. It must efficiently extract mycotoxins while minimizing the coextraction of other food components that may interfere with the analysis.
Cleanup	Following extraction, a cleanup step is often necessary to remove unwanted compounds from the extract, such as fats, sugars, and proteins, which can interfere with the sensitivity and accuracy of detection methods.	Cleanup methods vary by analytical technique and food matrix:Solid-Phase Extraction (SPE): extracts pass through an adsorbent column that binds unwanted substances while allowing mycotoxins to pass.Immunoaffinity Columns (IAC): use antibodies to selectively bind and isolate mycotoxins, which are then eluted with a solvent.Liquid–Liquid Partitioning: separates mycotoxins based on solubility in two immiscible phases, typically an organic solvent and water [103].	Cleanup must be optimized for different food types, as excessive removal of matrix components can result in the loss of mycotoxins, reducing the sensitivity of the analysis.
Concentration	After extraction and cleanup, mycotoxin levels may be too low for accurate detection. Concentration improves these levels, making quantification easier.	Concentration is typically achieved by evaporating the solvent used during extraction, leaving behind a more concentrated sample of mycotoxins. This is usually performed under reduced pressure or using rotary evaporation to avoid degradation of the mycotoxins [98].	Avoid over-concentration, as it can cause matrix effects or unwanted compound precipitation that interferes with analysis.

Preparing food samples for mycotoxin analysis involves carefully executing steps to ensure accurate detection and quantification. Each stage, from sampling and homogenization to extraction, cleanup, and concentration, must be tailored to the specific type of food and mycotoxin being analyzed [102]. Proper sample preparation is essential for reducing interference, enhancing extraction efficiency, and improving analytical techniques' reliability for mycotoxin detection.

4. Strategies for Mycotoxin Control

Controlling mycotoxin contamination is a pressing issue that requires a comprehensive approach. This approach should address pre- and post-harvest stages and incorporate effective regulatory and monitoring strategies. By implementing comprehensive control measures, we can significantly reduce the risk of mycotoxin contamination in food and feed, protecting public health and ensuring food safety [104].

4.1. Pre-Harvest Control Measures

Pre-harvest control measures focus on preventing fungal contamination and mycotoxin production before crops are harvested. These strategies involve various agricultural practices and interventions minimizing the conditions favoring fungal growth (Table 9) [62,75].

Table 9. Strategies involve various agricultural practices and interventions minimizing the conditions favoring fungal growth.

	Principle	Implementation	Challenges
Crop Rotation	Crop rotation alternates crops in a field across seasons, reducing fungal populations that target specific crops and lowering mycotoxin contamination risk [105].	By rotating crops, such as alternating cereals with legumes or other non-host plants, the life cycle of fungal pathogens is disrupted, reducing their ability to infect subsequent crops.	Effective crop rotation requires careful planning to prevent new crops from hosting pathogens and adapting practices to local conditions.
Use of Resistant Varieties	Plant breeding programs focus on developing crop varieties resistant to specific fungal pathogens. Resistant varieties can reduce fungal infection and subsequent mycotoxin production [106].	Farmers can select and plant varieties of crops that have been genetically modified or selectively bred for resistance to mycotoxin-producing fungi, such as maize varieties resistant to *Fusarium* species.	Regional factors may limit the availability of resistant varieties, and continuous breeding efforts are needed to address evolving fungal strains. Moreover, resistance does not always guarantee complete protection, so it should be used in conjunction with other measures.
Proper Irrigation and Field Management	Fungal pathogens thrive in warm, humid conditions, making proper irrigation and field management critical in preventing fungal contamination [107].	Techniques include optimizing irrigation, ensuring good drainage, and avoiding over-fertilization to reduce fungal growth. Managing plant residue and minimizing mechanical damage also help prevent fungal infection.	Effective field management requires monitoring weather conditions, soil moisture levels, and crop health, which may be resource-intensive. Farmers need access to proper tools and training to implement these practices effectively.

4.2. Post-Harvest Control Measures

Post-harvest control measures are crucial for minimizing mycotoxin contamination after harvesting. Proper storage conditions, such as maintaining cool and dry environments, are essential to inhibit fungal growth and mycotoxin production. Effective processing techniques, including cleaning, sorting, and milling, help reduce mycotoxin levels by removing contaminated parts and diluting mycotoxins. Chemical treatments, like ammonization and ozone treatment, can also detoxify mycotoxins in food products. Implementing these measures is vital to ensure food safety and protect public health.

4.3. Regulatory and Monitoring Approaches

Regulatory and monitoring approaches are essential for managing mycotoxin contamination and ensuring food safety. Regulatory bodies, such as the U.S. Food and Drug Administration (FDA) and the European Food Safety Authority (EFSA), establish guidelines and standards for maximum allowable levels of mycotoxins in food and feed [108]. These regulations protect public health by limiting exposure to mycotoxins through food and animal feed. Monitoring programs play a critical role in enforcing these standards. Regular testing of food and feed samples for mycotoxin contamination helps to identify and mitigate risks before they reach consumers. Techniques such as high-performance liquid chromatography (HPLC) and enzyme-linked immunosorbent assays (ELISAs) are commonly used to ensure compliance with regulatory limits. Additionally, government agencies and industry organizations often collaborate to conduct surveillance and train producers on best practices for managing mycotoxin risks [109].

Mycotoxin Regulation Framework

Global regulatory limits for various mycotoxins are set by different countries and international bodies, highlighting the variation in food safety standards. Aflatoxins, known for their carcinogenic effects, are regulated most strictly in the European Union (EU) at 2 ppb, while the United States (FDA) allows up to 20 ppb. Ochratoxin A, a nephrotoxic mycotoxin, has a 5 ppb limit in most regions, though it is unregulated in the U.S. [14]. For fumonisins, the limits range widely, from 1000 ppb in China to 4000 ppb in the U.S. Zearalenone, which affects hormone regulation, has limits varying from 60 ppb in China to 200 ppb in Japan, with no regulation in some countries. Lastly, deoxynivalenol (DON), known for its gastrointestinal effects, is regulated between 1000 ppb in the U.S. and 2000 ppb in Australia/New Zealand [110]. These data provide a basis for comparing the regulatory frameworks, emphasizing the need for harmonized global standards to ensure food safety.

5. Current Challenges and Limitations

The effective management of mycotoxin contamination in food and feed presents several challenges and limitations, particularly concerning detection methods and control strategies. Resolving these issues is crucial for improving food safety and mitigating health risks associated with mycotoxins.

5.1. Detection Challenges

Detection of mycotoxins in food involves sophisticated analytical techniques that face several challenges, including sensitivity, specificity, and cost (Table 10), while Table 11 illustrates the various regulatory and monitoring approaches. One significant challenge is achieving the required sensitivity to detect low levels of mycotoxins, especially when they are present in complex food matrices. Techniques such as high-performance liquid chromatography (HPLC) and gas chromatography–mass spectrometry (GC-MS) offer high sensitivity but can be expensive and require extensive sample preparation [111]. In addition, the specificity of these methods must be high to accurately distinguish between mycotoxins and similar compounds that may interfere with results.

Table 10. Post-harvest control measures.

Measure	Principle	Implementation	Challenges
Storage Conditions	Proper storage conditions are vital for preventing fungal growth and mycotoxin production post-harvest [112].	Critical practices include maintaining dry, cool, and ventilated storage to prevent fungal growth. Grains should be stored below 14% moisture in ventilated containers, with regular inspections of storage facilities.	Maintaining optimal storage conditions requires ongoing monitoring and control, which can be challenging in regions with limited infrastructure or resources.

Table 10. Cont.

Measure	Principle	Implementation	Challenges
Processing Techniques	Processing methods can help reduce mycotoxin levels in food products and remove contaminated portions [113].	Techniques like cleaning and sorting can remove contaminated food parts. Heat treatments may degrade some mycotoxins, while good manufacturing practices (GMPs) and hazard analysis and critical control point (HACCP) systems help manage contamination risks.	The effectiveness of processing techniques depends on the type of mycotoxin and the food matrix. Not all mycotoxins are easily removed or degraded by processing methods.
Chemical Treatments	Chemical treatments can help neutralize or remove mycotoxins from food and feed [114].	Adsorbents like activated carbon or clay can be added to animal feed to bind mycotoxins and reduce their bioavailability. Chemical decontamination agents, such as ozone or ammonia, can be used to treat contaminated grains.	Chemical treatments should be used cautiously to prevent new contaminants and preserve nutritional quality, as their effectiveness varies by mycotoxin type and treatment method.

Table 11. Regulatory and monitoring approaches.

Approach	Principle	Implementation	Challenges
Regulatory Guidelines and Standards	Regulatory guidelines establish permissible mycotoxin levels in food and feed to safeguard public health.	International bodies like Codex Alimentarius set global mycotoxin limits, while national authorities establish regulations based on these guidelines [115].	Ensuring compliance with regulations requires robust enforcement mechanisms and regular updates to guidelines based on new scientific data. Regulation variation between countries can also complicate international trade and food safety efforts.
Food Safety Monitoring	Monitoring involves regularly testing food and feed samples to detect and quantify mycotoxin contamination [116].	Food safety authorities monitor mycotoxin levels using analytical techniques and surveillance programs targeting high-risk products and regions.	Effective monitoring requires access to reliable and sensitive analytical methods and resources for sample collection and testing. Ensuring consistent testing quality and managing large volumes of samples can be resource-intensive.
Enforcement and Compliance	Enforcement ensures compliance with mycotoxin regulations and prompts corrective actions for contamination [117].	Regulatory authorities inspect and enforce mycotoxin standards, with non-compliance leading to fines, recalls, or closures.	Effective enforcement requires coordination and adequate resources to ensure compliance among all stakeholders, especially in regions with limited infrastructure.

Another challenge is the cost of advanced detection methods, which may be prohibitive for routine testing in low-resource settings. While immunoassays like enzyme-linked immunosorbent assays (ELISAs) are more cost-effective, they may lack the sensitivity and specificity required for detecting low levels of mycotoxins in complex samples [118]. Furthermore, developing and validating new detection methods can be time-consuming and resource-intensive.

5.2. Control Measures Limitations

While various control strategies are employed to manage mycotoxin contamination, each has its limitations in terms of effectiveness and feasibility.

Managing mycotoxin contamination involves addressing several challenges related to detection and control measures. Detection methods must be sensitive and specific, but high costs and technical limitations can restrict their use. Control measures, including pre-harvest practices, post-harvest treatments, and chemical decontamination, face limitations in effectiveness and feasibility. Regulatory and monitoring systems are crucial for ensuring compliance but may encounter consistency and resource availability challenges. To improve food safety and mitigate mycotoxin risks, ongoing research, innovation, and investment in more effective and accessible methods are needed [119].

6. Future Directions

The management and control of mycotoxins continue to evolve as new challenges and opportunities arise. Further research and the adoption of emerging technologies are essential to address these issues effectively. Identifying research gaps and staying abreast of emerging trends, such as advanced biosensors, machine learning models, and genome editing tools, can help enhance our ability to detect, control, and mitigate the risks associated with mycotoxins. This proactive approach ensures that food safety strategies remain dynamic and effective in the face of evolving threats [120].

6.1. Research Gaps

In recent years, significant progress has been made in understanding mycotoxins and their impact on food safety. However, several critical research gaps remain, hindering the development of more effective detection, control, and management strategies. Table 12 highlights these gaps, the needs they address, and the opportunities for future research to improve food safety and mitigate the risks posed by mycotoxins [121]

Table 12. Detection challenges.

	Issue	Challenges	Impacts
Sensitivity	Mycotoxins are often present at very low concentrations in food, making it challenging to detect them reliably. Analytical methods must be sensitive enough to identify trace amounts of mycotoxins to ensure accurate safety assessments [122].	Some mycotoxins have low natural abundance or are masked by matrix effects, which can interfere with detection. Analytical methods must be optimized to enhance sensitivity while minimizing false negatives.	Low sensitivity can result in underestimating contamination levels, potentially leading to unsafe food products reaching consumers.
Specificity	The specificity of detection methods is crucial to differentiate between mycotoxins and other compounds with similar chemical properties. Cross-reactivity with other substances can lead to false positives or inaccurate quantification [123].	Some methods, like immunoassays, may lack specificity and produce cross-reactivity with structurally similar compounds. Therefore, it is essential to ensure that methods can accurately target the specific mycotoxin of interest.	Lack of specificity can compromise the accuracy of results and lead to unnecessary regulatory actions or misinformed safety assessments.
Cost	Advanced analytical techniques, such as high-performance liquid chromatography–mass spectrometry (HPLC-MS), can be expensive due to equipment, reagents, and maintenance costs. This can limit their accessibility, especially in resource-limited settings [124].	High costs can restrict the frequency of testing and the number of samples analyzed, potentially leading to gaps in monitoring and an increased risk of undetected mycotoxin contamination.	Costs may limit the implementation of comprehensive testing programs, especially in developing regions with scarce resources.

6.2. Emerging Trends

As mycotoxin management evolves, several emerging trends are reshaping how mycotoxins are detected, controlled, and regulated. These advancements are driven by technological innovations, research into sustainable solutions, and enhanced global cooperation. Table 13 outlines the key emerging trends, highlighting their potential impact on food safety, agricultural practices, and regulatory frameworks [7]. In addition, Table 14 presents the research gaps, while Table 15 discusses key emerging trends.

Table 13. Control measure limitations.

Measures	Issue	Challenges	Impact
Pre-Harvest Control Measures	Pre-harvest measures, such as crop rotation and resistant varieties, are designed to reduce fungal contamination but may not always be practical or feasible [10].	Implementing crop rotation requires careful planning and may not be feasible for all crops. Resistant varieties may not always be available or fully protective.	Inadequate pre-harvest measures can lead to ongoing fungal contamination and mycotoxin production, especially in resource-limited regions.
Post-Harvest Control Measures	Post-harvest control measures, such as proper storage and processing techniques, are essential for managing mycotoxin contamination but may have limitations in effectiveness and feasibility [125].	Maintaining optimal storage conditions requires significant infrastructure, which may be lacking in some regions. Processing methods like heat treatments may not completely degrade all mycotoxins.	Inadequate post-harvest measures can lead to persistent mycotoxin contamination in stored food products, reducing the overall effectiveness of control strategies.
Chemical Treatments	Chemical treatments, such as the use of adsorbents or decontamination agents, can help reduce mycotoxin levels but are not consistently universally effective [126].	The effectiveness of chemical treatments can vary depending on the type of mycotoxin and the food matrix. Additionally, there may be concerns about potential residues or impacts on food quality and safety.	Variability in treatment effectiveness can limit the reliability of chemical methods in ensuring food safety and may require additional validation for different mycotoxins and food types.
Regulatory and Monitoring Approaches	Regulatory and monitoring systems are essential for ensuring compliance with mycotoxin limits but face challenges related to implementation and enforcement [127].	Regulatory guidelines vary between countries, causing inconsistencies in standards. Monitoring programs need substantial resources for sampling and testing, which may pose challenges in resource-limited settings.	Inconsistent regulations and inadequate monitoring can result in gaps in food safety oversight and an increased risk of mycotoxin contamination in the food supply.

Table 14. Research gaps.

Research Gaps	Need	Opportunity
Development of New Analytical Methods	While current analytical techniques such as HPLC, GC-MS, and ELISA are widely used, more advanced methods that offer improved sensitivity, specificity, and cost efficiency are needed.	Research into novel analytical techniques, such as portable sensors, lab-on-a-chip devices, or advanced mass spectrometry methods, could provide faster, more accurate, and cost-effective mycotoxin detection. Improved methods that can analyze multiple mycotoxins simultaneously or in real time would greatly enhance monitoring capabilities [126].
Innovative Control Strategies	Existing control strategies, including pre-harvest and post-harvest measures, are limited in effectiveness. Innovative approaches that can provide more reliable and scalable solutions are needed.	Research into biocontrol agents, such as beneficial microorganisms that inhibit fungal growth, or the development of novel chemical treatments that are both effective and safe, could offer new solutions for mycotoxin management. Additionally, integrating intelligent agriculture technologies, such as precision farming and remote sensing, may provide real-time data to optimize control measures [128].
Understanding Mycotoxin Interactions and Synergistic Effects	Many studies focus on individual mycotoxins, but there is limited research on the interactions and synergistic effects of multiple mycotoxins present in food.	Investigating how different mycotoxins interact and their combined effects on health could improve risk assessments and lead to more comprehensive control strategies. This includes studying the potential for additive or synergistic effects on toxicity and health outcomes [129].

Table 14. Cont.

Research Gaps	Need	Opportunity
Impact of Climate Change	Climate change can influence fungal growth and mycotoxin production, but the specific impacts on mycotoxin contamination and food safety are not fully understood.	Research into how changing climate conditions affect fungal populations, mycotoxin production, and crop susceptibility can help develop adaptive strategies and predictive models to better manage risks in varying environmental conditions [130].

Table 15. Key emerging trends.

	Trend	Impact
Advancements in Analytical Technology	New technologies, such as portable and field-deployable sensors, are emerging for rapid on-site mycotoxin detection. These devices can provide real-time results and are becoming increasingly affordable [131].	Portable sensors and devices can facilitate more frequent and widespread monitoring of mycotoxin contamination, particularly in developing regions or during critical stages of the food supply chain. These advancements improve the ability to detect contamination early and take appropriate actions.
Integration of Machine Learning and Artificial Intelligence	Machine learning (ML) and artificial intelligence (AI) enhance mycotoxin analysis by predicting contamination risks and optimizing control measures [132].	AI and ML algorithms can enhance predictive models for mycotoxin contamination, improve risk assessments, and optimize agricultural practices and monitoring systems. This integration can lead to more effective and targeted interventions.
Biocontrol and Natural Remedies	Research on biocontrol agents, like beneficial microbes and plant extracts, is increasing for environmentally friendly fungal inhibition and mycotoxin degradation [133].	The development and application of biocontrol agents offer a sustainable approach to managing mycotoxins. These natural remedies can reduce reliance on chemical treatments and contribute to more eco-friendly agricultural practices.
Enhanced Food Safety Regulations	Regulatory bodies continually update and refine guidelines and standards for mycotoxin levels in food and feed based on new research and emerging risks [134].	Enhanced regulations and standards can improve food safety and consumer protection. Continuous updates to regulatory frameworks ensure that they reflect the latest scientific knowledge and address emerging threats.
Global Collaboration and Data Sharing	Increased global collaboration and data sharing among researchers, regulatory bodies, and industry stakeholders are becoming more prevalent [135].	Collaborative efforts and shared data can improve the understanding of mycotoxin risks, enhance monitoring and control strategies, and promote the development of global best practices for mycotoxin management.

Addressing the challenges associated with mycotoxins requires ongoing research and adopting new technologies. Identifying research gaps, such as the need for advanced analytical methods and innovative control strategies, is crucial for improving mycotoxin management [113]. Emerging trends, including advancements in analytical technology, the integration of AI and ML, and the development of biocontrol agents, offer promising opportunities to enhance detection and control efforts [136]. By staying informed about these developments and investing in research and innovation, we can better manage the risks associated with mycotoxins [137] and protect public health.

7. Conclusions

This review examined the complex issue of mycotoxins in food, emphasizing their origin, significance, and impact on public health. Mycotoxins, toxic secondary metabolites produced by fungi such as Aspergillus, Penicillium, and Fusarium, represent a severe threat to food safety and public health. Key mycotoxins of concern include aflatoxins, ochratoxins, fumonisins, zearalenone, and patulin, each with specific sources and affected foods. The cancer risk associated with mycotoxins is notably significant, with aflatoxins being particularly carcinogenic and linked to liver cancer. Epidemiological evidence highlights a heightened risk in areas with substantial mycotoxin exposure, often exacerbated by inadequate control measures and regulatory oversight. Detection of mycotoxins remains challenging due to analytical methods' sensitivity, specificity, and cost issues. While techniques such as chromatography and immunoassays are effective, they have limitations that affect their reliability and accessibility. Control measures, including pre-harvest and post-harvest strategies, along with regulatory frameworks, are crucial but have limitations regarding feasibility and effectiveness. Emerging trends, such as advancements in analytical technology, the integration of AI and machine learning, and the development of biocontrol agents, offer promising opportunities for enhancing mycotoxin management. However, research gaps persist, particularly in developing new methods, understanding mycotoxin interactions, and adapting to climate change. To improve the management of mycotoxin contamination and ensure food safety, it is recommended to invest in advanced analytical technologies, develop and implement innovative control strategies, address research gaps and emerging risks, strengthen regulatory and monitoring frameworks, and promote global collaboration and data sharing. Addressing these challenges requires a coordinated effort that includes advancing technology, enhancing control measures, and fostering international cooperation to manage mycotoxin risks and protect public health effectively.

Author Contributions: Conceptualization, A.N.M. and D.B.; methodology, A.N.M.; software, A.N.M.; validation, D.B.; investigation, A.N.M.; writing—original draft preparation, A.N.M.; writing—review and editing, A.N.M. and D.B.; visualization, A.N.M. and D.B.; supervision, D.B.; project administration, D.B.; funding acquisition, D.B. All authors have read and agreed to the published version of the manuscript.

Funding: This work was supported by a National Priorities Research Program grant NPRP 14S-0311-210033; awarded to Dietrich Büsselberg, January 2023–Current) from the Qatar National Research Fund (QNRF, a member of Qatar Foundation). Publication costs for this work were covered by the Biomedical Research Program at Weill Cornell Medicine-Qatar, a program funded by Qatar Foundation. The statements made herein are solely the responsibility of the authors.

Data Availability Statement: No new data were created or analyzed in this study.

Conflicts of Interest: The authors declare no conflicts of interest.

References

1. El-Sayed, R.A.; Jebur, A.B.; Kang, W.; El-Demerdash, F.M. An overview on the major mycotoxins in food products: Characteristics, toxicity, and analysis. *J. Future Foods* **2022**, *2*, 91–102. [CrossRef]
2. Yu, J.; Pedroso, I.R. Mycotoxins in Cereal-Based Products and Their Impacts on the Health of Humans, Livestock Animals and Pets. *Toxins* **2023**, *15*, 480. [CrossRef] [PubMed]
3. Santos, A.R.; Carreiró, F.; Freitas, A.; Barros, S.; Brites, C.; Ramos, F.; Sanches Silva, A. Mycotoxins Contamination in Rice: Analytical Methods, Occurrence and Detoxification Strategies. *Toxins* **2022**, *14*, 647. [CrossRef] [PubMed]
4. Smaoui, S.; D'Amore, T.; Tarapoulouzi, M.; Agriopoulou, S.; Varzakas, T. Aflatoxins Contamination in Feed Commodities: From Occurrence and Toxicity to Recent Advances in Analytical Methods and Detoxification. *Microorganisms* **2023**, *11*, 2614. [CrossRef] [PubMed]
5. Pandey, A.K.; Samota, M.K.; Kumar, A.; Silva, A.S.; Dubey, N.K. Fungal mycotoxins in food commodities: Present status and future concerns. *Front. Sustain. Food Syst.* **2023**, *7*, 1162595. [CrossRef]
6. Ekwomadu, T.I.; Akinola, S.A.; Mwanza, M. Fusarium Mycotoxins, Their Metabolites (Free, Emerging, and Masked), Food Safety Concerns, and Health Impacts. *Int. J. Environ. Res. Public Health* **2021**, *18*, 11741. [CrossRef]
7. Khan, R.; Anwar, F.; Ghazali, F.M. A comprehensive review of mycotoxins: Toxicology, detection, and effective mitigation approaches. *Heliyon* **2024**, *10*, e28361. [CrossRef]
8. Omotayo, O.P.; Omotayo, A.O.; Mwanza, M.; Babalola, O.O. Prevalence of Mycotoxins and Their Consequences on Human Health. *Toxicol. Res.* **2019**, *35*, 1–7. [CrossRef]
9. Kappes, A.; Tozooneyi, T.; Shakil, G.; Railey, A.F.; McIntyre, K.M.; Mayberry, D.E.; Rushton, J.; Pendell, D.L.; Marsh, T.L. Livestock health and disease economics: A scoping review of selected literature. *Front. Vet. Sci.* **2023**, *10*, 1168649. [CrossRef]

10. Zhang, C.; Qu, Z.; Hou, J.; Yao, Y. Contamination and Control of Mycotoxins in Grain and Oil Crops. *Microorganisms* **2024**, *12*, 567. [CrossRef]
11. Mukhtar, K.; Nabi, B.G.; Ansar, S.; Bhat, Z.F.; Aadil, R.M.; Mousavi Khaneghah, A. Mycotoxins and consumers' awareness: Recent progress and future challenges. *Toxicon* **2023**, *232*, 107227. [CrossRef] [PubMed]
12. Faour-Klingbeil, D.; Todd, E. A Review on the Rising Prevalence of International Standards: Threats or Opportunities for the Agri-Food Produce Sector in Developing Countries, with a Focus on Examples from the MENA Region. *Foods* **2018**, *7*, 33. [CrossRef] [PubMed]
13. Magnoli, A.P.; Poloni, V.L.; Cavaglieri, L. Impact of mycotoxin contamination in the animal feed industry. *Curr. Opin. Food Sci.* **2019**, *29*, 99–108. [CrossRef]
14. Sirma, A.J.; Lindahl, J.F.; Makita, K.; Senerwa, D.; Mtimet, N.; Kang'ethe, E.K.; Grace, D. The impacts of aflatoxin standards on health and nutrition in sub-Saharan Africa: The case of Kenya. *Glob. Food Sec.* **2018**, *18*, 57–61. [CrossRef]
15. Lamm, K.W.; Randall, N.L.; Diez-Gonzalez, F. Critical Food Safety Issues Facing the Food Industry: A Delphi Analysis. *J. Food Prot.* **2021**, *84*, 680–687. [CrossRef]
16. Latham, R.L.; Boyle, J.T.; Barbano, A.; Loveman, W.G.; Brown, N.A. Diverse mycotoxin threats to safe food and feed cereals. *Essays Biochem.* **2023**, *67*, 797–809. [CrossRef]
17. Nji, Q.N.; Babalola, O.O.; Ekwomadu, T.I.; Nleya, N.; Mwanza, M. Six Main Contributing Factors to High Levels of Mycotoxin Contamination in African Foods. *Toxins* **2022**, *14*, 318. [CrossRef]
18. Ekwomadu, T.; Mwanza, M.; Musekiwa, A. Mycotoxin-Linked Mutations and Cancer Risk: A Global Health Issue. *Int. J. Environ. Res. Public Health* **2022**, *19*, 7754. [CrossRef]
19. Huong, B.T.M.; Tuyen, L.D.; Madsen, H.; Brimer, L.; Friis, H.; Dalsgaard, A. Total Dietary Intake and Health Risks Associated with Exposure to Aflatoxin B1, Ochratoxin A and Fuminisins of Children in Lao Cai Province, Vietnam. *Toxins* **2019**, *11*, 638. [CrossRef]
20. Soni, P.; Gangurde, S.S.; Ortega-Beltran, A.; Kumar, R.; Parmar, S.; Sudini, H.K.; Lei, Y.; Ni, X.; Huai, D.; Fountain, J.C.; et al. Functional Biology and Molecular Mechanisms of Host-Pathogen Interactions for Aflatoxin Contamination in Groundnut (*Arachis hypogaea* L.) and Maize (*Zea mays* L.). *Front. Microbiol.* **2020**, *11*, 227. [CrossRef]
21. Goessens, T.; Mouchtaris-Michailidis, T.; Tesfamariam, K.; Truong, N.N.; Vertriest, F.; Bader, Y.; De Saeger, S.; Lachat, C.; De Boevre, M. Dietary mycotoxin exposure and human health risks: A protocol for a systematic review. *Environ. Int.* **2024**, *184*, 108456. [CrossRef] [PubMed]
22. Luo, Y.; Liu, X.; Yuan, L.; Li, J. Complicated interactions between bio-adsorbents and mycotoxins during mycotoxin adsorption: Current research and future prospects. *Trends Food Sci. Technol.* **2020**, *96*, 127–134. [CrossRef]
23. Tolosa, J.; Rodríguez-Carrasco, Y.; Ruiz, M.J.; Vila-Donat, P. Multi-mycotoxin occurrence in feed, metabolism and carry-over to animal-derived food products: A review. *Food Chem. Toxicol.* **2021**, *158*, 112661. [CrossRef] [PubMed]
24. Gallage, S.; García-Beccaria, M.; Szydlowska, M.; Rahbari, M.; Mohr, R.; Tacke, F.; Heikenwalder, M. The therapeutic landscape of hepatocellular carcinoma. *Med* **2021**, *2*, 505–552. [CrossRef]
25. Malvandi, A.M.; Shahba, S.; Mehrzad, J.; Lombardi, G. Metabolic Disruption by Naturally Occurring Mycotoxins in Circulation: A Focus on Vascular and Bone Homeostasis Dysfunction. *Front. Nutr.* **2022**, *9*, 915681. [CrossRef]
26. Vörösházi, J.; Neogrády, Z.; Mátis, G.; Mackei, M. Pathological consequences, metabolism and toxic effects of trichothecene T-2 toxin in poultry. *Poult. Sci.* **2024**, *103*, 103471. [CrossRef]
27. Gómez-Osorio, L.-M.; Vasiljevic, M.; Raj, J.; Chaparro-Gutierréz, J.J.; López-Osorio, S. Mycotoxins and coccidiosis in poultry—co-occurrence, interaction, and effects. *Front. Vet. Sci.* **2024**, *11*, 1387856. [CrossRef]
28. Abraham, N.; Chan, E.T.S.; Zhou, T.; Seah, S.Y.K. Microbial detoxification of mycotoxins in food. *Front. Microbiol.* **2022**, *13*, 957148. [CrossRef]
29. Kortei, N.K.; Badzi, S.; Nanga, S.; Wiafe-Kwagyan, M.; Amon, D.N.K.; Odamtten, G.T. Survey of knowledge, and attitudes to storage practices preempting the occurrence of filamentous fungi and mycotoxins in some Ghanaian staple foods and processed products. *Sci. Rep.* **2023**, *13*, 8710. [CrossRef]
30. Liguori, J.; Trübswasser, U.; Pradeilles, R.; Le Port, A.; Landais, E.; Talsma, E.F.; Lundy, M.; Béné, C.; Bricas, N.; Laar, A.; et al. How do food safety concerns affect consumer behaviors and diets in low- and middle-income countries? A systematic review. *Glob. Food Sec.* **2022**, *32*, 100606. [CrossRef]
31. Nada, S.; Nikola, T.; Bozidar, U.; Ilija, D.; Andreja, R. Prevention and practical strategies to control mycotoxins in the wheat and maize chain. *Food Control* **2022**, *136*, 108855. [CrossRef]
32. Agriopoulou, S.; Stamatelopoulou, E.; Varzakas, T. Advances in Occurrence, Importance, and Mycotoxin Control Strategies: Prevention and Detoxification in Foods. *Foods* **2020**, *9*, 137. [CrossRef]
33. Kępińska-Pacelik, J.; Biel, W. Mycotoxins—Prevention, Detection, Impact on Animal Health. *Processes* **2021**, *9*, 2035. [CrossRef]
34. Awuchi, C.G.; Ondari, E.N.; Nwozo, S.; Odongo, G.A.; Eseoghene, I.J.; Twinomuhwezi, H.; Ogbonna, C.U.; Upadhyay, A.K.; Adeleye, A.O.; Okpala, C.O.R. Mycotoxins' Toxicological Mechanisms Involving Humans, Livestock and Their Associated Health Concerns: A Review. *Toxins* **2022**, *14*, 167. [CrossRef]
35. Alameri, M.M.; Kong, A.S.-Y.; Aljaafari, M.N.; Ali, H.A.; Eid, K.; Al Sallagi, M.; Cheng, W.-H.; Abushelaibi, A.; Lim, S.-H.E.; Loh, J.-Y.; et al. Aflatoxin Contamination: An Overview on Health Issues, Detection and Management Strategies. *Toxins* **2023**, *15*, 246. [CrossRef]

36. Ding, L.; Han, M.; Wang, X.; Guo, Y. Ochratoxin A: Overview of Prevention, Removal, and Detoxification Methods. *Toxins* **2023**, *15*, 565. [CrossRef]
37. Li, T.; Li, J.; Wang, J.; Xue, K.S.; Su, X.; Qu, H.; Duan, X.; Jiang, Y. The occurrence and management of fumonisin contamination across the food production and supply chains. *J. Adv. Res.* **2024**, *60*, 13–26. [CrossRef]
38. Balló, A.; Busznyákné Székvári, K.; Czétány, P.; Márk, L.; Török, A.; Szántó, Á.; Máté, G. Estrogenic and Non-Estrogenic Disruptor Effect of Zearalenone on Male Reproduction: A Review. *Int. J. Mol. Sci.* **2023**, *24*, 1578. [CrossRef]
39. Bacha, S.A.S.; Li, Y.; Nie, J.; Xu, G.; Han, L.; Farooq, S. Comprehensive review on patulin and Alternaria toxins in fruit and derived products. *Front. Plant Sci.* **2023**, *14*, 1139757. [CrossRef]
40. Hamad, G.M.; Mehany, T.; Simal-Gandara, J.; Abou-Alella, S.; Esua, O.J.; Abdel-Wahhab, M.A.; Hafez, E.E. A review of recent innovative strategies for controlling mycotoxins in foods. *Food Control* **2022**, *144*, 109350. [CrossRef]
41. Shabeer, S.; Asad, S.; Jamal, A.; Ali, A. Aflatoxin Contamination, Its Impact and Management Strategies: An Updated Review. *Toxins* **2022**, *14*, 307. [CrossRef] [PubMed]
42. Li, X.; Ma, W.; Ma, Z.; Zhang, Q.; Li, H. The Occurrence and Contamination Level of Ochratoxin A in Plant and Animal-Derived Food Commodities. *Molecules* **2021**, *26*, 6928. [CrossRef] [PubMed]
43. Yli-Mattila, T.; Sundheim, L. Fumonisins in African Countries. *Toxins* **2022**, *14*, 419. [CrossRef] [PubMed]
44. Ropejko, K.; Twarużek, M. Zearalenone and Its Metabolites—General Overview, Occurrence, and Toxicity. *Toxins* **2021**, *13*, 35. [CrossRef]
45. Umereweneza, D.; Kamizikunze, T.; Muhizi, T. Assessment of mycotoxins types in some foodstuff consumed in Rwanda. *Food Control* **2018**, *85*, 432–436. [CrossRef]
46. Peivasteh-roudsari, L.; Barzegar-bafrouei, R.; Sharifi, K.A.; Azimisalim, S.; Karami, M.; Abedinzadeh, S.; Asadinezhad, S.; Tajdar-oranj, B.; Mahdavi, V.; Alizadeh, A.M.; et al. Origin, dietary exposure, and toxicity of endocrine-disrupting food chemical contaminants: A comprehensive review. *Heliyon* **2023**, *9*, e18140. [CrossRef]
47. Godebo, T.R.; Stoner, H.; Kodsup, P.; Bases, B.; Marzoni, S.; Weil, J.; Frey, M.; Daley, P.; Earnhart, A.; Ellias, G.; et al. Occurrence of heavy metals coupled with elevated levels of essential elements in chocolates: Health risk assessment. *Food Res. Int.* **2024**, *187*, 114360. [CrossRef]
48. Lambré, C.; Barat Baviera, J.M.; Bolognesi, C.; Chesson, A.; Cocconcelli, P.S.; Crebelli, R.; Gott, D.M.; Grob, K.; Lampi, E.; Mengelers, M.; et al. Re-evaluation of the risks to public health related to the presence of bisphenol A (BPA) in foodstuffs. *EFSA J.* **2023**, *21*, e06857. [CrossRef]
49. Wang, X.; Nag, R.; Brunton, N.P.; Siddique, M.A.B.; Harrison, S.M.; Monahan, F.J.; Cummins, E. Human health risk assessment of bisphenol A (BPA) through meat products. *Environ. Res.* **2022**, *213*, 113734. [CrossRef]
50. Kovač, M.; Bulaić, M.; Nevistić, A.; Rot, T.; Babić, J.; Panjičko, M.; Kovač, T.; Šarkanj, B. Regulated Mycotoxin Occurrence and Co-Occurrence in Croatian Cereals. *Toxins* **2022**, *14*, 112. [CrossRef]
51. Casado, N.; Morante-Zarcero, S.; Sierra, I. The concerning food safety issue of pyrrolizidine alkaloids: An overview. *Trends Food Sci. Technol.* **2022**, *120*, 123–139. [CrossRef]
52. Warnatzsch, E.A.; Reay, D.S.; Camardo Leggieri, M.; Battilani, P. Climate Change Impact on Aflatoxin Contamination Risk in Malawi's Maize Crops. *Front. Sustain. Food Syst.* **2020**, *4*, 591792. [CrossRef]
53. Schrögel, P.; Wätjen, W. Insects for Food and Feed-Safety Aspects Related to Mycotoxins and Metals. *Foods* **2019**, *8*, 288. [CrossRef] [PubMed]
54. Açar, Y.; Akbulut, G. Evaluation of Aflatoxins Occurrence and Exposure in Cereal-Based Baby Foods: An Update Review. *Curr. Nutr. Rep.* **2024**, *13*, 59–68. [CrossRef]
55. Dai, C.; Tian, E.; Hao, Z.; Tang, S.; Wang, Z.; Sharma, G.; Jiang, H.; Shen, J. Aflatoxin B1 Toxicity and Protective Effects of Curcumin: Molecular Mechanisms and Clinical Implications. *Antioxidants* **2022**, *11*, 2031. [CrossRef]
56. Ben Miri, Y.; Benabdallah, A.; Chentir, I.; Djenane, D.; Luvisi, A.; De Bellis, L. Comprehensive Insights into Ochratoxin A: Occurrence, Analysis, and Control Strategies. *Foods* **2024**, *13*, 1184. [CrossRef]
57. Abrehame, S.; Manoj, V.R.; Hailu, M.; Chen, Y.-Y.; Lin, Y.-C.; Chen, Y.-P. Aflatoxins: Source, Detection, Clinical Features and Prevention. *Processes* **2023**, *11*, 204. [CrossRef]
58. Oke, O.E.; Akosile, O.A.; Oni, A.I.; Opowoye, I.O.; Ishola, C.A.; Adebiyi, J.O.; Odeyemi, A.J.; Adjei-Mensah, B.; Uyanga, V.A.; Abioja, M.O. Oxidative stress in poultry production. *Poult. Sci.* **2024**, *103*, 104003. [CrossRef]
59. Lumsangkul, C.; Tso, K.-H.; Fan, Y.-K.; Chiang, H.-I.; Ju, J.-C. Mycotoxin Fumonisin B1 Interferes Sphingolipid Metabolisms and Neural Tube Closure during Early Embryogenesis in Brown Tsaiya Ducks. *Toxins* **2021**, *13*, 743. [CrossRef]
60. Sun, Y.; Song, Y.; Long, M.; Yang, S. Immunotoxicity of Three Environmental Mycotoxins and Their Risks of Increasing Pathogen Infections. *Toxins* **2023**, *15*, 187. [CrossRef]
61. Ghazi, T.; Arumugam, T.; Foolchand, A.; Chuturgoon, A.A. The Impact of Natural Dietary Compounds and Food-Borne Mycotoxins on DNA Methylation and Cancer. *Cells* **2020**, *9*, 2004. [CrossRef] [PubMed]
62. Mahato, D.K.; Devi, S.; Pandhi, S.; Sharma, B.; Maurya, K.K.; Mishra, S.; Dhawan, K.; Selvakumar, R.; Kamle, M.; Mishra, A.K.; et al. Occurrence, Impact on Agriculture, Human Health, and Management Strategies of Zearalenone in Food and Feed: A Review. *Toxins* **2021**, *13*, 92. [CrossRef] [PubMed]
63. Ráduly, Z.; Szabó, L.; Madar, A.; Pócsi, I.; Csernoch, L. Toxicological and Medical Aspects of Aspergillus-Derived Mycotoxins Entering the Feed and Food Chain. *Front. Microbiol.* **2020**, *10*, 2908. [CrossRef] [PubMed]

64. Han, X.; Huangfu, B.; Xu, T.; Xu, W.; Asakiya, C.; Huang, K.; He, X. Research Progress of Safety of Zearalenone: A Review. *Toxins* **2022**, *14*, 386. [CrossRef]
65. Yang, L.; Liao, W.; Dong, J.; Chen, X.; Huang, L.; Yang, W.; Jiang, S. Zearalenone Promotes Uterine Hypertrophy through AMPK/mTOR Mediated Autophagy. *Toxins* **2024**, *16*, 73. [CrossRef]
66. Kościelecka, K.; Kuć, A.; Kubik-Machura, D.; Męcik-Kronenberg, T.; Włodarek, J.; Radko, L. Endocrine Effect of Some Mycotoxins on Humans: A Clinical Review of the Ways to Mitigate the Action of Mycotoxins. *Toxins* **2023**, *15*, 515. [CrossRef]
67. Yang, X.; Zheng, H.; Niu, J.; Chen, X.; Li, H.; Rao, Z.; Guo, Y.; Zhang, W.; Wang, Z. Curcumin alleviates zearalenone-induced liver injury in mice by scavenging reactive oxygen species and inhibiting mitochondrial apoptosis pathway. *Ecotoxicol. Environ. Saf.* **2024**, *277*, 116343. [CrossRef]
68. Saleh, I.; Goktepe, I. The characteristics, occurrence, and toxicological effects of patulin. *Food Chem. Toxicol.* **2019**, *129*, 301–311. [CrossRef]
69. Kobets, T.; Smith, B.P.C.; Williams, G.M. Food-Borne Chemical Carcinogens and the Evidence for Human Cancer Risk. *Foods* **2022**, *11*, 2828. [CrossRef]
70. Bardelčíková, A.; Šoltys, J.; Mojžiš, J. Oxidative Stress, Inflammation and Colorectal Cancer: An Overview. *Antioxidants* **2023**, *12*, 901. [CrossRef]
71. Cui, J.; Yin, S.; Zhao, C.; Fan, L.; Hu, H. Combining Patulin with Cadmium Induces Enhanced Hepatotoxicity and Nephrotoxicity In Vitro and In Vivo. *Toxins* **2021**, *13*, 221. [CrossRef] [PubMed]
72. Mulisa, G.; Pero-Gascon, R.; McCormack, V.; Bisanz, J.E.; Talukdar, F.R.; Abebe, T.; De Boevre, M.; De Saeger, S. Multiple mycotoxin exposure assessment through human biomonitoring in an esophageal cancer case-control study in the Arsi-Bale districts of Oromia region of Ethiopia. *Int. J. Hyg. Environ. Health* **2025**, *263*, 114466. [CrossRef] [PubMed]
73. Benkerroum, N. Chronic and Acute Toxicities of Aflatoxins: Mechanisms of Action. *Int. J. Environ. Res. Public Health* **2020**, *17*, 423. [CrossRef]
74. Andrews-Trevino, J.Y.; Webb, P.; Shively, G.; Kablan, A.; Baral, K.; Davis, D.; Paudel, K.; Shrestha, R.; Pokharel, A.; Acharya, S.; et al. Aflatoxin exposure and child nutrition: Measuring anthropometric and long-bone growth over time in Nepal. *Am. J. Clin. Nutr.* **2021**, *113*, 874–883. [CrossRef]
75. Mupunga, I.; Mngqawa, P.; Katerere, D. Peanuts, Aflatoxins and Undernutrition in Children in Sub-Saharan Africa. *Nutrients* **2017**, *9*, 1287. [CrossRef]
76. Ivanovics, B.; Gazsi, G.; Reining, M.; Berta, I.; Poliska, S.; Toth, M.; Domokos, A.; Nagy, B.; Staszny, A.; Cserhati, M.; et al. Embryonic exposure to low concentrations of aflatoxin B1 triggers global transcriptomic changes, defective yolk lipid mobilization, abnormal gastrointestinal tract development and inflammation in zebrafish. *J. Hazard. Mater.* **2021**, *416*, 125788. [CrossRef]
77. Morales, F.; Montserrat-de la Paz, S.; Leon, M.J.; Rivero-Pino, F. Effects of Malnutrition on the Immune System and Infection and the Role of Nutritional Strategies Regarding Improvements in Children's Health Status: A Literature Review. *Nutrients* **2023**, *16*, 1. [CrossRef]
78. Salas, R.; Acosta, N.; Garza, A.d.J.; Tijerina, A.; Dávila, R.; Jiménez-Salas, Z.; Otero, L.; Santos, M.; Trujillo, A.-J. Levels of Aflatoxin M1 in Breast Milk of Lactating Mothers in Monterrey, Mexico: Exposure and Health Risk Assessment of Newborns. *Toxins* **2022**, *14*, 194. [CrossRef]
79. Lukwago, F.B.; Mukisa, I.M.; Atukwase, A.; Kaaya, A.N.; Tumwebaze, S. Mycotoxins contamination in foods consumed in Uganda: A 12-year review (2006–18). *Sci. Afr.* **2019**, *3*, e00054. [CrossRef]
80. Anato, A.; Headey, D.; Hirvonen, K.; Pokharel, A.; Tessema, M.; Wu, F.; Baye, K. Feed handling practices, aflatoxin awareness and children's milk consumption in the Sidama region of southern Ethiopia. *One Health* **2024**, *18*, 100672. [CrossRef]
81. Visser, M.E.; Schoonees, A.; Ezekiel, C.N.; Randall, N.P.; Naude, C.E. Agricultural and nutritional education interventions for reducing aflatoxin exposure to improve infant and child growth in low- and middle-income countries. *Cochrane Database Syst. Rev.* **2020**, *2020*. [CrossRef] [PubMed]
82. Enespa; Chandra, P. Aflatoxins: Food Safety, Human Health Hazards and Their Prevention. In *Aflatoxins—Occurrence, Detoxification, Determination and Health Risks*; IntechOpen: London, UK, 2022; p. 2.
83. Boshra, M.H.; El-Housseiny, G.S.; Farag, M.M.S.; Aboshanab, K.M. Innovative approaches for mycotoxin detection in various food categories. *AMB Express* **2024**, *14*, 7. [CrossRef] [PubMed]
84. Keskin, E.; Eyupoglu, O.E. Determination of mycotoxins by HPLC, LC-MS/MS and health risk assessment of the mycotoxins in bee products of Turkey. *Food Chem.* **2023**, *400*, 134086. [CrossRef] [PubMed]
85. Janik, E.; Niemcewicz, M.; Podogrocki, M.; Ceremuga, M.; Gorniak, L.; Stela, M.; Bijak, M. The Existing Methods and Novel Approaches in Mycotoxins' Detection. *Molecules* **2021**, *26*, 3981. [CrossRef]
86. Agriopoulou, S.; Stamatelopoulou, E.; Varzakas, T. Advances in Analysis and Detection of Major Mycotoxins in Foods. *Foods* **2020**, *9*, 518. [CrossRef]
87. Malachová, A.; Stránská, M.; Václavíková, M.; Elliott, C.T.; Black, C.; Meneely, J.; Hajšlová, J.; Ezekiel, C.N.; Schuhmacher, R.; Krska, R. Advanced LC–MS-based methods to study the co-occurrence and metabolization of multiple mycotoxins in cereals and cereal-based food. *Anal. Bioanal. Chem.* **2018**, *410*, 801–825. [CrossRef]
88. Liew, W.-P.-P.; Sabran, M.-R. Recent advances in immunoassay-based mycotoxin analysis and toxicogenomic technologies. *J. Food Drug Anal.* **2022**, *30*, 549–561. [CrossRef]

89. Liang, M.; Zhang, Q.; Li, P. Advances in Visual Immunoassays for Sensitive Detection of Mycotoxins in Food—A Review. *Chem. Proc.* **2021**, *5*, 25. [CrossRef]
90. Shahjahan, T.; Javed, B.; Sharma, V.; Tian, F. Overview of Various Components of Lateral-Flow Immunochromatography Assay for the Monitoring of Aflatoxin and Limit of Detection in Food Products: A Systematic Review. *Chemosensors* **2023**, *11*, 520. [CrossRef]
91. Inglis, A.; Parnell, A.C.; Subramani, N.; Doohan, F.M. Machine Learning Applied to the Detection of Mycotoxin in Food: A Systematic Review. *Toxins* **2024**, *16*, 268. [CrossRef]
92. Hafez, E.; Abd El-Aziz, N.M.; Darwish, A.M.G.; Shehata, M.G.; Ibrahim, A.A.; Elframawy, A.M.; Badr, A.N. Validation of New ELISA Technique for Detection of Aflatoxin B1 Contamination in Food Products versus HPLC and VICAM. *Toxins* **2021**, *13*, 747. [CrossRef] [PubMed]
93. Wang, Y.; Wang, X.; Wang, S.; Fotina, H.; Wang, Z. A Novel Lateral Flow Immunochromatographic Assay for Rapid and Simultaneous Detection of Aflatoxin B1 and Zearalenone in Food and Feed Samples Based on Highly Sensitive and Specific Monoclonal Antibodies. *Toxins* **2022**, *14*, 615. [CrossRef] [PubMed]
94. Sadhasivam, S.; Barda, O.; Zakin, V.; Reifen, R.; Sionov, E. Rapid Detection and Quantification of Patulin and Citrinin Contamination in Fruits. *Molecules* **2021**, *26*, 4545. [CrossRef] [PubMed]
95. Guo, R.; Ji, Y.; Chen, J.; Ye, J.; Ni, B.; Li, L.; Yang, Y. Multicolor Visual Detection of Deoxynivalenol in Grain Based on Magnetic Immunoassay and Enzymatic Etching of Plasmonic Gold Nanobipyramids. *Toxins* **2023**, *15*, 351. [CrossRef]
96. Vargas Medina, D.A.; Bassolli Borsatto, J.V.; Maciel, E.V.S.; Lanças, F.M. Current role of modern chromatography and mass spectrometry in the analysis of mycotoxins in food. *TrAC Trends Anal. Chem.* **2021**, *135*, 116156. [CrossRef]
97. Di Nardo, F.; Chiarello, M.; Cavalera, S.; Baggiani, C.; Anfossi, L. Ten Years of Lateral Flow Immunoassay Technique Applications: Trends, Challenges and Future Perspectives. *Sensors* **2021**, *21*, 5185. [CrossRef]
98. Bian, Y.; Zhang, Y.; Zhou, Y.; Wei, B.; Feng, X. Recent Insights into Sample Pretreatment Methods for Mycotoxins in Different Food Matrices: A Critical Review on Novel Materials. *Toxins* **2023**, *15*, 215. [CrossRef]
99. Alahmad, W.; Kaya, S.I.; Cetinkaya, A.; Varanusupakul, P.; Ozkan, S.A. Green chemistry methods for food analysis: Overview of sample preparation and determination. *Adv. Sample Prep.* **2023**, *5*, 100053. [CrossRef]
100. Donnelly, R.; Elliott, C.; Zhang, G.; Baker, B.; Meneely, J. Understanding Current Methods for Sampling of Aflatoxins in Corn and to Generate a Best Practice Framework. *Toxins* **2022**, *14*, 819. [CrossRef]
101. Zhang, K.; Banerjee, K. A Review: Sample Preparation and Chromatographic Technologies for Detection of Aflatoxins in Foods. *Toxins* **2020**, *12*, 539. [CrossRef]
102. Colombo, R.; Papetti, A. Pre-Concentration and Analysis of Mycotoxins in Food Samples by Capillary Electrophoresis. *Molecules* **2020**, *25*, 3441. [CrossRef] [PubMed]
103. Delaunay, N.; Combès, A.; Pichon, V. Immunoaffinity Extraction and Alternative Approaches for the Analysis of Toxins in Environmental, Food or Biological Matrices. *Toxins* **2020**, *12*, 795. [CrossRef]
104. Fumagalli, F.; Ottoboni, M.; Pinotti, L.; Cheli, F. Integrated Mycotoxin Management System in the Feed Supply Chain: Innovative Approaches. *Toxins* **2021**, *13*, 572. [CrossRef] [PubMed]
105. Habschied, K.; Krstanović, V.; Zdunić, Z.; Babić, J.; Mastanjević, K.; Šarić, G.K. Mycotoxins Biocontrol Methods for Healthier Crops and Stored Products. *J. Fungi* **2021**, *7*, 348. [CrossRef] [PubMed]
106. Zadravec, M.; Markov, K.; Lešić, T.; Frece, J.; Petrović, D.; Pleadin, J. Biocontrol Methods in Avoidance and Downsizing of Mycotoxin Contamination of Food Crops. *Processes* **2022**, *10*, 655. [CrossRef]
107. Martín, I.; Gálvez, L.; Guasch, L.; Palmero, D. Fungal Pathogens and Seed Storage in the Dry State. *Plants* **2022**, *11*, 3167. [CrossRef]
108. Bartholomew, H.P.; Bradshaw, M.; Jurick, W.M.; Fonseca, J.M. The Good, the Bad, and the Ugly: Mycotoxin Production During Postharvest Decay and Their Influence on Tritrophic Host–Pathogen–Microbe Interactions. *Front. Microbiol.* **2021**, *12*, 611881. [CrossRef]
109. Loi, M.; Logrieco, A.F.; Pusztahelyi, T.; Leiter, É.; Hornok, L.; Pócsi, I. Advanced mycotoxin control and decontamination techniques in view of an increased aflatoxin risk in Europe due to climate change. *Front. Microbiol.* **2023**, *13*, 1085891. [CrossRef]
110. Abou Dib, A.; Assaf, J.C.; El Khoury, A.; El Khatib, S.; Koubaa, M.; Louka, N. Single, Subsequent, or Simultaneous Treatments to Mitigate Mycotoxins in Solid Foods and Feeds: A Critical Review. *Foods* **2022**, *11*, 3304. [CrossRef]
111. Chatterjee, S.; Dhole, A.; Krishnan, A.A.; Banerjee, K. Mycotoxin Monitoring, Regulation and Analysis in India: A Success Story. *Foods* **2023**, *12*, 705. [CrossRef]
112. Imade, F.; Ankwasa, E.M.; Geng, H.; Ullah, S.; Ahmad, T.; Wang, G.; Zhang, C.; Dada, O.; Xing, F.; Zheng, Y.; et al. Updates on food and feed mycotoxin contamination and safety in Africa with special reference to Nigeria. *Mycology* **2021**, *12*, 245–260. [CrossRef] [PubMed]
113. Chen, C.; Frank, K.; Wang, T.; Wu, F. Global wheat trade and Codex Alimentarius guidelines for deoxynivalenol: A mycotoxin common in wheat. *Glob. Food Sec.* **2021**, *29*, 100538. [CrossRef]
114. Muñoz-Solano, B.; Lizarraga Pérez, E.; González-Peñas, E. Monitoring Mycotoxin Exposure in Food-Producing Animals (Cattle, Pig, Poultry, and Sheep). *Toxins* **2024**, *16*, 218. [CrossRef] [PubMed]
115. Ngure, F.M.; Makule, E.; Mgongo, W.; Phillips, E.; Kassim, N.; Stoltzfus, R.; Nelson, R. Processing complementary foods to reduce mycotoxins in a medium scale Tanzanian mill: A hazard analysis critical control point (HACCP) approach. *Food Control* **2024**, *162*, 110463. [CrossRef]

116. Hao, W.; Li, A.; Wang, J.; An, G.; Guan, S. Mycotoxin Contamination of Feeds and Raw Materials in China in Year 2021. *Front. Vet. Sci.* **2022**, *9*, 929904. [CrossRef]
117. Hu, X.; Li, H.; Yang, J.; Wen, X.; Wang, S.; Pan, M. Nanoscale Materials Applying for the Detection of Mycotoxins in Foods. *Foods* **2023**, *12*, 3448. [CrossRef]
118. Bhatia, D.; Paul, S.; Acharjee, T.; Ramachairy, S.S. Biosensors and their widespread impact on human health. *Sens. Int.* **2024**, *5*, 100257. [CrossRef]
119. Kizis, D.; Vichou, A.-E.; Natskoulis, P.I. Recent Advances in Mycotoxin Analysis and Detection of Mycotoxigenic Fungi in Grapes and Derived Products. *Sustainability* **2021**, *13*, 2537. [CrossRef]
120. Papatheocharidou, C.; Samanidou, V. Two-Dimensional High-Performance Liquid Chromatography as a Powerful Tool for Bioanalysis: The Paradigm of Antibiotics. *Molecules* **2023**, *28*, 5056. [CrossRef]
121. Janssen, E.M.; Mourits, M.C.M.; van der Fels-Klerx, H.J.; Lansink, A.G.J.M.O. Pre-harvest measures against Fusarium spp. infection and related mycotoxins implemented by Dutch wheat farmers. *Crop Prot.* **2019**, *122*, 9–18. [CrossRef]
122. Odjo, S.; Alakonya, A.E.; Rosales-Nolasco, A.; Molina, A.L.; Muñoz, C.; Palacios-Rojas, N. Occurrence and postharvest strategies to help mitigate aflatoxins and fumonisins in maize and their co-exposure to consumers in Mexico and Central America. *Food Control* **2022**, *138*, 108968. [CrossRef]
123. Čolović, R.; Puvača, N.; Cheli, F.; Avantaggiato, G.; Greco, D.; Đuragić, O.; Kos, J.; Pinotti, L. Decontamination of Mycotoxin-Contaminated Feedstuffs and Compound Feed. *Toxins* **2019**, *11*, 617. [CrossRef] [PubMed]
124. Chilaka, C.A.; Obidiegwu, J.E.; Chilaka, A.C.; Atanda, O.O.; Mally, A. Mycotoxin Regulatory Status in Africa: A Decade of Weak Institutional Efforts. *Toxins* **2022**, *14*, 442. [CrossRef] [PubMed]
125. Eshelli, M.; Qader, M.M.; Jambi, E.J.; Hursthouse, A.S.; Rateb, M.E. Current Status and Future Opportunities of Omics Tools in Mycotoxin Research. *Toxins* **2018**, *10*, 433. [CrossRef]
126. Furlong, E.B.; Buffon, J.G.; Cerqueira, M.B.; Kupski, L. Mitigation of Mycotoxins in Food—Is It Possible? *Foods* **2024**, *13*, 1112. [CrossRef]
127. Magoke, G.Z.; Alders, R.G.; Krockenberger, M.; Bryden, W.L. Aflatoxin and Mycotoxin Analysis: An Overview Including Options for Resource-limited Settings. In *Aflatoxins—Occurrence, Detection and Novel Detoxification Strategies*; IntechOpen: London, UK, 2022.
128. Ayaz, M.; Li, C.-H.; Ali, Q.; Zhao, W.; Chi, Y.-K.; Shafiq, M.; Ali, F.; Yu, X.-Y.; Yu, Q.; Zhao, J.-T.; et al. Bacterial and Fungal Biocontrol Agents for Plant Disease Protection: Journey from Lab to Field, Current Status, Challenges, and Global Perspectives. *Molecules* **2023**, *28*, 6735. [CrossRef]
129. Lootens, O.; Vermeulen, A.; Croubels, S.; De Saeger, S.; Van Bocxlaer, J.; De Boevre, M. Possible Mechanisms of the Interplay between Drugs and Mycotoxins—Is There a Possible Impact? *Toxins* **2022**, *14*, 873. [CrossRef]
130. Zingales, V.; Taroncher, M.; Martino, P.A.; Ruiz, M.-J.; Caloni, F. Climate Change and Effects on Molds and Mycotoxins. *Toxins* **2022**, *14*, 445. [CrossRef]
131. Li, Y.; Zhang, D.; Zeng, X.; Liu, C.; Wu, Y.; Fu, C. Advances in Aptamer-Based Biosensors for the Detection of Foodborne Mycotoxins. *Molecules* **2024**, *29*, 3974. [CrossRef]
132. Camardo Leggieri, M.; Mazzoni, M.; Battilani, P. Machine Learning for Predicting Mycotoxin Occurrence in Maize. *Front. Microbiol.* **2021**, *12*, 661132. [CrossRef]
133. Makhuvele, R.; Naidu, K.; Gbashi, S.; Thipe, V.C.; Adebo, O.A.; Njobeh, P.B. The use of plant extracts and their phytochemicals for control of toxigenic fungi and mycotoxins. *Heliyon* **2020**, *6*, e05291. [CrossRef]
134. Stoev, S.D. Food Security and Foodborne Mycotoxicoses—What Should Be the Adequate Risk Assessment and Regulation? *Microorganisms* **2024**, *12*, 580. [CrossRef] [PubMed]
135. Susha, I.; Rukanova, B.; Zuiderwijk, A.; Gil-Garcia, J.R.; Gasco Hernandez, M. Achieving voluntary data sharing in cross sector partnerships: Three partnership models. *Inf. Organ.* **2023**, *33*, 100448. [CrossRef]
136. González-Rodríguez, V.E.; Izquierdo-Bueno, I.; Cantoral, J.M.; Carbú, M.; Garrido, C. Artificial Intelligence: A Promising Tool for Application in Phytopathology. *Horticulturae* **2024**, *10*, 197. [CrossRef]
137. Logrieco, A.F.; Miller, J.D.; Eskola, M.; Krska, R.; Ayalew, A.; Bandyopadhyay, R.; Battilani, P.; Bhatnagar, D.; Chulze, S.; De Saeger, S.; et al. The Mycotox Charter: Increasing Awareness of, and Concerted Action for, Minimizing Mycotoxin Exposure Worldwide. *Toxins* **2018**, *10*, 149. [CrossRef] [PubMed]

Disclaimer/Publisher's Note: The statements, opinions and data contained in all publications are solely those of the individual author(s) and contributor(s) and not of MDPI and/or the editor(s). MDPI and/or the editor(s) disclaim responsibility for any injury to people or property resulting from any ideas, methods, instructions or products referred to in the content.

Review

Impact of Metabolites from Foodborne Pathogens on Cancer

Alice N. Mafe [1] and Dietrich Büsselberg [2,*]

[1] Department of Biological Sciences, Faculty of Sciences, Taraba State University, Main Campus, Jalingo 660101, Taraba State, Nigeria; mafealice1991@gmail.com
[2] Weill Cornell Medicine-Qatar, Education City, Qatar Foundation, Doha Metropolitan Area P.O. Box 22104, Qatar
* Correspondence: dib2015@qatar-med.cornell.edu

Abstract: Foodborne pathogens are microorganisms that cause illness through contamination, presenting significant risks to public health and food safety. This review explores the metabolites produced by these pathogens, including toxins and secondary metabolites, and their implications for human health, particularly concerning cancer risk. We examine various pathogens such as *Salmonella* sp., *Campylobacter* sp., *Escherichia coli*, and *Listeria monocytogenes*, detailing the specific metabolites of concern and their carcinogenic mechanisms. This study discusses analytical techniques for detecting these metabolites, such as chromatography, spectrometry, and immunoassays, along with the challenges associated with their detection. This study covers effective control strategies, including food processing techniques, sanitation practices, regulatory measures, and emerging technologies in pathogen control. This manuscript considers the broader public health implications of pathogen metabolites, highlighting the importance of robust health policies, public awareness, and education. This review identifies research gaps and innovative approaches, recommending advancements in detection methods, preventive strategies, and policy improvements to better manage the risks associated with foodborne pathogens and their metabolites.

Keywords: metabolites; foodborne pathogens; cancer; detection methods; public health

Citation: Mafe, A.N.; Büsselberg, D. Impact of Metabolites from Foodborne Pathogens on Cancer. *Foods* **2024**, *13*, 3886. https://doi.org/10.3390/foods13233886

Academic Editors: Tamara Lazarević-Pašti and Nebojša Potkonjak

Received: 8 November 2024
Revised: 28 November 2024
Accepted: 29 November 2024
Published: 1 December 2024

Copyright: © 2024 by the authors. Licensee MDPI, Basel, Switzerland. This article is an open access article distributed under the terms and conditions of the Creative Commons Attribution (CC BY) license (https://creativecommons.org/licenses/by/4.0/).

1. Introduction

Foodborne pathogens encompass many microorganisms, including bacteria, viruses, fungi, and parasites, capable of contaminating food during production, processing, or storage [1]. When ingested, these microorganisms can cause foodborne diseases, a significant global public health challenge [2]. Pathogens enter food systems through various contamination pathways, such as inadequate hygiene, improper cooking, and cross-contamination [3]. The resulting illnesses range from mild gastrointestinal distress to severe, life-threatening conditions, emphasizing the importance of understanding their biology, transmission, and mitigation [4].

Salmonella sp., a genus of Gram-negative, rod-shaped bacteria, is one of the most significant contributors to foodborne illnesses globally. It is implicated in two primary health conditions [5]. The first is salmonellosis, a common gastrointestinal disease marked by symptoms such as diarrhea, fever, and abdominal cramps [6]. This condition often arises from consuming raw or undercooked eggs, poultry, or contaminated water. The second is typhoid fever, caused by *Salmonella typhi*, a more severe and systemic illness characterized by prolonged fever and potential complications, including intestinal perforation [7]. The pathogenicity of *Salmonella* sp. stems from its ability to produce lipopolysaccharides and effector proteins, which facilitate the invasion of host cells and the evasion of immune defenses, thereby posing significant health risks [8].

While most strains of *Escherichia coli* are harmless and play a commensal role in the human gut, certain pathogenic types can cause severe illnesses [9]. One such group is the Shiga toxin-producing *E. coli* (STEC), including the notorious *E. coli* O157 strain. These strains are associated with severe health outcomes such as hemolytic uremic syndrome (HUS), which is characterized by acute kidney failure, hemolytic anemia, and thrombocytopenia [10]. Infections often arise from consuming undercooked beef, raw vegetables, or unpasteurized milk. The Shiga toxin produced by STEC is a potent virulence factor that disrupts protein synthesis in host cells, leading to cell death and systemic complications [11]. This makes these strains particularly dangerous and underscores the need for stringent food safety measures [12].

Campylobacter species, particularly *C. jejuni* and *E. coli*, are among the most common bacterial causes of diarrheal diseases worldwide [13]. These spiral-shaped, Gram-negative bacteria are typically transmitted through the consumption of undercooked poultry, contaminated water, or raw milk [14]. Infections with *Campylobacter* sp. lead to gastrointestinal symptoms such as diarrhea, abdominal cramps, and fever [15]. Additionally, they can result in autoimmune complications like Guillain–Barré syndrome, a rare but serious condition which affects the nervous system. The bacteria's ability to colonize the intestines and evade the immune response is facilitated by their flagella and adherence factors, highlighting the importance of prevention strategies in food handling and preparation [16].

Listeria monocytogenes, a Gram-positive bacterium, is the causative agent of listeriosis, a severe foodborne illness which poses significant risks to vulnerable populations, including pregnant women, newborns, and immunocompromised individuals [17]. This pathogen is frequently associated with ready-to-eat foods like deli meats, soft cheeses, and raw produce. Listeriosis can manifest in severe forms, such as meningitis and septicemia, and may lead to pregnancy complications, including fetal loss [18]. Notably, *Listeria* sp. can survive and even grow at refrigeration temperatures, making it a persistent challenge in food safety management [19]. This unique trait underscores the critical need for stringent monitoring and control measures in the food supply chain [20].

Mycotoxins are toxic secondary metabolites produced by fungi, including *Aspergillus* sp., *Penicillium* sp., and *Fusarium* sp., which contaminate a wide range of food crops [21]. These compounds pose significant health risks to humans and animals. Among the most concerning mycotoxins are aflatoxins, including *Aspergillus flavus* and *A. parasiticus* [22]. These compounds are highly carcinogenic and are strongly associated with liver cancer, particularly in populations consuming contaminated grains and nuts. Fumonisins, produced by *Fusarium* sp., are linked to esophageal cancer and neural tube defects [23]. Another critical group, ochratoxins, produced by *Aspergillus* sp. and *Penicillium* sp., are nephrotoxic and suspected carcinogens. These mycotoxins often contaminate crops such as grains, nuts, and spices, underscoring the need for stringent food safety measures to mitigate their impact [24].

Mycotoxins, produced as secondary metabolites by certain fungi, have been well-documented for their pathogenic effects on humans and animals, primarily through chronic exposure and their potential carcinogenic, immunosuppressive, and hepatotoxic properties [25]. In contrast, foodborne pathogens secrete toxins that are typically proteinaceous in nature, such as enterotoxins and neurotoxins, [26]. These protein-based toxins operate via distinct pathogenic mechanisms, often causing acute illnesses by directly interacting with host cellular pathways, highlighting a fundamental difference in their modes of action [26]. Mycotoxins are secondary metabolites produced by specific fungi, such as *Aspergillus* sp., *Fusarium* sp., and *Penicillium* sp., often contaminating crops and stored foods under certain environmental conditions [27]. These compounds are chemically diverse, non-proteinaceous, and exhibit chronic toxic effects, including carcinogenicity, hepatotoxicity, immunosuppression, and neurotoxicity [28]. In contrast, toxins produced by foodborne pathogens, such as *Clostridium botulinum*, *Escherichia coli*, and *Staphylococcus aureus*, are primarily protein-based exotoxins [29]. These toxins are typically synthesized during pathogen growth in food or the host, leading to acute diseases through specific interactions with host cellular targets, such as disrupting membranes, inhibiting protein synthesis, or overstimulating immune responses [30]. Thus, mycotoxins and pathogen-derived toxins differ fundamentally in their origins, chemical nature, and mechanisms of pathogenicity [31].

Bacterial toxins are potent substances produced by certain pathogenic bacteria that exacerbate illness and contribute to systemic complications [32]. Shiga toxins, produced by Shiga toxin-producing *Escherichia coli* (STEC), inhibit protein synthesis in host cells, resulting in cell death and conditions such as hemolytic uremic syndrome (HUS), which can cause acute kidney failure [33]. Enterotoxins, secreted by bacteria like *Staphylococcus aureus* and *Bacillus cereus*, are common culprits in acute food poisoning [34]. These toxins lead to symptoms such as vomiting and diarrhea, typically caused by ingesting contaminated foods. Additionally, while not a traditional foodborne pathogen, *Helicobacter pylori* produces toxins that play a role in chronic gastritis and are implicated in gastric cancer, emphasizing the diverse impacts of bacterial toxins on human health [35].

Foodborne pathogens and spoilage organisms also produce other harmful metabolites that impact health. Biogenic amines, such as histamine, can accumulate in spoiled fish, cheese, and fermented foods, triggering toxic reactions in sensitive individuals [36]. The decarboxylation of amino acids forms these compounds during microbial activity. Another critical group of metabolites is lipopolysaccharides (LPSs), components of the outer membrane of Gram-negative bacteria [37]. LPSs can trigger systemic inflammation when they enter the bloodstream and are associated with chronic conditions, including cancer [38]. These metabolites highlight the complexity of food safety issues and the need for comprehensive monitoring and mitigation strategies [39].

Studying foodborne pathogens and their metabolites is critical for identifying health risks and developing mitigation strategies. Many metabolites, especially those with carcinogenic potential, persist in food products even after pathogens are eliminated [40]. This underscores the necessity for advanced detection technologies, such as mass spectrometry and molecular assays, and robust food safety practices [41]. Effective management strategies, including improved hygiene, thorough cooking, and regular food safety monitoring, are essential to reducing exposure and safeguarding public health [42].

The purpose and significance of this article lie in exploring the intricate relationship between these metabolites and cancer, particularly their role in inducing or exacerbating cancer in patients or causing severe complications. Understanding how metabolites like aflatoxins, fumonisins, and bacterial toxins interact with cellular mechanisms to promote carcinogenesis or hinder cancer treatment is vital for developing targeted interventions. Moreover, these metabolites not only pose direct threats to cancer patients but also represent a broader public health challenge by increasing the burden of cancer and related diseases globally.

This overview highlights the multifaceted risks posed by foodborne pathogens and their metabolites, emphasizing their potential to deteriorate the health of cancer patients and the far-reaching implications for public health. It underscores the importance of continued research and preventive measures to mitigate their global impact and improve food safety.

2. Metabolites from Foodborne Pathogens

2.1. Types of Pathogens and Their Metabolites

Foodborne pathogens such as *Salmonella*, *Campylobacter*, *Escherichia coli*, and *Listeria monocytogenes* are notorious for their ability to cause widespread foodborne illnesses [43]. These microorganisms produce various metabolites during their growth and metabolic processes that can affect human health [44]. For example, *Escherichia coli* (mainly *E. coli* O157) produce Shiga toxins, which cause severe gastrointestinal distress and can lead to hemolytic uremic syndrome (HUS) [45]. *Salmonella* sp. releases endotoxins, contributing to inflammation, fever, and diarrhea. *Listeria monocytogenes*, while primarily known for causing listeriosis, can also produce toxins that lead to significant immune system challenges [5]. *Campylobacter*, a leading cause of bacterial food poisoning, produces metabolites that can trigger gastrointestinal disorders and, in severe cases, neurological conditions like Guillain–Barré syndrome [46].

2.2. Metabolites of Concern

Several metabolites produced by these pathogens are associated with severe health conditions, including cancer [47]. For instance, mycotoxins produced by fungi, such as aflatoxins, are potent carcinogens linked to liver cancer [48]. Additionally, bacterial pathogens can generate harmful metabolites, such as the cytolethal distending toxin (CDT) produced by *Campylobacter*, which has been shown to damage DNA and promote tumor development [49]. *Helicobacter pylori*, a bacterium often associated with contaminated food and water, produces urease, which facilitates chronic inflammation in the stomach lining, eventually leading to gastric cancer [50]. Understanding these metabolites is crucial because their impact on health can be significant, contributing not only to acute illnesses but also to long-term, life-threatening diseases like cancer [51].

3. Emerging Metabolites and Their Carcinogenic Mechanisms

The field of foodborne pathogens is continuously evolving, and new evidence is surfacing on the roles metabolites from these microorganisms may play in human health, particularly in cancer development. Most research on foodborne pathogens has traditionally focused on gastrointestinal illnesses. Still, a growing body of work is uncovering how specific metabolites might act as carcinogens or co-factors in cancer progression. This is an exploration of metabolites from under-researched foodborne pathogens like *Cronobacter sakazakii* and *Aeromonas hydrophila* and their potential role in carcinogenesis [1]. This study also delves into how the combined effects of multiple pathogen metabolites can exacerbate cancer progression.

3.1. Metabolites from Emerging or Under-Researched Foodborne Pathogens

3.1.1. *Cronobacter sakazakii* Metabolites and Their Carcinogenic Mechanisms

Cronobacter sakazakii is a Gram-negative pathogen known for causing severe infections, particularly in infants. Traditionally linked to necrotizing enterocolitis, bacteremia, and meningitis in newborns, emerging research points to the ability of *C. sakazakii* to produce specific metabolites that may have carcinogenic effects, especially in gastrointestinal tissues [52].

Lipopolysaccharides (LPSs): LPS, an endotoxin found in the outer membrane of Gram-negative bacteria like *C. sakazakii*, can trigger chronic inflammation in the gastrointestinal tract. Chronic inflammation is a well-documented risk factor for colorectal cancer, as it promotes DNA mutations causing cellular damage and creates an environment conducive to tumorigenesis [53].

Toxins and Secondary Metabolites: Studies have suggested that certain toxins and secondary metabolites produced by *C. sakazakii* could have genotoxic effects, although the precise mechanisms remain largely unstudied. These metabolites may impair DNA repair mechanisms, allowing for the accumulation of mutations which can initiate cancerous growths [54].

Given the rising incidence of *C. sakazakii* infections in adults and its potential ability to colonize the intestines, exploring its metabolites' carcinogenic effects in adults, especially those with pre-existing inflammatory conditions, represents a novel area of research [55].

3.1.2. *Aeromonas hydrophila* Metabolites and Their Carcinogenic Mechanisms

Aeromonas hydrophila is another Gram-negative pathogen commonly found in aquatic environments and associated with foodborne infections from contaminated water, seafood, or produce. While primarily linked to gastrointestinal diseases, its metabolites are increasingly being studied for their role in cancer progression [56].

Aerolysin: *A. hydrophila* produces a pore-forming toxin called aerolysin, which disrupts the intestinal epithelium's integrity, causing damage and inflammation. Chronic damage to epithelial tissues is a recognized precursor to cancer, as it can lead to dysplasia and the eventual transformation of cells into a neoplastic state.

Siderophores and Reactive Oxygen Species (ROS): Another mechanism by which *A. hydrophila* metabolites may contribute to carcinogenesis is by producing siderophores and iron-chelating compounds. Siderophore activity can promote the generation of ROS, which leads to oxidative stress and DNA damage. Persistent DNA damage without efficient repair can result in mutations and cancer [57].

Endotoxins: Like *C. sakazakii*, *A. hydrophila* also produces endotoxins that can induce a pro-inflammatory response, particularly in the gastrointestinal tract. The relationship between chronic inflammation and cancer development, especially in the context of gastrointestinal cancers, underscores the importance of studying these endotoxins further [58].

Although the direct involvement of these metabolites in cancer remains underexplored, *A. hydrophila* is a promising subject for future investigations due to its capacity to disrupt gut homeostasis and promote a pro-inflammatory environment that could potentially contribute to cancer progression [59].

3.1.3. Synergistic Effects of Multiple Pathogen Metabolites on Cancer Progression

The human body is frequently exposed to multiple pathogens simultaneously, particularly in foodborne illnesses where various microorganisms may contaminate the same food source. Combined exposure to metabolites from different pathogens may exacerbate the carcinogenic potential of individual metabolites through synergistic mechanisms, including increased inflammation, oxidative stress, and DNA damage [60].

- Enhanced Inflammatory Responses

Metabolites from different pathogens, such as LPSs from *C. sakazakii* and aerolysin from *A. hydrophila*, could act in concert to amplify inflammatory responses in the host. Chronic inflammation is one of the critical drivers of cancer, especially in tissues like the gastrointestinal tract, where many foodborne pathogens exert their effects [61].

Cytokine Storms: The interaction between LPSs and other bacterial toxins can produce excessive pro-inflammatory cytokines, such as TNF-α, IL-6, and IL-1β. Sustained high levels of these cytokines promote a cycle of cellular injury, proliferation, and repair that can lead to mutations and malignant transformation [62].

Macrophage Polarization: Some pathogen metabolites influence the polarization of macrophages towards a pro-tumorigenic M2 phenotype. When exposed to multiple metabolites, macrophages may be more likely to adopt this tumor-supporting role, enhancing the growth and survival of cancer cells [63].

- Increased Oxidative Stress and DNA Damage

Oxidative stress is a critical factor in cancer development. Metabolites from foodborne pathogens can contribute to the accumulation of ROS, which causes oxidative damage to DNA, proteins, and lipids. When multiple pathogens are present, the total ROS burden can be significantly higher [64].

Cross-Reaction of Metabolites: For instance, siderophores produced by *A. hydrophila* can facilitate iron uptake in infected tissues, and iron is a well-known catalyst of ROS production. DNA damage can be more severe when combined with LPSs from *C. sakazakii*, inducing an immune response and further oxidative stress. The accumulation of mutations in key oncogenes and tumor suppressor genes (such as p53) can accelerate the initiation of cancer [65].

Metabolic Reprogramming: Certain pathogens may also alter the metabolic landscape of infected cells, making them more susceptible to oxidative stress. For instance, metabolite-induced metabolic shifts toward glycolysis (the Warburg effect) can increase the vulnerability of cells to ROS, compounding the risk of malignant transformation [66].

- Disruption of Gut Microbiota and Epithelial Barrier Integrity

Another synergistic effect arises from the combined impact of pathogen metabolites on the gut microbiota and the intestinal epithelial barrier. Many foodborne pathogens, including *C. sakazakii* and *A. hydrophila*, can alter the gut microbiome's composition, leading to dysbiosis, which is closely linked to cancer [67].

Microbiome Dysregulation: Pathogen metabolites can shift the balance of the gut microbiota towards pro-inflammatory and carcinogenic bacteria. When multiple pathogens are involved, the disruption is often more severe. This dysbiosis can promote conditions like colorectal cancer by creating a pro-carcinogenic microenvironment that supports tumor growth and suppresses anti-tumor immune responses [68].

Barrier Disruption: Metabolites from multiple pathogens may weaken the gut's epithelial barrier, increasing permeability (leaky gut). This allows harmful substances, including additional carcinogens from the diet or environment, to enter the bloodstream, where they can reach other tissues and potentially initiate tumorigenesis [69].

3.1.4. Future Directions in Research

Exploring pathogen metabolites and their role in cancer is still in its early stages. Future research should focus on the following:

- Characterizing new metabolites—advances in metabolomics can aid in identifying novel carcinogenic metabolites from under-studied pathogens;
- Longitudinal studies—tracking the long-term effects of chronic exposure to pathogen metabolites and their role in cancer development [70];
- Intervention strategies—investigating probiotic therapies, microbiome modulation, and dietary interventions to mitigate the carcinogenic effects of pathogen metabolites.

Metabolites from emerging or under-researched foodborne pathogens such as *Cronobacter sakazakii* and *Aeromonas hydrophila* hold significant potential to contribute to cancer development through mechanisms such as chronic inflammation, oxidative stress, and microbiome disruption [71]. The synergistic effects of combined pathogen metabolite exposure can further exacerbate these carcinogenic processes. Addressing this emerging risk requires a multifaceted approach involving advanced detection methods, improved food safety measures, and ongoing research into the long-term health impacts of these metabolites [72].

3.2. Metabolite Interaction with the Microbiome and Cancer

The gut microbiome, comprising trillions of microorganisms, is crucial in maintaining human health, including regulating immune responses, digestion and even influencing cancer development. When the gut microbiome is disrupted by factors such as metabolites from foodborne pathogens, it can lead to dysbiosis—a state of microbial imbalance, which has been increasingly linked to carcinogenesis. The metabolites produced by these pathogens may alter the composition and function of the gut microbiome, influencing cancer progression through various mechanisms, such as chronic inflammation, disruption of immune homeostasis, and the induction of genotoxicity [51]. This section explores the intricate relationship between foodborne pathogen metabolites and the gut microbiome, emphasizing how these metabolites alter microbial composition, induce dysbiosis, and contribute to cancer development [73].

3.2.1. Microbiome Interaction: How Pathogen Metabolites Disrupt the Gut Microbiota

The gut microbiome is a complex ecosystem; its balance is essential for maintaining health. However, metabolites produced by foodborne pathogens can interfere with the normal microbial flora, promoting an environment conducive to carcinogenesis [74].

- Metabolites and Gut Microbiota Composition

Pathogen metabolites can directly and indirectly affect the microbial community, disrupting the gut's balance between beneficial and harmful bacteria. Some metabolites produced by foodborne pathogens have the following potential:

They can promote the growth of pathogenic or opportunistic microbes: For instance, metabolites like lipopolysaccharides (LPSs) from Gram-negative bacteria (e.g., *Salmonella*, *Escherichia coli*) can suppress the growth of beneficial bacteria (such as *Lactobacillus* and *Bifidobacterium*) while promoting the proliferation of pathogenic bacteria. This shift in microbial composition can favor pro-inflammatory and tumor-promoting species [75].

They can also have direct antimicrobial activity: Some pathogen-derived metabolites possess antimicrobial properties that may selectively kill beneficial microbes, leading to the loss of microbiota diversity. Reduced diversity in the gut microbiome is a well-established risk factor for diseases such as colorectal cancer [76].

- Influence on Microbial Metabolic Functions

Pathogen metabolites not only alter the composition of the microbiome but also affect the metabolic functions of resident microbes.

Short-Chain Fatty Acid (SCFA) Production: SCFAs (such as butyrate, acetate, and propionate) are beneficial byproducts of the fermentation of dietary fibers by gut bacteria. SCFAs, particularly butyrate, have protective roles in preventing cancer by maintaining epithelial integrity, suppressing inflammation, and inducing apoptosis in cancer cells. Pathogen metabolites can inhibit SCFA-producing bacteria, resulting in lower SCFA levels, which compromises the protective functions of the gut [77].

Secondary Bile Acid Metabolism: The gut microbiota is critical in converting primary bile acids into secondary bile ones. Some secondary bile ones, such as deoxycholic acid (DCA), are carcinogenic and promote DNA damage in epithelial cells. Pathogen metabolites may induce dysbiosis that enhances the production of these carcinogenic bile acids, contributing to colorectal cancer risk [78].

3.2.2. Metabolite-Induced Dysbiosis: Mechanisms Contributing to Carcinogenesis

Dysbiosis, an imbalance in the microbial ecosystem, is a crucial factor linking pathogen metabolites to cancer. Metabolite-induced dysbiosis disrupts the normal protective functions of the gut microbiome, leading to several pathways of carcinogenesis [79].

- Chronic Inflammation and Immune Dysregulation

One primary mechanism by which metabolite-induced dysbiosis contributes to cancer is chronic inflammation. Inflammation is a double-edged sword: while it is essential for immune defense, chronic uncontrolled inflammation can promote cancer by damaging tissues, causing cellular mutations, and creating a tumor-promoting microenvironment [80].

Lipopolysaccharides (LPSs): LPS, produced by Gram-negative bacteria, is a potent inducer of inflammation by activating toll-like receptor 4 (TLR4) on immune cells. Pathogen-derived LPS induces dysbiosis, leading to an exaggerated immune response, including the release of pro-inflammatory cytokines such as IL-6, TNF-α, and IL-1β. Chronic exposure to these cytokines can promote epithelial damage and create a microenvironment that fosters cancer cell proliferation and survival [81].

Metabolite-induced immune evasion: Dysbiosis can shift immune responses in favor of tumor development by impairing immune surveillance mechanisms. For example, dysregulated gut microbiota resulting from pathogen metabolites can reduce the activity of anti-tumor immune cells, such as cytotoxic T cells, while promoting regulatory T cells (Tregs), which suppress anti-tumor immunity. This creates an immune-tolerant environment that allows cancer cells to evade immune destruction [82].

- Gut Barrier Dysfunction: A Pathway to Carcinogenesis

The gut epithelial barrier is the first defense against pathogenic bacteria and their metabolites. However, when dysbiosis compromises this barrier, pathogen metabolites can penetrate deeper into the tissues, promoting inflammation, immune dysregulation, and DNA damage [83].

Increased Gut Permeability (Leaky Gut): Dysbiosis caused by pathogen metabolites can disrupt tight-junction proteins between epithelial cells, leading to increased intestinal permeability. This condition, often called "leaky gut", allows toxins, metabolites, and even bacteria to enter the bloodstream. Once in circulation, these substances can affect distant organs, including the liver and colon, leading to systemic inflammation and promoting cancer in these tissues [84].

Genotoxic Effects: Some pathogen metabolites, such as colibactin (produced by certain strains of *Escherichia coli*), are directly genotoxic. Colibactin forms DNA adducts, leading to double-strand breaks in DNA. Dysbiosis that increases the abundance of colibactin-producing bacteria heightens the risk of mutagenesis and cancer development, especially in the colon [85].

- Oxidative Stress and DNA Damage

Pathogen metabolites can exacerbate oxidative stress in the gut, contributing to carcinogenesis through increased DNA damage.

Reactive Oxygen Species (ROS) Production: Certain metabolites from foodborne pathogens stimulate the production of ROS, either by the host immune response or through microbial activity. Chronic exposure to high levels of ROS leads to oxidative damage to DNA, proteins, and lipids, which can initiate carcinogenesis. For example, *Helicobacter pylori*, a well-known carcinogenic pathogen, produces metabolites that increase ROS production, leading to DNA damage and the initiation of gastric cancer [86].

Impaired DNA Repair Mechanisms: Dysbiosis induced by pathogen metabolites can impair the host's DNA repair mechanisms, making cells more susceptible to accumulating mutations. As a result, mutations in critical oncogenes or tumor suppressor genes (such as p53) can arise, promoting cancer initiation and progression [87].

- Tumor-Promoting Metabolites and Microbial Byproducts

Specific microbial metabolites, particularly those produced by dysbiotic microbiota, have been shown to promote tumor growth directly.

Secondary Bile Acids: Dysbiosis can enhance the production of carcinogenic secondary bile acids, such as deoxycholic acid (DCA), which induce DNA damage, promote chronic inflammation, and stimulate cancerous cell growth in the colon [88].

Polyamines: Pathogen-induced dysbiosis may also increase the production of polyamines, such as putrescine and spermidine, essential for cellular growth. Elevated polyamine levels have been associated with increased cancer cell proliferation, particularly in gastrointestinal cancers [89].

3.2.3. The Crosstalk Between Dysbiosis, Metabolites, and the Tumor Microenvironment

Once dysbiosis is established, pathogen metabolites may continue to interact with the tumor microenvironment, exacerbating cancer progression. These metabolites can modulate the behavior of not only cancer cells but also surrounding stromal cells, immune cells, and blood vessels, influencing tumor growth and metastasis [90].

- Impact on Cancer Cell Metabolism

Dysbiosis-induced pathogen metabolites can alter the metabolic programming of cancer cells, making them more aggressive and invasive.

Warburg Efffect: Dysbiosis can promote metabolic reprogramming in cancer cells, driving them towards glycolysis even in the presence of oxygen (the Warburg effect). This metabolic shift enhances the production of lactate, which acidifies the tumor microenvironment and promotes tumor invasion and immune evasion [91].

- Modulation of Angiogenesis

Specific metabolites from dysbiotic microbiota can also promote angiogenesis (forming new blood vessels), which is critical for tumor growth and metastasis. For instance, dysbiosis can lead to the increased production of pro-angiogenic factors like the vascular endothelial growth factor (VEGF), which supports the formation of blood vessels which supply nutrients to growing tumors [92].

3.2.4. Therapeutic Implications: Targeting Metabolite-Induced Dysbiosis

Given the critical role that metabolite-induced dysbiosis plays in cancer development, several therapeutic strategies are being explored to restore microbial balance and mitigate carcinogenesis [93].

- Probiotics and Prebiotics

Probiotics (live beneficial bacteria) and prebiotics (non-digestible fibers that promote beneficial bacterial growth) can help restore the gut microbiome's balance [94].

Butyrate-Producing Bacteria: Supplementation with butyrate-producing probiotics can restore SCFA levels, improving epithelial integrity and reducing inflammation. Probiotics such as *Lactobacillus* and *Bifidobacterium* have been shown to reduce tumor-promoting metabolites and restore gut homeostasis [95].

Dietary Interventions: Consuming prebiotics such as inulin and resistant starch can increase SCFA production and improve the gut's barrier function, counteracting the effects of dysbiosis [96].

- Fecal Microbiota Transplantation (FMT)

FMT involves transferring fecal material from a healthy donor to a patient with dysbiosis. This technique has shown promise in restoring microbial balance and reducing dysbiosis-associated cancer risks. However, the long-term efficacy of FMT in cancer prevention remains under investigation [97].

- Pharmacological Interventions

Emerging pharmacological interventions aim to target specific metabolites or pathways involved in dysbiosis-induced carcinogenesis.

Bile Acid Sequestrants: Drugs that bind and sequester carcinogenic bile acids like DCA may help reduce the risk of colorectal cancer in patients with dysbiosis [98].

Antioxidants: Given the role of ROS in cancer development, antioxidant therapy may help mitigate oxidative stress and prevent DNA damage in individuals with dysbiosis [99]. Table 1 summarizes case studies focusing on emerging metabolites from foodborne pathogens and their carcinogenic mechanisms, highlighting their public health implications.

Table 1. Case studies in emerging metabolites and their carcinogenic mechanisms.

Case Study	Emerging Metabolite	Pathogen	Carcinogenic Mechanism	Public Health Concern
Cronobacter sakazakii in Infant Formula	Cronobacter Metabolites	*Cronobacter sakazakii*	Alters gut permeability and induces inflammatory responses, potentially leading to carcinogenesis.	Increased cancer risk in infants due to compromised gut health [100].
Aeromonas hydrophila in Aquatic Foods	Aerolysin and Other Toxins	*Aeromonas hydrophila*	Induces apoptosis and disrupts cellular signaling pathways, promoting tumorigenesis.	Risk of gastrointestinal cancers from contaminated aquatic products [101].
Fusarium mycotoxins in Cereals	Zearalenone and Deoxynivalenol (DON)	*Fusarium* spp.	Estrogenic activity and DNA damage lead to hormonal cancers.	Long-term consumption of contaminated grains raises cancer risk.
Bacillus cereus in Rice and Grains	Cereulide and Other Enterotoxins	*Bacillus cereus*	Induces oxidative stress and DNA damage, potentially triggering carcinogenic pathways.	Associated risk of gastrointestinal cancers due to food poisoning incidents [102].
Clostridium botulinum in Canned Foods	Botulinum Neurotoxins	*Clostridium botulinum*	Neurotoxin-induced cell damage and inflammation can facilitate cancer development over time.	Severe public health risks in cases of foodborne botulism, with long-term effects [103].

Metabolite interaction with the microbiome is critical in cancer development, primarily through dysbiosis, chronic inflammation, immune dysregulation, and oxidative stress. Metabolites produced by foodborne pathogens disrupt the delicate balance of the gut microbiota, leading to an environment which promotes carcinogenesis [104]. Therapeutic strategies that restore microbial balance, such as probiotics, prebiotics, and pharmacological interventions, offer promising avenues for mitigating the cancer-promoting effects of dysbiosis. Understanding the intricate relationship between pathogen metabolites and the gut microbiome provides valuable insights into preventing and treating microbiome-associated cancers.

4. Cancer Risk Associated with Pathogen Metabolites

4.1. Mechanisms of Carcinogenicity

The metabolites produced by specific foodborne pathogens contribute to cancer development through various biochemical pathways [105]. Pathogens and their metabolites play significant roles in increasing cancer risks through multiple mechanisms. For example, *Aspergillus flavus* produces aflatoxins, mycotoxins which form DNA adducts in liver cells, leading to mutations which disrupt normal cellular processes [73]. These mutations accumulate over time, initiating carcinogenesis and significantly elevating the risk of liver cancer [106]. Similarly, *Helicobacter pylori* produces urease, an enzyme which causes chronic inflammation in the stomach lining, resulting in oxidative stress and the release of reactive oxygen species (ROS) [107]. This environment fosters DNA damage and promotes cell proliferation, thereby increasing the risk of gastric cancer. *Campylobacter jejuni* releases cytolethal distending toxin (CDT), which damages the DNA of host cells by causing double-strand breaks; if left unrepaired, these breaks lead to genomic instability, a critical factor in the development of colorectal cancer [108]. Also, *Salmonella typhi* produces typhoid toxin, inducing chronic inflammation in the gallbladder and bile ducts. This inflammation can cause cellular damage and mutations, predisposing cells to become cancerous, thereby heightening the risk of gallbladder cancer. Certain *Escherichia coli* strains (e.g., *E. coli* O157) produce colibactin, which forms DNA adducts and causes double-strand DNA breaks. If unrepaired, this damage promotes mutations and chromosomal instability, contributing to colorectal cancer development. *Fusobacterium nucleatum* employs FadA adhesin to adhere to and invade intestinal cells, triggering a pro-inflammatory response. This persistent inflammation leads to cellular proliferation and genetic instability, which can progress to colorectal cancer. Lastly, *Streptococcus bovis* produces carcinogenic metabolites that stimulate inflammatory responses in the colon and are associated with polyp formation. These metabolites disrupt normal cellular signaling pathways, increasing the likelihood of malignant transformation and colon cancer development (Figure 1). Another significant carcinogenic mechanism is the disruption of normal cell signaling pathways. For instance, some metabolites activate pathways involved in cell survival and proliferation, such as the NF-κB pathway, which is linked to inflammation and cancer progression [109]. Other metabolites interfere with tumor suppressor genes like p53, impairing the body's natural ability to prevent cancerous growth [110]. These biochemical disruptions are critical to understanding how pathogen-derived metabolites initiate or promote carcinogenesis [111].

4.2. Epidemiological Evidence

Numerous studies have linked exposure to pathogen-derived metabolites with an increased cancer risk [112]. One of the most extensively studied examples is the relationship between aflatoxins and liver cancer [113]. Research shows that populations in regions with high levels of aflatoxin contamination, such as parts of sub-Saharan Africa and Southeast Asia, have significantly higher rates of hepatocellular carcinoma [114]. Animal studies have also demonstrated the carcinogenic effects of aflatoxins, providing further evidence of their cancer-causing potential [115].

Helicobacter pylori is a significant focus of epidemiological research for bacterial metabolites. Chronic infection with this pathogen has been strongly associated with gastric cancer, and it has been classified as a Group 1 carcinogen by the International Agency for Research on Cancer (IARC) [116]. Studies in both human and animal models have confirmed the link between *H. pylori* infection, inflammation, and the development of gastric cancer [117]. In terms of *Campylobacter* sp. and its metabolites, although more research is needed, preliminary studies suggest that its CDT may play a role in colorectal cancer development, particularly in regions with high rates of foodborne infections [118].

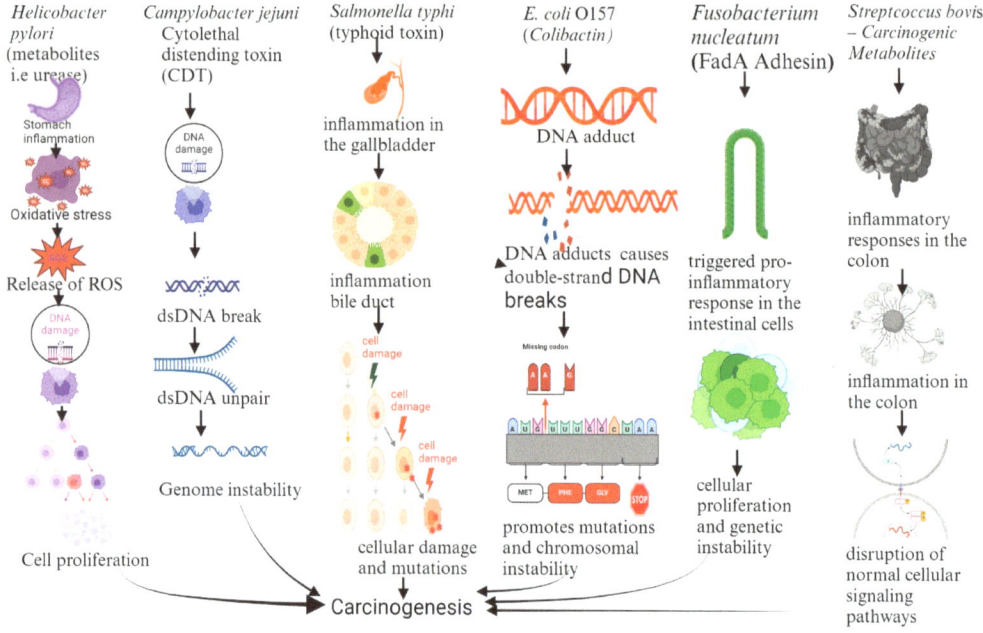

Figure 1. Mechanism of carcinogenicity (created using Bio Render). Legend: ROS: reactive oxygen species; DNA: deoxyribonucleic acid; dsDNA: double-stranded DNA, and CDT: Cytolethal distending toxin.

Together, these studies highlight the significant cancer risk posed by pathogen-derived metabolites, underscoring the need for better detection and control measures to reduce exposure and protect public health [119].

To better analyze and demonstrate the direct impact of foodborne pathogen metabolites on public health, several case studies can be further explored in terms of their emerging metabolites, carcinogenic mechanisms, and public health concerns:

Cronobacter sakazakii in Infant Formula: *Cronobacter sakazakii*, a pathogen commonly associated with powdered infant formula, produces metabolites that alter gut permeability and trigger inflammatory responses [120]. These changes can disrupt the gut's protective barrier, allowing harmful substances to enter the bloodstream and potentially lead to carcinogenesis [121]. In infants, whose immune systems and gut microbiomes are still developing, this disruption can have long-term health implications, increasing the risk of developing cancers later in life due to chronic inflammation and altered immune responses [122]. This case study highlights the vulnerability of infants to pathogen-induced gut dysfunction and the potential for increased cancer risk, especially in populations with compromised gut health.

Aeromonas hydrophila in Aquatic Foods: *Aeromonas hydrophila* is a pathogen found in aquatic foods, producing aerolysin and other toxins capable of inducing apoptosis (programmed cell death) and disrupting cellular signaling pathways [123]. These disruptions can promote tumorigenesis, particularly in tissues exposed to these toxins, such as those in the gastrointestinal system [124]. The ingestion of contaminated aquatic products can lead to gastrointestinal cancers, as the toxins alter normal cellular processes and trigger cancer-promoting pathways [125]. The public health concern here is the potential for widespread exposure to this pathogen through contaminated seafood, leading to a significant burden of gastrointestinal cancers [126].

Fusarium sp. Mycotoxins in Cereals: *Fusarium* species, such as *Fusarium graminearum*, produce mycotoxins like zearalenone and deoxynivalenol (DON) in contaminated cereals [127]. These mycotoxins have estrogenic properties, mimicking the effects of the hormone estrogen, and they can cause DNA damage in human cells [128]. The estrogenic activity of these metabolites has been linked to hormonal cancers, such as breast and ovarian cancer, by altering hormonal balances and inducing mutations in cells [129]. Long-term consumption of contaminated grains, especially in regions where food safety regulations may be less stringent, increases the risk of developing these cancers [130]. This case study emphasizes the need for the robust monitoring of food products to mitigate exposure to these harmful metabolites.

Bacillus cereus in Rice and Grains: *Bacillus cereus*, a pathogen commonly associated with foodborne illness from rice and grains, produces toxins like cereulide and enterotoxins which induce oxidative stress and DNA damage in human cells [131]. Oxidative stress is a known factor in the initiation of carcinogenic pathways, and the DNA damage caused by these toxins can lead to mutations that promote cancer development [125]. This pathogen poses a particular public health concern in areas where rice and grains are staple foods, and frequent food poisoning incidents increase the risk of gastrointestinal cancers over time [132]. Public health measures aimed at reducing contamination levels in these foods are essential to mitigate these risks [133].

Clostridium botulinum in Canned Foods: *Clostridium botulinum*, known for causing botulism through its production of neurotoxins in improperly canned foods, presents an unusual but significant carcinogenic mechanism [134]. The neurotoxins cause severe cell damage and inflammation in affected tissues, which, over time, can contribute to the development of cancer [135]. Although botulism itself is an acute condition, the prolonged effects of neurotoxin-induced inflammation and cellular damage can increase the risk of cancer in affected individuals [136]. This case study underscores the severe public health risks posed by foodborne botulism, highlighting the long-term effects of botulinum toxin exposure that go beyond immediate neurological symptoms and could lead to cancer development over time.

These case studies demonstrate how foodborne pathogen metabolites can contribute to cancer development, from inducing cellular damage and inflammation to disrupting hormonal and immune systems. Each case highlights the need for improved food safety measures, better detection of pathogen contamination, and increased public health awareness, especially in vulnerable populations exposed to these pathogens.

5. Detection Methods and Analytical Techniques

5.1. Detection Methods

Detection and control strategies for pathogen metabolites focus on identifying harmful metabolites produced by foodborne pathogens and implementing measures to mitigate their impact on human health. These metabolites, often linked to various health risks, including cancer, need to be carefully monitored and controlled to ensure food safety [1].

The detection of pathogen metabolites involves several advanced techniques. Analytical methods such as liquid chromatography–mass spectrometry (LC-MS), gas chromatography (GC), and enzyme-linked immunosorbent assay (ELISA) are widely used to detect and quantify these harmful substances in food samples. In addition, electrochemical and optical biosensors provide rapid and sensitive detection, allowing for the timely identification of specific metabolites. Spectroscopy techniques, such as nuclear magnetic resonance (NMR) and UV–visible (UV-Vis) spectroscopy, help analyze the chemical structures of these metabolites, further enhancing the detection process [137].

Control strategies aim to prevent or minimize the presence of pathogen metabolites in food. Traditional food preservation techniques, such as refrigeration, pasteurization, and drying, inhibit microbial growth and reduce the production of harmful metabolites. Using natural and synthetic antimicrobial agents and bacteriophages can further prevent or decrease the activity of pathogens responsible for producing these metabolites. Probiotics, beneficial microorganisms introduced into food systems, can also out-compete harmful pathogens, reducing their ability to produce toxic compounds [138].

In addition to these methods, maintaining good manufacturing practices (GMPs) is crucial. Ensuring the proper hygiene, handling, and storage of food products reduces contamination risks. In contrast, regulatory compliance with food safety standards, such as hazard analysis and critical control points (HACCPs), ensures consistent monitoring and the prevention of metabolite contamination. These comprehensive strategies are essential in safeguarding public health by reducing exposure to carcinogenic or harmful metabolites from foodborne pathogens (Figure 2) [139].

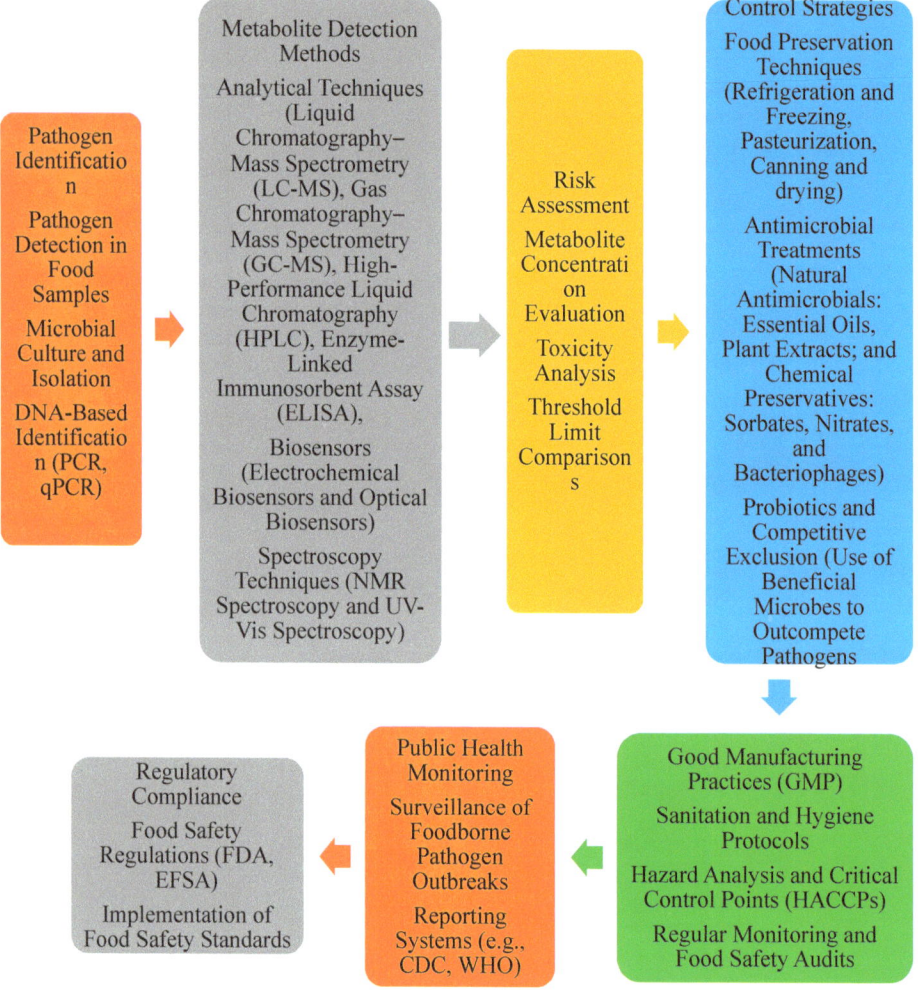

Figure 2. Comprehensive methods for pathogen detection and metabolite identification in food samples: from initial detection to regulatory compliance (created using Microsoft Word version 10.0).

5.2. Analytical Techniques

The detection and quantification of pathogen metabolites in food and biological samples rely on a variety of sophisticated analytical techniques [140] as presented in Table 2.

Table 2. Analytical techniques.

1.	Chromatography (HPLC, GC-MS)	High-performance liquid chromatography (HPLC) and gas chromatography–mass spectrometry (GC-MS) are widely used for detecting and separating pathogen metabolites in complex samples. HPLC is particularly effective for analyzing non-volatile metabolites, while GC-MS is valuable for volatile and semi-volatile compounds. Both techniques provide high sensitivity and specificity, allowing for the precise quantification of metabolites such as mycotoxins, bacterial toxins, and secondary metabolites from fungi. These methods are highly regarded for their accuracy in distinguishing different metabolite species.
2.	Spectrometry (MS)	Mass spectrometry (MS), often combined with chromatography (e.g., LC-MS or GC-MS), is a powerful tool for identifying and quantifying metabolites based on their mass-to-charge ratio. MS can analyze even trace levels of pathogen metabolites, making it ideal for detecting low-abundance carcinogenic compounds in food and biological samples. MS techniques can also provide structural information about metabolites, helping to identify specific toxins or degradation products linked to foodborne pathogens [141].
3.	Immunoassays	Immunoassays, such as enzyme-linked immunosorbent assays (ELISA), are widely used to rapidly detect pathogen metabolites, mainly in routine food safety testing. These assays are based on antigen–antibody interactions, where specific antibodies bind to target metabolites, allowing for quick, cost-effective, and large-scale screening. ELISA is commonly used to detect mycotoxins like aflatoxins and bacterial toxins like Shiga toxins in food products [142].

Advantages and Disadvantages of Analytical Techniques

Chromatography offers several advantages for analyzing pathogen metabolites in food and biological samples, particularly techniques like high-performance liquid chromatography (HPLC) and gas chromatography–mass spectrometry (GC-MS). These methods provide high sensitivity, specificity, and the ability to separate complex mixtures, making them ideal for detecting trace levels of carcinogenic metabolites produced by foodborne pathogens [143]. Chromatography also allows for the accurate quantification and identification of multiple metabolites simultaneously, providing valuable insights into their potential health impacts, including cancer risk [144]. However, the disadvantages include needing specialized equipment and trained personnel, which can be costly and time-consuming. Additionally, sample preparation is often labor-intensive, and the techniques may struggle with detecting volatile or low-abundance metabolites [145]. Despite these challenges, chromatography remains a powerful tool in assessing food safety and understanding the carcinogenic potential of pathogen metabolites, ultimately contributing to better public health protection by identifying and mitigating cancer risks [146].

Mass spectrometry (MS) offers significant advantages as an analytical technique for detecting pathogen metabolites in food and biological samples, particularly in assessing cancer risk [147]. It provides high sensitivity, precision, and the ability to identify metabolites at very low concentrations, making it suitable for detecting carcinogenic compounds produced by foodborne pathogens [148]. MS also allows for identifying complex metabolites, providing valuable insights into their biochemical pathways and potential health impacts [149]. However, the disadvantages include the high cost of equipment and maintenance, as well as the need for skilled operators and extensive sample preparation [150]. Mass spectrometry can be challenging in interpreting complex data, mainly when metabolites are present in trace amounts or have similar molecular weights [151]. Despite these challenges, MS remains an essential tool for assessing food safety and understanding the carcinogenic potential of pathogen metabolites, thereby aiding in cancer risk evaluation and public health protection [152].

Immunological assays, such as enzyme-linked immunosorbent assays (ELISA), are valuable for detecting pathogen metabolites in food and biological samples, particularly for assessing cancer risk [153]. One of the key advantages is their high specificity, as they can be designed to target specific metabolites or toxins produced by foodborne pathogens, enabling precise detection [154]. These assays are relatively cost-effective, easy to perform, and can be adapted for high-throughput screening, making them suitable for large-scale food safety testing [155]. However, the main disadvantages include their lower sensitivity compared to other techniques like mass spectrometry, which may limit their ability to detect metabolites at trace levels [156]. Additionally, immunological assays require the availability of high-quality antibodies, which can be challenging to produce for less-studied metabolites [157]. Furthermore, cross-reactivity with other substances in complex food matrices may lead to false positives or inaccurate results, impacting their reliability in assessing cancer-related metabolites [158]. Despite these limitations, immunological assays remain an essential tool for food safety monitoring and the evaluation of potential cancer risks posed by pathogen metabolites [159].

5.3. Novel Detection and Analytical Techniques

Detecting and analyzing carcinogenic metabolites from foodborne pathogens, particularly at trace levels, is crucial for the early diagnosis, prevention, and control of cancer development. Recent advances in technology have enabled researchers and clinicians to identify and quantify these harmful metabolites with higher sensitivity, accuracy, and efficiency [160]. Two significant areas of development in this field are advanced detection technologies and metabolomics. These approaches allow for the better detection of carcinogenic metabolites and facilitate biomarker discovery for early cancer diagnosis. This section provides an exhaustive overview of these emerging detection and analytical techniques, with a focus on biosensors, microfluidics, AI-assisted technologies, and the application of metabolomics for cancer biomarker discovery [161].

5.3.1. Advanced Detection Technologies: Emerging Tools for Trace-Level Metabolite Detection

Detecting carcinogenic metabolites from foodborne pathogens at trace levels is essential for identifying early risks associated with cancer development. Advanced detection technologies, including biosensors, microfluidic systems, and artificial intelligence (AI)-assisted methods, are revolutionizing how researchers detect these metabolites with high specificity and sensitivity. These technologies allow for real-time monitoring, rapid testing, and non-invasive sample analysis [162].

- Biosensors for Carcinogenic Metabolite Detection

Biosensors are analytical devices that convert biological interactions into measurable signals. They have become a powerful tool for detecting carcinogenic metabolites due to their sensitivity, specificity, and ability to detect even trace amounts of target compounds in food and biological samples. Biosensors combine a biological recognition element (such as enzymes, antibodies, or nucleic acids) with a transducer, which converts the biological interaction into an electrical, optical, or thermal signal [163].

- Electrochemical Biosensors

Due to their high sensitivity, electrochemical biosensors are among the most common types for detecting metabolites. They measure changes in electrical currents resulting from the interaction between a target metabolite and the biosensor's recognition element. These sensors can detect carcinogenic metabolites such as aflatoxins, mycotoxins, and nitrosamines, which are linked to cancer development. For instance, electrochemical biosensors have been developed to detect aflatoxin B1, a carcinogenic metabolite produced by the *Aspergillus* species, at extremely low concentrations in food samples [164].

- Optical Biosensors

These biosensors use light to detect metabolite interactions. Surface plasmon resonance (SPR) and fluorescence-based detection can provide real-time, label-free analysis. Optical biosensors are highly effective in detecting specific pathogen metabolites and can be applied to screen food samples for carcinogenic compounds such as polycyclic aromatic hydrocarbons (PAHs), which are byproducts of cooking processes and associated with an increased risk of cancer [165].

Biosensors have significant potential in food safety and public health surveillance. They are portable, cost-effective, and capable of providing rapid results, making them ideal for detecting carcinogenic metabolites in real-world settings such as food production facilities or healthcare clinics [166].

- Microfluidics for High-Sensitivity Detection

Microfluidic technology involves the manipulation of small volumes of fluids within micro-scale channels. It has emerged as a powerful tool for detecting metabolites due to its ability to integrate multiple analytical processes (e.g., separation, detection, and analysis) in a single platform, often called "lab-on-a-chip" systems. Microfluidic devices can perform complex analyses using minimal sample volumes and reagents, reducing the cost and time required for testing [167].

- Lab-on-a-Chip Systems

These miniaturized systems allow for the high-throughput screening of metabolites from foodborne pathogens. Microfluidic devices can incorporate biosensors, chromatography, or mass spectrometry to detect carcinogenic metabolites precisely. For instance, microfluidic chips have been developed to detect nitrosamines and carcinogenic compounds in preserved meats and processed foods. By using these systems, it is possible to detect multiple metabolites simultaneously, offering a comprehensive analysis of potential cancer risks from foodborne pathogens [168].

- Point-of-Care Diagnostics

Microfluidics also enable the development of point-of-care (POC) diagnostic tools for rapidly detecting carcinogenic metabolites in biological samples (such as blood, urine, or saliva). POC systems can be used in clinical settings to screen patients for early signs of cancer or monitor exposure to harmful metabolites. These devices offer a non-invasive, quick, and cost-effective alternative to traditional laboratory testing, making them highly valuable in both public health and clinical applications [169].

Microfluidics have significant potential for revolutionizing how carcinogenic metabolites are detected, allowing for real-time and on-site testing with high precision and minimal human intervention [170].

- Artificial Intelligence (AI)-Assisted Methods for Detection

Artificial intelligence (AI) and machine learning (ML) algorithms are increasingly integrated into metabolite detection systems to improve accuracy, enhance data analysis, and automate complex workflows. AI-assisted techniques can process large datasets generated from biosensors, microfluidic systems, and metabolomics platforms, identifying patterns and correlations which may not be immediately apparent through traditional analysis [171].

AI-Driven Biosensors: AI can optimize biosensor performance by enhancing signal interpretation and reducing noise. AI algorithms can detect subtle changes in biosensor signals and differentiate between specific metabolites, increasing detection sensitivity. For example, AI-assisted electrochemical biosensors have been used to detect cancer-associated metabolites in biological samples by learning from large datasets and refining detection parameters in real time [172].

Machine Learning in Microfluidics: Machine learning algorithms can process the complex datasets produced by microfluidic systems to improve the detection of carcinogenic metabolites. These algorithms can analyze high-throughput data from microfluidic assays, identifying key cancer-associated biomarkers. Machine learning has also been used to predict metabolite interactions with the gut microbiome, helping researchers understand how metabolite exposure may influence cancer risk [173].

AI-assisted methods allow for the more accurate and faster detection of carcinogenic metabolites, making them invaluable for large-scale public health monitoring and personalized medicine.

5.3.2. Metabolomics and Biomarker Discovery: Identifying Unique Cancer Biomarkers from Pathogen Metabolites

Metabolomics, the comprehensive study of metabolites in biological systems, has become essential for identifying biomarkers linked to diseases, including cancer. By analyzing the full spectrum of metabolites in a sample, metabolomics can provide insights into the metabolic pathways affected by pathogen-derived carcinogens and reveal unique biomarkers for early cancer detection [174].

- Metabolomics for Cancer Detection

Metabolomics involves using advanced analytical techniques, such as mass spectrometry (MS) and nuclear magnetic resonance (NMR) spectroscopy, to profile metabolites in food, biological fluids, or tissues. The application of metabolomics in cancer research has significantly advanced the identification of metabolite biomarkers associated with cancer development and progression [175].

- Mass Spectrometry (MS)-Based Metabolomics

MS is a highly sensitive and accurate method for identifying and quantifying metabolites in complex mixtures. MS-based metabolomics has been used to detect carcinogenic metabolites produced by foodborne pathogens, such as aflatoxins, nitrosamines, and PAHs, which are linked to gastrointestinal cancers. MS can detect these metabolites at trace levels, enabling researchers to track early exposure to these carcinogens and their potential links to cancer development [176].

- Nuclear Magnetic Resonance (NMR) Spectroscopy

NMR provides detailed information about the structure and concentration of metabolites in a sample. NMR-based metabolomics can identify unique metabolic signatures associated with cancer, helping to uncover early disease biomarkers. For example, NMR studies have revealed altered metabolite profiles in the urine of individuals exposed to aflatoxins, which are linked to liver cancer [177].

By leveraging metabolomics, researchers can identify early metabolic changes in individuals exposed to foodborne pathogen metabolites, offering a powerful tool for early cancer detection.

- Biomarker Discovery for Early Cancer Detection

One of the critical applications of metabolomics in cancer research is the discovery of biomarker molecules that indicate the presence of disease. Biomarkers can be used for early diagnosis, disease progression, monitoring, and treatment response prediction. Metabolites from foodborne pathogens can serve as biomarkers if they are uniquely associated with cancer development [178].

- Pathogen-Specific Metabolite Biomarkers

Metabolomics has identified several pathogen-specific metabolites that may serve as biomarkers for cancer. For example, mycotoxins, such as aflatoxin B1, have been found in the blood and urine of individuals with liver cancer. Detecting these metabolites in biological samples could allow for the early screening of populations at risk for cancer due to exposure to contaminated food [179].

- Host Metabolic Response Biomarkers

Besides detecting pathogen-derived metabolites, metabolomics can also reveal host metabolic changes in response to these metabolites. For instance, exposure to carcinogenic metabolites from pathogens may trigger oxidative stress, inflammation, or changes in lipid metabolism. These host responses can produce distinct metabolic signatures that serve as early indicators of cancer [180].

- Integrating Metabolomics with Other "Omics" for Comprehensive Biomarker Discovery

Metabolomics is often integrated with other "omics" approaches to enhance biomarker discovery, such as genomics, proteomics, and transcriptomics. This multi-omics strategy provides a more comprehensive understanding of the molecular mechanisms underlying cancer development and allows for the identification of robust biomarker panels [181].

Multi-Omics Approach: By combining metabolomics with genomics (DNA), transcriptomics (RNA), and proteomics (proteins), researchers can map the complete biological response to carcinogenic metabolites. This holistic approach improves the accuracy and reliability of biomarker discovery, enabling the identification of metabolite biomarkers which are specific to certain types of cancer [182].

Data Integration and AI: AI and machine learning are increasingly used to integrate multi-omics data, identifying complex biomarker signatures which could not be detected through a single approach. These AI-assisted methods allow researchers to discover novel biomarkers more efficiently, improving early cancer detection and personalized treatment strategies [183].

Emerging detection technologies and metabolomics are revolutionizing how carcinogenic metabolites from foodborne pathogens are detected and analyzed. Advanced tools like biosensors, microfluidic systems, and AI-assisted methods allow for the sensitive, rapid, and specific detection of these harmful compounds in food and biological samples [184]. Metabolomics provides a comprehensive platform for identifying unique cancer biomarkers associated with pathogen metabolites, offering critical insights into early disease detection and the molecular mechanisms driving cancer development. These novel analytical techniques have great potential for improving public health, food safety, and cancer prevention.

5.4. Challenges in Detection

Despite the effectiveness of these techniques, several challenges remain in detecting pathogen metabolites [185], as shown in Table 3.

Table 3. Challenges in detection.

Challenge	Description
Sensitivity and Specificity	One of the significant challenges in detecting pathogen metabolites is achieving high sensitivity and specificity. Many metabolites occur in extremely low concentrations, especially in complex food matrices or biological samples, making them difficult to detect without advanced methods [186]. False positives or negatives can also arise due to cross-reactivity in immunoassays or inadequate separation in chromatographic methods.
Sample Complexity	Food and biological samples often contain interfering substances, such as fats, proteins, or other organic compounds, which complicate detection. Sample preparation methods, including extraction and purification, must be carefully designed to minimize interference and concentrate the metabolites of interest. However, these processes can be time-consuming and require specialized expertise [187].
Cost and Accessibility	Advanced detection methods like GC-MS or LC-MS are expensive and require skilled technicians and sophisticated laboratory infrastructure. This limits their accessibility, especially in regions where foodborne pathogens are prevalent but resources are scarce [188]. Immunoassays, while more affordable and user-friendly, may lack the same level of precision as chromatographic or mass spectrometric techniques [189].

These challenges highlight the need for ongoing advancements in detection technologies to improve the accuracy, sensitivity, and cost-effectiveness of pathogen metabolite analysis in food safety and public health [40].

6. Control Strategies for Growth and Metabolite Production of Foodborne Pathogens

Foodborne pathogens significantly threaten public health, causing a wide range of illnesses. Understanding the mechanisms of pathogen growth and metabolite production is crucial for developing effective control strategies [190]. This section explores various techniques employed in food processing to minimize the risk of foodborne illness [191].

6.1. Prevention and Control Measures

Food processing techniques are vital in controlling foodborne pathogens' growth and metabolite production. The following table outlines vital techniques and their mechanisms of action [138] (Table 4).

Table 4. Food processing techniques.

Technique	Description
Thermal Processing (Pasteurization and Sterilization)	This involves heating foods to a specific temperature for a set period to kill or inactivate pathogens [192]. Pasteurization is standard in dairy products and juices, reducing pathogens like *Salmonella*, *Escherichia coli*, and *Listeria monocytogenes*. Sterilization, often used for canned goods, involves higher temperatures for complete microbial inactivation [193].
Cold Preservation (Refrigeration and Freezing)	Low temperatures slow or stop microbial growth. Refrigeration below 5 °C prevents pathogen growth, while freezing stops microbial activity. However, freezing does not kill all pathogens, so proper thawing and handling are essential.
Dehydration and Drying	Water activity is a critical factor for microbial growth. Drying methods (e.g., air drying and freeze drying) lower the water content of foods, inhibiting pathogens. They are often used for grains, fruits, and meats [194].
Fermentation	Controlled fermentation using lactic acid bacteria or yeasts creates acidic conditions that inhibit pathogens like *Clostridium botulinum* [195]. Producing organic acids, bacteriocins, and alcohol during fermentation can enhance food safety [196].
Irradiation	Ionizing radiation, such as gamma rays or electron beams, kills pathogens by damaging their DNA. This method is effective for spices, meats, and some produce without raising the temperature.
High-Pressure Processing (HPP)	HPP uses high pressure (up to 600 MPa) to inactivate pathogens without significantly altering the food's sensory properties [197]. It is effective against bacteria like *Listeria* and *E. coli* in foods like juices and meats [198].
Chemical Preservatives	Organic acids (lactic, acetic) and salts (nitrates, sulfites) are used to control pathogen growth [199]. These agents can directly inhibit microbial enzymes or alter the pH to levels unsuitable for pathogen survival [200].

6.2. Sanitation Practices

Maintaining a clean and sanitary environment is crucial for preventing foodborne illnesses. Adequate sanitation includes personal hygiene, equipment cleaning, and environmental monitoring. The following table highlights essential sanitation practices [201] (Table 5).

Table 5. Sanitation practices.

Practices	Description
Personal Hygiene	Proper hand washing and the use of protective gear like gloves and masks are essential to prevent food handler contamination. This is especially important in high-risk areas like kitchens and food processing plants and during food preparation [202].
Cleaning and Sanitizing Equipment	Equipment used in food production must be regularly cleaned and sanitized to prevent cross-contamination. Effective sanitization agents include chlorine-based solutions, quaternary ammonium compounds, and peracetic acid [203].
Environmental Monitoring	Monitoring the production environment (e.g., floors, walls, and equipment surfaces) for microbial contamination is crucial. Swabbing and testing for pathogens like *Listeria* sp. can prevent cross-contamination [204].
Segregation of Raw and Cooked Foods	Ensuring that raw and ready-to-eat foods are handled separately can reduce the risk of cross-contamination. This includes separate storage, preparation areas, and utensils for raw and cooked foods.

6.3. Regulatory Measures

Government regulations and industry standards play a vital role in ensuring food safety. These measures establish food production, handling, and processing guidelines to minimize the risk of foodborne illnesses. The following table outlines vital regulatory measures [205] (Table 6).

Table 6. Regulatory measures.

Measures	Description
Hazard Analysis and Critical Control Points (HACCPs)	This systematic approach identifies potential hazards in the production process and establishes critical control points (CCPs) to reduce or eliminate risks. HACCPs are mandatory for many food industries worldwide and are widely recognized for their effectiveness in preventing foodborne illnesses.
Good Manufacturing Practices (GMPs)	GMPs provide guidelines for producing, handling, and processing food products. These include proper facility design, sanitation, employee hygiene, and pest control measures to reduce contamination risks.
Food Safety Modernization Act (FSMA)	Enacted in the U.S., FSMA shifts the focus from responding to foodborne illness outbreaks to preventing them. It includes provisions for regular inspections, food safety plans, and increased oversight of imported foods.
Codex Alimentarius Standards	Developed by the FAO and WHO, the Codex provides international food standards, guidelines, and codes of practice to ensure food safety and fair-trade practices. These guidelines help harmonize food safety regulations across countries.

6.4. Emerging Technologies

Technological advancements offer promising solutions for enhancing food safety and controlling foodborne pathogens. These emerging technologies provide novel approaches to food preservation and pathogen inactivation [206]. The following table explores some of these technologies (Table 7).

Table 7. Emerging technologies.

Technologies	Description
Pulsed Electric Fields (PEFs)	PEF uses short bursts of high voltage to create pores in microbial cell membranes, effectively killing or inactivating pathogens without heating the food. This method is being explored for juices, milk, and liquid eggs [207].

Table 7. Cont.

Technologies	Description
Cold Plasma Technology	This non-thermal technology generates ionized gas (plasma) [208] that contains reactive oxygen and nitrogen species capable of killing bacteria, yeasts, and molds [209]. Cold plasma is studied in fresh produce, meats, and packaging materials [210].
Ultraviolet (UV) Light	UV light at specific wavelengths (particularly UV-C) damages the DNA of pathogens, [211], preventing their growth and reproduction. It is used in surface sanitation, water purification, and air treatment in food processing facilities [212].
Nanotechnology	Nanoparticles, particularly silver and copper [213], are being incorporated into packaging materials and coatings for antimicrobial purposes [214]. These materials can prevent pathogen growth on food surfaces and extend shelf life [215].
Phage Therapy	Bacteriophages (viruses which infect bacteria) are being explored as a targeted method to control particular pathogens like *Listeria* sp. or *Salmonella* sp. in foods [216]. Phages offer a natural and specific approach to pathogen control without affecting beneficial microorganisms [217,218].
Biocontrol Using Probiotics	Using beneficial microbes (probiotics) to outcompete or inhibit pathogens in food is gaining traction [219]. For example, *Lactobacillus* sp. species can inhibit *Listeria* sp. in fermented foods, while certain yeast strains are being investigated for pathogen control in alcoholic beverages [220].

The Potential and Challenges of Emerging Technologies in the Application of Food Detection and Safety Control

Pulsed Electric Fields (PEF): Pulsed Electric Field (PEF) technology is gaining attention for its ability to inactivate microorganisms and enzymes in food without significantly affecting its sensory and nutritional qualities [221]. The high-voltage electric pulses create pores in microbial cell membranes, effectively killing pathogens and spoilage organisms [222]. PEFs have potential for applications in liquid foods like juices and milk, ensuring microbial safety while retaining freshness [223]. However, challenges include the high initial investment costs and energy requirements and the limited applicability to solid foods [224]. Additionally, optimizing processing parameters for various food matrices and ensuring consistent microbial inactivation remain critical issues [225].

Cold Plasma Technology: Cold plasma technology is an emerging non-thermal method for microbial decontamination, offering a rapid and chemical-free alternative for food safety [226]. The reactive species generated by plasma can destroy bacteria, viruses, and fungi on surfaces and within food. This method holds promise for fresh produce, packaging materials, and ready-to-eat foods [227]. However, challenges include scalability for industrial use, the potential formation of undesirable byproducts, and limited knowledge of its long-term effects on food quality and human health [228]. Further research is needed to optimize plasma exposure conditions and assess regulatory concerns.

Ultraviolet (UV) Light: Ultraviolet (UV) light is widely recognized for its antimicrobial properties, among which is the remarkable UV-C light, which disrupts microbial DNA, rendering pathogens inactive [229]. It is effective for surface decontamination, liquid processing, and air purification in food facilities. The technology is cost-effective and environmentally friendly [230]. However, challenges include its limited penetration depth, which reduces efficacy for turbid liquids or opaque surfaces, and the potential for microbial resistance [231]. UV light can also cause oxidative effects, potentially altering the sensory and nutritional properties of certain foods [232].

Nanotechnology: Nanotechnology offers revolutionary potential in food safety through nanosensors for the real-time detection of pathogens, toxins, and spoilage indicators [233]. Nano-encapsulation enhances the stability and delivery of antimicrobial agents or preservatives [234]. However, challenges stem from the lack of standardized regulatory frameworks and uncertainties about the toxicity of nanomaterials [235]. Ensuring consumer acceptance and addressing public concerns about nanotechnology's safety and environmental impact on food systems is crucial for widespread adoption.

Phage Therapy: Phage therapy utilizes bacteriophages to target and destroy specific bacterial pathogens in foods. It is highly selective, minimizing disruptions to beneficial microorganisms [236]. This approach is particularly valuable for controlling multidrug-resistant pathogens. Despite its potential, challenges include the need for precise phage–host matching, the risk of bacterial resistance to phages, and the potential regulatory hurdles associated with introducing live viruses into food systems [237]. Ensuring phage stability during storage and distribution also requires further innovation.

Biocontrol Using Probiotics: The use of probiotics as biocontrol agents in food safety focuses on their ability to outcompete harmful microbes through mechanisms like competitive exclusion and the production of antimicrobial compounds [238]. This approach is promising for fermented foods, minimally processed products, and biofilms on food processing equipment. However, challenges include maintaining the viability and activity of probiotics during processing and storage, ensuring strain-specific safety and efficacy, and navigating regulatory approval [239]. Further research is needed to understand the long-term implications of probiotic use in diverse food systems.

These emerging technologies present transformative possibilities for enhancing food safety and quality. However, their practical application requires overcoming significant challenges, including cost, scalability, regulatory approval, and consumer acceptance. A multidisciplinary approach involving research, policy development, and industry collaboration is essential to unlock their full potential.

6.5. Control Strategies Using Biotechnology

- Probiotic Intervention

Engineered probiotics or microbiota-based interventions can neutralize the harmful metabolites produced by foodborne pathogens. By restoring gut microbiota balance, these probiotics reduce inflammation, bind to toxins (such as aflatoxins), and prevent the absorption of carcinogenic compounds, thereby mitigating their role in cancer development [240].

- Phage Therapy

Phage therapy uses bacteriophages to target and eliminate foodborne pathogens like *Salmonella* sp. or *Cronobacter sakazakii*. By controlling these pathogens, phage therapy reduces the production of carcinogenic metabolites, lowering the risk of cancer from chronic exposure [241]. Table 8 shows case studies and demonstrates how biotechnology can be applied to mitigate the effects of foodborne pathogen metabolites on cancer development. Probiotic interventions and phage therapy offer innovative, targeted strategies to neutralize or control harmful metabolites, contributing to improved public health outcomes.

Table 8. Case studies of control strategies using biotechnology.

Control Strategy	Case Study	Mechanism of Action	Impact on Carcinogenic Metabolites	Outcome
Probiotic Intervention	Engineered Probiotics for Neutralizing Aflatoxins	Genetically engineered probiotics (*Lactobacillus rhamnosus*) are designed to bind and detoxify aflatoxins in the gastrointestinal tract.	The probiotics bind aflatoxin B1, a carcinogenic metabolite produced by *Aspergillus* sp., neutralizing its effect and preventing absorption into the bloodstream.	Reduction in aflatoxin B1 bioavailability and minimized risk of liver cancer from aflatoxin exposure [242].

Table 8. Cont.

Control Strategy	Case Study	Mechanism of Action	Impact on Carcinogenic Metabolites	Outcome
Probiotic Intervention	Microbiota-Based Interventions to Prevent Dysbiosis and Carcinogenesis	The introduction of beneficial strains (*Lactobacillus plantarum* and *Bifidobacterium bifidum*) is needed to restore gut microbiome balance and prevent dysbiosis triggered by pathogen metabolites.	These probiotics improve gut integrity, reduce inflammation, and prevent the overgrowth of harmful bacteria that produce carcinogenic metabolites like nitrosamines.	Reduced inflammation and lower risk of gastrointestinal cancer due to balanced microbiota and suppression of harmful metabolite production [243].
Phage Therapy	Phage Treatment to Control *Salmonella* sp. in Food Production	Bacteriophages specific to *Salmonella* sp. are used to target and eliminate the pathogen in food processing environments.	By targeting Salmonella, phage therapy prevents the production of endotoxins and other carcinogenic metabolites produced during infection.	Significant reduction in *Salmonella* sp. contamination, decreasing the risk of cancer from chronic exposure to pathogen-associated toxins [244].
Phage Therapy	Phage-Based Control of *Cronobacter sakazakii* in Infant Formula	Bacteriophages specific to *Cronobacter sakazakii* are used to control contamination in powdered infant formula.	Phage therapy reduces the population of *C. sakazakii*, preventing the production of carcinogenic metabolites which may contribute to long-term health issues such as cancer in infants.	Lower contamination rates in infant formula, leading to reduced cancer risks from early exposure to pathogen metabolites [245].

A multifaceted approach combining traditional methods (thermal processing, sanitation) with emerging technologies (PEF, cold plasma) and regulatory frameworks (HACCP, FSMA) is essential for controlling foodborne pathogens and their metabolites. These strategies are critical to ensuring food safety and public health [246].

7. Public Health Implications of Pathogen Metabolites

Foodborne pathogens not only cause infections but also produce metabolites that can have significant adverse effects on human health [14]. These metabolites, particularly toxins, can exert a range of harmful effects, leading to acute and chronic illnesses [247]. Understanding the impact of these metabolites is crucial for public health interventions and strategies to mitigate the risks associated with foodborne pathogens.

7.1. Impact on Public Health

Pathogen metabolites, particularly toxins produced by foodborne pathogens, pose significant risks to human health [248]. These metabolites can be carcinogenic, hepatotoxic, neurotoxic, or immunosuppressive, leading to a range of acute and chronic health conditions [249] (Table 9) while Table 10 presents health policies and education.

Table 9. Impact on public health.

Impact	Description
Carcinogenic Effects	Aflatoxins (produced by *Aspergillus* sp.) are one of the most potent carcinogens found in food. Chronic exposure, particularly in developing countries where food storage conditions may promote fungal growth, is associated with liver cancer [22]. The burden of aflatoxin-related liver cancer is exceptionally high in sub-Saharan Africa and Southeast Asia, where hepatitis B virus infection is also prevalent, exacerbating cancer risk [250]. N-nitroso compounds (produced during the processing of meats) are linked to colorectal cancer. These metabolites are formed from nitrites and nitrates used in food preservation and are classified as probable human carcinogens by the International Agency for Research on Cancer (IARC) [251].
Hepatotoxicity	Pathogen-derived toxins, like aflatoxins and microcystins (produced by cyanobacteria), can cause severe liver damage. Acute aflatoxicosis can lead to liver failure, while chronic exposure leads to liver cirrhosis and increased susceptibility to liver cancer.
Neurological Disorders	Botulinum toxin (produced by *Clostridium botulinum*) is one of the most potent neurotoxins known. It can cause botulism, a life-threatening illness characterized by muscle paralysis, respiratory failure, and death if untreated [252]. Fumonisins, produced by *Fusarium* sp. in grains, are associated with neural tube defects in populations that consume contaminated maize. Animal studies also suggest a link between fumonisins and esophageal cancer [253].
Immunosuppression	Some pathogen metabolites, such as aflatoxins, have immunosuppressive effects, weakening the body's ability to fight infections [254]. This makes individuals more susceptible to other diseases, including HIV/AIDS and malaria, particularly in regions where these conditions are prevalent [255].
Burden of Foodborne Illnesses	Foodborne illnesses caused by bacterial pathogens (*Salmonella* sp., *E. coli*, and *Campylobacter* sp.) and their toxins lead to gastrointestinal diseases like diarrhea, which can be fatal in vulnerable populations such as children, the elderly, and immunocompromised individuals [256]. Chronic complications include post-infectious irritable bowel syndrome (IBS) and Guillain–Barré syndrome (a severe neurological disorder).

The global burden of diseases associated with pathogen metabolites emphasizes the need for improved food safety systems to prevent contamination [257].

Table 10. Health policies and education.

Roles	Policies	Description
Role of Health Policies	Health policies play a crucial role in regulating food safety, monitoring contaminant levels, and mitigating the risks of pathogen metabolites.	Regulatory Standards: National and international organizations like the World Health Organization (WHO), the Food and Agriculture Organization (FAO), and the Codex Alimentarius Commission have established maximum permissible levels for contaminants such as aflatoxins, fumonisins, and nitrates in food. These guidelines help ensure that food products meet safety standards before they reach consumers [258]. Surveillance and Monitoring: National agencies, such as the U.S. Food and Drug Administration (FDA) and the European Food Safety Authority (EFSA), monitor the presence of pathogen metabolites in food products and agricultural commodities. Early detection through food surveillance systems allows for a rapid response, including product recalls and public warnings. Food Safety Modernization Act (FSMA): The FSMA in the U.S. emphasizes preventive measures over reactive ones. It mandates hazard analysis, supply chain monitoring, and strict adherence to hygiene protocols to prevent contamination at the source. International Collaborations: Global cooperation through platforms like the Global Foodborne Infections Network (GFN) helps countries share information on foodborne disease outbreaks, improving response times and control strategies [259].

Table 10. *Cont.*

Roles	Policies	Description
Public Awareness and Education	Education plays a critical role in reducing the risks associated with pathogen metabolites. Increasing public awareness helps prevent and respond early to potential foodborne threats.	Regulatory Standards: National and international organizations like the World Health Organization (WHO), the Food and Agriculture Organization (FAO), and the Codex Alimentarius Commission have established maximum permissible levels for contaminants such as aflatoxins, fumonisins, and nitrates in food. These guidelines help ensure that food products meet safety standards before they reach consumers. Surveillance and Monitoring: National agencies, such as the U.S. Food and Drug Administration (FDA) and the European Food Safety Authority (EFSA), monitor the presence of pathogen metabolites in food products and agricultural commodities. Early detection through food surveillance systems allows for a rapid response, including product recalls and public warnings [258]. Food Safety Modernization Act (FSMA): The FSMA in the U.S. emphasizes preventive measures over reactive ones. It mandates hazard analysis, supply chain monitoring, and strict adherence to hygiene protocols to prevent contamination at the source. International Collaborations: Global cooperation through platforms like the Global Foodborne Infections Network (GFN) helps countries share information on foodborne disease outbreaks, improving response times and control strategies.

The Role of Policies and Regulations in Controlling Foodborne Pathogens and Ensuring Food Safety

Effective policies and regulations are critical for managing foodborne pathogens and ensuring food safety. These frameworks establish standards for food production, processing, distribution, and consumption, helping to reduce the incidence of foodborne illnesses [260]. However, the approach to food safety varies significantly across countries and regions due to differences in regulatory priorities, resources, and enforcement mechanisms [261].

- Developed Nations: Comprehensive and Stringent Frameworks

In developed nations, such as the United States, the European Union, and Japan, food safety is governed by comprehensive regulatory systems [262]. For instance, the U.S. Food and Drug Administration (FDA) and the U.S. Department of Agriculture (USDA) enforce stringent standards through programs like the Food Safety Modernization Act (FSMA), which emphasizes preventive controls and risk-based approaches. Similarly, the European Union's General Food Law mandates traceability across the food chain, ensuring rapid responses to food safety incidents. These policies are supported by advanced technologies, robust laboratory networks, and strict enforcement measures, resulting in better control of foodborne pathogens [260].

- Emerging Economies: Balancing Growth and Safety

In emerging economies, regulatory systems are often less developed, leading to inconsistent enforcement of food safety laws [263]. Countries like India, China, and Brazil are making significant strides by implementing food safety reforms [264]. For instance, China's Food Safety Law emphasizes improved risk assessment and supervision of high-risk foods [265]. However, challenges such as inadequate infrastructure, limited resources for inspections, and fragmented supply chains hinder effective pathogen control [266].

- Low-Income Nations: Resource Constraints and Informal Markets

Low-income nations face significant challenges in controlling foodborne pathogens due to limited resources, weak enforcement, and a high reliance on informal food markets [267]. Many countries in sub-Saharan Africa and South Asia lack standardized food safety frameworks, increasing vulnerability to foodborne diseases [268]. International organizations like the World Health Organization (WHO) and the Food and Agriculture Organization (FAO) work with these nations to build capacity, promote good agricultural practices, and implement hazard analysis and critical control point (HACCP) systems [269].

- Global Harmonization Efforts

Efforts to harmonize food safety standards globally are essential for managing cross-border food trade and mitigating risks of foodborne illnesses. Organizations such as the Codex Alimentarius Commission provide international guidelines and standards that countries can adopt to improve food safety [270]. Free trade agreements and regional bodies like the European Food Safety Authority (EFSA) and the Association of Southeast Asian Nations (ASEAN) Food Safety Policy also play a role in harmonizing standards and facilitating knowledge exchange.

- Challenges and Future Directions

Despite progress, several challenges remain in aligning food safety policies across regions. Variability in regulatory standards, differing levels of technological adoption, and cultural practices can create gaps in food safety. Furthermore, emerging risks such as climate change, antimicrobial resistance, and new foodborne pathogens require adaptive regulatory frameworks [271].

To address these challenges, governments must invest in infrastructure, enhance international collaboration, and engage stakeholders across the food system. Innovative tools like blockchain for traceability, rapid pathogen detection methods, and public–private partnerships can also strengthen food safety globally [272].

While policies and regulations significantly influence the control of foodborne pathogens, their effectiveness depends on enforcement, resource allocation, and international cooperation. Tailoring these frameworks to the specific needs and challenges of each region is vital for achieving safer food systems worldwide [273].

7.2. Public Health Implications in Developing Countries

- Focus on Vulnerable Populations

In developing countries, vulnerable populations face higher risks from carcinogenic metabolites due to poor food safety standards, limited access to clean water, and inadequate healthcare systems. Under-resourced settings often lack effective surveillance for foodborne pathogens and cancer, increasing the burden of disease, particularly among children, the elderly, and low-income communities [274].

- Policy Frameworks

Novel policy frameworks integrating food safety and cancer prevention are critical. These should prioritize strengthening food regulation, improving pathogen monitoring systems, and promoting public awareness. Policies could also support biotech interventions like probiotics and phage therapy, aiming to reduce pathogen exposure and related cancer risks in developing nations [275]. Table 11 shows case studies demonstrating the profound impact of foodborne pathogen metabolites on vulnerable populations in developing countries, where food safety and cancer surveillance are often inadequate. Policy frameworks aimed at improving food safety, integrating cancer prevention, and enhancing public health infrastructure can significantly reduce the risk of carcinogenic exposure and improve health outcomes.

Table 11. Case studies of public health implications in developing countries.

Case Study	Vulnerable Populations Affected	Metabolite Impact in Under-Resourced Settings	Policy Frameworks Proposed	Outcome
Aflatoxin Exposure in Rural Africa	Rural communities are dependent on maize and ground nuts as staple foods.	High levels of aflatoxins in improperly stored crops increase liver cancer risk, with limited cancer screening and food safety regulations.	Development of regional policies for improved crop storage, community education on mycotoxin risks, and integration of cancer surveillance programs.	Reduced aflatoxin-related liver cancer cases and improved early detection rates in rural areas [276].
Cronobacter sakazakii in Infant Formula in Southeast Asia	Infants in low-income households are reliant on formula feeding.	Contamination of powdered infant formula with *Cronobacter sakazakii* increases risks of infections and long-term cancer effects, with limited food safety oversight.	Strengthening regulations for infant formula production, implementing routine pathogen testing, and subsidizing safer alternatives for low-income families.	Lower contamination rates and reduced infant mortality and cancer risks from early-life pathogen exposure [100].
Hepatitis B and Mycotoxins in Sub-Saharan Africa	These affect low-income populations with poor access to vaccines and healthcare.	Co-exposure to hepatitis B virus and aflatoxins exacerbates liver cancer risks, with little access to vaccinations or mycotoxin control programs.	Integrating mycotoxin control into national cancer prevention policies, expanding hepatitis B vaccination, and improving food safety through regional cooperation.	Decreased liver cancer incidence due to improved vaccination and food safety measures [277].
Foodborne Bacterial Infections in Latin America	These affect children and the elderly in regions with poor sanitation and food handling practices.	Exposure to bacterial pathogens produces carcinogenic metabolites due to a lack of sanitation and food safety measures, which increase cancer risks.	Implement cross-border food safety policies, including stricter controls on imports and exports, and invest in sanitation infrastructure.	Enhanced food safety and reduced incidence of foodborne diseases leading to cancer [278].

Resolving the public health implications of pathogen metabolites requires coordinated efforts across health policy, education, and food safety regulation [279]. By ensuring that food products are safe for consumption and educating the public on proper handling and prevention practices, the risks posed by these harmful metabolites can be effectively mitigated, leading to improved public health outcomes [280].

7.3. Climate Change and Foodborne Pathogen Metabolites

Climate change intensifies the proliferation of foodborne pathogens by creating warmer, more humid conditions that promote microbial growth in food systems. This shift increases the production of carcinogenic metabolites like mycotoxins, nitrosamines, and endotoxins. Rising temperatures and unpredictable weather patterns compromise food storage and safety, leading to more significant contamination risks [281]. As a result, climate change is expected to escalate public health concerns related to foodborne pathogens, particularly in vulnerable regions, contributing to higher cancer risks through prolonged exposure to these harmful metabolites. Table 12 shows the case studies highlighting the link between climate change and the increased proliferation of foodborne pathogens and carcinogenic metabolites. They underscore the need for climate-adaptive food safety measures to prevent future public health crises [282].

Table 12. Case studies of climate change and foodborne pathogen metabolites.

Case Study	Climate Change Impact	Pathogen/Metabolite Affected	Public Health Concern	Outcome
Aflatoxin Contamination in Maize (Sub-Saharan Africa)	Increased droughts and heat stress cause higher fungal contamination in crops, especially maize.	*Aspergillus* sp. produces aflatoxins.	Elevated liver cancer risk due to increased aflatoxin levels in staple crops, particularly affecting rural populations.	This results in an increased incidence of liver cancer and the urgent need for climate-resilient crop storage and fungal monitoring systems [283].
Vibrio Infections in Seafood (Coastal Regions of North America and Europe)	Rising ocean temperatures promote the growth of Vibrio bacteria in shellfish, increasing the risk of bacterial infections.	*Vibrio* sp. produces harmful endotoxins.	Greater risk of gastrointestinal cancers linked to chronic exposure to Vibrio endotoxins through seafood consumption.	This results in a growing number of foodborne illnesses; stricter seafood safety regulations and monitoring are needed [284].
Mycotoxin Production in Wheat (Europe)	Warmer and wetter conditions during the growing season lead to increased fungal contamination in wheat.	*Fusarium* sp. produces deoxynivalenol (DON) and other mycotoxins.	Increased risk of cancers related to chronic mycotoxin exposure through wheat-based products.	Food products have a higher prevalence of mycotoxin contamination; enhanced monitoring and climate-adaptive farming practices are required [285].
Salmonella in Poultry (Global)	Rising temperatures accelerate the proliferation of *Salmonella* sp. in poultry farming and food processing environments.	*Salmonella* sp. produces endotoxins and other carcinogenic compounds.	Increased risk of foodborne illnesses and cancer from chronic *Salmonella* sp. exposure.	This results in higher contamination rates and foodborne illness and the need for enhanced cooling and sanitation measures in poultry farming [286].

7.4. Comparison of Metabolite Levels in Different Foods

Different foods harbor varying levels of harmful metabolites produced by pathogens. For instance, aflatoxins in maize and peanuts can reach up to 500 µg/kg and 3000 µg/kg, respectively, posing a significant risk for liver cancer. Wheat often contains deoxynivalenol (DON) from *Fusarium* sp., with levels between 200 and 1200 µg/kg, leading to gastrointestinal issues and potential cancer risk. Though present at lower levels, endotoxins from *Vibrio* sp. and *Salmonella* sp. in shellfish and poultry contribute to gastrointestinal and cancer risks [287]. Effective monitoring and food safety strategies are crucial in managing these risks. Table 13 illustrates varying metabolite levels in different foods and their potential health impacts, highlighting the significance of food safety measures in limiting carcinogenic exposure [12].

Table 13. Comparison of metabolite levels in different foods.

Food Item	Pathogen	Metabolite	Average Metabolite Level (µg/kg)	Health Impact
Maize	*Aspergillus* sp.	Aflatoxins	10–500	Liver cancer risk
Wheat	*Fusarium* sp.	Deoxynivalenol (DON)	200–1200	Gastrointestinal issues and cancer
Peanuts	*Aspergillus* sp.	Aflatoxins	20–3000	Hepatocellular carcinoma

Table 13. Cont.

Food Item	Pathogen	Metabolite	Average Metabolite Level (µg/kg)	Health Impact
Shellfish	*Vibrio* sp.	Endotoxins	0.5–5	Gastrointestinal cancers
Poultry	*Salmonella* sp.	Endotoxins and Nitrosamines	5–50	Increased cancer risk
Dairy Products	*Campylobacter* sp.	Toxins (*Campylobacter jejuni*)	1–10	Gastrointestinal disorders and cancer
Fruits (e.g., Apples)	*Penicillium* sp.	Patulin	10–50	Increased cancer risk
Soy Products	*Streptomyces* sp.	Streptomycin Residues	0.1–10	Disruption of gut microbiota and potential cancer

8. Future Directions in Pathogen Metabolite Control and Public Health

While significant strides have been made in understanding and mitigating the risks associated with foodborne pathogens, the threat posed by their metabolites remains a paramount public health concern [12]. These toxins, produced by bacteria and fungi, can cause a range of acute and chronic illnesses, impacting both individual health and global food security [288]. To effectively protect public health, we must continue to invest in research and innovation to address the challenges posed by pathogen metabolites [289]. This section explores critical research gaps and innovative approaches that hold promise for the future of pathogen metabolite control [290].

Research Gaps

Despite advances in food safety, there are several areas where further research is needed to enhance the detection and control of pathogen metabolites [291] (Table 14) while the innovative approaches are presented on Table 15.

Table 14. Research gaps.

Research Gap	Description
Improved Detection Methods	Rapid, Field-Deployable Detection Tools: Current detection methods for pathogen metabolites, like aflatoxins and fumonisins, often require sophisticated lab equipment (e.g., HPLC, ELISA), making them inaccessible in resource-limited settings. Portable, cost-effective, and easy-to-use diagnostic tools that can be deployed in the field or at small-scale production facilities are needed. This is especially critical in regions with high contamination risks, such as sub-Saharan Africa and Southeast Asia [99]. Biosensors: Developing biosensors that can quickly and accurately detect multiple metabolites (mycotoxins, bacterial toxins, etc.) in food matrices is an important research priority. These devices could provide real-time monitoring of contamination in food processing lines, helping to reduce outbreaks of foodborne illnesses [100]. Early Detection of Contaminants in Supply Chains: Better surveillance systems are needed to detect contaminants at early stages of the food supply chain, such as on farms or during storage. This includes better methods for detecting fungal growth or toxin production in grains before they reach consumers [101].

Table 14. Cont.

Research Gap	Description
Mechanisms of Metabolite Toxicity	Understanding Chronic Exposure: While the acute toxicity of some pathogen metabolites is well-documented [102], the long-term effects of chronic low-dose exposure (e.g., through diet) are not fully understood. More research is needed to investigate how prolonged exposure to aflatoxins, fumonisins, or nitrates may contribute to cancers, immune disorders, and developmental issues [103]. Interaction Between Pathogens and Metabolites: The interactions between different pathogens (bacteria, fungi) and the metabolites they produce in complex food matrices need more exploration. Understanding how environmental factors (e.g., humidity, temperature) influence these interactions could help develop better prevention and control measures [104].
Probiotic Interventions	The role of probiotics and the human microbiome in mitigating the effects of pathogen metabolites is an emerging research area [292]. Further investigation is needed into how beneficial microbes [293] (such as *Saccharomyces cerevisiae* var. *boulardii*) can detoxify harmful metabolites or inhibit the growth of toxin-producing pathogens [294].
Food Safety and Climate Change	Climate change alters environmental conditions to promote the growth of toxin-producing pathogens, particularly fungi. Research into how shifting climate patterns affect the production of pathogen metabolites in crops and food products is needed to predict and mitigate future risks [295].

Table 15. Innovative approaches.

Approaches	Description
Predictive Analytics and AI in Food Safety	Machine Learning and Artificial Intelligence: AI and machine learning models are being developed to predict foodborne pathogen outbreaks and contamination in supply chains. These models can analyze large datasets (weather patterns, food processing conditions, and shipment routes) to identify trends and risks. Predictive analytics could help food safety regulators and producers proactively prevent contamination before it reaches consumers [109]. Blockchain for Supply Chain Transparency: Blockchain technology enhances traceability and transparency in the food supply chain. It enables the real-time tracking of food products from farm to table, helping detect contamination sources quickly and ensuring accountability in handling practices.
Microbiome-Based Solutions	Recent research into the human microbiome has revealed that certain probiotic strains can help mitigate the effects of pathogen metabolites. For example, some probiotic strains have been shown to bind and neutralize aflatoxins, preventing their absorption in the gastrointestinal tract. This presents a promising avenue for developing functional foods and supplements designed to detoxify harmful metabolites in the body [296].
CRISPR and Gene Editing	Gene editing tools like CRISPR are being explored to target and eliminate toxin-producing genes in pathogens. By removing or disabling specific genes responsible for metabolite production, safer fungi or bacteria may be possible for food production. CRISPR technology could also be used to engineer plants resistant to contamination by pathogens or mycotoxins [297].
Natural Antimicrobials and Toxin Binders	Plant-Based Compounds: Plant-derived compounds such as essential oils (e.g., from oregano and thyme) and bioactive peptides are being researched for their antimicrobial and antifungal properties. These natural antimicrobials could be incorporated into food packaging or coatings to prevent pathogen growth and toxin production during storage [112]. Toxin Binders: Another innovative area of research is the development of bioactive compounds that can bind and neutralize harmful metabolites in foods. Certain clays, for instance, have been found to bind aflatoxins, preventing their absorption when consumed. These binders could be used as food additives or in feed to protect both animals and humans from toxin exposure.

Table 15. *Cont.*

Approaches	Description
Smart Packaging and Sensors	Smart packaging with embedded sensors that monitor the condition of food products (e.g., temperature, humidity, and microbial activity) is an emerging trend [298]. These sensors can detect changes in the food environment that may promote the growth of pathogens and the production of harmful metabolites. For example, sensors that detect volatile compounds associated with fungal growth could trigger alerts before spoilage occurs [299].
Biopreservation	Biopreservation involves using natural or controlled microbial flora and their metabolites to extend the shelf life of food and inhibit the growth of spoilage and pathogenic microorganisms [300]. Lactic acid bacteria and yeasts can produce organic acids, hydrogen peroxide, and bacteriocins that inhibit harmful microbes. This approach aligns with consumer demand for more "natural" food preservation methods without synthetic additives [301].

Resolving the challenges of pathogen metabolites in food safety requires a combination of innovative technologies, better detection methods, and an expanded understanding of how these toxins interact with human health [302]. Ongoing research into probiotics, gene editing, natural antimicrobials, and AI-based systems offers promising solutions to reduce the risks posed by these harmful compounds.

9. Conclusions

This review underscores the critical public health challenges posed by pathogen metabolites, which can lead to severe health issues such as cancer, liver damage, neurological disorders, and immunosuppression. The need for advanced detection methods, a better understanding of long-term exposure effects, and effective control strategies is evident. Recommendations include prioritizing research into rapid, field-deployable diagnostic tools and innovative approaches like probiotics and gene editing. Strengthening food safety regulations, enhancing global surveillance, and increasing public education on safe food handling practices are essential to mitigate these risks. We can better protect public health and improve food safety by resolving these gaps and implementing targeted policies.

Author Contributions: Conceptualization, A.N.M. and D.B.; methodology, A.N.M.; software, A.N.M.; validation, D.B.; investigation, A.N.M.; writing—original draft preparation, A.N.M.; writing—review and editing, A.N.M. and D.B.; visualization, A.N.M. and D.B.; supervision, D.B.; project administration, D.B.; and funding acquisition, D.B. All authors have read and agreed to the published version of the manuscript.

Funding: This work was supported by the National Priorities Research Program grant NPRP 14S0311-210033, awarded to Dietrich Büsselberg (January 2023–Current) from the Qatar National Research Fund (QNRF, a member of the Qatar Foundation). The publication costs for this work were covered by the Biomedical Research Program at Weill Cornell Medicine-Qatar, a program funded by the Qatar Foundation. The statements made herein are solely the responsibility of the authors.

Data Availability Statement: No new data were created or analyzed in this study. Data sharing is not applicable to this article.

Conflicts of Interest: The authors declare no conflicts of interest.

References

1. Elbehiry, A.; Abalkhail, A.; Marzouk, E.; Elmanssury, A.E.; Almuzaini, A.M.; Alfheeaid, H.; Alshahrani, M.T.; Huraysh, N.; Ibrahem, M.; Alzaben, F.; et al. An Overview of the Public Health Challenges in Diagnosing and Controlling Human Foodborne Pathogens. *Vaccines* **2023**, *11*, 725. [CrossRef] [PubMed]
2. Awad, D.A.; Masoud, H.A.; Hamad, A. Climate changes and food-borne pathogens: The impact on human health and mitigation strategy. *Clim. Chang.* **2024**, *177*, 92. [CrossRef]

3. Augustin, J.-C.; Kooh, P.; Bayeux, T.; Guillier, L.; Meyer, T.; Jourdan-Da Silva, N.; Villena, I.; Sanaa, M.; Cerf, O. Contribution of Foods and Poor Food-Handling Practices to the Burden of Foodborne Infectious Diseases in France. *Foods* **2020**, *9*, 1644. [CrossRef] [PubMed]
4. Piret, J.; Boivin, G. Pandemics Throughout History. *Front. Microbiol.* **2021**, *11*, 631736. [CrossRef]
5. Lamichhane, B.; Mawad, A.M.M.; Saleh, M.; Kelley, W.G.; Harrington, P.J.; Lovestad, C.W.; Amezcua, J.; Sarhan, M.M.; El Zowalaty, M.E.; Ramadan, H.; et al. Salmonellosis: An Overview of Epidemiology, Pathogenesis, and Innovative Approaches to Mitigate the Antimicrobial Resistant Infections. *Antibiotics* **2024**, *13*, 76. [CrossRef]
6. Zha, L.; Garrett, S.; Sun, J. *Salmonella* Infection in Chronic Inflammation and Gastrointestinal Cancer. *Diseases* **2019**, *7*, 28. [CrossRef]
7. Marchello, C.S.; Birkhold, M.; Crump, J.A. Complications and mortality of typhoid fever: A global systematic review and meta-analysis. *J. Infect.* **2020**, *81*, 902–910. [CrossRef]
8. Khan, M.; Shamim, S. Understanding the Mechanism of Antimicrobial Resistance and Pathogenesis of *Salmonella enterica* Serovar Typhi. *Microorganisms* **2022**, *10*, 2006. [CrossRef]
9. Ramos, S.; Silva, V.; Dapkevicius, M.d.L.E.; Caniça, M.; Tejedor-Junco, M.T.; Igrejas, G.; Poeta, P. *Escherichia coli* as Commensal and Pathogenic Bacteria among Food-Producing Animals: Health Implications of Extended Spectrum β-Lactamase (ESBL) Production. *Animals* **2020**, *10*, 2239. [CrossRef]
10. Liu, Y.; Thaker, H.; Wang, C.; Xu, Z.; Dong, M. Diagnosis and Treatment for Shiga Toxin-Producing *Escherichia coli* Associated Hemolytic Uremic Syndrome. *Toxins* **2022**, *15*, 10. [CrossRef]
11. Bagel, A.; Sergentet, D. Shiga Toxin-Producing *Escherichia coli* and Milk Fat Globules. *Microorganisms* **2022**, *10*, 496. [CrossRef]
12. Onyeaka, H.; Ghosh, S.; Obileke, K.; Miri, T.; Odeyemi, O.A.; Nwaiwu, O.; Tamasiga, P. Preventing chemical contaminants in food: Challenges and prospects for safe and sustainable food production. *Food Control* **2024**, *155*, 110040. [CrossRef]
13. Liu, F.; Lee, S.A.; Xue, J.; Riordan, S.M.; Zhang, L. Global epidemiology of campylobacteriosis and the impact of COVID-19. *Front. Cell. Infect. Microbiol.* **2022**, *12*, 979055. [CrossRef] [PubMed]
14. Musa Moi, I.; Ibrahim, Z.; Mohammed Abubakar, B.; Mohammed Katagum, Y.; Abdullahi, A.; Ajibji Yiga, G.; Abdullahi, B.; Mustapha, I.; Ali, J.; Mahmud, Z.; et al. Properties of Foodborne Pathogens and Their Diseases. In *Foodborne Pathogens—Recent Advances in Control and Detection*; IntechOpen: Rijeka, Croatia, 2023.
15. Metreveli, M.; Bulia, S.; Shalamberidze, I.; Tevzadze, L.; Tsanava, S.; Goenaga, J.C.; Stingl, K.; Imnadze, P. Campylobacteriosis, Shigellosis and Salmonellosis in Hospitalized Children with Acute Inflammatory Diarrhea in Georgia. *Pathogens* **2022**, *11*, 232. [CrossRef] [PubMed]
16. Lin, Q.; Lin, S.; Fan, Z.; Liu, J.; Ye, D.; Guo, P. A Review of the Mechanisms of Bacterial Colonization of the Mammal Gut. *Microorganisms* **2024**, *12*, 1026. [CrossRef]
17. Kraus, V.; Čižmárová, B.; Birková, A. Listeria in Pregnancy—The Forgotten Culprit. *Microorganisms* **2024**, *12*, 2102. [CrossRef] [PubMed]
18. Ribeiro, A.C.; Almeida, F.A.d.; Medeiros, M.M.; Miranda, B.R.; Pinto, U.M.; Alves, V.F. *Listeria monocytogenes*: An Inconvenient Hurdle for the Dairy Industry. *Dairy* **2023**, *4*, 316–344. [CrossRef]
19. Osek, J.; Lachtara, B.; Wieczorek, K. *Listeria monocytogenes*—How This Pathogen Survives in Food-Production Environments? *Front. Microbiol.* **2022**, *13*, 866462. [CrossRef]
20. Duan, K.; Onyeaka, H.; Pang, G. Leveraging blockchain to tackle food fraud: Innovations and obstacles. *J. Agric. Food Res.* **2024**, *18*, 101429. [CrossRef]
21. Awuchi, C.G.; Ondari, E.N.; Ogbonna, C.U.; Upadhyay, A.K.; Baran, K.; Okpala, C.O.R.; Korzeniowska, M.; Guiné, R.P.F. Mycotoxins Affecting Animals, Foods, Humans, and Plants: Types, Occurrence, Toxicities, Action Mechanisms, Prevention, and Detoxification Strategies—A Revisit. *Foods* **2021**, *10*, 1279. [CrossRef]
22. Benkerroum, N. Aflatoxins: Producing-Molds, Structure, Health Issues and Incidence in Southeast Asian and Sub-Saharan African Countries. *Int. J. Environ. Res. Public Health* **2020**, *17*, 1215. [CrossRef] [PubMed]
23. Lumsangkul, C.; Tso, K.-H.; Fan, Y.-K.; Chiang, H.-I.; Ju, J.-C. Mycotoxin Fumonisin B1 Interferes Sphingolipid Metabolisms and Neural Tube Closure during Early Embryogenesis in Brown Tsaiya Ducks. *Toxins* **2021**, *13*, 743. [CrossRef] [PubMed]
24. Mafe, A.N.; Büsselberg, D. Mycotoxins in Food: Cancer Risks and Strategies for Control. *Foods* **2024**, *13*, 3502. [CrossRef] [PubMed]
25. Khan, R.; Anwar, F.; Ghazali, F.M. A comprehensive review of mycotoxins: Toxicology, detection, and effective mitigation approaches. *Heliyon* **2024**, *10*, e28361. [CrossRef] [PubMed]
26. Popoff, M.R. Overview of Bacterial Protein Toxins from Pathogenic Bacteria: Mode of Action and Insights into Evolution. *Toxins* **2024**, *16*, 182. [CrossRef]
27. Greeff-Laubscher, M.R.; Beukes, I.; Marais, G.J.; Jacobs, K. Mycotoxin production by three different toxigenic fungi genera on formulated abalone feed and the effect of an aquatic environment on fumonisins. *Mycology* **2020**, *11*, 105–117. [CrossRef]
28. Janik, E.; Niemcewicz, M.; Ceremuga, M.; Stela, M.; Saluk-Bijak, J.; Siadkowski, A.; Bijak, M. Molecular Aspects of Mycotoxins—A Serious Problem for Human Health. *Int. J. Mol. Sci.* **2020**, *21*, 8187. [CrossRef]
29. Forbes, J.D. Clinically Important Toxins in Bacterial Infection: Utility of Laboratory Detection. *Clin. Microbiol. Newsl.* **2020**, *42*, 163–170. [CrossRef]

30. Ghazaei, C. Advances in the Study of Bacterial Toxins, Their Roles and Mechanisms in Pathogenesis. *Malays. J. Med. Sci.* **2022**, *29*, 4–17. [CrossRef] [PubMed]
31. Awuchi, C.G.; Ondari, E.N.; Nwozo, S.; Odongo, G.A.; Eseoghene, I.J.; Twinomuhwezi, H.; Ogbonna, C.U.; Upadhyay, A.K.; Adeleye, A.O.; Okpala, C.O.R. Mycotoxins' Toxicological Mechanisms Involving Humans, Livestock and Their Associated Health Concerns: A Review. *Toxins* **2022**, *14*, 167. [CrossRef]
32. Soni, J.; Sinha, S.; Pandey, R. Understanding bacterial pathogenicity: A closer look at the journey of harmful microbes. *Front. Microbiol.* **2024**, *15*, 1370818. [CrossRef] [PubMed]
33. Joseph, A.; Cointe, A.; Mariani Kurkdjian, P.; Rafat, C.; Hertig, A. Shiga Toxin-Associated Hemolytic Uremic Syndrome: A Narrative Review. *Toxins* **2020**, *12*, 67. [CrossRef] [PubMed]
34. Dietrich, R.; Jessberger, N.; Ehling-Schulz, M.; Märtlbauer, E.; Granum, P.E. The Food Poisoning Toxins of *Bacillus cereus*. *Toxins* **2021**, *13*, 98. [CrossRef] [PubMed]
35. Reyes, V.E. *Helicobacter pylori* and Its Role in Gastric Cancer. *Microorganisms* **2023**, *11*, 1312. [CrossRef] [PubMed]
36. Mafe, A.N.; Edo, G.I.; Makia, R.S.; Joshua, O.A.; Akpoghelie, P.O.; Gaaz, T.S.; Jikah, A.N.; Yousif, E.; Isoje, E.F.; Igbuku, U.A.; et al. A review on food spoilage mechanisms, food borne diseases and commercial aspects of food preservation and processing. *Food Chem. Adv.* **2024**, *5*, 100852. [CrossRef]
37. Saha Turna, N.; Chung, R.; McIntyre, L. A review of biogenic amines in fermented foods: Occurrence and health effects. *Heliyon* **2024**, *10*, e24501. [CrossRef]
38. Ebbensgaard, A.; Mordhorst, H.; Aarestrup, F.M.; Hansen, E.B. The Role of Outer Membrane Proteins and Lipopolysaccharides for the Sensitivity of *Escherichia coli* to Antimicrobial Peptides. *Front. Microbiol.* **2018**, *9*, 2153. [CrossRef] [PubMed]
39. Pedrosa, M.C.; Lima, L.; Heleno, S.; Carocho, M.; Ferreira, I.C.F.R.; Barros, L. Food Metabolites as Tools for Authentication, Processing, and Nutritive Value Assessment. *Foods* **2021**, *10*, 2213. [CrossRef]
40. Aladhadh, M. A Review of Modern Methods for the Detection of Foodborne Pathogens. *Microorganisms* **2023**, *11*, 1111. [CrossRef]
41. Taiwo, O.R.; Onyeaka, H.; Oladipo, E.K.; Oloke, J.K.; Chukwugozie, D.C. Advancements in Predictive Microbiology: Integrating New Technologies for Efficient Food Safety Models. *Int. J. Microbiol.* **2024**, *2024*, 6612162. [CrossRef]
42. Insfran-Rivarola, A.; Tlapa, D.; Limon-Romero, J.; Baez-Lopez, Y.; Miranda-Ackerman, M.; Arredondo-Soto, K.; Ontiveros, S. A Systematic Review and Meta-Analysis of the Effects of Food Safety and Hygiene Training on Food Handlers. *Foods* **2020**, *9*, 1169. [CrossRef] [PubMed]
43. Warmate, D.; Onarinde, B.A. Food safety incidents in the red meat industry: A review of foodborne disease outbreaks linked to the consumption of red meat and its products, 1991 to 2021. *Int. J. Food Microbiol.* **2023**, *398*, 110240. [CrossRef]
44. Fouillaud, M.; Dufossé, L. Microbial Secondary Metabolism and Biotechnology. *Microorganisms* **2022**, *10*, 123. [CrossRef] [PubMed]
45. Mayer, C.L.; Leibowitz, C.S.; Kurosawa, S.; Stearns-Kurosawa, D.J. Shiga Toxins and the Pathophysiology of Hemolytic Uremic Syndrome in Humans and Animals. *Toxins* **2012**, *4*, 1261–1287. [CrossRef] [PubMed]
46. Igwaran, A.; Okoh, A.I. Human campylobacteriosis: A public health concern of global importance. *Heliyon* **2019**, *5*, e02814. [CrossRef]
47. Gou, H.; Zeng, R.; Lau, H.C.H.; Yu, J. Gut microbial metabolites: Shaping future diagnosis and treatment against gastrointestinal cancer. *Pharmacol. Res.* **2024**, *208*, 107373. [CrossRef]
48. Wenndt, A.; Mutua, F.; Grace, D.; Thomas, L.F.; Lambertini, E. Quantitative assessment of aflatoxin exposure and hepatocellular carcinoma (HCC) risk associated with consumption of select Nigerian staple foods. *Front. Sustain. Food Syst.* **2023**, *7*, 1128540. [CrossRef]
49. Malet-Villemagne, J.; Vidic, J. Extracellular vesicles in the pathogenesis of *Campylobacter jejuni*. *Microbes Infect.* **2024**, *26*, 105377. [CrossRef]
50. Ali, A.; AlHussaini, K.I. *Helicobacter pylori*: A Contemporary Perspective on Pathogenesis, Diagnosis and Treatment Strategies. *Microorganisms* **2024**, *12*, 222. [CrossRef]
51. Afzaal, M.; Saeed, F.; Shah, Y.A.; Hussain, M.; Rabail, R.; Socol, C.T.; Hassoun, A.; Pateiro, M.; Lorenzo, J.M.; Rusu, A.V.; et al. Human gut microbiota in health and disease: Unveiling the relationship. *Front. Microbiol.* **2022**, *13*, 999001. [CrossRef]
52. Ling, N.; Forsythe, S.; Wu, Q.; Ding, Y.; Zhang, J.; Zeng, H. Insights into *Cronobacter sakazakii* Biofilm Formation and Control Strategies in the Food Industry. *Engineering* **2020**, *6*, 393–405. [CrossRef]
53. Ling, N.; Zhang, X.; Forsythe, S.; Zhang, D.; Shen, Y.; Zhang, J.; Ding, Y.; Wang, J.; Wu, Q.; Ye, Y. Bacteroides fragilis ameliorates *Cronobacter malonaticus* lipopolysaccharide-induced pathological injury through modulation of the intestinal microbiota. *Front. Immunol.* **2022**, *13*, 931871. [CrossRef] [PubMed]
54. Cheng, C.; Yan, X.; Liu, B.; Jiang, T.; Zhou, Z.; Guo, F.; Zhang, Q.; Li, C.; Fang, T. SdiA Enhanced the Drug Resistance of *Cronobacter sakazakii* and Suppressed Its Motility, Adhesion and Biofilm Formation. *Front. Microbiol.* **2022**, *13*, 901912. [CrossRef] [PubMed]
55. Ke, A.; Parreira, V.R.; Goodridge, L.; Farber, J.M. Current and Future Perspectives on the Role of Probiotics, Prebiotics, and Synbiotics in Controlling Pathogenic *Cronobacter* Spp. in Infants Front. Microbiol. **2021**, *12*, 755083. [CrossRef] [PubMed]
56. Semwal, A.; Kumar, A.; Kumar, N. A review on pathogenicity of Aeromonas hydrophila and their mitigation through medicinal herbs in aquaculture. *Heliyon* **2023**, *9*, e14088. [CrossRef] [PubMed]
57. Khasheii, B.; Mahmoodi, P.; Mohammadzadeh, A. Siderophores: Importance in bacterial pathogenesis and applications in medicine and industry. *Microbiol. Res.* **2021**, *250*, 126790. [CrossRef]
58. Wu, H.; Wang, Y.; Li, H.; Meng, L.; Zheng, N.; Wang, J. Effect of Food Endotoxin on Infant Health. *Toxins* **2021**, *13*, 298. [CrossRef]

59. Zhou, X.; Kandalai, S.; Hossain, F.; Zheng, Q. Tumor microbiome metabolism: A game changer in cancer development and therapy. *Front. Oncol.* **2022**, *12*, 933407. [CrossRef]
60. Okaiyeto, S.A.; Sutar, P.P.; Chen, C.; Ni, J.-B.; Wang, J.; Mujumdar, A.S.; Zhang, J.-S.; Xu, M.-Q.; Fang, X.-M.; Zhang, C.; et al. Antibiotic resistant bacteria in food systems: Current status, resistance mechanisms, and mitigation strategies. *Agric. Commun.* **2024**, *2*, 100027. [CrossRef]
61. Ji, X.; Lu, P.; Xue, J.; Zhao, N.; Zhang, Y.; Dong, L.; Zhang, X.; Li, P.; Hu, Y.; Wang, J.; et al. The lipoprotein NlpD in *Cronobacter sakazakii* responds to acid stress and regulates macrophage resistance and virulence by maintaining membrane integrity. *Virulence* **2021**, *12*, 415–429. [CrossRef]
62. Karki, R.; Kanneganti, T.-D. The 'cytokine storm': Molecular mechanisms and therapeutic prospects. *Trends Immunol.* **2021**, *42*, 681–705. [CrossRef] [PubMed]
63. Saha, P.; Ettel, P.; Weichhart, T. Leveraging macrophage metabolism for anticancer therapy: Opportunities and pitfalls. *Trends Pharmacol. Sci.* **2024**, *45*, 335–349. [CrossRef] [PubMed]
64. Arfin, S.; Jha, N.K.; Jha, S.K.; Kesari, K.K.; Ruokolainen, J.; Roychoudhury, S.; Rathi, B.; Kumar, D. Oxidative Stress in Cancer Cell Metabolism. *Antioxidants* **2021**, *10*, 642. [CrossRef] [PubMed]
65. Singh, P.; Chauhan, P.K.; Upadhyay, S.K.; Singh, R.K.; Dwivedi, P.; Wang, J.; Jain, D.; Jiang, M. Mechanistic Insights and Potential Use of Siderophores Producing Microbes in Rhizosphere for Mitigation of Stress in Plants Grown in Degraded Land. *Front. Microbiol.* **2022**, *13*, 898979. [CrossRef] [PubMed]
66. Allen, C.N.S.; Arjona, S.P.; Santerre, M.; Sawaya, B.E. Hallmarks of Metabolic Reprogramming and Their Role in Viral Pathogenesis. *Viruses* **2022**, *14*, 602. [CrossRef]
67. Ke, A.; Parreira, V.R.; Farber, J.M.; Goodridge, L. Inhibition of *Cronobacter sakazakii* in an infant simulator of the human intestinal microbial ecosystem using a potential synbiotic. *Front. Microbiol.* **2022**, *13*, 947624. [CrossRef]
68. Wu, H.; Ma, W.; Wang, Y.; Wang, Y.; Sun, X.; Zheng, Q. Gut microbiome-metabolites axis: A friend or foe to colorectal cancer progression. *Biomed. Pharmacother.* **2024**, *173*, 116410. [CrossRef]
69. Aleman, R.S.; Moncada, M.; Aryana, K.J. Leaky Gut and the Ingredients That Help Treat It: A Review. *Molecules* **2023**, *28*, 619. [CrossRef]
70. Liu, B.; Li, Y.; Suo, L.; Zhang, W.; Cao, H.; Wang, R.; Luan, J.; Yu, X.; Dong, L.; Wang, W.; et al. Characterizing microbiota and metabolomics analysis to identify candidate biomarkers in lung cancer. *Front. Oncol.* **2022**, *12*, 1058436. [CrossRef]
71. Jiang, Y.; Li, Y. Nutrition Intervention and Microbiome Modulation in the Management of Breast Cancer. *Nutrients* **2024**, *16*, 2644. [CrossRef] [PubMed]
72. Li, X.; Shen, X.; Jiang, W.; Xi, Y.; Li, S. Comprehensive review of emerging contaminants: Detection technologies, environmental impact, and management strategies. *Ecotoxicol. Environ. Saf.* **2024**, *278*, 116420. [CrossRef] [PubMed]
73. Agus, A.; Clément, K.; Sokol, H. Gut microbiota-derived metabolites as central regulators in metabolic disorders. *Gut* **2021**, *70*, 1174–1182. [CrossRef] [PubMed]
74. Barathan, M.; Ng, S.L.; Lokanathan, Y.; Ng, M.H.; Law, J.X. The Profound Influence of Gut Microbiome and Extracellular Vesicles on Animal Health and Disease. *Int. J. Mol. Sci.* **2024**, *25*, 4024. [CrossRef] [PubMed]
75. Ma, T.; Shen, X.; Shi, X.; Sakandar, H.A.; Quan, K.; Li, Y.; Jin, H.; Kwok, L.-Y.; Zhang, H.; Sun, Z. Targeting gut microbiota and metabolism as the major probiotic mechanism—An evidence-based review. *Trends Food Sci. Technol.* **2023**, *138*, 178–198. [CrossRef]
76. Cao, C.; Yue, S.; Lu, A.; Liang, C. Host-Gut Microbiota Metabolic Interactions and Their Role in Precision Diagnosis and Treatment of Gastrointestinal Cancers. *Pharmacol. Res.* **2024**, *207*, 107321. [CrossRef]
77. Zhan, Z.; Tang, H.; Zhang, Y.; Huang, X.; Xu, M. Potential of gut-derived short-chain fatty acids to control enteric pathogens. *Front. Microbiol.* **2022**, *13*, 976406. [CrossRef]
78. Režen, T.; Rozman, D.; Kovács, T.; Kovács, P.; Sipos, A.; Bai, P.; Mikó, E. The role of bile acids in carcinogenesis. *Cell. Mol. Life Sci.* **2022**, *79*, 243. [CrossRef]
79. Goswami, M.; Bose, P.D. Gut microbial dysbiosis in the pathogenesis of leukemia: An immune-based perspective. *Exp. Hematol.* **2024**, *133*, 104211. [CrossRef] [PubMed]
80. Khatun, S.; Appidi, T.; Rengan, A.K. The role played by bacterial infections in the onset and metastasis of cancer. *Curr. Res. Microb. Sci.* **2021**, *2*, 100078. [CrossRef] [PubMed]
81. Kalyan, M.; Tousif, A.H.; Sonali, S.; Vichitra, C.; Sunanda, T.; Praveenraj, S.S.; Ray, B.; Gorantla, V.R.; Rungratanawanich, W.; Mahalakshmi, A.M.; et al. Role of Endogenous Lipopolysaccharides in Neurological Disorders. *Cells* **2022**, *11*, 4038. [CrossRef] [PubMed]
82. Shakhpazyan, N.K.; Mikhaleva, L.M.; Bedzhanyan, A.L.; Gioeva, Z.V.; Mikhalev, A.I.; Midiber, K.Y.; Pechnikova, V.V.; Biryukov, A.E. Exploring the Role of the Gut Microbiota in Modulating Colorectal Cancer Immunity. *Cells* **2024**, *13*, 1437. [CrossRef] [PubMed]
83. Stolfi, C.; Maresca, C.; Monteleone, G.; Laudisi, F. Implication of Intestinal Barrier Dysfunction in Gut Dysbiosis and Diseases. *Biomedicines* **2022**, *10*, 289. [CrossRef] [PubMed]
84. Di Vincenzo, F.; Del Gaudio, A.; Petito, V.; Lopetuso, L.R.; Scaldaferri, F. Gut microbiota, intestinal permeability, and systemic inflammation: A narrative review. *Intern. Emerg. Med.* **2024**, *19*, 275–293. [CrossRef] [PubMed]
85. Sadeghi, M.; Mestivier, D.; Sobhani, I. Contribution of pks+ *Escherichia coli* (*E. coli*) to Colon Carcinogenesis. *Microorganisms* **2024**, *12*, 1111. [CrossRef] [PubMed]

86. Pawłowska, M.; Mila-Kierzenkowska, C.; Szczegielniak, J.; Woźniak, A. Oxidative Stress in Parasitic Diseases—Reactive Oxygen Species as Mediators of Interactions between the Host and the Parasites. *Antioxidants* **2023**, *13*, 38. [CrossRef]
87. Kwao-Zigah, G.; Bediako-Bowan, A.; Boateng, P.A.; Aryee, G.K.; Abbang, S.M.; Atampugbire, G.; Quaye, O.; Tagoe, E.A. Microbiome Dysbiosis, Dietary Intake and Lifestyle-Associated Factors Involve in Epigenetic Modulations in Colorectal Cancer: A Narrative Review. *Cancer Control* **2024**, *31*. [CrossRef]
88. Liu, Y.; Zhang, S.; Zhou, W.; Hu, D.; Xu, H.; Ji, G. Secondary Bile Acids and Tumorigenesis in Colorectal Cancer. *Front. Oncol.* **2022**, *12*, 813745. [CrossRef]
89. Bui, T.I.; Britt, E.A.; Muthukrishnan, G.; Gill, S.R. Probiotic induced synthesis of microbiota polyamine as a nutraceutical for metabolic syndrome and obesity-related type 2 diabetes. *Front. Endocrinol.* **2023**, *13*, 1094258. [CrossRef] [PubMed]
90. Chen, Z.; Han, F.; Du, Y.; Shi, H.; Zhou, W. Hypoxic microenvironment in cancer: Molecular mechanisms and therapeutic interventions. *Signal Transduct. Target. Ther.* **2023**, *8*, 70. [CrossRef]
91. Mathew, M.; Nguyen, N.; Bhutia, Y.; Sivaprakasam, S.; Ganapathy, V. Metabolic Signature of Warburg Effect in Cancer: An Effective and Obligatory Interplay between Nutrient Transporters and Catabolic/Anabolic Pathways to Promote Tumor Growth. *Cancers* **2024**, *16*, 504. [CrossRef]
92. Ngaha, T.Y.S.; Zhilenkova, A.V.; Essogmo, F.E.; Uchendu, I.K.; Abah, M.O.; Fossa, L.T.; Sangadzhieva, Z.D.; Sanikovich, V.D.; Rusanov, A.S.; Pirogova, Y.N.; et al. Angiogenesis in Lung Cancer: Understanding the Roles of Growth Factors. *Cancers* **2023**, *15*, 4648. [CrossRef] [PubMed]
93. Bhatt, A.P.; Redinbo, M.R.; Bultman, S.J. The role of the microbiome in cancer development and therapy. *CA. Cancer J. Clin.* **2017**, *67*, 326–344. [CrossRef] [PubMed]
94. Zhou, P.; Chen, C.; Patil, S.; Dong, S. Unveiling the therapeutic symphony of probiotics, prebiotics, and postbiotics in gut-immune harmony. *Front. Nutr.* **2024**, *11*, 1355542. [CrossRef] [PubMed]
95. Flaig, B.; Garza, R.; Singh, B.; Hamamah, S.; Covasa, M. Treatment of Dyslipidemia through Targeted Therapy of Gut Microbiota. *Nutrients* **2023**, *15*, 228. [CrossRef] [PubMed]
96. Yoo, S.; Jung, S.-C.; Kwak, K.; Kim, J.-S. The Role of Prebiotics in Modulating Gut Microbiota: Implications for Human Health. *Int. J. Mol. Sci.* **2024**, *25*, 4834. [CrossRef]
97. Sahle, Z.; Engidaye, G.; Shenkute Gebreyes, D.; Adenew, B.; Abebe, T.A. Fecal microbiota transplantation and next-generation therapies: A review on targeting dysbiosis in metabolic disorders and beyond. *SAGE Open Med.* **2024**, *12*. [CrossRef]
98. Bernstein, H.; Bernstein, C. Bile acids as carcinogens in the colon and at other sites in the gastrointestinal system. *Exp. Biol. Med.* **2023**, *248*, 79–89. [CrossRef] [PubMed]
99. Muscolo, A.; Mariateresa, O.; Giulio, T.; Mariateresa, R. Oxidative Stress: The Role of Antioxidant Phytochemicals in the Prevention and Treatment of Diseases. *Int. J. Mol. Sci.* **2024**, *25*, 3264. [CrossRef]
100. Haston, J.C.; Miko, S.; Cope, J.R.; McKeel, H.; Walters, C.; Joseph, L.A.; Griswold, T.; Katz, L.S.; Andújar, A.A.; Tourdot, L.; et al. *Cronobacter sakazakii* Infections in Two Infants Linked to Powdered Infant Formula and Breast Pump Equipment—United States, 2021 and 2022. *MMWR. Morb. Mortal. Wkly. Rep.* **2023**, *72*, 223–226. [CrossRef] [PubMed]
101. Pessoa, R.B.G.; Oliveira, W.F.d.; Correia, M.T.d.S.; Fontes, A.; Coelho, L.C.B.B. Aeromonas and Human Health Disorders: Clinical Approaches. *Front. Microbiol.* **2022**, *13*, 868890. [CrossRef]
102. Jovanovic, J.; Tretiak, S.; Begyn, K.; Rajkovic, A. Detection of Enterotoxigenic Psychrotrophic Presumptive *Bacillus cereus* and Cereulide Producers in Food Products and Ingredients. *Toxins* **2022**, *14*, 289. [CrossRef] [PubMed]
103. Rawson, A.M.; Dempster, A.W.; Humphreys, C.M.; Minton, N.P. Pathogenicity and virulence of Clostridium botulinum. *Virulence* **2023**, *14*, 2205251. [CrossRef] [PubMed]
104. Huang, J.-T.; Mao, Y.-Q. The impact of the microbiome in cancer: Targeting metabolism of cancer cells and host. *Front. Oncol.* **2022**, *12*, 1029033. [CrossRef] [PubMed]
105. Zhang, Y.; Tang, N.; Zhou, H.; Zhu, Y. The role of microbial metabolites in endocrine tumorigenesis: From the mechanistic insights to potential therapeutic biomarkers. *Biomed. Pharmacother.* **2024**, *172*, 116218. [CrossRef] [PubMed]
106. Li, Y.; Zhu, R.; Jin, J.; Guo, H.; Zhang, J.; He, Z.; Liang, T.; Guo, L. Exploring the Role of Clustered Mutations in Carcinogenesis and Their Potential Clinical Implications in Cancer. *Int. J. Mol. Sci.* **2024**, *25*, 6744. [CrossRef]
107. Han, L.; Shu, X.; Wang, J. *Helicobacter pylori*-Mediated Oxidative Stress and Gastric Diseases: A Review. *Front. Microbiol.* **2022**, *13*, 811258. [CrossRef] [PubMed]
108. Gu, J.; Lin, Y.; Wang, Z.; Pan, Q.; Cai, G.; He, Q.; Xu, X.; Cai, X. *Campylobacter jejuni* Cytolethal Distending Toxin Induces GSDME-Dependent Pyroptosis in Colonic Epithelial Cells. *Front. Cell. Infect. Microbiol.* **2022**, *12*, 853204. [CrossRef] [PubMed]
109. Jha, N.K.; Arfin, S.; Jha, S.K.; Kar, R.; Dey, A.; Gundamaraju, R.; Ashraf, G.M.; Gupta, P.K.; Dhanasekaran, S.; Abomughaid, M.M.; et al. Re-establishing the comprehension of phytomedicine and nanomedicine in inflammation-mediated cancer signaling. *Semin. Cancer Biol.* **2022**, *86*, 1086–1104. [CrossRef]
110. Lacroix, M.; Riscal, R.; Arena, G.; Linares, L.K.; Le Cam, L. Metabolic functions of the tumor suppressor p53: Implications in normal physiology, metabolic disorders, and cancer. *Mol. Metab.* **2020**, *33*, 2–22. [CrossRef]
111. Rahman, M.M.; Islam, M.R.; Shohag, S.; Ahasan, M.T.; Sarkar, N.; Khan, H.; Hasan, A.M.; Cavalu, S.; Rauf, A. Microbiome in cancer: Role in carcinogenesis and impact in therapeutic strategies. *Biomed. Pharmacother.* **2022**, *149*, 112898. [CrossRef]
112. Zhang, Y.; Chen, R.; Zhang, D.; Qi, S.; Liu, Y. Metabolite interactions between host and microbiota during health and disease: Which feeds the other? *Biomed. Pharmacother.* **2023**, *160*, 114295. [CrossRef] [PubMed]

113. Marchese, S.; Polo, A.; Ariano, A.; Velotto, S.; Costantini, S.; Severino, L. Aflatoxin B1 and M1: Biological Properties and Their Involvement in Cancer Development. *Toxins* **2018**, *10*, 214. [CrossRef] [PubMed]
114. Rushing, B.R.; Selim, M.I. Aflatoxin B1: A review on metabolism, toxicity, occurrence in food, occupational exposure, and detoxification methods. *Food Chem. Toxicol.* **2019**, *124*, 81–100. [CrossRef] [PubMed]
115. Peles, F.; Sipos, P.; Győri, Z.; Pfliegler, W.P.; Giacometti, F.; Serraino, A.; Pagliuca, G.; Gazzotti, T.; Pócsi, I. Adverse Effects, Transformation and Channeling of Aflatoxins Into Food Raw Materials in Livestock. *Front. Microbiol.* **2019**, *10*, 2861. [CrossRef]
116. Piscione, M.; Mazzone, M.; Di Marcantonio, M.C.; Muraro, R.; Mincione, G. Eradication of *Helicobacter pylori* and Gastric Cancer: A Controversial Relationship. *Front. Microbiol.* **2021**, *12*, 630852. [CrossRef]
117. Díaz, P.; Valenzuela Valderrama, M.; Bravo, J.; Quest, A.F.G. *Helicobacter pylori* and Gastric Cancer: Adaptive Cellular Mechanisms Involved in Disease Progression. *Front. Microbiol.* **2018**, *9*, 5. [CrossRef] [PubMed]
118. Heimesaat, M.M.; Backert, S.; Alter, T.; Bereswill, S. Molecular Targets in Campylobacter Infections. *Biomolecules* **2023**, *13*, 409. [CrossRef] [PubMed]
119. Weir, T.L.; Trikha, S.R.J.; Thompson, H.J. Diet and cancer risk reduction: The role of diet-microbiota interactions and microbial metabolites. *Semin. Cancer Biol.* **2021**, *70*, 53–60. [CrossRef]
120. Ling, N.; Jiang, X.; Forsythe, S.; Zhang, D.; Shen, Y.; Ding, Y.; Wang, J.; Zhang, J.; Wu, Q.; Ye, Y. Food Safety Risks and Contributing Factors of *Cronobacter* spp. *Engineering* **2022**, *12*, 128–138. [CrossRef]
121. Sun, Y.; Wang, X.; Li, L.; Zhong, C.; Zhang, Y.; Yang, X.; Li, M.; Yang, C. The role of gut microbiota in intestinal disease: From an oxidative stress perspective. *Front. Microbiol.* **2024**, *15*, 1328324. [CrossRef]
122. Buonocore, G. Microbiota and gut immunity in infants and young children. *Glob. Pediatr.* **2024**, *9*, 100202. [CrossRef]
123. Fernández-Bravo, A.; Figueras, M.J. An Update on the Genus Aeromonas: Taxonomy, Epidemiology, and Pathogenicity. *Microorganisms* **2020**, *8*, 129. [CrossRef] [PubMed]
124. Bonfiglio, R.; Sisto, R.; Casciardi, S.; Palumbo, V.; Scioli, M.P.; Palumbo, A.; Trivigno, D.; Giacobbi, E.; Servadei, F.; Melino, G.; et al. The impact of toxic metal bioaccumulation on colorectal cancer: Unravelling the unexplored connection. *Sci. Total Environ.* **2024**, *906*, 167667. [CrossRef] [PubMed]
125. Wang, F.; Wu, F.; Chen, H.; Tang, B. The effect of probiotics in the prevention of atopic dermatitis in children: A systematic review and meta-analysis. *Transl. Pediatr.* **2023**, *12*, 731–748. [CrossRef] [PubMed]
126. Lanrewaju, A.A.; Enitan-Folami, A.M.; Sabiu, S.; Edokpayi, J.N.; Swalaha, F.M. Global public health implications of human exposure to viral contaminated water. *Front. Microbiol.* **2022**, *13*, 981896. [CrossRef]
127. Mielniczuk, E.; Skwaryło-Bednarz, B. Fusarium Head Blight, Mycotoxins and Strategies for Their Reduction. *Agronomy* **2020**, *10*, 509. [CrossRef]
128. Balló, A.; Busznyákné Székvári, K.; Czétány, P.; Márk, L.; Török, A.; Szántó, Á.; Máté, G. Estrogenic and Non-Estrogenic Disruptor Effect of Zearalenone on Male Reproduction: A Review. *Int. J. Mol. Sci.* **2023**, *24*, 1578. [CrossRef]
129. Kozieł, M.J.; Piastowska-Ciesielska, A.W. Estrogens, Estrogen Receptors and Tumor Microenvironment in Ovarian Cancer. *Int. J. Mol. Sci.* **2023**, *24*, 14673. [CrossRef]
130. Rai, P.K.; Lee, S.S.; Zhang, M.; Tsang, Y.F.; Kim, K.-H. Heavy metals in food crops: Health risks, fate, mechanisms, and management. *Environ. Int.* **2019**, *125*, 365–385. [CrossRef]
131. Choi, W.; Kim, S.-S. Outbreaks, Germination, and Inactivation of *Bacillus cereus* in Food Products: A Review. *J. Food Prot.* **2020**, *83*, 1480–1487. [CrossRef]
132. Leong, S.S.; Korel, F.; King, J.H. *Bacillus cereus*: A review of "fried rice syndrome" causative agents. *Microb. Pathog.* **2023**, *185*, 106418. [CrossRef] [PubMed]
133. Todd, E. Food-Borne Disease Prevention and Risk Assessment. *Int. J. Environ. Res. Public Health* **2020**, *17*, 5129. [CrossRef]
134. Munir, M.T.; Mtimet, N.; Guillier, L.; Meurens, F.; Fravalo, P.; Federighi, M.; Kooh, P. Physical Treatments to Control Clostridium botulinum Hazards in Food. *Foods* **2023**, *12*, 1580. [CrossRef] [PubMed]
135. Was, H.; Borkowska, A.; Bagues, A.; Tu, L.; Liu, J.Y.H.; Lu, Z.; Rudd, J.A.; Nurgali, K.; Abalo, R. Mechanisms of Chemotherapy-Induced Neurotoxicity. *Front. Pharmacol.* **2022**, *13*, 750507. [CrossRef] [PubMed]
136. Witmanowski, H.; Błochowiak, K. The whole truth about botulinum toxin—A review. *Adv. Dermatol. Allergol.* **2020**, *37*, 853–861. [CrossRef] [PubMed]
137. De Girolamo, A.; Lippolis, V.; Pascale, M. Overview of Recent Liquid Chromatography Mass Spectrometry-Based Methods for Natural Toxins Detection in Food Products. *Toxins* **2022**, *14*, 328. [CrossRef]
138. Abdelhamid, A.G.; El-Dougdoug, N.K. Controlling foodborne pathogens with natural antimicrobials by biological control and antivirulence strategies. *Heliyon* **2020**, *6*, e05020. [CrossRef]
139. Lee, J.C.; Daraba, A.; Voidarou, C.; Rozos, G.; Enshasy, H.A.E.; Varzakas, T. Implementation of Food Safety Management Systems along with Other Management Tools (HAZOP, FMEA, Ishikawa, Pareto). The Case Study of *Listeria monocytogenes* and Correlation with Microbiological Criteria. *Foods* **2021**, *10*, 2169. [CrossRef]
140. Kabiraz, M.P.; Majumdar, P.R.; Mahmud, M.M.C.; Bhowmik, S.; Ali, A. Conventional and advanced detection techniques of foodborne pathogens: A comprehensive review. *Heliyon* **2023**, *9*, e15482. [CrossRef] [PubMed]
141. Hernández-Mesa, M.; Moreno-González, D. Current Role of Mass Spectrometry in the Determination of Pesticide Residues in Food. *Separations* **2022**, *9*, 148. [CrossRef]

142. Lei, H.; Wang, Z.; Eremin, S.A.; Liu, Z. Application of Antibody and Immunoassay for Food Safety. *Foods* **2022**, *11*, 826. [CrossRef] [PubMed]
143. Gould, O.; Nguyen, N.; Honeychurch, K.C. New Applications of Gas Chromatography and Gas Chromatography-Mass Spectrometry for Novel Sample Matrices in the Forensic Sciences: A Literature Review. *Chemosensors* **2023**, *11*, 527. [CrossRef]
144. Zhou, J.; Zhong, L. Applications of liquid chromatography-mass spectrometry based metabolomics in predictive and personalized medicine. *Front. Mol. Biosci.* **2022**, *9*, 1049016. [CrossRef] [PubMed]
145. Kiseleva, O.; Kurbatov, I.; Ilgisonis, E.; Poverennaya, E. Defining Blood Plasma and Serum Metabolome by GC-MS. *Metabolites* **2021**, *12*, 15. [CrossRef]
146. Madia, F.; Worth, A.; Whelan, M.; Corvi, R. Carcinogenicity assessment: Addressing the challenges of cancer and chemicals in the environment. *Environ. Int.* **2019**, *128*, 417–429. [CrossRef]
147. Son, A.; Kim, W.; Park, J.; Park, Y.; Lee, W.; Lee, S.; Kim, H. Mass Spectrometry Advancements and Applications for Biomarker Discovery, Diagnostic Innovations, and Personalized Medicine. *Int. J. Mol. Sci.* **2024**, *25*, 9880. [CrossRef]
148. Muthusamy, G.; Karthikeyan, S.; Arun Giridhari, V.; Alhimaidi, A.R.; Balachandar, D.; Ammari, A.A.; Paranidharan, V.; Maruthamuthu, T. Identification of Potential Biomarkers and Spectral Fingerprinting for Detection of Foodborne Pathogens in Raw Chicken Meat Matrix Using GCMS and FTIR. *Foods* **2024**, *13*, 3416. [CrossRef] [PubMed]
149. Jha, B.K.; Sherpa, M.L.; Imran, M.; Mohammed, Y.; Jha, L.A.; Paudel, K.R.; Jha, S.K. Progress in Understanding Metabolic Syndrome and Knowledge of Its Complex Pathophysiology. *Diabetology* **2023**, *4*, 134–159. [CrossRef]
150. Hamasha, M.M.; Bani-Irshid, A.H.; Al Mashaqbeh, S.; Shwaheen, G.; Al Qadri, L.; Shbool, M.; Muathen, D.; Ababneh, M.; Harfoush, S.; Albedoor, Q.; et al. Strategical selection of maintenance type under different conditions. *Sci. Rep.* **2023**, *13*, 15560. [CrossRef] [PubMed]
151. Amara, A.; Frainay, C.; Jourdan, F.; Naake, T.; Neumann, S.; Novoa-del-Toro, E.M.; Salek, R.M.; Salzer, L.; Scharfenberg, S.; Witting, M. Networks and Graphs Discovery in Metabolomics Data Analysis and Interpretation. *Front. Mol. Biosci.* **2022**, *9*, 841373. [CrossRef]
152. Fan, M.; Xu, X.; Lang, W.; Wang, W.; Wang, X.; Xin, A.; Zhou, F.; Ding, Z.; Ye, X.; Zhu, B. Toxicity, formation, contamination, determination and mitigation of acrylamide in thermally processed plant-based foods and herbal medicines: A review. *Ecotoxicol. Environ. Saf.* **2023**, *260*, 115059. [CrossRef]
153. Venbrux, M.; Crauwels, S.; Rediers, H. Current and emerging trends in techniques for plant pathogen detection. *Front. Plant Sci.* **2023**, *14*, 1120968. [CrossRef] [PubMed]
154. Quintela, I.A.; Vasse, T.; Lin, C.-S.; Wu, V.C.H. Advances, applications, and limitations of portable and rapid detection technologies for routinely encountered foodborne pathogens. *Front. Microbiol.* **2022**, *13*, 1054782. [CrossRef] [PubMed]
155. Bibi Sadeer, N.; Montesano, D.; Albrizio, S.; Zengin, G.; Mahomoodally, M.F. The Versatility of Antioxidant Assays in Food Science and Safety—Chemistry, Applications, Strengths, and Limitations. *Antioxidants* **2020**, *9*, 709. [CrossRef] [PubMed]
156. Flasch, M.; Koellensperger, G.; Warth, B. Comparing the sensitivity of a low- and a high-resolution mass spectrometry approach for xenobiotic trace analysis: An exposome-type case study. *Anal. Chim. Acta* **2023**, *1279*, 341740. [CrossRef]
157. Register, A.C.; Tarighat, S.S.; Lee, H.Y. Bioassay Development for Bispecific Antibodies—Challenges and Opportunities. *Int. J. Mol. Sci.* **2021**, *22*, 5350. [CrossRef]
158. Pratiwi, R.; Ramadhanti, S.P.; Amatulloh, A.; Megantara, S.; Subra, L. Recent Advances in the Determination of Veterinary Drug Residues in Food. *Foods* **2023**, *12*, 3422. [CrossRef]
159. Chandrasekar, V.; Panicker, A.J.; Dey, A.K.; Mohammad, S.; Chakraborty, A.; Samal, S.K.; Dash, A.; Bhadra, J.; Suar, M.; Khare, M.; et al. Integrated approaches for immunotoxicity risk assessment: Challenges and future directions. *Discov. Toxicol.* **2024**, *1*, 9. [CrossRef]
160. Saini, R.V.; Vaid, P.; Saini, N.K.; Siwal, S.S.; Gupta, V.K.; Thakur, V.K.; Saini, A.K. Recent Advancements in the Technologies Detecting Food Spoiling Agents. *J. Funct. Biomater.* **2021**, *12*, 67. [CrossRef]
161. Das, S.; Dey, M.K.; Devireddy, R.; Gartia, M.R. Biomarkers in Cancer Detection, Diagnosis, and Prognosis. *Sensors* **2023**, *24*, 37. [CrossRef]
162. Noor, J.; Chaudhry, A.; Batool, S. Microfluidic Technology, Artificial Intelligence, and Biosensors As Advanced Technologies in Cancer Screening: A Review Article. *Cureus* **2023**, *15*. [CrossRef] [PubMed]
163. Ramesh, M.; Janani, R.; Deepa, C.; Rajeshkumar, L. Nanotechnology-Enabled Biosensors: A Review of Fundamentals, Design Principles, Materials, and Applications. *Biosensors* **2022**, *13*, 40. [CrossRef] [PubMed]
164. Shanbhag, M.M.; Manasa, G.; Mascarenhas, R.J.; Mondal, K.; Shetti, N.P. Fundamentals of bio-electrochemical sensing. *Chem. Eng. J. Adv.* **2023**, *16*, 100516. [CrossRef]
165. Singh, A.K.; Mittal, S.; Das, M.; Saharia, A.; Tiwari, M. Optical biosensors: A decade in review. *Alexandria Eng. J.* **2023**, *67*, 673–691. [CrossRef]
166. Szelenberger, R.; Cichoń, N.; Zajaczkowski, W.; Bijak, M. Application of Biosensors for the Detection of Mycotoxins for the Improvement of Food Safety. *Toxins* **2024**, *16*, 249. [CrossRef]
167. Preetam, S.; Nahak, B.K.; Patra, S.; Toncu, D.C.; Park, S.; Syväjärvi, M.; Orive, G.; Tiwari, A. Emergence of microfluidics for next generation biomedical devices. *Biosens. Bioelectron. X* **2022**, *10*, 100106. [CrossRef]
168. Jiang, Z.; Zhuang, Y.; Guo, S.; Sohan, A.S.M.M.F.; Yin, B. Advances in Microfluidics Techniques for Rapid Detection of Pesticide Residues in Food. *Foods* **2023**, *12*, 2868. [CrossRef]

169. Pittman, T.W.; Decsi, D.B.; Punyadeera, C.; Henry, C.S. Saliva-based microfluidic point-of-care diagnostic. *Theranostics* **2023**, *13*, 1091–1108. [CrossRef]
170. Bakhshi, A.; Pandey, A.; Kharaba, Z.; Razlansari, M.; Sargazi, S.; Behzadmehr, R.; Rahdar, A.; Díez-Pascual, A.M.; Fathi-karkan, S. Microfluidic-based nanoplatforms for cancer theranostic applications: A mini-review on recent advancements. *OpenNano* **2024**, *15*, 100197. [CrossRef]
171. Cernat, A.; Groza, A.; Tertis, M.; Feier, B.; Hosu-Stancioiu, O.; Cristea, C. Where artificial intelligence stands in the development of electrochemical sensors for healthcare applications-A review. *TrAC Trends Anal. Chem.* **2024**, *181*, 117999. [CrossRef]
172. Nashruddin, S.N.A.B.M.; Salleh, F.H.M.; Yunus, R.M.; Zaman, H.B. Artificial intelligence−powered electrochemical sensor: Recent advances, challenges, and prospects. *Heliyon* **2024**, *10*, e37964. [CrossRef] [PubMed]
173. Rodoplu Solovchuk, D. Advances in AI-assisted biochip technology for biomedicine. *Biomed. Pharmacother.* **2024**, *177*, 116997. [CrossRef] [PubMed]
174. Liang, L.; Sun, F.; Wang, H.; Hu, Z. Metabolomics, metabolic flux analysis and cancer pharmacology. *Pharmacol. Ther.* **2021**, *224*, 107827. [CrossRef] [PubMed]
175. Moco, S. Studying Metabolism by NMR-Based Metabolomics. *Front. Mol. Biosci.* **2022**, *9*, 882487. [CrossRef]
176. Alseekh, S.; Aharoni, A.; Brotman, Y.; Contrepois, K.; D'Auria, J.; Ewald, J.; Ewald, J.C.; Fraser, P.D.; Giavalisco, P.; Hall, R.D.; et al. Mass spectrometry-based metabolomics: A guide for annotation, quantification and best reporting practices. *Nat. Methods* **2021**, *18*, 747–756. [CrossRef] [PubMed]
177. Nagana Gowda, G.A.; Raftery, D. NMR-Based Metabolomics. In *Advances in Experimental Medicine and Biology*; Springer: Cham, Switzerland, 2021; pp. 19–37. [CrossRef]
178. Lin, C.; Tian, Q.; Guo, S.; Xie, D.; Cai, Y.; Wang, Z.; Chu, H.; Qiu, S.; Tang, S.; Zhang, A. Metabolomics for Clinical Biomarker Discovery and Therapeutic Target Identification. *Molecules* **2024**, *29*, 2198. [CrossRef]
179. Owolabi, I.O.; Siwarak, K.; Greer, B.; Rajkovic, A.; Dall'asta, C.; Karoonuthaisiri, N.; Uawisetwathana, U.; Elliott, C.T.; Petchkongkaew, A. Applications of Mycotoxin Biomarkers in Human Biomonitoring for Exposome-Health Studies: Past, Present, and Future. *Expo. Health* **2024**, *16*, 837–859. [CrossRef]
180. Rahman, M.; Schellhorn, H.E. Metabolomics of infectious diseases in the era of personalized medicine. *Front. Mol. Biosci.* **2023**, *10*, 1120376. [CrossRef]
181. Babu, M.; Snyder, M. Multi-Omics Profiling for Health. *Mol. Cell. Proteom.* **2023**, *22*, 100561. [CrossRef]
182. Magro, D.; Venezia, M.; Rita Balistreri, C. The omics technologies and liquid biopsies: Advantages, limitations, applications. *Med. Omi.* **2024**, *11*, 100039. [CrossRef]
183. Ozaki, Y.; Broughton, P.; Abdollahi, H.; Valafar, H.; Blenda, A.V. Integrating Omics Data and AI for Cancer Diagnosis and Prognosis. *Cancers* **2024**, *16*, 2448. [CrossRef] [PubMed]
184. Ahuja, V.; Singh, A.; Paul, D.; Dasgupta, D.; Urajová, P.; Ghosh, S.; Singh, R.; Sahoo, G.; Ewe, D.; Saurav, K. Recent Advances in the Detection of Food Toxins Using Mass Spectrometry. *Chem. Res. Toxicol.* **2023**, *36*, 1834–1863. [CrossRef] [PubMed]
185. Nnachi, R.C.; Sui, N.; Ke, B.; Luo, Z.; Bhalla, N.; He, D.; Yang, Z. Biosensors for rapid detection of bacterial pathogens in water, food and environment. *Environ. Int.* **2022**, *166*, 107357. [CrossRef]
186. Alonso, N.; Zelzer, S.; Eibinger, G.; Herrmann, M. Vitamin D Metabolites: Analytical Challenges and Clinical Relevance. *Calcif. Tissue Int.* **2023**, *112*, 158–177. [CrossRef] [PubMed]
187. Malcangi, S.; Romagnoli, M.; Beccaria, M.; Catani, M.; Chenet, T.; De Luca, C.; Felletti, S.; Pasti, L.; Cavazzini, A.; Franchina, F.A. Modern sample preparation approaches for small metabolite elucidation to support biomedical research. *Adv. Sample Prep.* **2022**, *2*, 100017. [CrossRef]
188. Jafari, S.; Guercetti, J.; Geballa-Koukoula, A.; Tsagkaris, A.S.; Nelis, J.L.D.; Marco, M.-P.; Salvador, J.-P.; Gerssen, A.; Hajslova, J.; Elliott, C.; et al. ASSURED Point-of-Need Food Safety Screening: A Critical Assessment of Portable Food Analyzers. *Foods* **2021**, *10*, 1399. [CrossRef] [PubMed]
189. Geballa-Koukoula, A.; Gerssen, A.; Nielen, M.W.F. Direct analysis of lateral flow immunoassays for deoxynivalenol using electrospray ionization mass spectrometry. *Anal. Bioanal. Chem.* **2020**, *412*, 7547–7558. [CrossRef] [PubMed]
190. Ge, H.; Wang, Y.; Zhao, X. Research on the drug resistance mechanism of foodborne pathogens. *Microb. Pathog.* **2022**, *162*, 105306. [CrossRef]
191. Uçar, A.; Yilmaz, M.V.; Çakiroglu, F.P. Food Safety—Problems and Solutions. In *Significance, Prevention and Control of Food Related Diseases*; InTech: London, UK, 2016.
192. Chiozzi, V.; Agriopoulou, S.; Varzakas, T. Advances, Applications, and Comparison of Thermal (Pasteurization, Sterilization, and Aseptic Packaging) against Non-Thermal (Ultrasounds, UV Radiation, Ozonation, High Hydrostatic Pressure) Technologies in Food Processing. *Appl. Sci.* **2022**, *12*, 2202. [CrossRef]
193. Silva, F.V.M.; Evelyn, E. Pasteurization of Food and Beverages by High Pressure Processing (HPP) at Room Temperature: Inactivation of *Staphylococcus aureus*, *Escherichia coli*, *Listeria monocytogenes*, *Salmonella*, and Other Microbial Pathogens. *Appl. Sci.* **2023**, *13*, 1193. [CrossRef]
194. Nowak, D.; Jakubczyk, E. The Freeze-Drying of Foods—The Characteristic of the Process Course and the Effect of Its Parameters on the Physical Properties of Food Materials. *Foods* **2020**, *9*, 1488. [CrossRef] [PubMed]
195. Cai, H.; Tao, L.; Zhou, X.; Liu, Y.; Sun, D.; Ma, Q.; Yu, Z.; Jiang, W. Lactic acid bacteria in fermented fish: Enhancing flavor and ensuring safety. *J. Agric. Food Res.* **2024**, *16*, 101206. [CrossRef]

196. Aguirre-Garcia, Y.L.; Nery-Flores, S.D.; Campos-Muzquiz, L.G.; Flores-Gallegos, A.C.; Palomo-Ligas, L.; Ascacio-Valdés, J.A.; Sepúlveda-Torres, L.; Rodríguez-Herrera, R. Lactic Acid Fermentation in the Food Industry and Bio-Preservation of Food. *Fermentation* **2024**, *10*, 168. [CrossRef]
197. Xia, Q.; Liu, Q.; Denoya, G.I.; Yang, C.; Barba, F.J.; Yu, H.; Chen, X. High Hydrostatic Pressure-Based Combination Strategies for Microbial Inactivation of Food Products: The Cases of Emerging Combination Patterns. *Front. Nutr.* **2022**, *9*, 878904. [CrossRef]
198. Usaga, J.; Acosta, Ó.; Churey, J.J.; Padilla-Zakour, O.I.; Worobo, R.W. Evaluation of high pressure processing (HPP) inactivation of *Escherichia coli* O157:H7, *Salmonella enterica*, and *Listeria monocytogenes* in acid and acidified juices and beverages. *Int. J. Food Microbiol.* **2021**, *339*, 109034. [CrossRef] [PubMed]
199. Ji, Q.-Y.; Wang, W.; Yan, H.; Qu, H.; Liu, Y.; Qian, Y.; Gu, R. The Effect of Different Organic Acids and Their Combination on the Cell Barrier and Biofilm of *Escherichia coli*. *Foods* **2023**, *12*, 3011. [CrossRef] [PubMed]
200. Dey, S.; Nagababu, B.H. Applications of food color and bio-preservatives in the food and its effect on the human health. *Food Chem. Adv.* **2022**, *1*, 100019. [CrossRef]
201. Pakdel, M.; Olsen, A.; Bar, E.M.S. A Review of Food Contaminants and Their Pathways Within Food Processing Facilities Using Open Food Processing Equipment. *J. Food Prot.* **2023**, *86*, 100184. [CrossRef]
202. Guennouni, M.; Admou, B.; Bourrhouat, A.; El Khoudri, N.; Zkhiri, W.; Talha, I.; Hazime, R.; Hilali, A. Knowledge and practices of food safety among health care professionals and handlers working in the kitchen of a Moroccan university hospital. *J. Food Prot.* **2022**, *85*, 676–685. [CrossRef]
203. Mendoza, I.C.; Luna, E.O.; Pozo, M.D.; Vásquez, M.V.; Montoya, D.C.; Moran, G.C.; Romero, L.G.; Yépez, X.; Salazar, R.; Romero-Peña, M.; et al. Conventional and non-conventional disinfection methods to prevent microbial contamination in minimally processed fruits and vegetables. *LWT* **2022**, *165*, 113714. [CrossRef] [PubMed]
204. Bourdichon, F.; Betts, R.; Dufour, C.; Fanning, S.; Farber, J.; McClure, P.; Stavropoulou, D.A.; Wemmenhove, E.; Zwietering, M.H.; Winkler, A. Processing environment monitoring in low moisture food production facilities: Are we looking for the right microorganisms? *Int. J. Food Microbiol.* **2021**, *356*, 109351. [CrossRef] [PubMed]
205. Shabir Ahmad, R.; Munawar, H.; Saima, H.; Siddique, F. Introductory Chapter: Food Safety. In *Food Safety—New Insights*; IntechOpen: Rijeka, Croatia, 2023.
206. Jadhav, H.B.; Choudhary, P. Emerging techniques for the processing of food to ensure higher food safety with enhanced food quality: A review. *Discov. Food* **2024**, *4*, 20. [CrossRef]
207. Ghoshal, G. Comprehensive review on pulsed electric field in food preservation: Gaps in current studies for potential future research. *Heliyon* **2023**, *9*, e17532. [CrossRef] [PubMed]
208. Chacha, J.S.; Zhang, L.; Ofoedu, C.E.; Suleiman, R.A.; Dotto, J.M.; Roobab, U.; Agunbiade, A.O.; Duguma, H.T.; Mkojera, B.T.; Hossaini, S.M.; et al. Revisiting Non-Thermal Food Processing and Preservation Methods—Action Mechanisms, Pros and Cons: A Technological Update (2016–2021). *Foods* **2021**, *10*, 1430. [CrossRef] [PubMed]
209. Pipliya, S.; Kumar, S.; Babar, N.; Srivastav, P.P. Recent trends in non-thermal plasma and plasma activated water: Effect on quality attributes, mechanism of interaction and potential application in food & agriculture. *Food Chem. Adv.* **2023**, *2*, 100249. [CrossRef]
210. Safwa, S.M.; Ahmed, T.; Talukder, S.; Sarker, A.; Rana, M.R. Applications of non-thermal technologies in food processing Industries-A review. *J. Agric. Food Res.* **2023**, 100917. [CrossRef]
211. Duering, H.; Westerhoff, T.; Kipp, F.; Stein, C. Short-Wave Ultraviolet-Light-Based Disinfection of Surface Environment Using Light-Emitting Diodes: A New Approach to Prevent Health-Care-Associated Infections. *Microorganisms* **2023**, *11*, 386. [CrossRef] [PubMed]
212. Tchonkouang, R.D.; Lima, A.R.; Quintino, A.C.; Cristofoli, N.L.; Vieira, M.C. UV-C Light: A Promising Preservation Technology for Vegetable-Based Nonsolid Food Products. *Foods* **2023**, *12*, 3227. [CrossRef] [PubMed]
213. Onyeaka, H.; Passaretti, P.; Miri, T.; Al-Sharify, Z.T. The safety of nanomaterials in food production and packaging. *Curr. Res. Food Sci.* **2022**, *5*, 763–774. [CrossRef]
214. Adeyemi, J.O.; Fawole, O.A. Metal-Based Nanoparticles in Food Packaging and Coating Technologies: A Review. *Biomolecules* **2023**, *13*, 1092. [CrossRef]
215. Fadiji, T.; Rashvand, M.; Daramola, M.O.; Iwarere, S.A. A Review on Antimicrobial Packaging for Extending the Shelf Life of Food. *Processes* **2023**, *11*, 590. [CrossRef]
216. Lavilla, M.; Domingo-Calap, P.; Sevilla-Navarro, S.; Lasagabaster, A. Natural Killers: Opportunities and Challenges for the Use of Bacteriophages in Microbial Food Safety from the One Health Perspective. *Foods* **2023**, *12*, 552. [CrossRef] [PubMed]
217. Endersen, L.; Coffey, A. The use of bacteriophages for food safety. *Curr. Opin. Food Sci.* **2020**, *36*, 1–8. [CrossRef]
218. Bai, J.; Kim, Y.-T.; Ryu, S.; Lee, J.-H. Biocontrol and Rapid Detection of Food-Borne Pathogens Using Bacteriophages and Endolysins. *Front. Microbiol.* **2016**, *7*, 474. [CrossRef] [PubMed]
219. Sionek, B.; Szydłowska, A.; Küçükgöz, K.; Kołożyn-Krajewska, D. Traditional and New Microorganisms in Lactic Acid Fermentation of Food. *Fermentation* **2023**, *9*, 1019. [CrossRef]
220. Zapaśnik, A.; Sokołowska, B.; Bryła, M. Role of Lactic Acid Bacteria in Food Preservation and Safety. *Foods* **2022**, *11*, 1283. [CrossRef]
221. Martínez, J.M.; Abad, V.; Quílez, J.; Raso, J.; Cebrián, G.; Álvarez-Lanzarote, I. Pulsed Electric Fields (PEF) applications in the inactivation of parasites in food. *Trends Food Sci. Technol.* **2023**, *138*, 470–479. [CrossRef]

222. Nowosad, K.; Sujka, M.; Pankiewicz, U.; Kowalski, R. The application of PEF technology in food processing and human nutrition. *J. Food Sci. Technol.* **2021**, *58*, 397–411. [CrossRef]
223. Cavalcanti, R.N.; Balthazar, C.F.; Margalho, L.P.; Freitas, M.Q.; Sant'Ana, A.S.; Cruz, A.G. Pulsed electric field-based technology for microbial inactivation in milk and dairy products. *Curr. Opin. Food Sci.* **2023**, *54*, 101087. [CrossRef]
224. Lv, Y. Transitioning to sustainable energy: Opportunities, challenges, and the potential of blockchain technology. *Front. Energy Res.* **2023**, *11*, 1258044. [CrossRef]
225. Lima, V.; Pinto, C.A.; Saraiva, J.A. The dependence of microbial inactivation by emergent nonthermal processing technologies on pH and water activity. *Innov. Food Sci. Emerg. Technol.* **2023**, *89*, 103460. [CrossRef]
226. Yawut, N.; Mekwilai, T.; Vichiansan, N.; Braspaiboon, S.; Leksakul, K.; Boonyawan, D. Cold plasma technology: Transforming food processing for safety and sustainability. *J. Agric. Food Res.* **2024**, *18*, 101383. [CrossRef]
227. Ranjan, R.; Gupta, A.K.; Pandiselvam, R.; Chauhan, A.K.; Akhtar, S.; Jha, A.K.; Pratiksha; Ghosh, T.; Purohit, S.R.; Rather, M.A.; et al. Plasma treatment: An alternative and sustainable green approach for decontamination of mycotoxin in dried food products. *J. Agric. Food Res.* **2023**, *14*, 100867. [CrossRef]
228. Rațu, R.N.; Velescu, I.D.; Stoica, F.; Usturoi, A.; Arsenoaia, V.N.; Crivei, I.C.; Postolache, A.N.; Lipșa, F.D.; Filipov, F.; Florea, A.M.; et al. Application of Agri-Food By-Products in the Food Industry. *Agriculture* **2023**, *13*, 1559. [CrossRef]
229. Lorenzo-Leal, A.C.; Tam, W.; Kheyrandish, A.; Mohseni, M.; Bach, H. Antimicrobial Activity of Filtered Far-UVC Light (222 nm) against Different Pathogens. *BioMed Res. Int.* **2023**, *2023*, 2085140. [CrossRef] [PubMed]
230. Ansari, A.; Parmar, K.; Shah, M. A comprehensive study on decontamination of food-borne microorganisms by cold plasma. *Food Chem. Mol. Sci.* **2022**, *4*, 100098. [CrossRef] [PubMed]
231. Khan, J.; Tarar, S.M.; Gul, I.; Nawaz, U.; Arshad, M. Challenges of antibiotic resistance biofilms and potential combating strategies: A review. *3 Biotech* **2021**, *11*, 169. [CrossRef]
232. Juan-Polo, A.; Beltrán Sanahuja, A.; Monedero Prieto, M.; Sánchez Reig, C.; Valdés García, A.; Maestre Pérez, S.E. Impact of UV-light irradiation on sensory properties, volatile, fatty acid, and tocopherol composition of peanuts (*Arachis hypogaea* L.). *LWT* **2023**, *173*, 114247. [CrossRef]
233. Primožič, M.; Knez, Ž.; Leitgeb, M. (Bio)Nanotechnology in Food Science—Food Packaging. *Nanomaterials* **2021**, *11*, 292. [CrossRef]
234. Liao, W.; Badri, W.; Dumas, E.; Ghnimi, S.; Elaissari, A.; Saurel, R.; Gharsallaoui, A. Nanoencapsulation of Essential Oils as Natural Food Antimicrobial Agents: An Overview. *Appl. Sci.* **2021**, *11*, 5778. [CrossRef]
235. Allan, J.; Belz, S.; Hoeveler, A.; Hugas, M.; Okuda, H.; Patri, A.; Rauscher, H.; Silva, P.; Slikker, W.; Sokull-Kluettgen, B.; et al. Regulatory landscape of nanotechnology and nanoplastics from a global perspective. *Regul. Toxicol. Pharmacol.* **2021**, *122*, 104885. [CrossRef] [PubMed]
236. Zalewska-Piątek, B. Phage Therapy—Challenges, Opportunities and Future Prospects. *Pharmaceuticals* **2023**, *16*, 1638. [CrossRef]
237. Principi, N.; Silvestri, E.; Esposito, S. Advantages and Limitations of Bacteriophages for the Treatment of Bacterial Infections. *Front. Pharmacol.* **2019**, *10*, 513. [CrossRef] [PubMed]
238. Tomé, A.R.; Carvalho, F.M.; Teixeira-Santos, R.; Burmølle, M.; Mergulhão, F.J.M.; Gomes, L.C. Use of Probiotics to Control Biofilm Formation in Food Industries. *Antibiotics* **2023**, *12*, 754. [CrossRef] [PubMed]
239. Abouelela, M.E.; Helmy, Y.A. Next-Generation Probiotics as Novel Therapeutics for Improving Human Health: Current Trends and Future Perspectives. *Microorganisms* **2024**, *12*, 430. [CrossRef]
240. Wang, Z.; Li, L.; Wang, S.; Wei, J.; Qu, L.; Pan, L.; Xu, K. The role of the gut microbiota and probiotics associated with microbial metabolisms in cancer prevention and therapy. *Front. Pharmacol.* **2022**, *13*, 1025860. [CrossRef] [PubMed]
241. Bumunang, E.W.; Zaheer, R.; Niu, D.; Narvaez-Bravo, C.; Alexander, T.; McAllister, T.A.; Stanford, K. Bacteriophages for the Targeted Control of Foodborne Pathogens. *Foods* **2023**, *12*, 2734. [CrossRef]
242. Zolfaghari, H.; Khezerlou, A.; Ehsani, A.; Khosroushahi, A.Y. Detoxification of Aflatoxin B1 by Probiotic Yeasts and Bacteria Isolated From Dairy Products of Iran. *Adv. Pharm. Bull.* **2020**, *10*, 482–487. [CrossRef] [PubMed]
243. Dahiya, D.; Nigam, P.S. Biotherapy Using Probiotics as Therapeutic Agents to Restore the Gut Microbiota to Relieve Gastrointestinal Tract Inflammation, IBD, IBS and Prevent Induction of Cancer. *Int. J. Mol. Sci.* **2023**, *24*, 5748. [CrossRef] [PubMed]
244. Khan, M.A.S.; Rahman, S.R. Use of Phages to Treat Antimicrobial-Resistant *Salmonella* Infections in Poultry. *Vet. Sci.* **2022**, *9*, 438. [CrossRef] [PubMed]
245. Andrezal, M.; Oravcova, L.; Kadličekova, V.; Ozaee, E.; Elnwrani, S.; Bugala, J.; Markuskova, B.; Kajsik, M.; Drahovska, H. Characterization and the host specificity of Pet-CM3-4, a new phage infecting *Cronobacter* and *Enterobacter* strains. *Virus Res.* **2023**, *324*, 199025. [CrossRef] [PubMed]
246. Pasdar, N.; Mostashari, P.; Greiner, R.; Khelfa, A.; Rashidinejad, A.; Eshpari, H.; Vale, J.M.; Gharibzahedi, S.M.T.; Roohinejad, S. Advancements in Non-Thermal Processing Technologies for Enhancing Safety and Quality of Infant and Baby Food Products: A Review. *Foods* **2024**, *13*, 2659. [CrossRef] [PubMed]
247. Mitra, S.; Chakraborty, A.J.; Tareq, A.M.; Emran, T.B.; Nainu, F.; Khusro, A.; Idris, A.M.; Khandaker, M.U.; Osman, H.; Alhumaydhi, F.A.; et al. Impact of heavy metals on the environment and human health: Novel therapeutic insights to counter the toxicity. *J. King Saud Univ.—Sci.* **2022**, *34*, 101865. [CrossRef]

248. Hernández-Cortez, C.; Palma-Martínez, I.; Gonzalez-Avila, L.U.; Guerrero-Mandujano, A.; Solís, R.C.; Castro-Escarpulli, G. Food Poisoning Caused by Bacteria (Food Toxins). In *Poisoning—From Specific Toxic Agents to Novel Rapid and Simplified Techniques for Analysis*; IntechOpen: Rijeka, Croatia, 2017.
249. Esheli, M.; Thissera, B.; El-Seedi, H.R.; Rateb, M.E. Fungal Metabolites in Human Health and Diseases—An Overview. *Encyclopedia* **2022**, *2*, 1590–1601. [CrossRef]
250. Williams, J.H.; Phillips, T.D.; Jolly, P.E.; Stiles, J.K.; Jolly, C.M.; Aggarwal, D. Human aflatoxicosis in developing countries: A review of toxicology, exposure, potential health consequences, and interventions. *Am. J. Clin. Nutr.* **2004**, *80*, 1106–1122. [CrossRef] [PubMed]
251. Xie, Y.; Geng, Y.; Yao, J.; Ji, J.; Chen, F.; Xiao, J.; Hu, X.; Ma, L. N-nitrosamines in processed meats: Exposure, formation and mitigation strategies. *J. Agric. Food Res.* **2023**, *13*, 100645. [CrossRef]
252. Corsalini, M.; Inchingolo, F.; Dipalma, G.; Wegierska, A.E.; Charitos, I.A.; Potenza, M.A.; Scarano, A.; Lorusso, F.; Inchingolo, A.D.; Montagnani, M.; et al. Botulinum Neurotoxins (BoNTs) and Their Biological, Pharmacological, and Toxicological Issues: A Scoping Review. *Appl. Sci.* **2021**, *11*, 8849. [CrossRef]
253. Poulain, B.; Popoff, M.R. Why Are Botulinum Neurotoxin-Producing Bacteria So Diverse and Botulinum Neurotoxins So Toxic? *Toxins* **2019**, *11*, 34. [CrossRef]
254. Navale, V.; Vamkudoth, K.R.; Ajmera, S.; Dhuri, V. Aspergillus derived mycotoxins in food and the environment: Prevalence, detection, and toxicity. *Toxicol. Rep.* **2021**, *8*, 1008–1030. [CrossRef] [PubMed]
255. Kumar, P.; Mahato, D.K.; Kamle, M.; Mohanta, T.K.; Kang, S.G. Aflatoxins: A Global Concern for Food Safety, Human Health and Their Management. *Front. Microbiol.* **2017**, *7*, 2170. [CrossRef]
256. Kreling, V.; Falcone, F.H.; Kehrenberg, C.; Hensel, A. *Campylobacter* sp.: Pathogenicity factors and prevention methods—New molecular targets for innovative antivirulence drugs? *Appl. Microbiol. Biotechnol.* **2020**, *104*, 10409–10436. [CrossRef]
257. Hamaideh, S.; Olaimat, A.; Al-Holy, M.; Ababneh, A.; Shahbaz, H.; Abughoush, M.; Al-Nabulsi, A.; Osaili, T.; Ayyash, M.; Holley, R. The Influence of Technological Shifts in the Food Chain on the Emergence of Foodborne Pathogens: An Overview. *Appl. Microbiol.* **2024**, *4*, 594–606. [CrossRef]
258. Omojokun, J. Regulation and Enforcement of Legislation on Food Safety in Nigeria. In *Mycotoxin and Food Safety in Developing Countries*; IntechOpen: Rijeka, Croatia, 2013.
259. De Oliveira Mota, J.; Boué, G.; Prévost, H.; Maillet, A.; Jaffres, E.; Maignien, T.; Arnich, N.; Sanaa, M.; Federighi, M. Environmental monitoring program to support food microbiological safety and quality in food industries: A scoping review of the research and guidelines. *Food Control* **2021**, *130*, 108283. [CrossRef]
260. Faour-Klingbeil, D.; Todd, E.C.D. Prevention and Control of Foodborne Diseases in Middle-East North African Countries: Review of National Control Systems. *Int. J. Environ. Res. Public Health* **2019**, *17*, 70. [CrossRef] [PubMed]
261. Dama, T.I.; Loki, O.; Fitawek, W.; Mpuzu, S.M. Food policy analyses and prioritisation of food systems to achieve safer food for South Africa. *Appl. Food Res.* **2024**, *4*, 100476. [CrossRef]
262. Thapliyal, D.; Karale, M.; Diwan, V.; Kumra, S.; Arya, R.K.; Verros, G.D. Current Status of Sustainable Food Packaging Regulations: Global Perspective. *Sustainability* **2024**, *16*, 5554. [CrossRef]
263. Adaku, A.A.; Egyir, I.S.; Gadegbeku, C.; Kunadu, A.P.-H.; Amanor-Boadu, V.; Laar, A. Barriers to ensuring and sustaining street food safety in a developing economy. *Heliyon* **2024**, *10*, e32190. [CrossRef]
264. Rahman, T.; Deb, N.; Alam, M.Z.; Moniruzzaman, M.; Miah, M.S.; Horaira, M.A.; Kamal, R. Navigating the contemporary landscape of food waste management in developing countries: A comprehensive overview and prospective analysis. *Heliyon* **2024**, *10*, e33218. [CrossRef] [PubMed]
265. Yu, H.; Song, Y.; Lv, W.; Liu, D.; Huang, H. Food safety risk assessment and countermeasures in China based on risk matrix method. *Front. Sustain. Food Syst.* **2024**, *8*, 1351826. [CrossRef]
266. Moyo, P.; Moyo, E.; Mangoya, D.; Mhango, M.; Mashe, T.; Imran, M.; Dzinamarira, T. Prevention of antimicrobial resistance in sub-Saharan Africa: What has worked? What still needs to be done? *J. Infect. Public Health* **2023**, *16*, 632–639. [CrossRef] [PubMed]
267. Leahy, E.; Mutua, F.; Grace, D.; Lambertini, E.; Thomas, L.F. Foodborne zoonoses control in low- and middle-income countries: Identifying aspects of interventions relevant to traditional markets which act as hurdles when mitigating disease transmission. *Front. Sustain. Food Syst.* **2022**, *6*, 913560. [CrossRef]
268. Tohonon, A.C.; Ouétchéhou, R.; Hounsou, M.; Zannou, O.; Dabadé, D.S. Food hygiene in Sub-Saharan Africa: A focus on catering services. *Food Control* **2024**, *168*, 110938. [CrossRef]
269. Wu, D.; Elliott, C.; Wu, Y. Food Safety Strategies: The One Health Approach to Global Challenges and China's Actions. *China CDC Wkly.* **2021**, *3*, 507–513. [CrossRef] [PubMed]
270. Lee, J.-G.; Lee, Y.; Kim, C.S.; Han, S.B. Codex Alimentarius commission on ensuring food safety and promoting fair trade: Harmonization of standards between Korea and codex. *Food Sci. Biotechnol.* **2021**, *30*, 1151–1170. [CrossRef] [PubMed]
271. Vågsholm, I.; Arzoomand, N.S.; Boqvist, S. Food Security, Safety, and Sustainability—Getting the Trade-Offs Right. *Front. Sustain. Food Syst.* **2020**, *4*, 16. [CrossRef]
272. Ellahi, R.M.; Wood, L.C.; Bekhit, A.E.-D.A. Blockchain-Driven Food Supply Chains: A Systematic Review for Unexplored Opportunities. *Appl. Sci.* **2024**, *14*, 8944. [CrossRef]
273. Radu, E.; Dima, A.; Dobrota, E.M.; Badea, A.-M.; Madsen, D.Ø.; Dobrin, C.; Stanciu, S. Global trends and research hotspots on HACCP and modern quality management systems in the food industry. *Heliyon* **2023**, *9*, e18232. [CrossRef] [PubMed]

274. Njoagwuani, E.I.; Onyeaka, H.; Mazi, I.M.; Akegbe, H.; Oladunjoye, I.O.; Ochulor, C.E.; Omotosho, A.D.; Odeyemi, O.A.; Nwaiwu, O.; Tamasiga, P. Food safety in vulnerable populations: A perspective on the challenges and solutions. *FASEB J.* **2023**, *37*, e22872. [CrossRef] [PubMed]
275. Farber, J.M.; Zwietering, M.; Wiedmann, M.; Schaffner, D.; Hedberg, C.W.; Harrison, M.A.; Hartnett, E.; Chapman, B.; Donnelly, C.W.; Goodburn, K.E.; et al. Alternative approaches to the risk management of *Listeria monocytogenes* in low risk foods. *Food Control* **2021**, *123*, 107601. [CrossRef]
276. Gichohi-Wainaina, W.N.; Kimanya, M.; Muzanila, Y.C.; Kumwenda, N.C.; Msere, H.; Rashidi, M.; Mponda, O.; Okori, P. Aflatoxin Contamination, Exposure among Rural Smallholder Farming Tanzanian Mothers and Associations with Growth among Their Children. *Toxins* **2023**, *15*, 257. [CrossRef]
277. Githang'a, D.; Wangia, R.N.; Mureithi, M.W.; Wandiga, S.O.; Mutegi, C.; Ogutu, B.; Agweyu, A.; Wang, J.-S.; Anzala, O. The effects of aflatoxin exposure on Hepatitis B-vaccine induced immunity in Kenyan children. *Curr. Probl. Pediatr. Adolesc. Health Care* **2019**, *49*, 117–130. [CrossRef] [PubMed]
278. Pakbin, B.; Amani, Z.; Rahimi, Z.; Najafi, S.; Familsatarian, B.; Khakpoor, A.; Brück, W.M.; Brück, T.B. Prevalence of Foodborne Bacterial Pathogens and Antibiotic Resistance Genes in Sweets from Local Markets in Iran. *Foods* **2023**, *12*, 3645. [CrossRef] [PubMed]
279. Barlow, S.M.; Boobis, A.R.; Bridges, J.; Cockburn, A.; Dekant, W.; Hepburn, P.; Houben, G.F.; König, J.; Nauta, M.J.; Schuermans, J.; et al. The role of hazard- and risk-based approaches in ensuring food safety. *Trends Food Sci. Technol.* **2015**, *46*, 176–188. [CrossRef]
280. Nordhagen, S.; Lamberlini, E.; DeWaal, C.S.; McClafferty, B.; Neufeld, L.M. Integrating nutrition and food safety in food systems policy and programming. *Glob. Food Sec.* **2022**, *32*, 100593. [CrossRef]
281. Smith, B.; Fazil, A. How will climate change impact microbial foodborne disease in Canada? *Can. Commun. Dis. Rep.* **2019**, *45*, 108–113. [CrossRef] [PubMed]
282. Opoku, S.K.; Filho, W.L.; Hubert, F.; Adejumo, O. Climate change and health preparedness in Africa: Analysing trends in six African countries. *Int. J. Environ. Res. Public Health* **2021**, *18*, 4672. [CrossRef]
283. Massomo, S.M.S. Aspergillus flavus and aflatoxin contamination in the maize value chain and what needs to be done in Tanzania. *Sci. Afr.* **2020**, *10*, e00606. [CrossRef]
284. Abdalla, T.; Al-Rumaithi, H.; Osaili, T.M.; Hasan, F.; Obaid, R.S.; Abusheleibi, A.; Ayyash, M.M. Prevalence, Antibiotic-Resistance, and Growth Profile of *Vibrio* spp. Isolated From Fish and Shellfish in Subtropical-Arid Area. *Front. Microbiol.* **2022**, *13*, 861547. [CrossRef]
285. Kos, J.; Anić, M.; Radić, B.; Zadravec, M.; Janić Hajnal, E.; Pleadin, J. Climate Change—A Global Threat Resulting in Increasing Mycotoxin Occurrence. *Foods* **2023**, *12*, 2704. [CrossRef] [PubMed]
286. Shaji, S.; Selvaraj, R.K.; Shanmugasundaram, R. *Salmonella* Infection in Poultry: A Review on the Pathogen and Control Strategies. *Microorganisms* **2023**, *11*, 2814. [CrossRef]
287. Mahato, D.K.; Lee, K.E.; Kamle, M.; Devi, S.; Dewangan, K.N.; Kumar, P.; Kang, S.G. Aflatoxins in Food and Feed: An Overview on Prevalence, Detection and Control Strategies. *Front. Microbiol.* **2019**, *10*, 2266. [CrossRef] [PubMed]
288. Simões, D.; de Andrade, E.; Sabino, R. Fungi in a One Health Perspective. *Encyclopedia* **2023**, *3*, 900–918. [CrossRef]
289. Salam, M.A.; Al-Amin, M.Y.; Salam, M.T.; Pawar, J.S.; Akhter, N.; Rabaan, A.A.; Alqumber, M.A.A. Antimicrobial Resistance: A Growing Serious Threat for Global Public Health. *Healthcare* **2023**, *11*, 1946. [CrossRef] [PubMed]
290. Oon, Y.-L.; Oon, Y.-S.; Ayaz, M.; Deng, M.; Li, L.; Song, K. Waterborne pathogens detection technologies: Advances, challenges, and future perspectives. *Front. Microbiol.* **2023**, *14*, 1286923. [CrossRef] [PubMed]
291. Choi, J.R.; Yong, K.W.; Choi, J.Y.; Cowie, A.C. Emerging Point-of-care Technologies for Food Safety Analysis. *Sensors* **2019**, *19*, 817. [CrossRef]
292. Maftei, N.-M.; Raileanu, C.R.; Balta, A.A.; Ambrose, L.; Boev, M.; Marin, D.B.; Lisa, E.L. The Potential Impact of Probiotics on Human Health: An Update on Their Health-Promoting Properties. *Microorganisms* **2024**, *12*, 234. [CrossRef]
293. Petrariu, O.-A.; Barbu, I.C.; Niculescu, A.-G.; Constantin, M.; Grigore, G.A.; Cristian, R.-E.; Mihaescu, G.; Vrancianu, C.O. Role of probiotics in managing various human diseases, from oral pathology to cancer and gastrointestinal diseases. *Front. Microbiol.* **2024**, *14*, 1296447. [CrossRef] [PubMed]
294. Średnicka, P.; Juszczuk-Kubiak, E.; Wójcicki, M.; Akimowicz, M.; Roszko, M.Ł. Probiotics as a biological detoxification tool of food chemical contamination: A review. *Food Chem. Toxicol.* **2021**, *153*, 112306. [CrossRef]
295. Duchenne-Moutien, R.A.; Neetoo, H. Climate Change and Emerging Food Safety Issues: A Review. *J. Food Prot.* **2021**, *84*, 1884–1897. [CrossRef]
296. Pop, O.L.; Suharoschi, R.; Gabbianelli, R. Biodetoxification and Protective Properties of Probiotics. *Microorganisms* **2022**, *10*, 1278. [CrossRef]
297. Rafeeq, H.; Afsheen, N.; Rafique, S.; Arshad, A.; Intisar, M.; Hussain, A.; Bilal, M.; Iqbal, H.M.N. Genetically engineered microorganisms for environmental remediation. *Chemosphere* **2023**, *310*, 136751. [CrossRef] [PubMed]
298. Bhatlawande, A.R.; Ghatge, P.U.; Shinde, G.U.; Anushree, R.K.; Patil, S.D. Unlocking the future of smart food packaging: Biosensors, IoT, and nano materials. *Food Sci. Biotechnol.* **2024**, *33*, 1075–1091. [CrossRef] [PubMed]
299. Rodrigues, C.; Souza, V.G.L.; Coelhoso, I.; Fernando, A.L. Bio-Based Sensors for Smart Food Packaging—Current Applications and Future Trends. *Sensors* **2021**, *21*, 2148. [CrossRef] [PubMed]

300. Sionek, B.; Szydłowska, A.; Kołożyn-Krajewska, D. The Role of Microorganisms and Their Antibacterial Compounds in Food Biopreservation. *Appl. Sci.* **2024**, *14*, 5557. [CrossRef]
301. Barcenilla, C.; Ducic, M.; López, M.; Prieto, M.; Álvarez-Ordóñez, A. Application of lactic acid bacteria for the biopreservation of meat products: A systematic review. *Meat Sci.* **2022**, *183*, 108661. [CrossRef] [PubMed]
302. Bergwerff, A.A.; Debast, S.B. Modernization of Control of Pathogenic Micro-Organisms in the Food-Chain Requires a Durable Role for Immunoaffinity-Based Detection Methodology—A Review. *Foods* **2021**, *10*, 832. [CrossRef]

Disclaimer/Publisher's Note: The statements, opinions and data contained in all publications are solely those of the individual author(s) and contributor(s) and not of MDPI and/or the editor(s). MDPI and/or the editor(s) disclaim responsibility for any injury to people or property resulting from any ideas, methods, instructions or products referred to in the content.

Article

Ultrasound-Assisted Extraction Followed by Inductively Coupled Plasma Mass Spectrometry and Multivariate Profiling of Rare Earth Elements in Coffee

Aleksandra Savić [1], Jelena Mutić [2], Milica Lučić [3], Jelena Vesković [4], Andrijana Miletić [4] and Antonije Onjia [4,*]

1. Anahem Laboratory, Mocartova 10, 11160 Belgrade, Serbia
2. Department of Analytical Chemistry, Faculty of Chemistry, University of Belgrade, 11158 Belgrade, Serbia
3. Innovation Center of the Faculty of Technology and Metallurgy, 11120 Belgrade, Serbia
4. Faculty of Technology and Metallurgy, University of Belgrade, 11120 Belgrade, Serbia
* Correspondence: onjia@tmf.bg.ac.rs

Abstract: A rapid and efficient ultrasound-assisted extraction (UAE) procedure followed by inductively coupled plasma mass spectrometry (ICP-MS) was developed for the determination of 14 rare earth elements (REEs) (La, Ce, Pr, Nd, Sm, Eu, Gd, Tb, Dy, Ho, Er, Tm, Yb, Lu), along with yttrium (Y) and scandium (Sc), in coffee samples. The method was validated using certified reference material (NIST SRM 1547), recovery tests at four fortification levels, and comparisons with microwave-assisted digestion (MAD). Excellent accuracy and precision were achieved, with recovery rates ranging from 80.1% to 112% and relative standard deviations (RSD%) below 14%. Limits of detection (LODs) ranged from 0.2 ng/kg (Yb) to 0.16 µg/kg (Nd). Total REE concentrations varied between 8.3 µg/kg and 1.1 mg/kg, with the highest individual mean concentrations (µg/kg) observed for Ce (11.7), La (6.0), and Sc (4.7). The lowest individual mean concentrations (µg/kg) were for Ho (0.16), Lu (0.066), and Tm (0.063). Multivariate analysis of REE profiles from 92 coffee samples collected in Serbia revealed clear distinctions between ground roasted and instant coffees, as well as between different surrogate blends. This study indicated that the determination of coffee's geographical origin was not possible due to the diverse types, blends, and additives. However, differences in REE profiles suggest potential classification based on variety. REEs pose a negligible health risk to coffee consumers, with HI values ranging from 4.7×10^{-8} to 6.3×10^{-6} and TCR ranging from 2.6×10^{-14} to 3.5×10^{-12}.

Keywords: Arabica; Canephora; surrogate; REEs; inductively coupled plasma mass spectrometry; positive matrix factorization

1. Introduction

Coffee most likely originated from the northern regions of Africa, and today, it is commercially cultivated in approximately 70 countries worldwide. These countries span Central and South America, the Caribbean, Africa, and Asia [1–3]. The consumption and economic cultivation of coffee crops began in the 15th century [1,4]. Generally, two main species of coffee plants are grown: Arabica (*Coffea arabica*), which has a lower caffeine content and accounts for 70% of all coffee production, and Robusta (*Coffea canephora*), which has a significantly higher caffeine content and accounts for 30% of all coffee production [3,5]. The chemical profile and quality of coffee (including flavor, taste, aroma, and style) are influenced by several factors, including the coffee species, soil, field practices, climate, and processing methodology (roasting) [6–8].

Rare earth elements (REEs) include 17 elements that encompass 15 lanthanoids, as well as scandium (Sc), and yttrium (Y). The lanthanide series (from Z = 57 to Z = 71) includes the following elements: lanthanum (La), cerium (Ce), praseodymium (Pr), neodymium (Nd), promethium (Pm), samarium (Sm), europium (Eu), gadolinium (Gd), terbium (Tb), dysprosium (Dy), holmium (Ho), erbium (Er), thulium (Tm), ytterbium (Yb), and lutetium (Lu) [9,10]. These elements are non-essential to living organisms and have low to moderate toxicity [11,12]. REEs are usually classified as light REEs (LRREs) from La to Sm and heavy REEs (HREEs) from Eu to Lu, including Y [13,14]. Scandium is not classified into any sub-group because of its small ionic radius [15,16].

REEs have applications in different industries, as well as in medicine and agriculture. They are used not only for producing high-tech products but also in traditional industries. In agriculture, these elements are also used as fertilizers, plant growth promoters, and feed additives for livestock, poultry, and aquaculture [9,12,16]. In nature, they occur in the form of minerals, and 95% of these elements are found in bastnasite, monazite, and xenotime [10]. Under natural conditions, the availability of REEs in the environment (soil [17], water [18], atmosphere [19], and plants [20]) is low. However, REE mining processes and rapidly growing REE usage in modern industry and everyday life can increase REE concentrations and pose a threat to environmental and human health [9,21,22].

Most studies have shown that REEs are present at low concentrations in most food products [23,24]. The chemical composition of food is directly influenced by the soil in which it is grown. Therefore, REE content can serve as a means of authenticating various food products. In previous research, REEs have provided information on the geographic origins of wine [25], lentils [26], honey [27], and *Ruditapes philippinarum* (clam) [28]. To trace the origins of green and roasted coffee, several studies have investigated the content of REEs alongside other elements [3,29,30].

Concentrations of REEs in soils, water, and food are usually in traces, so analyses of these elements in food matrices are usually performed using inductively coupled plasma mass spectrometry (ICP-MS). This analytical technique is most commonly used for determining REEs in biological samples, including various foodstuffs, such as tomato plants [31], olive oils [32], meat and dairy products [33], pistachios [34], coffee [29,35], etc. Among the analytical methods for authenticating the geographical origin of food, ICP-MS stands out as a leading technique due to its robustness, accuracy, and sensitivity in detecting inorganic elements in food samples [36]. In general, food samples are usually prepared by acid digestion using concentrated acids. This type of sample preparation has several drawbacks, such as reducing sensitivity caused by a large volume dilution before instrumental analysis, and it can also influence the precision of element determination due to possible contamination. Ultrasound-assisted extraction (UAE) is a promising alternative to microwave-assisted digestion (MAD). The primary advantages of the UAE method are shorter sample preparation time (extraction), the use of diluted acids (which reduce waste and eliminate the need for dilution), and the application of lower temperatures and pressures during extraction. All these factors contribute to lower costs, increased safety, simplified sample preparation, and give more efficient and cleaner method [37–39]. Lima et al. (2000) compared UAE and MAD for the determination of Cd, Cu, and Pb in biological and sediment samples, finding that UAE required less solvent and energy, aligning with green chemistry principles [40]. Furthermore, Fotouh R. Mansour et al. (2024) utilized the Modified GAPI (MoGAPI) tool to evaluate the greenness of several analytical methods, including UAE and MAD methods [41]. Their findings indicated that the UAE method generally scores higher on greenness metrics such as the Analytical Eco-Scale (AES) and Analytical GREEnness Metric (AGREE) due to its lower solvent consumption and reduced waste generation. Although specific studies applying green metric scales to UAE and MAD are limited, the general principles suggest that methods requiring fewer hazardous reagents and less energy consumption score higher on green metric scales. The main novelty of this study lies in the application of UAE, in combination with ICP-MS, for determining REEs in various types of coffee.

The objective of the present study was to develop an accurate and precise method for determining REEs in commercial ground roasted and instant coffee. Additionally, the study aimed to assess the REE profiles of coffee originating from different regions worldwide. A total of 92 coffee samples were analyzed using two sample preparation procedures (MAD and UAE). These procedures were validated using certified reference material (CRM). To the best of our knowledge, this study is the first to investigate the REE concentrations in coffee samples by combining UAE and ICP-MS analysis and to assess REE patterns using multivariate methods.

2. Materials and Methods

2.1. Samples and Reagents

The samples of various coffees and coffee surrogates (n = 92) were obtained from the local market in Belgrade, Serbia, in May 2024. Samples included 37 ground roasted coffees, 12 ground roasted coffees in capsules, 24 instant coffees with additives, 15 instant coffees, and 4 coffee surrogates. A detailed description of the analyzed samples is provided in Supplementary Materials Table S1.

All reagents used for digesting coffee samples were of analytical reagent grade. Hydrogen peroxide (30%, w/w) was provided by Lachner (Neratovice, Czech Republic), and nitric acid (67–70%, v/v) was supplied by Fisher Chemical (Merelbeke, Belgium). All solutions were prepared and diluted using Milli-Q water (Millipore, Bedford, MA, USA) with a resistivity of 18.2 MΩ·cm. Internal standard (Rhodium, Rh) and multi-element solutions (10 mg/L each) were purchased from AccuStandard (New Haven, CN, USA). All glassware and polypropylene tubes used in the analysis were soaked in 10% HNO_3 overnight, rinsed with ultrapure water, and dried before use.

2.2. Samples Digestion and Instrument Analysis

In this study, coffee samples were prepared using two methods for element analysis: the conventional one, MAD, and the alternative one, UAE.

For MAD, a mass of 0.250 g was weighed and transferred into a polytetrafluoroethylene (PTFE) tube. This step was followed by the addition of 4 mL of 67–70% HNO_3 and 1 mL of 30% H_2O_2. The tubes containing the samples were then subjected to a digestion process using a microwave oven MARS 5 (CEM, Matthews, NC, USA). The digestion regime was 25 min to 120 psi and held for 10 min (operating parameters were power 1600 W, pressure sensor control). After cooling, the mixtures were added to a final volume of 25 mL ultrapure water. The solutions were centrifuged, diluted five times, and measured at ICP-MS.

The UAE procedure was performed following the method of Gohlke et al. (2024) [42] with modifications. A precise mass of 0.250 g was weighed and transferred to a 15 mL Falcon polypropylene (PP) tube featuring screw caps (Biosigma S.p.A, Cona, Italy). Then, 10 mL of the extraction solution (1 M HNO_3) was added. Prior to the sonication, the sample was vortexed for 1 min; afterward, sonication was conducted for 15 min in an ultrasonic bath (Elmasonic S15H, Elma Schmidbauer GmbH, Singen, Germany) operating at 95 W of effective ultrasonic power, a frequency of 37 kHz, and a temperature of 80 °C. The resulting mixtures were then centrifuged at 3000 rpm for 3 min. The obtained supernatants were diluted with ultrapure water (18.2 MΩ·cm) to 25 mL in a polymethyl-pentene (PMP) volumetric flask (class A, VITLAB GmbH, Großostheim, Germany). The solutions were stored at 4 °C until analysis.

The analysis of 14 REEs (La, Ce, Pr, Nd, Sm, Eu, Gd, Tb, Dy, Ho, Er, Tm, Yb, and Lu), Y, and Sc was performed using an inductively coupled plasma mass spectrometer (Thermo Scientific iCAP TQ ICP-MS, Bremen, Germany. The correction for interferences caused by polyatomic ions and other unwanted species was made using the KED (Kinetic Energy Discrimination) mode for all elements. The instrument operating parameters are presented in Table 1.

Table 1. Instrument operating conditions for ICP-MS.

Parameter	Value
Forward power	1550 W
Nebulizer gas (Ar)	0.99 L/min
Auxiliary gas (Ar)	0.8 L/min
Cool gas flow	14 L/min
Collision cell gas (He)	4.5 mL/min
Sample uptake/wash time	45 s each
Dwell time	0.01 s
Repeats per sample	3
Pump speed	40 rpm

All coffee samples and NIST SRM 1547 were prepared and analyzed in their original state, without prior drying. The moisture content was subsequently determined (oven-based drying at 103 °C until a constant weight was reached, as described in ISO 11294 [43]), and results were expressed on a dry weight basis. Moisture content ranged from 1.1% to 4.3% for roasted ground coffees and from 1.2% to 4.7% for instant coffees.

2.3. Analytical Method Validation

External standard calibration combined with an internal standard (Rh) was employed to mitigate matrix effects and compensate for instrumental drift. Just to note that in this work, internal standard calibration was not used. Instead, the internal standard served to trigger recalibration of the ICP-MS instrument when a significant deviation in the internal standard signal occurred, below 80% or above 120% of the initial value. The calibration standards were prepared using the multi-element standard solution (PE-MECAL2-ASL-1, Accu Standard, New Heaven, CN, USA) and simulated matrix solutions that represent simulated digestions, containing element concentrations typical of those found in coffee as a matrix. A matrix solution containing Ca (1000 mg/L), Mg (1500 mg/L), K (15,000 mg/L), Na (300 mg/L), Cl (1400 mg/L), C (3.3%) as acetic acid, P (300 mg/L), S (200 mg/L), and F (200 mg/L). An internal standard was then added to reach a final concentration of 10 µg/L of Rh [7,44]. The method's accuracy was confirmed by analyzing the CRM of botanical material obtained from the National Institute of Standards and Technology (NIST, Gaithersburg, MD, USA): Standard Reference Material 1547 (NIST SRM 1547) Peach Leaves. Analyses and measurements of all samples and CRM were carried out in triplicate. Additionally, the recovery of REEs was determined using four levels of fortification (0.5; 5.0; 40, and 200 µg/kg). Specifically, two coffee samples (one ground roasted coffee and one instant coffee) were spiked after digestion. Three separate subsamples of the same coffee were spiked and analyzed three times. The homogeneity of the spiked samples was ensured by vortexing for 1 min. The indicative values of the REE concentrations were obtained by analyzing the coffee samples ten times. The limit of detection (LOD) and limit of quantification (LOQ) were calculated as three and ten times the standard deviation (SD) of six measurements of analytical blank divided by the slope of the calibration curve.

2.4. Multivariate Statistics

Several multivariate statistical methods were employed to analyze the data. Pearson correlation analysis, principal component analysis (PCA), and hierarchical cluster analysis (HCA) utilizing the R software package, version 4.1.2 were performed to assess the linear relationships between REEs, to reduce dimensionality while preserving the variability within the data, and to classify the coffee samples into clusters based on similarity, allowing the grouping of REEs with shared characteristics. Additionally, the positive matrix factorization (PMF) method, using the EPA PMF version 5.0 software, was employed to decompose the data matrix into source profiles and contributions, providing insights into the underlying factors influencing the REE content in coffee samples.

2.5. Health Risks from REE in Coffee

The potential adverse human health effects of REEs present in coffee for consumers were evaluated using the Human Health Risk Assessment (HRA) model developed by the United States Environmental Protection Agency (USEPA) [45]. The model estimates the hazard index (HI) and target cancer risk (TCR) for assessing health risks. All model equations and exposure factor values used in the model are presented in Tables S2 and S3. The reference dose (RfD) and cancer slope factor (CSF) for the assessed REEs were set at $0.02~\text{mg}\cdot\text{kg}^{-1}\cdot\text{bw}\cdot\text{day}^{-1}$ and $3.2 \times 10^{-12}~\text{kg}\cdot\text{bw}\cdot\text{day}\cdot\text{mg}^{-1}$ [46].

3. Results and Discussion

3.1. Method Development and Validation

The limits of detection (LODs), correlation coefficient (R^2), and calibration equations are presented in Table 2. The linearity of the calibration curves was satisfactory for all elements, with an R^2 value above 0.995. The five point calibration curves covered the concentration range from 1 to 25 µg/L. LOD values were in the range of 0.002 µg/kg (Yb) to 0.41 µg/kg (Sc).

Table 2. Calibration parameters for ICP-MS of REEs.

m/z	Element	Polyatomic Interferences	LOD, ng/L	R^2	Linear Equation
141	Pr	/	0.19	0.9962	$f(x) = 48.6 \times 10^3 x + 1.67$
146	Nd	BaO, RuO$_3$	1.38	0.9970	$f(x) = 13.6 \times 10^3 x + 1.33$
147	Sm	RuO$_3$	0.52	0.9999	$f(x) = 68.8 \times 10^3 x + 1.45$
153	Eu	BaO	0.19	0.9986	$f(x) = 33.8 \times 10^3 x + 2.19$
157	Gd	CeO, PrO, LaO, BF	0.50	0.9989	$f(x) = 69.1 \times 10^3 x + 0.70$
159	Tb	NdO, PrO	0.14	0.9981	$f(x) = 45.9 \times 10^3 x + 3.27$
163	Dy	NdO, SmO	0.62	0.9998	$f(x) = 11.3 \times 10^3 x + 0.33$
165	Ho	SmO	0.16	0.9964	$f(x) = 44.0 \times 10^3 x + 1.45$
166	Er	SmO, NdO	0.48	0.9955	$f(x) = 14.9 \times 10^3 x + 1.01$
169	Tm	EuO	0.04	0.9977	$f(x) = 41.0 \times 10^3 x + 0.45$
172	Yb	DyO, SmO, GdO	0.02	0.9997	$f(x) = 19.6 \times 10^3 x + 0.79$
175	Lu	GdO, TbO	0.12	0.9982	$f(x) = 40.7 \times 10^3 x + 3.40$
45	Sc	CO$_2$, SiO, BO$_2$, AlO, CaH, CHO$_2$, SiOH, N$_2$OH	4.09	0.9982	$f(x) = 16.4 \times 10^3 x + 1.02$
89	Y		0.91	0.9985	$f(x) = 38.5 \times 10^3 x + 4.70$
140	Ce	/	0.44	0.9987	$f(x) = 38.0 \times 10^3 x + 20.7$
139	La	/	0.30	0.9988	$f(x) = 44.1 \times 10^3 x + 37.3$

f(x): instrument response in cps (counts per second); x: analyte concentration in µg/L.

Microwave-assisted digestion methods are widely used to extract elements from various foodstuffs [47–49]. This method has previously been used by other researchers for the digestion of coffee samples to analyze REEs [3,29,30]. However, these methods often require a large amount of reagents, which can affect the measurement of elements present at low concentrations [37,50]. Ultrasound-assisted extraction offers a promising alternative with several advantages over conventional acid digestion methods, including reduced extraction time, the use of diluted acids (resulting in less waste and improved safety), simplicity, and lower operating costs [38,51]. Ultrasound waves enhance the extraction process via acoustic cavitation, which induces rapid and localized fluctuations in temperature and pressure, resulting in cell wall rupture [52]. In this context, we developed a UAE-based method for extracting REEs from coffee samples. To validate this method, a comparison of the UAE with MAD was performed, and a CRM was analyzed. The linear regressions between REE concentrations in MAD and UAE coffee samples ($n = 10$) are shown in Figure 1, while the measured concentrations are presented in Supplementary Materials Table S4. The results showed that the UAE procedure was an efficient and valid alternative to MAD, except for Sc and Ce, which had slightly lower extraction efficiency using the UAE procedure.

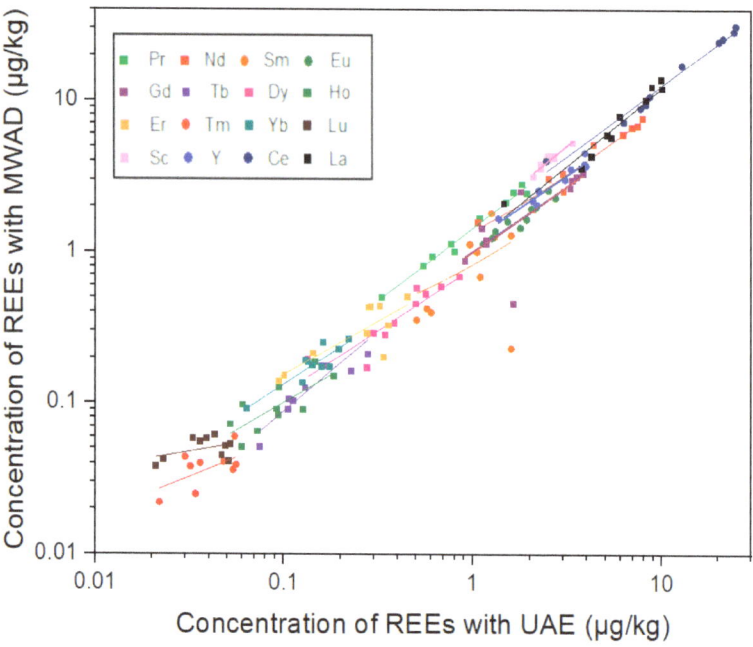

Figure 1. Linear regressions between REE concentrations in coffee after MAD and UAE ($n = 10$).

Figure 1 presents a linear regression analysis of REEs in coffee samples. The plot compares the concentration of REEs obtained using two different methods: MAD and UAE. The results of the linear regression analysis reveal varying degrees of correlation (Table S5) between concentrations. High correlation coefficients ($R^2 > 0.9$) were obtained for Pr, Nd, Sc, Y, Ce, and La, indicating strong predictability and reliability of the method. Moderate to strong correlations ($0.7 \leq R^2 \leq 0.9$) were observed for Eu, Gd, Tb, Dy, and Yb, suggesting that while the UAE method is fairly reliable, some variability remains unexplained. Lower correlations ($R^2 < 0.7$) were found for Ho, Er, Tm, Sm, and Lu, indicating weaker predictability, particularly in cases where the element concentrations were very low. These findings suggest that while UAE is a suitable alternative to MAD for most REEs, further refinement may be required for elements with low concentration levels to enhance the accuracy and reliability of the method.

The accuracy of the measurements was verified by preparing and analyzing the CRM in the same way as the coffee samples. Although NIST SRM 1547 is not coffee-specific, it provides a plant-based matrix similar to coffee. The CRM data are presented in Table S6. The percentage recovery was obtained in the range of 93.5–112%. The precision of the method was determined by the relative standard deviation (RSD), which ranged from 0.63% to 11.9%. The method validation showed that the obtained concentrations of Eu, Gd, Nd, Sm, Sc, Tb, and Yb were not significantly different from the CRM values, based on the 95% confidence intervals (CIs, Table S6). Significant differences were observed for Ce and La. Additionally, the accuracy of the method is confirmed by spiking experiments. The initial concentrations of spiked coffee samples (ground roasted coffee and instant coffee) are presented in Table S7. The recoveries of spiked coffee samples ranged from 80.1 to 108% for UAE experiments, while those for MAD experiments ranged from 79.3 to 113%. The relative standard deviation (RSD%) of spiking experiments ranged from 1.8 to 13.0% for UAE experiments and from 1.5 to 12.3% for MAD experiments (Supplementary Materials Table S8a–d).

3.2. REE Content in Coffee

The descriptive statistics for REEs in the studied coffee samples are presented in Figure 2 and Supplementary Table S9. Box plots were used to display the distribution of REEs, with concentrations presented on a logarithmic scale.

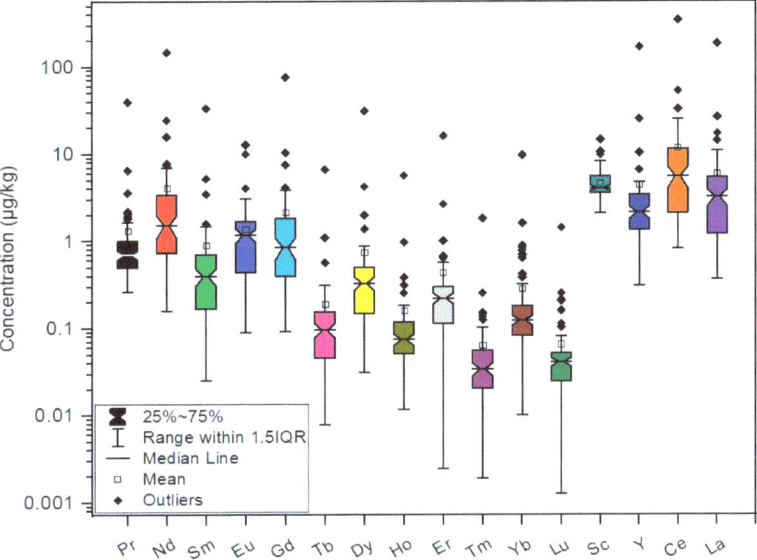

Figure 2. Box plots of REE concentrations in analyzed coffee samples.

The total REE content ranged from 8.35 to 1097 µg/kg. The highest total content was measured in the surrogate sample (100% chicory). Other samples containing surrogates also exhibited higher total REE contents than the pure coffee samples. The lowest REE content was observed in instant coffee with additives (Cappuccino Irish Cream). Generally, instant coffees and instant coffees with additives had lower total REE contents than other analyzed samples. Two samples from Rwanda and Peru had significantly lower total REE content than other ground roasted coffees. This indicates that geographical origin may influence the REE content.

The mean concentrations of REEs followed the order Ce (11.7 µg/kg) > La (5.96 µg/kg) > Sc (4.68 µg/kg) > Y (4.43 µg/kg) > Nd (4.11 µg/kg) > Gd (2.45 µg/kg) > Pr (1.33 µg/kg) > Eu (1.28 µg/kg) > Sm (1.09 µg/kg) > Dy (0.740 µg/kg) > Er (0.438 µg/kg) > Yb (0.286 µg/kg) > Tb (0.191 µg/kg) > Ho (0.161 µg/kg) > Lu (0.066 µg/kg) > Tm (0.063 µg/kg). REEs (Tb, Sm, Ho, Er, Dy, Tm, and Lu) were below the corresponding LOD values in 23 coffee samples (3 ground roasted, 18 instant, and 2 ground roasted coffees in capsules). The minimum and maximum concentrations of REEs align with the concentrations reported by Vezzulli et al. (2023) [30], except for the maximum concentration of Eu, which was higher in our study (Table 3). The maximum concentrations of Eu are in good agreement with the results of Barbosa et al. (2014) [29] and Messaoudi et al. (2018) [53]. On the other hand, the maximum Ce and La concentrations reported for coffee samples from India [30] and Brazil [35] were significantly higher than the maximum values obtained for our coffee samples. Our results for pure coffee samples agree with those of Santato et al. (2012) [3] for 62 raw green coffee samples originating from Central America, South America, Africa, and Asia. Compared with pure coffee samples, surrogates and coffee with surrogates (2% chickpeas) exhibited higher minimum and maximum values for all REEs. For all elements, skewness and kurtosis values were greater than 2, indicating that the distribution of data was not normal; the data were highly skewed to the right and too peaked. Before further analyses of the normalized data, the logarithm transformation was applied.

Table 3. The range of REE concentrations (µg/kg) in the coffee samples from other studies compared to the present study.

REEs	Vezzulli et al. (2023) [30]	de Oliveira Costa et al. (2024) [35] [1]	Barbosa et al. (2014) [29]	Messaoudi et al. (2018) [53]	Santato et al. (2012) [3]	Present Study $	Present Study #
Pr	<2.12–2.74 [2]; 8.96 [3]; 2.71 [5]; 13.61 [6]	3–200	-	-	0.16–3.38	0.27–3.58	0.89–39.7
Nd	<3.12–11.9 [1]; 38.2 [3]; 3.5 [4]; 13.2 [5]; 76.8 [6]; 3.9 [7]	10–700	-	-	0.38–12.1	<LOD–15.9	2.13–148
Sm	<3.99–9.26 [3]; 8.8 [6]	3–400	3.5–62	7.8–8.3 [a]; 4.0–4.3 [b]	0.20–2.36	<LOD–3.46	0.62–33.8
Eu	<3.93–4.3 [6]	1–10	1.3–15.1	nd [a]; nd–11.7 [b]	0.13–1.02	0.090–12.9	0.83–10.1
Gd	<3.71–7.57 [3]; 6.9 [6]	3–60	0.9–24.9	-	0.18–2.39	0.092–7.59	1.18–76.2
Tb	<3.25	1–10	0.07–10.6	-	-	<LOD–0.57	0.096–6.68
Dy	<3.94–7.4 [3]; 9.9 [6]	2–30	0.4–39.3	-	0.17–1.84	<LOD–1.98	0.42–31.3
Ho	<2.79	1–10	0.11–6.5	-	-	<LOD–0.38	0.072–5.69
Er	<2.66–4.0 [3]; 3.4 [6]	2–20	0.3–11.1	-	0.21–1.07	<LOD–1.00	0.25–16.1
Tm	<2.23	1–8	-	-	<LOD–0.21	<LOD–0.15	0.043–1.83
Yb	<3.82	2–10	0.1–9.2	-	<LOD–0.94	0.010–0.89	0.17–9.74
Lu	<2.46	3–20	0.13–0.72	-	-	<LOD–0.26	0.040–1.42
Ce	<2.55–23.2 [1]; 4.0 [2]; 64.2 [3]; 6.8 [4]; 28.5 [5]; 114.5 [6]; 9.4 [7]	10–2100	9.5–361	74–89 [a]; 490–510 [b]	0.73–36.2	0.81–32.9	7.58–347
La	<2.40–23.1 [1]; 4.9 [2]; 34.8 [3]; 5.3 [4]; 15.2 [5]; 85.9 [6]; 7.1 [7]	10–1000	3.2–122	35–36 [a]; 85–91 [b]	0.59–15.5	0.37–17.1	3.45–186
Sc	-	10–40	-	37–38 [a]; 66–68 [b]	-	2.09–10.6	3.81–14.5
Y	-	3–100	-	-	-	0.31–10.4	2.33–169

[1]—Brazil; [2]—Columbia; [3]—Panama; [4]—Costa Rica; [5]—Ethiopia; [6]—India; [7]—Indonesia; [a]—Arabica; [b]—Robusta. $—coffee samples; #—surrogates and coffee with the addition of surrogates (2% chickpeas or 10% barley).

3.3. Multivariate Analyses

Pearson correlation analysis was employed to investigate the relationships among the analyzed REEs (Figure 3, Supplementary Materials Table S10). Positive correlation coefficients indicate a positive relationship, whereas negative values signify a negative relationship. Typically, a strong correlation exists between two parameters if the coefficient falls within the range of 0.7 to 1.0, a moderate correlation between 0.5 and 0.7, and a weak correlation if the coefficient is less than 0.5. Most REEs exhibited positive correlations, except for Sc with Nd (r = −0.247), Sc with Sm (r = −0.172), Sc with Eu (r = −0.218), Sc with Tb (r = −0.036), Sc with Ce (r = −0.260), and Sc with La (r = −0.266). The correlation coefficient of Sc with other elements indicated weak positive correlations. Additionally, Eu, Lu, Tm, and Yb generally exhibited moderate or weak correlations with other elements, suggesting that they originated from sources different than the rest of the REEs. However, all other REEs showed strong (or closely to strong) correlations with each other.

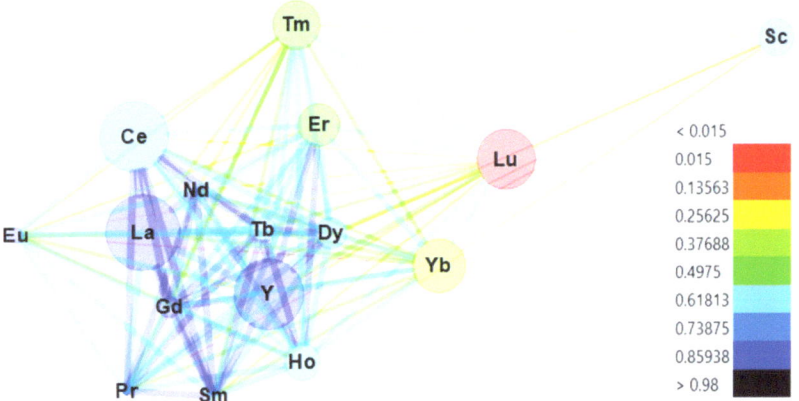

Figure 3. Pearson correlation network among concentrations of 16 different REEs in coffee samples (*n* = 92). Only positive correlations are represented: blue lines indicate strong positive correlations, light blue lines indicate moderate correlations, and the red line indicates weak positive correlations.

To differentiate the coffee samples, principal component analysis (PCA) was applied. The model extracted two components with eigenvalues greater than 1, which explained 83.6% of the total variance (Supplementary Materials Table S11). The first principal component was loaded with Pr, Nd, Sm, Gd, Tb, Dy, Ho, Er, Y, Ce, and La, accounting for 68.9% of the total variance. The second principal component explained 14.7% of the total variance and had a strong loading of Eu, Tm, Yb, Lu, and Sc. The score plots for the first two components are presented in Figure 4. The score plot shows the separation of coffees with additives on one side (orange and violet samples) and roasted ground coffees and roasted ground coffees in capsules on the other side (pink and blue samples). Two samples of coffees with 2% chickpeas (red star) were closest to the roasted ground coffee samples. On the other hand, the surrogates for coffee (red dots) were separated from all other coffee samples. The PCA results are important in distinguishing coffee samples from surrogate samples and identifying coffee with added surrogates. This analysis not only underscores the variations in composition but also aids in identifying potential adulteration in coffee products.

Hierarchical cluster analysis (HCA) revealed three clusters of coffee samples (Figure 5). The first cluster (blue) comprised 34.8% of the analyzed samples. This cluster mostly represented instant coffees and instant coffees with additives, which were separated into two sub-clusters. The second (green) and third clusters (red) comprised 38.0% and 27.2% of the analyzed samples, respectively. The second cluster mostly included ground roasted coffee and ground roasted coffee in capsules but also included several instant coffees and instant coffees with additives.

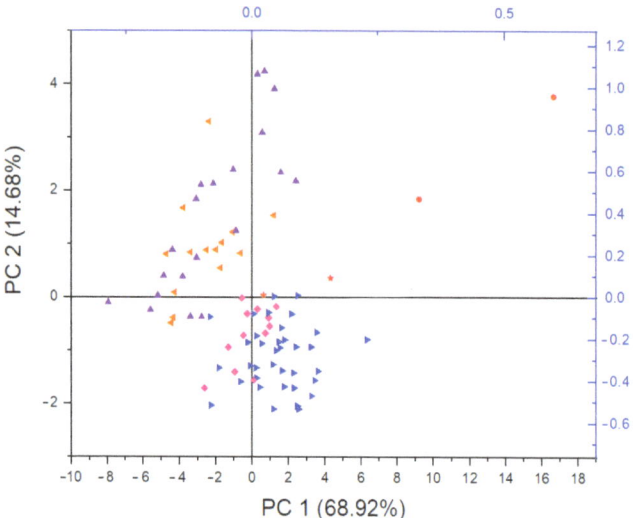

Figure 4. Principal component analysis (PCA). Score plot for 92 coffee samples based on rare earth element (REE) concentrations: blue ▶—ground roasted coffees; orange ◀—instant coffees; violet ▲—instant coffees with additives; pink ◆—ground roasted coffees in capsules; red ●—surrogates for coffee; red star—coffees with surrogates (2% chickpeas).

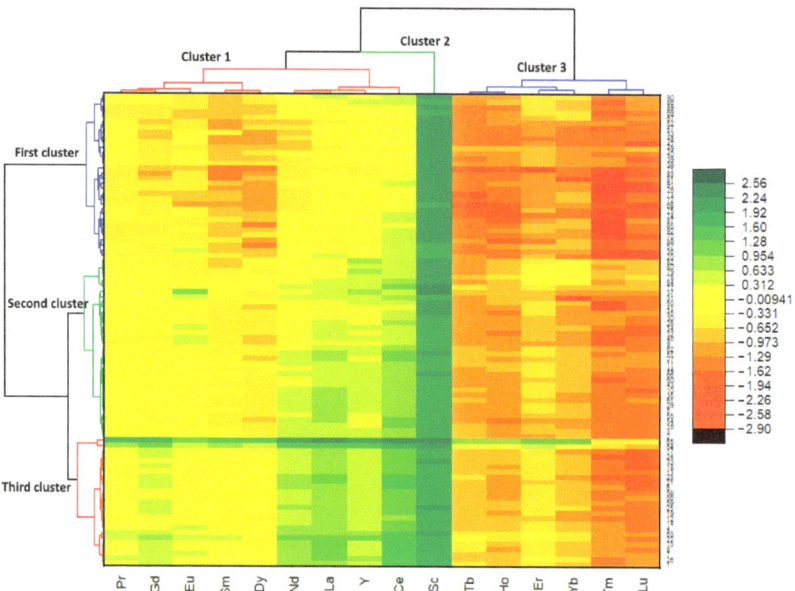

Figure 5. Heatmap dendrogram for the coffee samples and rare earth elements (REEs).

Instant coffees with additives from the second cluster generally contained chocolate as an ingredient. Additionally, 28.6% of the coffee in the second cluster was labeled with 100% Arabica, while 20% had a geographic origin. The last data are likely consistent with the fact that these samples generally have lower REE content, which is why they are positioned in the same cluster. The third cluster consisted of ground roasted coffee and two samples of ground roasted coffee with surrogates, one with 2% chickpeas and the other with 10% barley. Most samples in this cluster represented a blend of two coffee species, Arabica and Robusta, even in the case of coffees containing substitutes. This cluster also had a right sub-cluster that included only two samples that were coffee substitutes (100% chicory and 70% barley + 30% chicory).

The HCA identified three REE clusters. Cluster 1: Pr, Gd, Eu, Sm, Dy; Cluster 2: Nd, La, Y, Ce, Sc; Cluster 3: Tb, Ho, Er, Yb, Tm, Lu. Most samples showed concentrations in the order of Cluster 2 > Cluster 1 > Cluster 3, meaning Ce, La, Sc, Nd, and Y were most abundant, followed by Pr, Gd, Eu, Sm, and Dy, with Tb, Ho, Er, Yb, Tm, and Lu being the least abundant.

The positive matrix factorization method (PMF) is often used in environmental studies to characterize natural and anthropogenic pollution sources in soil [54], sediment [55], and water [18]. Alongside correlation analysis, PCA, HCA, and PMF can help identify different behaviors and potential sources of REEs in coffee samples. In our study, the PMF model was run 20 times, with the number of factors set between 3 and 5. The best results were obtained for three factors (Figure 6), yielding the smallest Q value. However, the coefficients of the observed and predicted values (R^2) varied. Most REEs had R^2 values above 0.81, except for Yb, Eu, and Sc, which had R^2 values of 0.69, 0.52, and 0.041, respectively.

Figure 6. PMF results illustrate the contribution of REEs to three different factors.

Factor 1 was characterized by Tm, Tb, and Er; Factor 2 by Lu, Sc, and Yb; and Factor 3 was associated with the remaining elements (Pr, Nd, Sm, Eu, Gd, Dy, Ho, Y, Ce, and La). Our classification of REEs based on the identified factors is quite similar to the classification of REEs using PCA in a study by de Oliveira Costa et al. (2024) [35]. The formation of a distinct group comprising Lu, Tm, and Yb was observed. Elements that stand out are the most significant for sample separation. Additionally, in Vezzulli et al.'s study (2023) [30], which explored the elemental profiling and origin of specialty and high-quality coffees, Tm was identified as the significant discriminant element among others. Previous research has noted that Lu and Yb exhibit the same trend in plants as in terrestrial matrices, meaning that plants reflect the soil composition [56]. Artificial fertilizers lead to an increase in REEs in soil and, consequently, in coffee. Special coffee varieties are generally expected to have lower REE concentrations due to careful cultivation and better control of artificial fertilizer addition [35]. Furthermore, an increase in REEs in final coffee samples can occur due to the addition of surrogates. A study from 2019 that investigated the REE content in tap water and beverages from fast-food franchises revealed that Coca-Cola beverages are characterized by a strong enrichment of Yb and Lu derived from syrup [57].

Based on our analysis, it is not possible to unambiguously determine the origin of the coffee samples, as they consist of various coffee types, blends, and samples with different additives. However, clear distinctions were observed in cases involving surrogate additives (e.g., barley, chickpea) and between most of the different coffee types (ground versus instant). Our findings also suggest the potential for classifying coffee samples based on their origins or varieties (e.g., Arabica versus Robusta). This highlights that the analysis of rare earth elements (REEs), combined with multivariate analysis techniques, could be valuable for authenticating unmodified or minimally processed coffee samples, providing a foundation for future research. Furthermore, the assessment of REE content could serve as a tool for detecting potential adulteration in coffee products.

3.4. Potential REE Hazards to Human Health

The potential health effects of REEs in coffee were assessed using the model previously applied for heavy metal(loid)s. This model incorporates ingestion exposure pathways. The calculated hazard quotient (HQ) values follow a decreasing order (mean) as follows: Ce (6.77×10^{-8}) > La (3.43×10^{-8}) > Sc (2.70×10^{-8}) > Y (2.55×10^{-8}) > Nd (2.37×10^{-8}) > Gd (1.25×10^{-8}) > Eu (8.03×10^{-9}) > Pr (7.67×10^{-9}) > Sm (5.26×10^{-9}) > Dy (4.26×10^{-9}) > Er (2.52×10^{-9}) > Yb (1.65×10^{-9}) > Tb (1.10×10^{-9}) > Ho (9.25×10^{-10}) > Lu (3.78×10^{-10}) > Tm (3.64×10^{-10}). The min–max HQ values for REEs in coffee are provided in Table 4. Based on the resulting HQ, HI, and TCR values, one may assume that REEs pose a negligible health risk to coffee consumers. A previous study by Savić et al. (2024) [7] evidenced that the adverse health effects of coffee are predominantly driven by the presence of HMs rather than REEs. The contribution of REEs to HI and TCR is presented in Figure 7. Scandium and Ce were responsible for 25.5% and 24.8% of non-carcinogenic and carcinogenic risks.

Table 4. Hazard quotients (HQs), hazard index (HI), and target cancer risk (TCR) for REEs in coffee.

Health Risk Index	REE	Mean	Min	Max
HQ	Ce	6.77×10^{-8}	4.69×10^{-9}	2.00×10^{-6}
	La	3.43×10^{-8}	2.11×10^{-9}	1.07×10^{-6}
	Sc	2.70×10^{-8}	1.20×10^{-8}	8.38×10^{-8}
	Y	2.55×10^{-8}	1.80×10^{-9}	9.74×10^{-7}
	Nd	2.37×10^{-8}	9.29×10^{-10}	8.50×10^{-7}
	Gd	1.25×10^{-8}	5.31×10^{-10}	4.39×10^{-7}
	Eu	8.03×10^{-9}	5.17×10^{-10}	7.42×10^{-8}
	Pr	7.67×10^{-9}	1.56×10^{-9}	2.29×10^{-7}
	Sm	5.26×10^{-9}	1.45×10^{-10}	1.95×10^{-7}
	Dy	4.26×10^{-9}	1.82×10^{-10}	1.80×10^{-7}
	Er	2.52×10^{-9}	1.42×10^{-11}	9.30×10^{-8}
	Yb	1.65×10^{-9}	5.90×10^{-11}	5.61×10^{-8}
	Tb	1.10×10^{-9}	4.52×10^{-11}	3.85×10^{-8}
	Ho	9.25×10^{-10}	6.79×10^{-11}	3.27×10^{-8}
	Lu	3.78×10^{-10}	7.30×10^{-11}	8.20×10^{-9}
	Tm	3.64×10^{-10}	1.09×10^{-11}	1.06×10^{-8}
HI	ΣREEs	2.23×10^{-7}	4.70×10^{-8}	6.32×10^{-6}
TCR	ΣREEs	1.24×10^{-13}	2.61×10^{-14}	3.51×10^{-12}

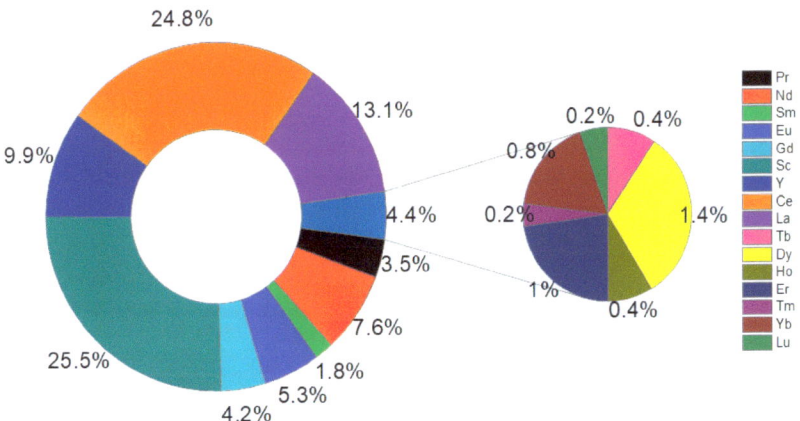

Figure 7. Contribution of REEs to the hazard index (HI) and target cancer risk (TCR) from coffee.

4. Conclusions

This study describes the levels of 14 rare earth elements (REEs) (La, Ce, Pr, Nd, Sm, Eu, Gd, Tb, Dy, Ho, Er, Tm, Yb, Lu), and Y and Sc in 92 coffee samples. The ultrasound-assisted extraction (UAE) procedure was developed for the extraction of REEs; subsequently, their determination was performed by ICP-MS. This study confirmed that the UAE procedure is an efficient and accurate method for preparing coffee samples for further determination of REEs. The most abundant REE was Ce, followed by La, Sc, Y, Nd, Gd, Eu, Pr, Sm, Dy, Er, Yb, Tb, Ho, Lu, and Tm. Among the analyzed elements, Tb was below the LOD value in 4 samples; Sm, Ho, and Er were below the LOD values in 5 samples; while Dy, Tm, and Lu were below the corresponding LOD values in 8, 9, and 10 samples, respectively. The surrogates had significantly higher total REE content than the pure coffee samples. The multivariate analysis techniques effectively distinguished between different types of coffee, clearly separating surrogates from 100% coffee samples. Principal component analysis (PCA) highlighted Eu, Tm, Yb, Lu, and Sc as significant elements of separation, which was similarly confirmed by the positive matrix factorization method (PMF), where there was a distinct separation of Tm, Tb, and Er, as well as Lu, Sc, and Yb from other elements. PMF analysis suggested that coffee reflects the soil composition. An increase in REEs in coffee can result from the addition of surrogates and the use of artificial fertilizers.

In conclusion, future research should investigate the potential for authenticating raw and minimally processed coffee based on the content of REEs. Furthermore, additional studies are needed to explore the applicability of this method to other types of food matrices and to evaluate the long-term health risks associated with REE exposure.

Supplementary Materials: The following supporting information can be downloaded at https://www.mdpi.com/article/10.3390/foods14020275/s1: Table S1: Coffee samples analyzed in this study; Table S2: Equations used in the risk assessment model; Table S3: Exposure parameters and their distributions in the risk assessment model; Table S4: Concentration of REEs in coffee samples ($n = 10$) when prepared by MAD and UAE procedures (µg/kg); Table S5: Correlation coefficients (R^2) of the linear regression (MAD vs. UAE); Table S6: Analysis of certified reference material (SRM 1547) for the content of REEs (mean ± SD, n = 3); Table S7: Initial concentrations (µg/kg) of REEs in spiked samples; Table S8a: Recovery and precision for spiking of UAE experiments (Spike 1: 0.5 µg/kg; Spike 2: 5.0 µg/kg); Table S8b: Recovery and precision for spiking of UAE experiments (Spike 3: 40 µg/kg; Spike 4: 200 µg/kg); Table S8c: Recovery and precision for spiking of MAD experiments (Spike 1: 0.5 µg/kg; Spike 2: 5.0 µg/kg); Table S8d: Recovery and precision for spiking of MAD experiments (Spike 3: 40 µg/kg; Spike 4: 200 µg/kg); Table S9: Descriptive statistics of the REE concentrations in coffee; Table S10: Pearson correlation matrix for REEs in coffee samples; Table S11: Principal component analysis showing two extracted components with eigenvalues greater than one, and their loadings.

Author Contributions: Conceptualization, M.L; methodology, M.L.; software, A.M.; validation, A.M.; formal analysis, J.M.; investigation, A.S.; resources, A.S.; data curation, J.V.; writing—original draft preparation, A.S.; writing—review and editing, A.O.; visualization, J.V.; supervision, J.M.; project administration, A.O.; funding acquisition, A.O. All authors have read and agreed to the published version of the manuscript.

Funding: This research was funded by the Ministry of Science, Technological Development and Innovation of the Republic of Serbia (Contracts 451–03-66/2024–03/200287, 451–03-65/2024–03/200135, and 451–03-66/2024–03/200168).

Institutional Review Board Statement: Not applicable.

Informed Consent Statement: Not applicable.

Data Availability Statement: The original contributions presented in the study are included in the article/Supplementary Materials, further inquiries can be directed to the corresponding author.

Conflicts of Interest: The authors declare no conflicts of interest.

References

1. Butt, M.S.; Sultan, M.T. Coffee and Its Consumption: Benefits and Risks. *Crit. Rev. Food Sci. Nutr.* **2011**, *51*, 363–373. [CrossRef] [PubMed]
2. George, S.E.; Ramalakshmi, K.; Rao, L.J.M. A Perception on Health Benefits of Coffee. *Crit. Rev. Food Sci. Nutr.* **2008**, *48*, 464–486. [CrossRef] [PubMed]
3. Santato, A.; Bertoldi, D.; Perini, M.; Camin, F.; Larcher, R. Using Elemental Profiles and Stable Isotopes to Trace the Origin of Green Coffee Beans on the Global Market. *J. Mass Spectrom.* **2012**, *47*, 1132–1140. [CrossRef] [PubMed]
4. Samoggia, A.; Riedel, B. Consumers' Perceptions of Coffee Health Benefits and Motives for Coffee Consumption and Purchasing. *Nutrients* **2019**, *11*, 653. [CrossRef]
5. Markos, M.U.; Tola, Y.; Kebede, B.T.; Ogah, O. Metabolomics: A Suitable Foodomics Approach to the Geographical Origin Traceability of Ethiopian Arabica Specialty Coffees. *Food Sci. Nutr.* **2023**, *11*, 4419–4431. [CrossRef]
6. Debastiani, R.; Iochims dos Santos, C.E.; Maciel Ramos, M.; Sobrosa Souza, V.; Amaral, L.; Yoneama, M.L.; Ferraz Dias, J. Elemental Analysis of Brazilian Coffee with Ion Beam Techniques: From Ground Coffee to the Final Beverage. *Food Res. Int.* **2019**, *119*, 297–304. [CrossRef]
7. Savić, A.; Mutić, J.; Lučić, M.; Onjia, A. Dietary Intake of Minerals and Potential Human Exposure to Toxic Elements via Coffee Consumption. *Biol. Trace Elem. Res.* **2024**. [CrossRef]
8. Tian, L.; Guo, Y.; Zhang, A.; Zhong, H. Simultaneously Verifying the Original Region of Green and Roasted Coffee Beans by Stable Isotopes and Elements Combined with Random Forest. *J. Food Qual.* **2022**, *2022*, 1308645. [CrossRef]
9. Migaszewski, Z.M.; Gałuszka, A. The Characteristics, Occurrence, and Geochemical Behavior of Rare Earth Elements in the Environment: A Review. *Crit. Rev. Environ. Sci. Technol.* **2015**, *45*, 429–471. [CrossRef]
10. Ramos, S.J.; Dinali, G.S.; Oliveira, C.; Martins, G.C.; Moreira, C.G.; Siqueira, J.O.; Guilherme, L.R.G. Rare Earth Elements in the Soil Environment. *Curr. Pollut. Rep.* **2016**, *2*, 28–50. [CrossRef]
11. Jenkins, J.A.; Musgrove, M.; White, S.J.O. Outlining Potential Biomarkers of Exposure and Effect to Critical Minerals: Nutritionally Essential Trace Elements and the Rare Earth Elements. *Toxics* **2023**, *11*, 188. [CrossRef] [PubMed]
12. Jiang, D.G.; Yang, J.; Zhang, S.; Yang, D.J. A Survey of 16 Rare Earth Elements in the Major Foods in China. *Biomed. Environ. Sci.* **2012**, *25*, 267–271. [CrossRef] [PubMed]
13. Tao, Y.; Shen, L.; Feng, C.; Yang, R.; Qu, J.; Ju, H.; Zhang, Y. Distribution of Rare Earth Elements (REEs) and Their Roles in Plant Growth: A Review. *Environ. Pollut.* **2022**, *298*, 118540. [CrossRef] [PubMed]
14. Kanazawa, Y.; Kamitani, M. Rare Earth Minerals and Resources in the World. *J. Alloys Compd.* **2006**, *408–412*, 1339–1343. [CrossRef]
15. Dushyantha, N.; Batapola, N.; Ilankoon, I.M.S.K.; Rohitha, S.; Premasiri, R.; Abeysinghe, B.; Ratnayake, N.; Dissanayake, K. The Story of Rare Earth Elements (REEs): Occurrences, Global Distribution, Genesis, Geology, Mineralogy and Global Production. *Ore Geol. Rev.* **2020**, *122*, 103521. [CrossRef]
16. Turra, C. Sustainability of Rare Earth Elements Chain: From Production to Food—A Review. *Int. J. Environ. Health Res.* **2017**, *28*, 23–42. [CrossRef]

17. Khan, A.M.; Bakar, N.K.A.; Bakar, A.F.A.; Ashraf, M.A. Chemical Speciation and Bioavailability of Rare Earth Elements (REEs) in the Ecosystem: A Review. *Environ. Sci. Pollut. Res.* **2017**, *24*, 22764–22789. [CrossRef]
18. Vesković, J.; Bulatović, S.; Ražić, S.; Lučić, M.; Miletić, A.; Nastasović, A.; Onjia, A. Arsenic-Contaminated Groundwater of the Western Banat (Pannonian Basin): Hydrogeochemical Appraisal, Pollution Source Apportionment, and Monte Carlo Simulation of Source-Specific Health Risks. *Water Environ. Res.* **2024**, *96*, e11087. [CrossRef]
19. Yan, Y.; Chi, H.; Liu, J.; Hu, G.; Yu, R.; Huang, H.; Lin, C. Provenance and Bioaccessibility of Rare Earth Elements in Atmospheric Particles in Areas Impacted by the Optoelectronic Industry. *Environ. Pollut.* **2020**, *263*, 114349. [CrossRef]
20. Miclean, M.; Levei, E.A.; Tanaselia, C.; Cadar, O. Rare Earth Elements Transfer from Soil to Vegetables and Health Risks Associated with Vegetable Consumption in a Former Mining Area. *Agronomy* **2023**, *13*, 1399. [CrossRef]
21. Arciszewska, Z.; Gama, S.; Lesniewska, B.; Malejko, J.; Nalewajko-Sieliwoniuk, E.; Zambrzycka-Szelewa, E.; Godlewska-Zyłkiewicz, B. The Translocation Pathways of Rare Earth Elements from the Environment to the Food Chain and Their Impact on Human Health. *Process Saf. Environ. Prot.* **2022**, *168*, 205–223. [CrossRef]
22. Shi, Z.; Yong, L.; Liu, Z.; Wang, Y.; Sui, H.; Mao, W.; Zhang, L.; Li, Y.; Liu, J.; Wei, S.; et al. Risk Assessment of Rare Earth Elements in Fruits and Vegetables from Mining Areas in China. *Environ. Sci. Pollut. Res.* **2022**, *29*, 48694–48703. [CrossRef]
23. Kollander, B.; Rodushkin, I.; Sundström, B. Multi-Element Assessment of Potentially Toxic and Essential Elements in New and Traditional Food Varieties in Sweden. *Foods* **2023**, *12*, 1831. [CrossRef] [PubMed]
24. Nolasco, A.; Squillante, J.; Esposito, F.; Velotto, S.; Romano, R.; Aponte, M.; Giarra, A.; Toscanesi, M.; Montella, E.; Cirillo, T. Coffee Silverskin: Chemical and Biological Risk Assessment and Health Profile for Its Potential Use in Functional Foods. *Foods* **2022**, *11*, 2834. [CrossRef]
25. Temerdashev, Z.; Bolshov, M.; Abakumov, A.; Khalafyan, A.; Kaunova, A.; Vasilyev, A.; Sheludko, O.; Ramazanov, A. Can Rare Earth Elements Be Considered as Markers of the Varietal and Geographical Origin of Wines? *Molecules* **2023**, *28*, 4319. [CrossRef]
26. Lymperopoulou, T.; Balta-Brouma, K.; Tsakanika, L.A.; Tzia, C.; Tsantili-Kakoulidou, A.; Tsopelas, F. Identification of Lentils (Lens Culinaris Medik) from Eglouvi (Lefkada, Greece) Based on Rare Earth Elements Profile Combined with Chemometrics. *Food Chem.* **2024**, *447*, 138965. [CrossRef]
27. Drivelos, S.A.; Danezis, G.P.; Halagarda, M.; Popek, S.; Georgiou, C.A. Geographical Origin and Botanical Type Honey Authentication through Elemental Metabolomics via Chemometrics. *Food Chem.* **2021**, *338*, 127936. [CrossRef]
28. Tanaka, K.; Zhao, L.; Tazoe, H.; Iizuka, T.; Murakami-Sugihara, N.; Toyama, K.; Yamamoto, T.; Yorisue, T.; Shirai, K. Using Neodymium Isotope Ratio in Ruditapes Philippinarum Shells for Tracking the Geographical Origin. *Food Chem.* **2022**, *382*, 131914. [CrossRef]
29. Barbosa, R.M.; Batista, B.L.; Varrique, R.M.; Coelho, V.A.; Campiglia, A.D.; Barbosa, F., Jr. The Use of Advanced Chemometric Techniques and Trace Element Levels for Controlling the Authenticity of Organic Coffee. *Food Res. Int.* **2014**, *61*, 246–251. [CrossRef]
30. Vezzulli, F.; Fontanella, M.C.; Lambri, M.; Beone, G.M. Specialty and High-Quality Coffee: Discrimination through Elemental Characterization via ICP-OES, ICP-MS, and ICP-MS/MS of Origin, Species, and Variety. *J. Sci. Food Agric.* **2023**, *103*, 4303–4316. [CrossRef]
31. Spalla, S.; Baffi, C.; Barbante, C.; Turretta, C.; Cozzi, G.; Beone, G.M.; Bettinelli, M. Determination of Rare Earth Elements in Tomato Plants by Inductively Coupled Plasma Mass Spectrometry Techniques. *Rapid Commun. Mass Spectrom.* **2009**, *23*, 3285–3292. [CrossRef] [PubMed]
32. Chiaudani, A.; Flamminii, F.; Consalvo, A.; Bellocci, M.; Pizzi, A.; Passamonti, C.; Cichelli, A. Rare Earth Element Variability in Italian Extra Virgin Olive Oils from Abruzzo Region. *Foods* **2024**, *13*, 141. [CrossRef]
33. Bandoniene, D.; Walkner, C.; Ringdorfer, F.; Meisel, T. Authentication of Meat and Dairy Products Using Rare Earth Element Labeling and Detection by Solution Based and Laser Ablation ICP-MS. *Food Res. Int.* **2020**, *132*, 109106. [CrossRef] [PubMed]
34. Kalogiouri, N.P.; Manousi, N.; Klaoudatos, D.; Spanos, T.; Topi, V.; Zachariadis, G.A. Rare Earths as Authenticity Markers for the Discrimination of Greek and Turkish Pistachios Using Elemental Metabolomics and Chemometrics. *Foods* **2021**, *10*, 349. [CrossRef] [PubMed]
35. de Oliveira Costa, T.; Rangel Botelho, J.; Cassago Nascimento, M.H.; Krause, M.; Weitzel Dias Carneiro, M.T.; Coelho Ferreira, D.; Filgueiras, P.R.; de Oliveira Souza, M. A One-Class Classification Approach for Authentication of Specialty Coffees by Inductively Coupled Plasma Mass Spectroscopy (ICP-MS). *Food Chem.* **2024**, *442*, 138268. [CrossRef]
36. Mazarakioti, E.C.; Zotos, A.; Thomatou, A.A.; Kontogeorgos, A.; Patakas, A.; Ladavos, A. Inductively Coupled Plasma-Mass Spectrometry (ICP-MS), a Useful Tool in Authenticity of Agricultural Products' and Foods' Origin. *Foods* **2022**, *11*, 3705. [CrossRef]
37. Gamela, R.R.; Costa, V.C.; Pereira-Filho, E.R. Multivariate Optimization of Ultrasound-Assisted Extraction Procedure for the Determination of Ca, Fe, K, Mg, Mn, P, and Zn in Pepper Samples by ICP OES. *Food Anal. Methods* **2020**, *13*, 69–77. [CrossRef]
38. Saleem, M.; Durani, A.I.; Alharthy, R.D.; Ahmed, M.; Shafiq, M.I.; Hameed, A.; Mubasher Hussain, S.; Bashir, W. Ultrasound-Assisted Extraction of Micro- and Macroelements in Fruit Peel Powder Mineral Supplement for Osteoporosis Patients and Their Determination by Flame Atomic Absorption Spectrometry. *J. Chem.* **2021**, *2021*, 5151560. [CrossRef]

39. Ul-Haq, I.; Ahmed, E.; Sharif, A.; Ahmed, M.; Ahmad, W. Optimization of Ultrasound-Assisted Extraction of Essential and Non-Essential/Toxic Trace Metals in Vegetables and Their Determination by FAAS and ICP-OES: An Evaluation of Human Health Risk. *Food Anal. Methods* **2021**, *14*, 2262–2275. [CrossRef]
40. Lima, E.C.; Barbosa, O.F.; Krug, F.J.; Silva, M.M.; Vale, M.G.R. Comparison of Ultrasound-Assisted Extraction, Slurry Sampling and Microwave-Assisted Digestion for Cadmium, Copper and Lead Determination in Biological and Sediment Samples by Electrothermal Atomic Absorption Spectrometry. *J. Anal. At. Spectrom.* **2000**, *15*, 995–1000. [CrossRef]
41. Mansour, F.R.; Płotka-Wasylka, J.; Locatelli, M. Modified GAPI (MoGAPI) Tool and Software for the Assessment of Method Greenness: Case Studies and Applications. *Analytica* **2024**, *5*, 451–457. [CrossRef]
42. Gohlke, G.; Cauduro, V.H.; Frozi, E.; Rocha, L.F.; Machado, G.R.; Henn, A.S.; Tao, Y.; Mesko, M.F.; Flores, E.M.M. Low Cost Sample Preparation Method Using Ultrasound for the Determination of Environmentally Critical Elements in Seaweed. *Ultrason. Sonochem.* **2024**, *103*, 106788. [CrossRef] [PubMed]
43. *ISO 11294:1994*; ISO International Standard—Roasted Ground Coffee—Determination of Moisture Content—Method by Determination of Loss in Mass at 103 °C (Routine Method). ISO: Geneva, Switzerland, 1994.
44. *EN 17851:2023*; Comite Europeen de Normalisation—Foodstuffs—Determination of Elements and Their Chemical Species—Determination of Ag, As, Cd, Co, Cr, Cu, Mn, Mo, Ni, Pb, Se, Tl, U and Zn in Foodstuffs by Inductively Coupled Plasma Mass Spectrometry (ICP-MS) after Pressure Digestion. Comite Europeen de Normalisation: Brussels, Belgium, 2023.
45. USEPA Human Health Risk Assessment I US EPA. Available online: https://www.epa.gov/risk/human-health-risk-assessment (accessed on 15 October 2024).
46. Slavković-Beškoski, L.; Ignjatović, L.; Ćujić, M.; Vesković, J.; Trivunac, K.; Stojaković, J.; Perić-Grujić, A.; Onjia, A. Ecological and Health Risks Attributed to Rare Earth Elements in Coal Fly Ash. *Toxics* **2024**, *12*, 71. [CrossRef] [PubMed]
47. Lee, J.; Park, Y.S.; Lee, H.J.; Koo, Y.E. Microwave-Assisted Digestion Method Using Diluted Nitric Acid and Hydrogen Peroxide for the Determination of Major and Minor Elements in Milk Samples by ICP-OES and ICP-MS. *Food Chem.* **2022**, *373*, 131483. [CrossRef]
48. Telloli, C.; Cicconi, F.; Manzi, E.; Borgognoni, F.; Salvi, S.; Iapalucci, M.C.; Rizzo, A. Multi-Elemental Analysis of Commercial Wheat Flours by ICP-MS Triple Quadrupole in Function of the Milling Degree. *Food Chem.* **2024**, *450*, 139370. [CrossRef]
49. Škrbić, B.; Živančev, J.; Mrmoš, N. Concentrations of Arsenic, Cadmium and Lead in Selected Foodstuffs from Serbian Market Basket: Estimated Intake by the Population from the Serbia. *Food Chem. Toxicol.* **2013**, *58*, 440–448. [CrossRef]
50. Naicker, K.; Mahlambi, P.; Mahlambi, M. Comparison of Ultrasonic and Microwave Assisted Digestion Methods for the Determination of Heavy Metals in Soil and Sediment: The Effect of Seasonal Variations on Metal Concentrations and Risk Assessment. *Soil Sediment Contam. Int. J.* **2023**, *32*, 320–336. [CrossRef]
51. Alves, V.; de Andrade, J.K.; Felsner, M.L. Green and Fast Ultrasound-Assisted Extraction Procedures for Fe, Mn, Mg and Ca Analysis in Cane Syrups by FAAS. *J. Food Compos. Anal.* **2023**, *123*, 105495. [CrossRef]
52. Biondić Fučkar, V.; Nutrizio, M.; Grudenić, A.; Djekić, I.; Režek Jambrak, A. Sustainable Ultrasound Assisted Extractions and Valorization of Coffee Silver Skin (CS). *Sustainability* **2023**, *15*, 8198. [CrossRef]
53. Messaoudi, M.; Begaa, S.; Hamidatou, L.; Salhi, M.; Ouakouak, H.; Mouzai, M.; Hassani, A. Neutron Activation Analysis of Major and Trace Elements in Arabica and Robusta Coffee Beans Samples Consumed in Algeria. *Radiochim. Acta* **2018**, *106*, 525–533. [CrossRef]
54. Miletić, A.; Vesković, J.; Lučić, M.; Onjia, A. Monte Carlo Simulation of Source-Specific Risks of Soil at an Abandoned Lead-Acid Battery Recycling Site. *Stoch. Environ. Res. Risk Assess.* **2024**, *38*, 3313–3329. [CrossRef]
55. Yazman, M.M.; Yüksel, B.; Ustaoğlu, F.; Şen, N.; Tepe, Y.; Tokatlı, C. Investigation of Groundwater Quality in the Southern Coast of the Black Sea: Application of Computational Health Risk Assessment in Giresun, Türkiye. *Environ. Sci. Pollut. Res.* **2024**, *31*, 52306–52325. [CrossRef] [PubMed]
56. Squadrone, S.; Brizio, P.; Stella, C.; Mantia, M.; Battuello, M.; Nurra, N.; Sartor, R.M.; Orusa, R.; Robetto, S.; Brusa, F.; et al. Rare Earth Elements in Marine and Terrestrial Matrices of Northwestern Italy: Implications for Food Safety and Human Health. *Sci. Total Environ.* **2019**, *660*, 1383–1391. [CrossRef] [PubMed]
57. Schmidt, K.; Bau, M.; Merschel, G.; Tepe, N. Anthropogenic Gadolinium in Tap Water and in Tap Water-Based Beverages from Fast-Food Franchises in Six Major Cities in Germany. *Sci. Total Environ.* **2019**, *687*, 1401–1408. [CrossRef]

Disclaimer/Publisher's Note: The statements, opinions and data contained in all publications are solely those of the individual author(s) and contributor(s) and not of MDPI and/or the editor(s). MDPI and/or the editor(s) disclaim responsibility for any injury to people or property resulting from any ideas, methods, instructions or products referred to in the content.

Article

Influence of Ultrasonic and Chemical Pretreatments on Quality Attributes of Dried Pepper (*Capsicum annuum*)

Milica Lučić [1], Nebojša Potkonjak [2,*], Ivana Sredović Ignjatović [3], Steva Lević [3], Zora Dajić-Stevanović [3], Stefan Kolašinac [3], Miona Belović [4], Aleksandra Torbica [4], Ivan Zlatanović [5], Vladimir Pavlović [3] and Antonije Onjia [6,*]

[1] Innovation Center of the Faculty of Technology and Metallurgy, Karnegijeva 4, 11120 Belgrade, Serbia
[2] Vinča Institute of Nuclear Sciences–National Institute of Republic of Serbia, University of Belgrade, Mike Petrovića Alasa 12-14, Vinča, 11351 Belgrade, Serbia
[3] Faculty of Agriculture, University of Belgrade, Nemanjina 6, 11080 Belgrade, Serbia
[4] Institute of Food Technology, University of Novi Sad, Bulevar Cara Lazara 1, 21000 Novi Sad, Serbia
[5] Faculty of Mechanical Engineering, University of Belgrade, Kraljice Marije 16, 11120 Belgrade, Serbia
[6] Faculty of Technology and Metallurgy, University of Belgrade, Karnegijeva 4, 11120 Belgrade, Serbia
* Correspondence: npotkonjak@vin.bg.ac.rs (N.P.); onjia@tmf.bg.ac.rs (A.O.)

Abstract: This study investigates the effects of ultrasound, in combination with chemical pretreatments, on the quality attributes (total phenolic and carotenoid content, antioxidant activity (2,2-Diphenyl-1-picrylhydrazyl assay (DPPH)), ferric-reducing ability (FRAP), CIE L* a* b* color, non-enzymatic browning, rehydration ratio, textural and morphological properties) of red pepper subjected to drying (hot air drying or freeze drying). The fractional factorial design was used to assess the impact of factors. The global Derringer desirability function was used to determine the optimal conditions for the best quality attributes of dried pepper. The drying method influenced total phenolic content, a* (redness), and initial rehydration ratio; pretreatment time significantly affected FRAP antiradical activity, a*, chroma and non-browning index, while pH-value had a significant effect on the texture of dried pepper. Non-enzymatic browning was reduced to 72.6%, while the DPPH antioxidant capacity of freeze-dried peppers was enhanced from 4.2% to 71.9%. Ultrasonic pretreatment led to changes in the pepper morphology, while potassium metabisulfite (KMS) was a more effective additive than citric acid.

Keywords: Derringer desirability function; factorial design; ultrasound; drying; citric acid; metabisulfite; food quality; antioxidant activity; color

1. Introduction

Pepper (*Capsicum annuum*) is a marketable vegetable and a part of culinary practices worldwide [1]. Peppers are grown on all continents, where China is the largest producer of fresh peppers and India of dried peppers. In 2021, the production of fresh pepper in the world reached 36 million tons, while the production of dried pepper reached 4.8 million tons. China dominates the worldwide production of fresh peppers, with over 16 million tons in 2021. Turkey is in second place with 3.1 million tons, followed by Indonesia (2.7 million tons) and Mexico (2.6 million tons). India is the largest producer of dried pepper (over 2.0 million tons), followed by Thailand (over 0.336 million tons) and China (over 0.312 million tons) (FAOSTAT, 2021) [2]. It is consumed fresh, processed, or in the form of dehydrated products [1,3]. Dehydrated pepper products are whole dried pepper, pepper flakes, and spice [4,5]. Paprika is a non-pungent red pepper powder obtained by grinding dried fruits of different varieties of *C. annuum* [6]. It is used as a colorant and a flavor in preparing different dishes (soups, sauces, stews, processed meats, dairy products, snacks, pizzas, etc.) [1,6,7].

The fruits of fresh peppers are a good source of many compounds with significant antioxidant potential. Antioxidants found in the fruits of sweet *C. annuum* are phenolic compounds, carotenoids, ascorbic acid, capsinoids, vitamin E, and other nutritional components [1,8]. All these nutritional components have a beneficial effect on human health. A diet rich in fruits and vegetables can reduce a person's risk of developing numerous chronic diseases, such as heart disease, diabetes, cancer, and other diseases [9,10]. Compared with other vegetables, these have the highest amount of vitamin C, carotenoids, and phenolics [11]. In most cases, eating ~60–80 g of fresh pepper is enough to meet the recommended daily intake for vitamin C [10]. Carotenoids are the compounds that are the most accountable for the color (and also influence the price of paprika) of yellow-orange and red varieties [12,13]. While hot capsaicinoid compounds are present in pungent varieties, non-pungent capsinoids are identified in sweet varieties. These compounds also have significant antioxidant properties [14].

Tunnel hot air-drying and sun drying are the most commonly used ways for fruit and vegetable dehydration, although they have certain disadvantages. The main downsides of sun drying are long drying time and risk of fungi proliferation, while high temperature during tunnel drying can result in significant degradation of valuable antioxidant compounds. Furthermore, high-temperature changes the color of the dried product due to oxidation and degradation of carotenoids, as well as the occurrence of Maillard reaction products [6,15,16].

The primary role of food drying is to extend the shelf life of perishable foods [3]. Blanching and chemical pretreatments are often used to decrease drying time, improve color, and better preserve the nutritional components. Ultrasonic pretreatment is one of the possible ways that can be used to produce dried products of better quality [17]. Previous studies indicate that it shortens the drying time and improves the rehydration of mushrooms, Brussels sprouts, and cauliflower [18,19]. Ultrasound pretreatment improves the retention of bioactive compounds and physical properties of fresh-cut quince fruit [20] and the antioxidant properties of ultrasonicated apple–grape juice compared to other treatments [21]. However, data on its influence on the chemical composition and antioxidant properties are scarce. To our knowledge, there is no data on the impact of simultaneous blanching, chemical pretreatments, and ultrasound on dried products. The findings of this study may help to improve the quality of dried peppers and other dried foods. To our knowledge, this is the first study utilizing the experimental design to simultaneously investigate the influence of individual factors (drying and applied pretreatments) and their interactions on the examined quality parameters and overall quality of dried red pepper.

The main goal of this study was to investigate the influence of different processing parameters (the mass of the sample subjected to pretreatment while the volume of the solution is kept constant, pretreatment time, the temperature of the pretreatment solution, application of ultrasound during pretreatment, application of different additives during pretreatment, the impact of pH value, the drying method) and their interactions on physicochemical properties and antioxidant activities of dried sweet red pepper. The fractional factorial design (FFD) was used to study different factors simultaneously. Additionally, the overall quality of dried pepper is studied by combining several responses using the derringer desirability function.

2. Materials and Methods

2.1. Plant Material, Reagents, and Standards

The fruits of the sweet red peppers (*C. annuum*) cultivar "Horgoš sweet 6" were purchased from a local farmer in Bački Petrovac, Serbia. Trolox (6-hydroxy-2,5,7,8-tetramethylchroman-2-carboxylic acid) and TPTZ (2,4,6-Tris(2-pyridyl)-s-triazine) were purchased from Acros Organics (Fair Lawn, NJ, USA), DPPH (2,2-diphenyl-1-picrylhydrazyl) was purchased from Sigma Aldrich (Darmstadt, Germany), gallic acid and Folin–Ciocalteu's reagent were purchased from Carlo Erba Reagents S.A.S. (Val de Reuil Cedex, France), HPLC grade methanol was purchased from J.T. Baker (Gliwice, Poland). Citric acid monohydrate, potassium metabisulfite, HCl, acetone, NaOH, FeCl$_3$·6H$_2$O, sodium carbonate anhydrous (p.a. > 99%), sodium acetate trihydrate (p.a. > 99%) were of analytical grade.

2.2. Pretreatment Procedure

Soon after collecting fresh pepper fruits, pretreatments were performed according to the experimental design (Table 1 and Table S1). After the peppers washing, they were cut in half longitudinally and cleaned of seeds, stalks, and placenta. One-half of the experiments were conducted in an ultrasonic bath (Elmasonic S15H, Elma Schmidbauer GmbH, Singen, Germany) at constant power (95 W) and frequency (37 kHz), while another half were done in the same ultrasonic bath but without ultrasonic power (Table S1). The pretreatment solution was filled to the recommended point to achieve uniform ultrasound pretreatment. It is clear that complete uniformity of ultrasound effect is not possible, but sandwich transducer systems, as a part of used ultrasonic bath, enable high homogeneity of ultrasound transmission. Based on the specifications given by the manufacturer, a sweep function was used. According to the manufacturer, the sweep function provides an optimized sound field distribution in the liquid. Additionally, used ultrasound bath unit was equipped with an additional mixing device that assures the optimum mixing of the treated content during the pretreatment. Pretreatment solutions were made with a constant volume of 1 L at mass concentrations of 0.25% citric acid (CA), 0.25% K$_2$S$_2$O$_5$ (KMS), or a mixture of citric acid and K$_2$S$_2$O$_5$ (CA/KMS), each salt having a concentration of 0.25%. Desired pH values for pretreatment solutions were adjusted using 10M NaOH or concentrated HCl.

Table 1. Experimental values and coded levels of the factors used for FFD.

No.	Effect of Factor	Factor	Level −1	Level 0	Level +1
1	A	Drying method	HD	-	FD
2	B	pH	3	6.5	10
3	C	Additive (0.25%)	CA	CA/KMS	KMS
4	D	US	Off	-	On
5	E	T (°C)	20	50	80
6	F	t (min)	1	3	5
7	G	m (g)	30	100	170

US—ultrasound; T—pretreatment temperature; t—pretreatment time; m—sample mass; HD—hot air-dried; FD—freeze-dried; CA—citric acid; CA/KMS—citric acid/potassium metabisulfite; KMS—potassium metabisulfite.

2.3. Drying by Experimental Design

After pretreatments, obtained pepper slices were dried according to the experimental design (Table S1). Tunnel hot air-drying (HD) was carried out as described in the study of Milanović et al. [22] at 60 °C and a constant air velocity of 2 m/s until water content in the final product decreased below 10%. Pepper samples were placed on a perforated tray to allow free circulation of hot air. The relative humidity in the dryer was an unregulated parameter, and its values ranged from 4% to 9%. For freeze-drying (FD) pretreated peppers were frozen at −20 °C and vacuum freeze-dried, maintaining collector temperature at −40 °C and chamber pressure 13.3 Pa for 24 h, using Labconco FreeZone® 18 freeze-dry system (Labconco Corporation, Kansas, MO, USA). The final sample temperature was 25 °C. After dying, all samples were separately vacuum-packed and stored in a dark place at room temperature until analysis. Before analysis, except for the analysis of textural and morphology properties, all samples were grounded to a fine powder. To determine moisture content, the obtained powders were dried at 105 °C to a constant mass [23].

The influence of seven different factors on antioxidant activity, total phenolic content (TPC), total carotenoid content (TCC), the rehydration process, non-enzymatic browning index, surface color, texture and morphological characteristics of dried peppers were assessed using a 1/8 fractional factorial design (FFD) as detailed in Table 1. The design resolution was IV. The alias structure of the FFD is given in Table S2. Five factors were estimated at three levels (−1, 0, +1), and two factors were non-numeric, estimated at two levels: low (−1) and high (+1). Four central points were replicated three times. The experimental design consisted of twenty-eight combinations of seven independent variables (Table S1). Pareto chart, main effect plot, and interaction plot were used to interpret the results. The Pareto chart is a bar chart that ranks the absolute value of the standardized effects of studied factors from the largest to the smallest. The reference line (red line) indicates which effects are statistically significant (in our study level of significance was $\alpha = 0.05$). Terms A, B, C, D, E, F, and G mark the effect of individual factors, while two or three terms combined denote the effects of factors interactions. The main effect plot shows how factor affects the response. The horizontal line indicates no main effect present. The interaction plot is used to see interactions between factors.

Experimental design allows the study of the influence of several factors simultaneously [24]. Three graphs were used to interpret obtained results: Pareto chart, main effect plot, and interaction plot. Pareto charts provide information about the statistical significance of all variables, where the vertical line is calculated for $\alpha = 0.05$ and the confidence level 95%. The main effect and interaction plots give additional information about examined variables and their interactions.

2.4. Total Phenolic and Carotenoid Analysis

TPC, as gallic acid equivalent, was determined by Folin–Ciocalteu assay according to Dewanto et al. [25] at 760 nm. Approximately 200 ± 1 mg of ground dried pepper was mixed with 5 mL of 80% methanol, sonicated for 30 min at room temperature, and centrifuged for 5 min at $1000 \times g$. The supernatant was collected in 10 mL volumetric flask and combined with the next supernatant obtained by re-extraction of the residue under the same conditions. The extraction solution was used to fill up the volumetric flask to the given mark. Every extraction was carried out in triplicate. Obtained extracts were kept at $-20\ °C$ until analyses and were analyzed within five days after the extraction. Pepper powders were extracted with acetone until colorless residue and obtained extracts were used to analyze TCC. The absorbance of this solution was measured at 662, 644, and 440 nm [26].

2.5. Antioxidant Properties

Antioxidant activity was determined by two assays using the same extracts as for TPC analysis. The results were expressed on dry mass as Trolox equivalent g/kg.

The DPPH assay was done following the method of Thaipong et al. [27] with minor modifications. First, the working solution was prepared by diluting 9 mL of stock solution (25 mg of DPPH in 100 mL of methanol) up to 50 mL with methanol. Then, a reaction mixture was made by mixing 150 µL of extract or standard with 2850 µL of working DPPH solution. The absorbance of the reaction mixture was measured at 517 nm after 30 min incubation at room temperature.

The ferric-reducing ability (FRAP) was assessed using the method of Taipong et al. [27]. The FRAP reagent contained 50 mL of sodium acetate buffer (pH 3.6), 5 mL of 10 mM TPTZ (2,4,6-Tris(2-pyridyl)-s-triazine) solution in 40 mM HCl, and 5 mL of 20 mM $FeCl_3 \cdot 6H_2O$. The mixture was heated in a water bath to $37\ °C$ and, immediately after that, added to the extract (150 µL of extract or standard + 2850 µL of FRAP reagent). The absorbance of the reaction mixture was measured at 593 nm after 30 min incubation. The buffer solution was used as blank.

2.6. Color Analysis

2.6.1. Surface Color Measurement

The color characteristics of powder samples were measured in CIE $L^*\ a^*\ b^*$ color space with a Chroma Meter (Model CR-400, Konica Minolta Inc., Tokyo, Japan), using D65 illuminating condition at $2°$ observed angle. Parameters L^* (darkness/whiteness), a^* (greenness/redness), b^* (blueness/yellowness), chroma C^*, and hue angle h^* were measured directly. Calibration of the instrument was performed with standard white tile.

Three readings were measured for each sample, and an average value was used for data analysis [28].

2.6.2. The Non-Enzymatic Browning Index (NBI)

Extraction of pepper samples was performed according to [29]. The absorbance of the resulting supernatants was measured at 420 nm after four-fold dilution. The results were expressed per kg of dry mass, taking into account the moisture content of the sample (Table S1).

2.7. Rehydration Analysis

The pepper flakes were rehydrated in distilled water at $20 \pm 1\ °C$. The solid-to-liquid ratio was 1:50. The rehydration kinetics was followed in time intervals by measuring the mass of slices after 15, 30, 60, 90, 120, 180, 240, 300, 360, 420, and 480 min. Before measuring, the flakes were taken out from the water, drained, and blotted with a paper towel for 20 s. All measurements were carried out in triplicate. The rehydration ratio (RR) was estimated as a ratio of m_t to m_0, where m_t is the mass of rehydrated sample at interval t, and m_0 is the mass of the dried sample before rehydration [30].

2.8. Analysis of Textural and Morphological Properties

2.8.1. Texture

The puncture force of dried samples was measured by using a TA.XT Plus Texture Analyser (Stable Micro Systems, Godalming, UK), equipped with a 5 kg load cell. Puncture force was obtained by 1 penetration in each sample (3 strips per treatment), with a 2 mm diameter stainless steel needle probe (P/2N) and a travel distance of 12 mm.

2.8.2. Scanning Electron Microscopy (SEM)

The surface characteristics of dried peppers were analyzed by scanning electron microscope (JEOL JSM6390LV). Sample coating with a layer of Au was performed using a sputter coater (Baltec scd 005) [31].

2.9. Desirability Function

The desirability function [32] was used to find the optimal conditions of the examined factors to define the optimal quality levels for different responses. The desirability function is a quick transformation of different responses to one objective function [33]. The desirability function has two steps: (1) transformation of every individual response to an individual desirability function (d_i) that ranges from 0 to 1 ($d_i = 0$ undesirable response; $d_i = 1$ desirable response) and (2) calculating of overall desirability (D) by taking the geometric average of all individual desirability values (Equation (1)).

$$D = \left(d_1^{r_1} * d_2^{r_2} * d_3^{r_3} * \ldots * d_n^{r_n}\right)^{1/\Sigma r_i} \quad (1)$$

where d_i is the individual desirability of response y_i ($i = 1, 2, 3, \ldots, n$), n is the number of responses, and r_i is the importance of every variable relative to others. In our work, we chose weights (r_i) equal to 1 for all twelve responses. The outcome of the overall desirability D depends on r_i values that offer users flexibility in the definition of desirability functions. If any of the responses are undesirable, overall desirability will become zero.

Individual desirability is defined by Equation (2) if a response is to be maximized.

$$d_i(\hat{y}_i(x)) = \begin{bmatrix} 0 & \text{if } \hat{y}_i(x) < L_i \\ \left(\frac{\hat{y}_i(x)-L_i}{U_i-L_i}\right)^s & \text{if } L_i \leq \hat{y}_i(x) \leq U_i \\ 1 & \text{if } \hat{y}_i(x) > U_i \end{bmatrix} \quad (2)$$

Individual desirability is defined by Equation (3) if a response is to be minimized.

$$d_i(\hat{y}_i(x)) = \begin{bmatrix} 1 & if\, \hat{y}_i(x) < L_i \\ \left(\frac{U_i - \hat{y}_i(x)}{U_i - L_i}\right)^t & if\, L_i \leq \hat{y}_i(x) \leq U_i \\ 0 & if\, \hat{y}_i(x) > U_i \end{bmatrix} \quad (3)$$

The exponents s and t are the weights assigned to individual responses that determine how important it is for d_i to be close to maximum or minimum, respectively. In our study, s and t were chosen to be 1. U_i and L_i are upper and lower acceptable values for the response, respectively [34]. In our study, L_i and U_i are the lowest and the highest values obtained for the response, respectively.

The desired responses for TPC, TCC, antioxidant activity obtained by DPPH and FRAP assays, lightness (L*), redness (a*), chroma (C*), hue (h*), rehydration ratio, texture were set to be maximized, while the desired responses for yellowness (b*) and non-enzymatic browning index were set to be minimized. The importance of all responses was the same.

3. Results and Discussion

3.1. Total Phenolic Content and Total Carotenoid Content

Among examined parameters, the drying method significantly affected the TPC in pepper (Table S3, Figure S1a). Other parameters and their interactions also influence the TPC, although it is not statistically significant. Tunnel hot air drying was a better method than freeze-drying to preserve TPC (Table S4, Figure S1b). This can be caused by a higher degree of cell destruction throughout hot air-drying at 60 °C compared to freeze-drying, so these compounds are more available for extraction [35] or by the emergence of new phenolic substances due to non-enzymatic interconversion between phenolic molecules [36]. The same trend in TPC was found in control samples produced from fresh pepper (higher TPC in HD than FD samples).

Contrary to this, FD samples produced from water-blanched (WB) pepper had slightly higher TPC than complementary HD samples. For HD samples, all pretreatments had a positive effect compared to water blanching. TPC increases ranged from 2.3% to 91.3%. For FD samples, pretreatments led to both decreases and increases compared to water blanching (from −21.4% to +26.9%). Most of the investigated pretreatments did not contribute to the better preservation of TPC if compared to the control produced from fresh peppers. Slight enhancement in TPC was observed by raising the pretreatment temperature from 20 °C to 50 °C and more intense by increasing the pretreatment time by up to 3 min. A further temperature rise and prolonged pretreatment reduced TPC, probably due to the leaching and degradation of phenolic compounds [37]. The decomposition of phenolic compounds during extraction will likely occur at higher temperatures [37]. The highest retention of TPC was in HD peppers samples with the following pretreatment: pH = 6.5; CA/KMS; without applying ultrasound; T = 50 °C; time = 3 min and mass 100 g in 1 L (experiment 27, Table S1).

This study indicated that most applied pretreatments did not contribute to better preservation of TCC. Control samples revealed that freeze-drying is better for the preservation of TCC (65.45% higher TCC than in HD samples), while water blanching improved the preservation of TCC in the final dried HD sample by 28.64%, but not in FD samples. Contrary to these results, FFD showed that none of the examined parameters have a statistically significant influence on TCC. Additional information on TCC is given in supplementary material S1.

3.2. Antioxidant Activity

Two different assays based on radical scavenging capacity (DPPH and FRAP) were used to measure antioxidant activity. The FRAP assay showed higher antioxidant capacity values (from 10.03 ± 1.12 to 20.2 ± 0.03 Trolox equivalent g/kg) compared to the DPPH assay (from 4.0 ± 0.30 to 15.7 ± 0.80 Trolox equivalent g/kg). Also, the FRAP test showed that antioxidant capacity significantly depends on pretreatment time (Table S3, Figure S2a, Pareto chart); for the DPPH test, none of the examined parameters significantly influence antioxidant capacity. Nevertheless, the influence of many individual parameters is the same or similar between these tests (pH value, additive, pretreatment temperature, and mass of treated sample). For FRAP assay, prolonged pretreatment, up to 3 min, enhanced antioxidant capacity (Figure S2b). Further prolongation of pretreatment time reduced antioxidant activity, probably due to the loss of antioxidants that pass into the solution. Both tests indicate that for antioxidant capacity, the best additives were in the following order CA/KSA > KMS > CA. KMS itself acts as an antioxidant and can preserve and stabilize carotenoids that contribute to antioxidant potential [17].

For both tests raising the temperature to 50 °C had a weak positive effect, while higher temperatures led to a sharp decrease in antioxidant capacity. It appears that pretreatment temperature and time had the same effect on TPC (see above). The higher temperatures and prolonged pretreatment caused lower antioxidant capacity due to the leaching and degradation of water-soluble phenolic and other antioxidant compounds responsible for the radical scavenging activity. The reduction of antioxidant capacity during hot water blanching was reported by other authors [38,39]. Our results indicate that the pH value of pretreatment solutions somewhat affected antioxidant capacity (Figure S2a,b). The best results were achieved at a pH value of 6.5.

These two antioxidant tests indicate that activity also depends on certain interactions between parameters. The FRAP assay showed an interaction between the drying method and pretreatment temperature; an interaction between pH and ultrasound pretreatment, while the DPPH assay showed an interaction between the drying method and additive. Better antioxidant activity was achieved at lower pretreatment temperatures, 20 °C and 50 °C for FD samples and 50 °C for HD samples (Figure S2c). Ultrasonic pretreatment positively affected the antioxidant capacity for pH values 3 and 6.5 and had a negative effect at pH value 10 (Figure S2c). The DPPH assay showed that a mixture of CA/KMS was the best additive for both drying methods. Pretreatments with citric acid have yielded the lowest results obtained from antioxidant activity tests when considering HD samples.

Water blanching negatively affected the antioxidant capacity of HD peppers (reduced by 42.2% and 40.9% for DPPH and FRAP assay, respectively) compared to drying without WB. However, WB improved the antioxidant activity of FD peppers for both tests (41.5% and 1.3% for DPPH and FRAP assay, respectively). All pretreatments from the FFD positively affected the ferric reduction ability of HD samples (increments ranged from 5.4% to 105%) compared to WB samples. Our findings are that all of the applied pretreatments from FFD are better for preserving the antioxidant capacity of HD samples than water blanching. Pretreatments from the FFD also improved the DPPH radical scavenging ability of FD samples (increscent ranged from 4.2% to 71.9%) compared to the FD control produced from fresh pepper.

The highest antioxidant capacity, measured by both assays, was obtained for the HD sample with the following pretreatment: pH = 6.5; CA/KMS; applied ultrasound; T = 50 °C; t = 3 min; mass 100 g in 1 L.

3.3. Color Analysis

The lightness (L*), yellowness (b*), and hue (h*) of samples were not significantly affected by any of examined parameters. Contrarily, the redness (a*) was found to be significantly dependent on the drying method and the pretreatment time, while the chroma (C*) was significantly affected only by pretreatment time. Better preservation of red pigments was achieved with freeze-drying than tunnel air-drying, probably due to less degradation of red carotenoids (capsanthin and capsorubin) [16], which occurs at higher drying temperatures. A longer pretreatment time positively affected the redness (a*) and color saturation (C*) of paprika. Longer pretreatments, probably due to better absorption of applied additives, gave the final product a more vivid color. In this study, the ultrasound pretreatment did not significantly affect color parameters. The best color characteristic of dried red pepper was obtained for the FD sample with the following pretreatment: pH 3; CA; applied ultrasound; T = 80 °C; t = 5 min; mass 30 g in 1 L.

All pretreatments from FFD, except experiment no. 15, reduce the NBI compared to corresponding controls without pretreatments. Reduction in NBI ranges from 30.7% to 72.6% for HD samples and from 19.6% to 58.5% for FD samples. Non-enzymatic browning was also reduced by water blanching (16.3% and 25.3% in HD and FD samples, respectively). Our results indicate that almost all pretreatments reduced non-enzymatic browning compared to drying without pretreatment. Although longer pretreatments improved the retention of red color, they also significantly affected the browning index due to the formation of brown compounds. Pretreatments with additives were undoubtedly better for reducing non-enzymatic browning than water bleaching. KMS and CA pretreatment solutions enhance the quality of dried foods. Inhibition of non-enzymatic browning with sulfite pretreatment was observed in dried peppers [40], while KMS, CA, and KMS/CA pretreatments improved the color characteristics of sweet bell-pepper powder [41]. In addition, citric acid proved to be a better additive to prevent browning, which was considered safer than a KMS due to some health problems (e.g., asthmatic reactions) [17]. The highest reduction of non-enzymatic browning was obtained for the following pretreatment: pH 3; KMS; applied ultrasound; T = 20 °C; t = 1 min; mass 30 g in 1 L.

3.4. Rehydration

The drying method affected the rehydration ratio in the initial period (30 min), where FD samples were rehydrated faster (Figure 1: experiments 1, 10, 13 and control X1b). A higher water absorption rate in the early phase of the rehydration process is observed for freeze-dried peppers compared to hot air-dried peppers (40 °C, 50 °C, 60 °C) by Kheto et al. (2021) [42]. In another study, freeze-dried tomato slices also exhibited higher rehydration ratios during 20 min of rehydration than slices dried using other drying methods [43]. In our study, at the equilibrium point (8 h), the drying method was still the parameter that had the most significant impact on rehydration. However, its impact is not statistically significant and had the opposite effect compared to the onset of rehydration, i.e., samples dried in the air-dryer showed better rehydration (Figure 1 experiment 2, 4, 14 and control X1a). This finding is similar to the results of Kheto et al. (2021) for green, yellow, and red bell peppers [42]. During the rehydration procedure, it was noticed that the HD samples preserved the structure better and that they did not break down during rehydration. FD samples were more brittle after they had been packed into vacuum bags. Fen et al. (2021) [44] and Zheng et al. (2023) [45] reported that during the rehydration of freeze-dried garlic, the potential of water was insufficient to exhaust all intercellular air left behind by the freeze-drying process. This phenomenon may also be the reason for the low final rehydration of freeze-dried pepper samples. Additionally, the vacuum freezing technique can cause structural deformations of the freeze-dried samples [1].

3.5. Texture

The pH value of the pretreatment solution had a statistically significant effect on the texture of dried pepper (Table S3, Figure S3a). Higher pH values of the pretreatment solution gave products that are firmer (Figure S3b). This change in the texture is probably induced due to the gelation of the pepper pectin under the influence of monovalent Na^+ ions, which were added as NaOH to adjust the pH value of the pretreatment solution. Pepper fruits can be a good source of pectin [46]. According to the degree of methylation, there are two groups of pectin, highly methylated pectin (HMP) with a degree of methylation of more than 50% and low methylated pectin (LMP) with a degree of methylation less than 50% [47]. Numerous studies examined the effect of divalent cations (Ca^{2+}, Cu^{2+}, Fe^{2+}) on LMP gelling, but it has been found that monovalent cations can also induce gelling of LMP and HMP [47–49]. Alkaline conditions lead to pectin demethylation, after which gel formation can occur under the influence of monovalent cations [49], increasing fruit and vegetable firmness [50]. Wang et al. (2019) [48] found that Na+ and K+ cations in alkaline solutions can lead to HMP gelling, while Pan et al. (2021) [47] found that Na$^+$ can lead to LMP gelling.

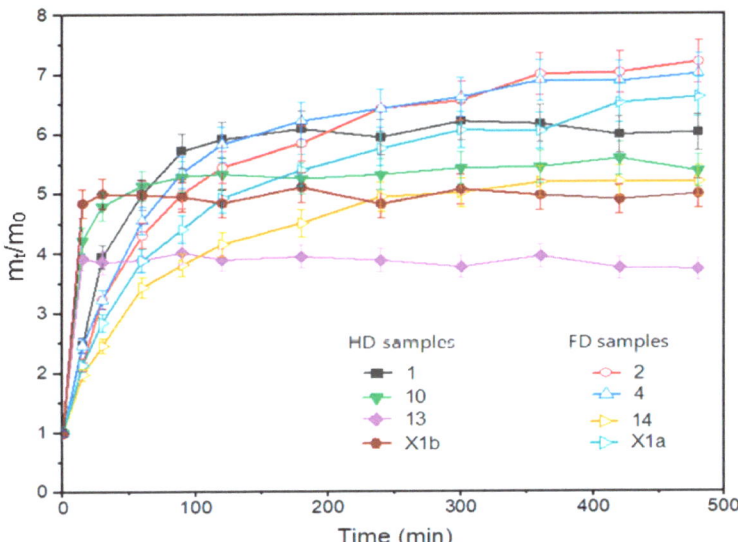

Figure 1. Rehydration ratio of pepper samples, freeze-dried: 1, 10, 13 and hot air-dried: 2, 4, 14 from FFD (Table S1); X1a—hot air-dried control without pretreatment; X1b—freeze-dried control without pretreatment.

Other parameters also influence the texture of dried peppers: pretreatment temperature, the interaction between drying and pH value, drying and applied additive, the influence of drying method, and mass of the treated sample (Figure S3c). The main effect plot (Figure S3b) shows that higher temperatures positively affected the texture of dried samples. Blanching processes can activate the enzyme pectin–methylesterase (PME), which de-esterifies pectin, whereby the newly formed product, more precisely its free carboxyl groups, can react with cations present in the solution, resulting in a gelling process [50].

This study also found an interaction between the drying method and pH value and between the drying method and additive. For FD samples, increasing the pH value also increases the strength of the final product, while for HD samples, the best results are achieved at pH 3 and pH 10, where pH 6.5 gives the weakest texture. Moreira et al. (2014) noticed that pectin degradation occurs at pH values of 5.35 and higher, and no gel formation is possible [51]. The pretreatment that gave the firmest peppers was: pH 10; CA; applied ultrasound; T = 80 °C; t = 1 min; mass 170 g in 1 L and hot air-drying.

3.6. Morphological Properties of Pretreated Dried Sweet Red Pepper

The SEM analysis was used to examine the influences of different factors on pepper surface properties. Further, the variations in surface properties may indicate potential tissue damage, which could cause the nutritional value to decrease. Morphological properties of controls, without pretreatment, indicate that hot air-drying caused the formation of furrows on the outer surface (Figure 2(A1)), while freeze-drying caused greater changes on the inside of the fruit, i.e., cracking of the inner surface (Figure 2(B2)). Wang et al. [52] also observed similar cracks to those of the hot air-dried control in hot air-dried pepper samples, previously blanched. Higher drying temperatures may result in more damage to the cellular structure [53]. Most pretreatments positively influenced the morphological properties of the dried product compared to controls without pretreatments. Pretreatments reduced (Figure 2(C1)) or completely stopped (Figure 2(D1,E1); Figure S4A,B) the formation of furrows at the outer surface of HD peppers. Vega-Gálvez et al. [54] also observed that sodium metabisulfite pretreated pepper samples (dried at 70 °C) suffered less structural damage than non-pretreated samples. Most of the examined pretreatments did not significantly affect the inner surface of HD peppers (Figure 2(C2,D2); Figure S4C), except the pretreatment No. 14 (Table S1; Figure 2(E2)), which led to considerable cracking of the inner surface compared to the HD control. Changes on the inner surface of dried pepper shown in Figure 2(E2) may result from ultrasonic and temperature (80 °C, 1 min) pretreatment [55,56]. Most of the examined pretreatments did not prevent breakage of the inner surface of freeze-dried samples (Figure 2(F2,H2)), except pretreatment No. 11 (Table S1; Figure 2(G2)). Pretreatment from experiment No. 16 (Table S1; Figure 2(H1,H2)) enhanced the cracking of the outer and inner surfaces. The variation in morphology of pretreated samples may indicate that the ultrasound pretreatment led to changes on the inside surface of pepper fruits (Figure 2(E2,F2)) regardless of whether samples were hot air or freeze-dried.

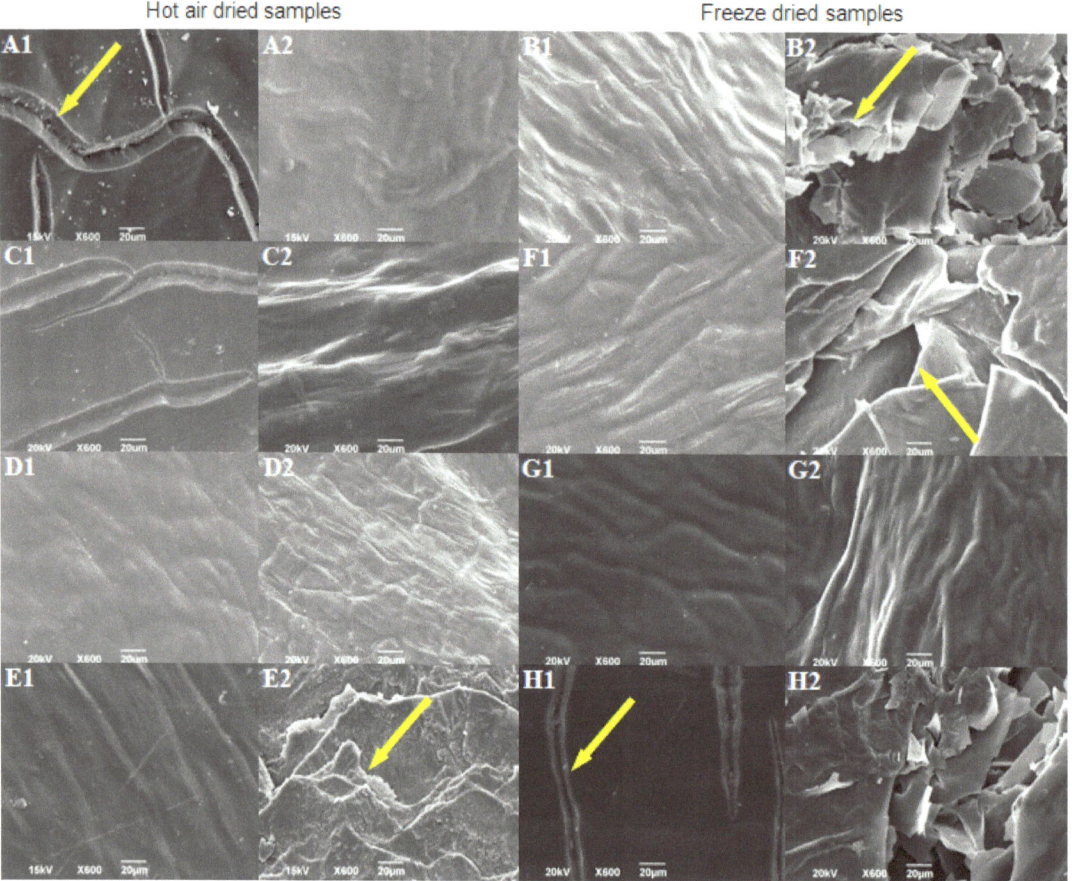

Figure 2. SEM micrographs of the dried sweet pepper: (**A1**)—outer and (**A2**)—the inner surface of HD sample without pretreatment; (**B1**)—outer and (**B2**)—the inner surface of FD sample without pretreatment; (**C1**)—outer and (**C2**)—the inner surface of sample 2 from FFD; (**D1**)—outer and (**D2**)—the inner surface of sample 4 from FFD; (**E1**)—outer and (**E2**)—the inner surface of sample 14 from FFD; (**F1**)—outer and (**F2**)—the inner surface of sample 6 from FFD; (**G1**)—outer and (**G2**)—the inner surface of sample 11 from FFD; (**H1**)—outer and (**H2**)—the inner surface of sample 16 from FFD. Arrows indicate cracks formed during drying.

3.7. Overall Desirability

By analyzing the experimental results with overall desirability function, it was found that the best quality of dried red pepper is significantly influenced by two pretreatment parameters, i.e., ultrasonic pretreatment and type of applied additive (Figure 3). The analysis of variance for the desirability function is presented in Table 2. The better overall quality of dried pepper, when considering 12 responses, is achieved with the following pretreatment: pH value 6.5, KMS pretreatment without applied ultrasound during 3 min at 50 °C, mass to volume ratio 100:1 (g:L), and final freeze-drying (Figure 4). For the overall quality of dried red pepper, ultrasonic pretreatment had a negative effect, probably due to cell wall raptures and the leaching of different bioactive compounds. Applying ultrasound waves in a liquid medium produces cavitations that cause sudden and localized changes in temperature and pressure [57,58]. As a result, bubbles form, rapidly grow, and collapse. When a solid is present in the liquid medium, the acoustic wave can form a microjet in the bubble. This microjet moves through the bubble, leaves it, and passes into the solid, changing the solid structure [1,57,58]. This further leads to the liquid extraction from the solid and the fluid penetration from the outside. The formation of microscopic channels may also occur and facilitate mass transport [1,58]. An interaction plot for 7 observed variables and 12 responses is given in Figure 4.

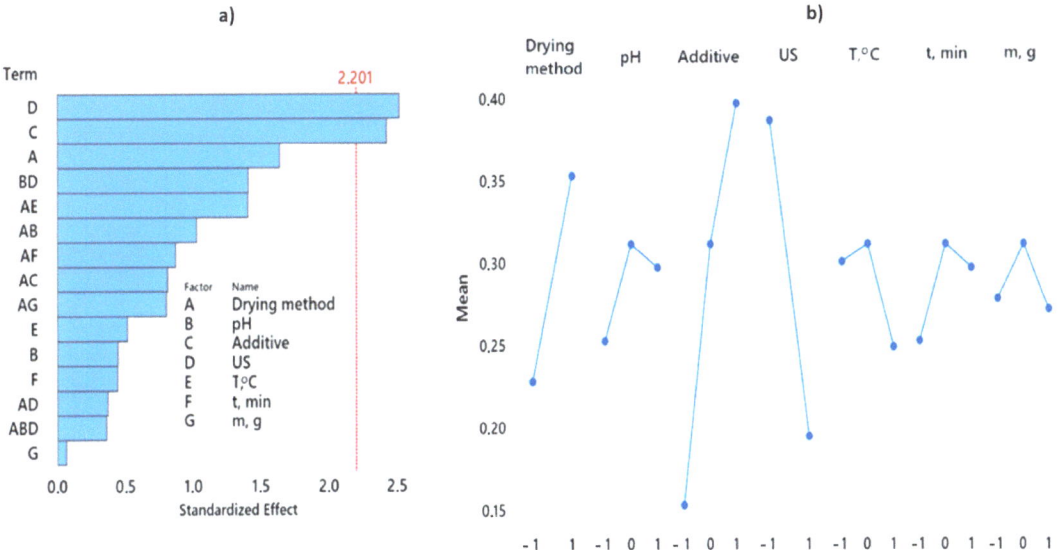

Figure 3. Pareto chart and Main effects plot obtained for overall desirability. (**a**) Pareto chart showing the standardized effect of independent variables and their interactions on overall desirability. (**b**) Main effects plot showing the effect of independent variables. Twelve responses get combined in one desirability function.

Table 2. Analysis of variance for desirability function.

Source	DF	Adj SS	Adj MS	F-Value	p-Value
Model	16	0.93981	0.058738	1.45	0.271
Linear	7	0.63253	0.090361	2.23	0.114
Drying method	1	0.10933	0.109333	2.69	0.129
pH	1	0.00805	0.008050	0.20	0.665
Additive	1	0.23870	0.238696	5.88	0.034

Table 2. Cont.

Source	DF	Adj SS	Adj MS	F-Value	p-Value
US	1	0.25751	0.257508	6.34	0.029
T, °C	1	0.01078	0.010783	0.27	0.617
t, min	1	0.00797	0.007974	0.20	0.666
m, g	1	0.00018	0.000181	0.00	0.948
2-Way Interactions	7	0.29276	0.041823	1.03	0.463
Drying method*pH	1	0.04277	0.042775	1.05	0.327
Drying method*Additive	1	0.02684	0.026838	0.66	0.433
Drying method*US	1	0.00554	0.005539	0.14	0.719
Drying method*T, C	1	0.08010	0.080101	1.97	0.188
Drying method*t, min	1	0.03090	0.030902	0.76	0.402
Drying method*m, g	1	0.02626	0.026262	0.65	0.438
pH*US	1	0.08034	0.080342	1.98	0.187
3-Way Interactions	1	0.00530	0.005301	0.13	0.725
Drying method*pH*US	1	0.00530	0.005301	0.13	0.725
Curvature	1	0.00922	0.009221	0.23	0.643
Error	11	0.44673	0.040611		
Lack-of-Fit	3	0.28896	0.096321	4.88	0.032
Pure Error	8	0.15776	0.019720		
Total	27	1.38653			

*—indicate interactions between factors.

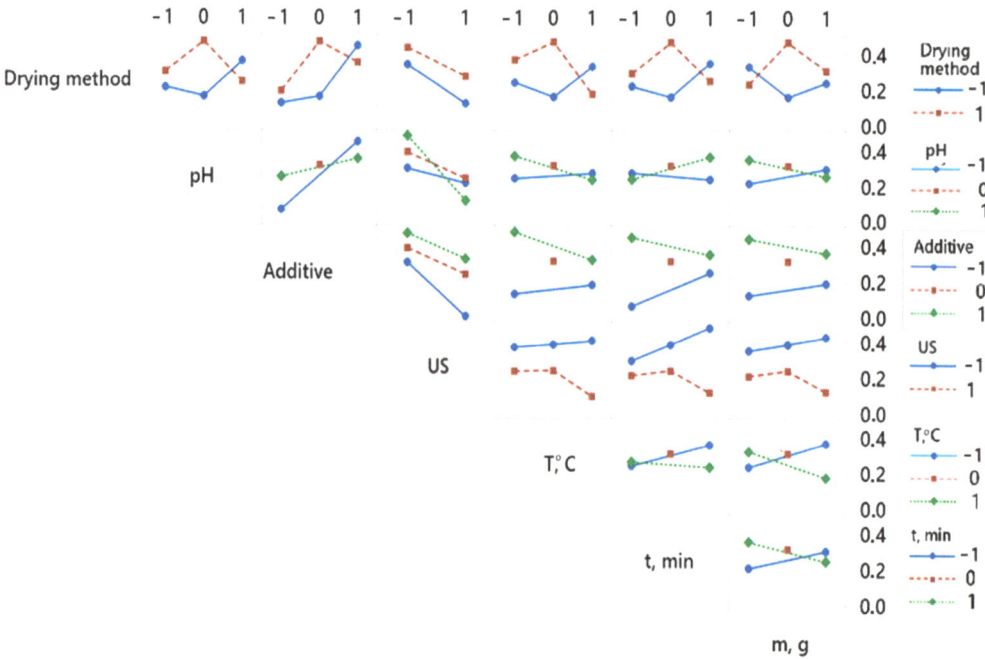

Figure 4. Interaction plot showing the effect of variable's interactions obtained for overall desirability. Twelve responses get combined in one desirability function.

4. Conclusions

The ultrasound and applied additive significantly affected the overall quality of dried red pepper. Regarding the best quality, ultrasound negatively affected physicochemical properties and antioxidant activities, i.e., the overall quality. The best additive was KMS, followed by CA/KMS, while CA exhibited poor results as a pretreatment additive. On the other hand, the drying method, pretreatment time, and pH value significantly influenced individual quality parameters. Hot air-drying provided better results than freeze-drying for retention of TPC and antioxidant capacity while freeze-drying provided better preservation of TCC. The pretreatment time is an important parameter that affected color parameters a*, chroma, and non-enzymatic browning. While prolonged pretreatment positively affected color characteristics, it also influenced higher non-enzymatic browning, even though most of the applied pretreatments reduced non-enzymatic browning (up to 72.6%). The texture was affected by pH value, where higher pH values gave firmer dried peppers.

The relevance of the responses was assumed to be the same; therefore, the same weight was given to each response (equal to 1), which is a limitation of this study. Additional research is required to evaluate the significance of individual responses in the overall quality of dried red pepper.

Supplementary Materials: The following supporting information can be downloaded at: https://www.mdpi.com/article/10.3390/foods12132468/s1, Figure S1: Total phenolic content (TPC): (a) Pareto chart of standardized effect (response in g/kg GAE, $\alpha = 0.05$); (b) Main effect plot for g/kg GAE; (c) Interaction plot for g/kg GAE; Figure S2: FRAP assay: (a) Pareto chart of standardized effect (response in equivalent Trolox g/kg, $\alpha = 0.05$); (b) Main effect plot for equivalent Trolox g/kg ; (c) Interaction plot equivalent Trolox g/kg; Figure S3: Texture analysis / Skin puncture force (g): (a) Pareto chart for standardized effect; (b) Main effect plot; (c) Interaction plot; Figure S4: SEM micrographs of the dried sweet pepper: (A) outer surface of sample 3 from FFD, (B) outer surface of sample 24 from FFD, (C) inner surface of sample 24 from FFD; Table S1: Experimental values for the fractional factorial design (FFD) and results obtained for all the independent variables; Table S2: Alias Structure for 1/8 FFD, 7 factors, 28 runs, resolution IV; Table S3: Main and interaction effects for each of the independent variables.; Table S4: Results for control pepper samples g/kg dry basis [59,60].

Author Contributions: Conceptualization, I.S.I.; methodology, S.L.; software, M.B.; validation, S.K.; formal analysis, V.P.; investigation, Z.D.-S.; resources, M.B.; data curation, I.Z.; writing—original draft preparation, M.L.; writing—review and editing, A.O.; visualization, A.T.; supervision, A.O.; project administration, N.P.; and funding acquisition, N.P. All authors have read and agreed to the published version of the manuscript.

Funding: This work was financially supported by the Ministry of Education, Science and Technological Development of the Republic of Serbia (Contract No. 451-03-47/2023-01/200287, Contract No. 451-03-47/2023-01/200017, No. 451-03-47/2023-01/200116, Contract No. 451-03-47/2023-01/200135, No. 451-03-47/2023-01/200105 and No. 451-03-47/2023-01/200222).

Data Availability Statement: The data that support the findings of this study are available from the corresponding authors upon reasonable request.

Conflicts of Interest: The authors have declared no conflict of interest for this article.

References

1. Rybak, K.; Wiktor, A.; Witrowa-Rajchert, D.; Parniakov, O.; Nowacka, M. The Quality of Red Bell Pepper Subjected to Freeze-Drying Preceded by Traditional and Novel Pretreatment. *Foods* **2021**, *10*, 226. [CrossRef]
2. FAO. FAOSTAT. Available online: https://www.fao.org/faostat/en/#data (accessed on 29 May 2023).
3. Geng, Z.; Huang, X.; Wang, J.; Xiao, H.; Yang, X.; Zhu, L.; Qi, X.; Zhang, Q.; Hu, B. Pulsed Vacuum Drying of Pepper (*Capsicum annuum* L.): Effect of High-Humidity Hot Air Impingement Blanching Pretreatment on Drying Kinetics and Quality Attributes. *Foods* **2022**, *11*, 318. [CrossRef]
4. Łechtańska, J.M.; Szadzinska, J.; Kowalski, S.J. Microwave- and Infrared-Assisted Convective Drying of Green Pepper: Quality and Energy Considerations. *Chem. Eng. Process.* **2015**, *98*, 155–164. [CrossRef]
5. Finger, F.L.; Pereira, G.M. Physiology and Postharvest of Pepper Fruits. In *Production and Breeding of Chilli Peppers (Capsicum spp.)*; do Rêgo, E.R., do Rêgo, M.M., Finger, F.L., Eds.; Springer: Cham, Switzerland, 2016; pp. 27–40. ISBN 9783319065328.
6. Topuz, A.; Feng, H.; Kushad, M. The Effect of Drying Method and Storage on Color Characteristics of Paprika. *LWT—Food Sci. Technol.* **2009**, *42*, 1667–1673. [CrossRef]
7. Galvin-King, P.; Haughey, S.A.; Elliott, C.T. The Detection of Substitution Adulteration of Paprika Spectroscopy Tools. *Foods* **2020**, *9*, 944. [CrossRef]
8. Ropelewska, E.; Sabanci, K.; Aslan, M.F. The Changes in Bell Pepper Flesh as a Result of Lacto-Fermentation Evaluated Using Image Features and Machine Learning. *Foods* **2022**, *11*, 2956. [CrossRef]

9. Del Río-Celestino, M.; Font, R. The Health Benefits of Fruits and Vegetables. *Foods* **2020**, *9*, 369. [CrossRef]
10. Jin, L.; Jin, N.; Wang, S.; Meng, X.; Xie, Y.; Li, Z.; Zhang, G.; Yin, X.; Lyu, J.; Zhong, Y.; et al. Comparing the Morphological Characteristics and Nutritional Composition of 23 Pepper (*Capsicum annuum* L.) Varieties. *Eur. Food Res. Technol.* **2022**, *249*, 963–974. [CrossRef]
11. Lučić, M.; Miletić, A.; Savić, A.; Lević, S.; Sredović Ignjatović, I.; Onjia, A. Dietary Intake and Health Risk Assessment of Essential and Toxic Elements in Pepper (*Capsicum annuum*). *J. Food Compos. Anal.* **2022**, *111*, 104598. [CrossRef]
12. Campos, M.R.S.; Gómez, K.R.; Ordoñez, Y.M.; Ancona, D.B. Polyphenols, Ascorbic Acid and Carotenoids Contents and Antioxidant Properties of Habanero Pepper (*Capsicum chinense*) Fruit. *Food Nutr. Sci.* **2013**, *4*, 47–54. [CrossRef]
13. Ergüneş, G.; Tarhan, S. Color Retention of Red Peppers by Chemical Pretreatments during Greenhouse and Open Sun Drying. *J. Food Eng.* **2006**, *76*, 446–452. [CrossRef]
14. Rosa, A.; Deiana, M.; Casu, V.; Paccagnini, S.; Appendino, G.; Ballero, M.; Dessí, M.A. Antioxidant Activity of Capsinoids. *J. Agric. Food Chem.* **2002**, *50*, 7396–7401. [CrossRef]
15. Vega-Gálvez, A.; Di Scala, K.; Rodríguez, K.; Lemus-Mondaca, R.; Miranda, M.; López, J.; Perez-Won, M. Effect of Air-Drying Temperature on Physico-Chemical Properties, Antioxidant Capacity, Colour and Total Phenolic Content of Red Pepper (*Capsicum annuum* L. Var. Hungarian). *Food Chem.* **2009**, *117*, 647–653. [CrossRef]
16. Yang, X.H.; Deng, L.Z.; Mujumdar, A.S.; Xiao, H.W.; Zhang, Q.; Kan, Z. Evolution and Modeling of Colour Changes of Red Pepper (*Capsicum annuum* L.) during Hot Air Drying. *J. Food Eng.* **2018**, *231*, 101–108. [CrossRef]
17. Deng, L.Z.; Mujumdar, A.S.; Zhang, Q.; Yang, X.H.; Wang, J.; Zheng, Z.A.; Gao, Z.J.; Xiao, H.W. Chemical and Physical Pretreatments of Fruits and Vegetables: Effects on Drying Characteristics and Quality Attributes—A Comprehensive Review. *Crit. Rev. Food Sci. Nutr.* **2017**, *59*, 1408–1432. [CrossRef]
18. Çakmak, R.Ş.; Tekeoğlu, O.; Bozkir, H.; Ergün, A.R.; Baysal, T. Effects of Electrical and Sonication Pretreatments on the Drying Rate and Quality of Mushrooms. *LWT—Food Sci. Technol.* **2016**, *69*, 197–202. [CrossRef]
19. Jambrak, A.R.; Mason, T.J.; Paniwnyk, L.; Lelas, V. Accelerated Drying of Button Mushrooms, Brussels Sprouts and Cauliflower by Applying Power Ultrasound and Its Rehydration Properties. *J. Food Eng.* **2007**, *81*, 88–97. [CrossRef]
20. Yildiz, G.; Izli, G.; Muhammad, R. Comparison of Chemical, Physical, and Ultrasound Treatments on the Shelf Life of Fresh-Cut Quince Fruit (*Cydonia oblonga* Mill.). *J. Food Process. Preserv.* **2019**, *44*, e14366. [CrossRef]
21. Aadil, R.M.; Khalil, A.A.; Rehman, A.; Khalid, A.; Inam-ur-Raheem, M.; Karim, A.; Gill, A.A.; Abid, M.; Afraz, M.T. Assessing the Impact of Ultra-Sonication and Thermo-Ultrasound on Antioxidant Indices and Polyphenolic Profile of Apple-Grape Juice Blend. *J. Food Process. Preserv.* **2019**, *44*, e14406. [CrossRef]
22. Milanović, M.; Komatina, M.; Zlatanović, I.; Manić, N.; Antonijević, D. Kinetic Parameters Identification of Conductive Enhanced Hot Air Drying Process of Food Waste. *Therm. Sci.* **2021**, *25*, 1795–1807. [CrossRef]
23. Salević, A.; Stojanović, D.; Lević, S.; Pantić, M.; Đorđević, V.; Pešić, R.; Bugarski, B.; Pavlović, V.; Uskoković, P.; Nedović, V. The Structuring of Sage (*Salvia officinalis* L.) Extract-Incorporating Edible Zein-Based Materials with Antioxidant and Antibacterial Functionality by Solvent Casting versus Electrospinning. *Foods* **2022**, *11*, 390. [CrossRef]
24. Lučić, M.; Sredović Ignjatović, I.; Lević, S.; Pećinar, I.; Antić, M.; Đurđić, S.; Onjia, A. Ultrasound-Assisted Extraction of Essential and Toxic Elements from Pepper in Different Ripening Stages Using Box–Behnken Design. *J. Food Process. Preserv.* **2022**, *46*, e16493. [CrossRef]
25. Dewanto, V.; Xianzhong, W.; Adom, K.K.; Liu, R.H. Thermal Processing Enhances the Nutritional Value of Tomatoes by Increasing Total Antioxidant Activity. *J. Agric. Food Chem.* **2002**, *50*, 3010–3014. [CrossRef]
26. Ilić, S.Z.; Milenković, L.; Dimitrijević, A.; Stanojević, L.; Cvetković, D.; Kevrešan, Ž.; Fallik, E.; Mastilović, J. Light Modification by Color Nets Improve Quality of Lettuce from Summer Production. *Sci. Hortic.* **2017**, *226*, 389–397. [CrossRef]
27. Thaipong, K.; Boonprakob, U.; Crosby, K.; Cisneros-Zevallos, L.; Hawkins Byrne, D. Comparison of ABTS, DPPH, FRAP, and ORAC Assays for Estimating Antioxidant Activity from Guava Fruit Extracts. *J. Food Compos. Anal.* **2006**, *19*, 669–675. [CrossRef]
28. Kojić, J.; Belović, M.; Krulj, J.; Pezo, L.; Teslić, N.; Kojić, P.; Tukuljac, L.P.; Šeregelj, V.; Ilić, N. Textural, Color and Sensory Features of Spelt Wholegrain Snack Enriched with Betaine. *Foods* **2022**, *11*, 475. [CrossRef]
29. Delgado-Andrade, C.; Morales, F.J.; Seiquer, I.; Pilar Navarro, M. Maillard Reaction Products Profile and Intake from Spanish Typical Dishes. *Food Res. Int.* **2010**, *43*, 1304–1311. [CrossRef]
30. Szadzińska, J.; Łechtańska, J.; Kowalski, S.J.; Stasiak, M. The Effect of High Power Airborne Ultrasound and Microwaves on Convective Drying Effectiveness and Quality of Green Pepper. *Ultrason. Sonochem.* **2017**, *34*, 531–539. [CrossRef]
31. Kalušević, A.M.; Lević, S.M.; Čalija, B.R.; Milić, J.R.; Pavlović, V.B.; Bugarski, B.M.; Nedović, V.A. Effects of Different Carrier Materials on Physicochemical Properties of Microencapsulated Grape Skin Extract. *J. Food Sci. Technol.* **2017**, *54*, 3411–3420. [CrossRef] [PubMed]
32. Derringer, G.; Suich, R. Simultaneous Optimization of Several Response Variables. *J. Qual. Technol.* **1980**, *12*, 214–219. [CrossRef]
33. Lukić, J.; Radulović, J.; Lučić, M.; Đurkić, T.; Onjia, A. Chemometric Optimization of Solid-Phase Extraction Followed by Liquid Chromatography-Tandem Mass Spectrometry and Probabilistic Risk Assessment of Ultraviolet Filters in an Urban Recreational Lake. *Front. Environ. Sci.* **2022**, *10*, 911. [CrossRef]
34. Vera Candioti, L.; De Zan, M.M.; Cámara, M.S.; Goicoechea, H.C. Experimental Design and Multiple Response Optimization. Using the Desirability Function in Analytical Methods Development. *Talanta* **2014**, *124*, 123–138. [CrossRef]

35. Campos-Hernández, N.; Jaramillo-Flores, M.E.; Téllez-Medina, D.I.; Alamilla-Beltrán, L. Effect of Traditional Dehydration Processing of Pepper Jalapeno Rayado (*Capsicum annuum*) on Secondary Metabolites with Antioxidant Activity. *CYTA—J. Food* **2018**, *16*, 316–324. [CrossRef]
36. Que, F.; Mao, L.; Fang, X.; Wu, T. Comparison of Hot Air-Drying and Freeze-Drying on the Physicochemical Properties and Antioxidant Activities of Pumpkin (*Cucurbita moschata* Duch.) Flours. *Int. J. Food Sci. Technol.* **2008**, *43*, 1195–1201. [CrossRef]
37. Bouzari, A.; Holstege, D.; Barrett, D.M. Mineral, Fiber, and Total Phenolic Retention in Eight Fruits and Vegetables: A Comparison of Refrigerated and Frozen Storage. *J. Agric. Food Chem.* **2015**, *63*, 951–956. [CrossRef]
38. Norafida, A.; Aminah, A. Effect of Blanching Treatments on Antioxidant Activity of Frozen Green Capsicum (*Capsicum annuum* L. Var Bell Pepper). *Int. Food Res. J.* **2018**, *25*, 1427–1434.
39. Wang, J.; Yang, X.H.; Mujumdar, A.S.; Wang, D.; Zhao, J.H.; Fang, X.M.; Zhang, Q.; Xie, L.; Gao, Z.J.; Xiao, H.W. Effects of Various Blanching Methods on Weight Loss, Enzymes Inactivation, Phytochemical Contents, Antioxidant Capacity, Ultrastructure and Drying Kinetics of Red Bell Pepper (*Capsicum annuum* L.). *LWT—Food Sci. Technol.* **2017**, *77*, 337–347. [CrossRef]
40. Melgar-Lalanne, G.; Hernández-Álvarez, A.J.; Jiménez-Fernández, M.; Azuara, E. Oleoresins from *Capsicum* Spp.: Extraction Methods and Bioactivity. *Food Bioprocess Technol.* **2017**, *10*, 51–76. [CrossRef]
41. Sharma, R.; Joshi, V.K.; Kaushal, M. Effect of Pre-Treatments and Drying Methods on Quality Attributes of Sweet Bell-Pepper (*Capsicum annum*) Powder. *J. Food Sci. Technol.* **2015**, *52*, 3433–3439. [CrossRef]
42. Kheto, A.; Dhua, S.; Nema, P.K.; Sharanagat, V.S. Influence of Drying Temperature on Quality Attributes of Bell Pepper (*Capsicum annuum* L.): Drying Kinetics and Modeling, Rehydration, Color, and Antioxidant Analysis. *J. Food Process Eng.* **2021**, *44*, e13880. [CrossRef]
43. Gaware, T.J.; Sutar, N.; Thorat, B.N. Drying of Tomato Using Different Methods: Comparison of Dehydration and Rehydration Kinetics. *Dry. Technol.* **2010**, *28*, 651–658. [CrossRef]
44. Feng, Y.; Xu, B.; ElGasim, A.; Yagoub, A.; Ma, H.; Sun, Y.; Xu, X.; Yu, X.; Zhou, C. Role of Drying Techniques on Physical, Rehydration, Flavor, Bioactive Compounds and Antioxidant Characteristics of Garlic. *Food Chem.* **2021**, *343*, 128404. [CrossRef] [PubMed]
45. Zheng, Z.A.; Wang, S.Y.; Wang, H.; Xiao, H.; Liu, Z.L.; Pan, Y.H.; Gao, L. Comparative Study on the Influence of Various Drying Techniques on Drying Characteristics and Physicochemical Quality of Garlic Slices. *Foods* **2023**, *12*, 1314. [CrossRef] [PubMed]
46. do Nascimento, G.E.; Iacomini, M.; Cordeiro, L.M.C. New Findings on Green Sweet Pepper (*Capsicum annum*) Pectins: Rhamnogalacturonan and Type I and II Arabinogalactans. *Carbohydr. Polym.* **2017**, *171*, 292–299. [CrossRef] [PubMed]
47. Pan, M.K.; Zhou, F.F.; Liu, Y.; Wang, J.H. Na+-Induced Gelation of a Low-Methoxyl Pectin Extracted from *Premna microphylla* Turcz. *Food Hydrocoll.* **2021**, *110*, 106153. [CrossRef]
48. Wang, H.; Wan, L.; Chen, D.; Guo, X.; Liu, F.; Pan, S. Unexpected Gelation Behavior of Citrus Pectin Induced by Monovalent Cations under Alkaline Conditions. *Carbohydr. Polym.* **2019**, *212*, 51–58. [CrossRef]
49. Wehr, J.B.; Menzies, N.W.; Blamey, F.P.C. Alkali Hydroxide-Induced Gelation of Pectin. *Food Hydrocoll.* **2004**, *18*, 375–378. [CrossRef]
50. Castro, S.M.; Van Loey, A.; Saraiva, J.A.; Smout, C.; Hendrickx, M. Inactivation of Pepper (*Capsicum annuum*) Pectin Methylesterase by Combined High-Pressure and Temperature Treatments. *J. Food Eng.* **2006**, *75*, 50–58. [CrossRef]
51. Moreira, H.R.; Munarin, F.; Gentilini, R.; Visai, L.; Granja, P.L.; Tanzi, M.C.; Petrini, P. Injectable Pectin Hydrogels Produced by Internal Gelation: PH Dependence of Gelling and Rheological Properties. *Carbohydr. Polym.* **2014**, *103*, 339–347. [CrossRef]
52. Wang, J.; Fang, X.M.; Mujumdar, A.S.; Qian, J.Y.; Zhang, Q.; Yang, X.H.; Liu, Y.H.; Gao, Z.J.; Xiao, H.W. Effect of High-Humidity Hot Air Impingement Blanching (HHAIB) on Drying and Quality of Red Pepper (*Capsicum annuum* L.). *Food Chem.* **2017**, *220*, 145–152. [CrossRef]
53. Deng, L.Z.; Yang, X.H.; Mujumdar, A.S.; Zhao, J.H.; Wang, D.; Zhang, Q.; Wang, J.; Gao, Z.J.; Xiao, H.W. Red Pepper (*Capsicum annuum* L.) Drying: Effects of Different Drying Methods on Drying Kinetics, Physicochemical Properties, Antioxidant Capacity, and Microstructure. *Dry. Technol.* **2018**, *36*, 893–907. [CrossRef]
54. Vega-Gálvez, A.; Lemus-Mondaca, R.; Bilbao-Sáinz, C.; Fito, P.; Andrés, A. Effect of Air Drying Temperature on the Quality of Rehydrated Dried Red Bell Pepper (Var. Lamuyo). *J. Food Eng.* **2008**, *85*, 42–50. [CrossRef]
55. Dias, A.L.B.; Arroio Sergio, C.S.; Santos, P.; Barbero, G.F.; Rezende, C.A.; Martínez, J. Effect of Ultrasound on the Supercritical CO_2 Extraction of Bioactive Compounds from Dedo de Moça Pepper (*Capsicum baccatum* L. Var. Pendulum). *Ultrason. Sonochem.* **2016**, *31*, 284–294. [CrossRef] [PubMed]
56. Nowacka, M.; Wiktor, A.; Śledź, M.; Jurek, N.; Witrowa-Rajchert, D. Drying of Ultrasound Pretreated Apple and Its Selected Physical Properties. *J. Food Eng.* **2012**, *113*, 427–433. [CrossRef]
57. Montalvo-González, E.; Anaya-Esparza, L.M.; Abraham Domínguez-Avila, J.; González-Aguilar, G.A. Chapter 5—Ultrasonic Processing Technology for Postharvest Disinfection. In *Postharvest Disinfection of Fruits and Vegetables*; Siddiqui, M.W., Ed.; Academic Press: Cambridge, MA, USA, 2018; pp. 101–119. ISBN 9780128126981.
58. Rodríguez, Ó.; Eim, V.; Rosselló, C.; Femenia, A.; Cárcel, J.A.; Simal, S. Application of Power Ultrasound on the Convective Drying of Fruits and Vegetables: Effects on Quality. *J. Sci. Food Agric.* **2018**, *98*, 1660–1673. [CrossRef] [PubMed]

59. Kaur, R.; Kaur, K.; Ahluwalia, P. Effect of Drying Temperatures and Storage on Chemical and Bioactive Attributes of Dried Tomato and Sweet Pepper. *LWT—Food Sci. Technol.* **2020**, *117*, 108604. [CrossRef]
60. Bechoff, A.; Westby, A.; Menya, G.; Tomlins, K.I. Effect of Pretreatments for Retaining Total Carotenoids in Dried and Stored Orange-Fleshed-Sweet Potato Chips. *J. Food Qual.* **2011**, *34*, 259–267. [CrossRef]

Disclaimer/Publisher's Note: The statements, opinions and data contained in all publications are solely those of the individual author(s) and contributor(s) and not of MDPI and/or the editor(s). MDPI and/or the editor(s) disclaim responsibility for any injury to people or property resulting from any ideas, methods, instructions or products referred to in the content.

Article

The Effect of Temperature and Storage Duration on the Quality and Attributes of the Breast Meat of Hens after Their Laying Periods

Anna Augustyńska-Prejsnar [1], Paweł Hanus [2], Małgorzata Ormian [1], Miroslava Kačániová [3], Zofia Sokołowicz [1] and Jadwiga Topczewska [1,*]

[1] Department of Animal Production and Poultry Products Evaluation, Institute of Food and Nutrition Technology, University of Rzeszow, 35-959 Rzeszow, Poland; aaugustynska@ur.edu.pl (A.A.-P.); mormian@ur.edu.pl (M.O.); zsokolowicz@ur.edu.pl (Z.S.)

[2] Department of Food Technology and Human Nutrition, Institute of Food and Nutrition Technology, University of Rzeszow, 35-959 Rzeszow, Poland; phanus@ur.edu.pl

[3] Institute of Horticulture, Faculty of Horticulture and Landscape Engineering, Slovak University of Agriculture, 949 76 Nitra, Slovakia; miroslava.kacaniova@gmail.com

* Correspondence: jtopczewska@ur.edu.pl

Abstract: The purpose of this study was to evaluate the effect of temperature (2 °C and 6 °C) and storage duration on the quality and attributes of hens' breast meat after their laying periods. The study included physicochemical characteristics (pH, drip loss, colour, shear force), microbiological quality (total Enterobacteriaceae family and Pseudomonas count), and sensory quality. Bacterial identification was performed using matrix-assisted laser desorption/ionisation time-of-flight mass spectrometry. The increased meat pH and drip loss was greater at 6 than 2 °C ($p < 0.05$). An increase in the tenderness of the meat stored at 6 °C was found as early as day 4, as well as at 2 °C on day 8 of storage ($p < 0.05$). On day 4 of storage, the meat was characterised by a darker colour than on the first day, but the darkening was greater at 6 °C than at 2 °C ($p < 0.05$). At 6 °C, on day 4 of storage, there was an increase in yellow saturation (b*) of the meat, which was higher at 6 °C than at 2 °C ($p < 0.05$). At 2 °C, the total bacterial count and number of *Pseudomonas* spp. in the meat gradually increased along with increasing storage duration, reaching 4.64 log cfu/g and 4.48 log cfu/g, respectively, on the 8th day of storage. At 6 °C, on the sixth day of storage, the total bacterial count in the meat exceeded 7 log cfu/g, considered the limit of microbiological safety. The meat stored at 2 °C had an acceptable sensory quality until the 8th day of storage. The study shows that storage at 2 °C preserves the sensory characteristics and microbiological safety of the hen meat longer at an acceptable level after the laying period. Extended storage life may be of importance to consumers and the meat industry.

Keywords: post-laying hen; storage time; storage temperature; MALDI-TOF MS; physical and chemical characteristics; sensory quality

1. Introduction

Laying-hens' meat, after termination of the egg-laying period, is most often disposed of or used in animal fodder [1]. The use of meat from poultry production after the end of the laying period for consumer purposes is marginal, mainly due to its less favourable technological and sensory characteristics [2]. However, the increased demand for local and regional food products observed in recent years offers new opportunities and perspectives for using the meat of native and local breeds, e.g., RIR hens of free-range/organic breeding [3]. At the same time, it should be emphasised that the use of such raw material is part of the need to reduce food waste. According to the concept contained in the European Green Deal, a 10% reduction in food processing waste should be sought by 2030 [4,5]. The most common method of preserving poultry meat for short-term storage is refrigeration. This, however, requires periodic storage of larger amounts of hen meat. A significant number of studies were conducted concerning the effects of time and storage conditions on the quality of broiler chicken meat [6–10]. However, there is a lack of research concerning the influence of time and storage conditions on the quality of hen meat after termination of the laying period, as well as a lack of estimates on the length of its shelf life. Shelf life is the recommended maximum storage time allowing for safe food consumption. Poultry meat is perishable and a loss/drop in quality occurs quickly during its storage [11,12]. Inappropriate storage conditions result in the development of microorganisms and enzymatic changes that cause meat spoilage [13–18], deterioration of its sensory characteristics [12,19,20] and physicochemical characteristics [21]. In order to delay spoilage and lengthen the shelf life, poultry meat is most often stored in cool conditions, at a temperature of 0 °C to 7 °C [22–24]. Storage in cool conditions limits the development of microorganisms and slows down the intensity of chemical transformations [25–27]. The total number of microorganisms and *Pseudomonas* spp. is most frequently used to assess the microbiological quality of meat stored under aerobic conditions [10,14,15,18,28]. *Pseudomonas* spp. bacteria are considered specific microorganisms that spoil fresh meat during chilled storage. Poultry meat is particularly sensitive to the activity of these bacteria. Their impact involves the enzymatic acceleration of proteolysis, and oxidative and hydrolytic processes of fat tissues, generating unpleasant flavour and odour, improper colouring, and mucus [14,15,18,29,30]. The shelf life for consumption of poultry meat is most often determined on the basis of its sensory characteristics and the overall number of microorganisms, which should not exceed 7 log cfu/g [31]. Also, according to Gratta et al. [32], the termination of storage life of poultry meat is 7 log cfu/g for the total number of microorganisms and 7.3 log cfu/g for *Pseudomonas* bacteria, which are characteristic for spoiled poultry meat [15,16]. The shelf life of poultry meat depends mainly part of the initial microbiological load [15,33,34]. During the successive steps from poultry slaughtering to meat production, bacteria from the air and the environment can contaminate poultry meat. The skin of poultry carcasses and cuts is directly in contact with air and equipment surfaces, and is therefore easily contaminated. However, in fresh meat, bacteria are present on the surface rather than in the meat. Poultry meat is therefore exposed to a large number of bacteria that cause meat to spoil [17,18,35,36], and the initial level of bacterial contamination is greater than that of beef or pork [37]. An inappropriate storage temperature may be a factor that reduces the shelf life of cooled foodstuffs [6,16,38], especially when it undergoes fluctuations [39]. Traditional cooling temperatures in refrigeration equipment are usually between 0 °C and 7 °C, but often, consumers do not pay attention to the importance of the impact of storage temperature on the quality characteristics of meat without any treatment.

The purpose of this work was to assess the effect of temperature (2 °C and 6 °C) and storage time on the physicochemical, microbiological, and sensory characteristics of hen meat after the end of the laying period. The storage temperatures were chosen to account for the typical temperature range during refrigerated storage of poultry meat.

2. Materials and Methods

2.1. Research Material

The research material consisted of breast meat obtained from the carcasses of 56-week-old Rod Island Red (RIR) hens after the end of the laying period, of even body weights. RIRs are hens of dual-purpose breeds. The hens were on a certified ecological farm. During their entire period of use, the birds were raised according to the principles of ecological hen production, determined by the EU and national regulations [40–42]. The carcasses were subjected to a cooling process at a temperature of 4 °C. Maintaining the cooling chain, the carcasses were transported in isothermal containers to the laboratory. A simplified dissection according to the method of Ziołecki and Doruchowski [43] was conducted. During the dissection with a sterile knife, the breast meat (90 single units) was prepared, removing the external fat, which was used in further research. The evaluation of the material quality (day 0 of storage) was conducted on 10 single, randomly chosen meat breasts. The breast meat prepared for storage was divided into 2 experimental groups: the first group with a cool storage temperature of 2 °C; and the second group with a storage temperature of 6 °C. Each group included 40 randomly chosen meat breasts for the purpose of their further storage (for 2, 4, 6, 8 days) under stable temperature conditions in 2 refrigerators (FK v36/10 of Liebherr, Germany company). The samples (single meat breasts with a weight of 110 ± 5 g) were identified by weight, and were quickly and without delay transferred to separate marked containers for food storage. They were hand-wrapped in an air-permeable polyethylene foil, maintaining sterility. Each day, the samples were taken from 10 containers from the 2 refrigerators with the stable, monitored temperature conditions of 2 °C and 6 °C for further microbiological, physicochemical, and sensory evaluation.

2.2. Evaluation of the Physicochemical Characteristics of Meat

The pH level was measured with a digital pH meter-HI 99,163 (Hanna Instrument Company, Woonsocket, RI, USA) equipped with a combined electrode FC232 (Hanna Instrument Company, Woonsocket, RI, USA). Before measuring, the pH meter was calibrated using a two-point method towards the calibrated pH 4.01 and pH 7.01 buffers (Hanna Instrument Company, Salaj, Romania). The measures were conducted at the room temperature of 20–24 °C. The average pH value was determined based on 5 measurements of the same sample, and the procedures were the same for all samples. The surface colour of the cross section of the stored meat was determined with use of a colorimeter (CR-300; Minolta Camera, Osaka, Japan). Before measurement, the colorimeter was calibrated regarding the white standard to Y 94.2, x 31.63, y 33.30. A method of reflection was used with a standard lighting D65 observation angle of 2°. The product colour was shown as lightness (L*), redness (a*), and yellowness (b*), in accordance with the International Commission on Illumination (CIE) colour systems; the values were indicated from the average value of six randomly chosen readings. Tenderness was assessed on raw samples with the dimensions of (mm) 10 × 10 × 50, cut out parallelly to the course of muscle fibres. For measurement of the shear force (F max) use was made of a universal testing machine, Zwick/Roell BT1-FR1.OTH.D14 (Zwick CmbH&Co.KG., Ulm, Germany), with a Warner-Bratzler single-knife cutting system (one flat knife with a width of 1.2 mm with triangular incision at a 60° angle, of which the internal edge is also the working edge), with a head speed of 100 mm·min^{-1}, and pretension force of 0.2 N. The cutting was conducted perpendicularly to the course of muscle fibres, and the temperature of the samples was 20 °C. The water holding capacity (WHC) was determined based on the amount of juice pressed out using the method of Grau and Hamm [44]. For this purpose, samples with a weight of 0.25–0.35 g were placed on tissue paper to press out the meat juice, and were loaded with a mass of 2 kg. After pressing out the juice, the samples were weighed again. Water absorption (%) was calculated from the following formula: $M_1 - M_2/M_1 \times 100$, where: M_1—mass of test portion; and M_2—mass of the test portion after squeezing out the juice. The drip loss (%) was determined on the basis of the loss in weight of the meat before (day 0) and after the storage period.

2.3. Microbiological Analysis

The study determined the total number of aerobic microorganisms, including Enterobacteriaceae, *Pseudomonas* spp., and *Salmonella* bacteria. In addition, typical microbial colonies were identified using MALDI-TOF MS Biotyper mass spectrometry by measuring unique bacterial proteins. Microbiological analysis was performed at 0, 2, 6, and 8 days of refrigerated storage at 2 °C and 6 °C. Each test subject was analysed in 3 independent replicates. Serial dilutions from 10^{-1} to 10^{-5} were made from the samples obtained. Total aerobic microbial counts to calculate the colony-forming units (cfu/g) were performed on a TSA medium (Triticasein Soy Lab-agar, Biomaxima, Lublin, Poland) according to PN-EN ISO 11133:2014-07 [45]. An endo Agar (Merck KGaA, Darmstadt, Germany), according to ISO 9308-1:2014 [46], used to determine Enterobacteriaceae bacteria, *Pseudomonas* spp. bacteria on a Pseudomonas Isolation Agar (Merck KGaA, Darmstadt, Germany) according to ISO 13720:2010 [47], and Salmonella bacteria on an SS medium (Biomaxima, Lublin, Poland) according to EN ISO 6579-1:2017 [48].

Qualitative analysis of microbial isolates was performed using the method described by Shell et al. [49] using MALDI-TOF mass spectrometry (Bruker Daltonics, Bremen, Germany). The generated spectra were analysed on a MALDI-TOF Microflex LT instrument (Bruker Daltonics, Bremen, Germany) using Flex Control 3.4 software. The probability of a correct identification was expressed by the instrument in point form. For scores between 2300 and 3000, a reliable identification of microorganisms to the species level was obtained; for 2000–2299, reliable designation to the genus level and probable identification to the species level were achieved; scores between 1700 and 1999 indicated a likely identification result to the genus level; and scores below and/or equal to 1699 produced an unreliable identification result [50]. The presented results fall within the range of 2000–3000. The MaldI-TOF MS method (matrix-assisted laser desorption/ionisation time-of-flight mass spectrometry) method is based on the analysis of the protein profile of the organism [51]. This method has found its special place in meat microbiology, as a fast and inexpensive method, additionally characterised by high accuracy in the identification of bacteria. The identification of microorganisms is primarily based on the detection of ribosomal proteins, but also mitochondrial proteins that can be isolated [52–55].

2.4. Sensory Evaluation

Evaluation of the sensory parameters of samples in cold storage was conducted by a 10-person assessment team with tested sensory sensitivity and at least 4 years of experience in evaluation using a scaling method. The panellists were trained in sensory analysis according to ISO 8586-2 and familiarised themselves with the evaluation form prior to participating in the study [56].

The study made use of a specially prepared evaluation form.

A 5-point hedonic scale was used. The attributes subject to evaluation were: intensity and desirability of odour (5—very strong, typical; 4—strong, typical; 3—weak, unnoticeable, typical; 2—slightly changed, not strong, 1—changed, spoiled); colour of cross section and external colour (5—even, typical; 4—desired, less even, typical; 3—moderately desired, uneven, changed in places; 2—slightly undesired, changed in places, infiltrations, yellow; 1—very undesired, changed in places, yellow or green); texture (5—resilient muscle tissue, dense; 4—averagely resilient, deformation evens out after pressing; 3—muscle tissue after finger pressure remains deformed; 2—muscle tissue loosened, flattens after pressing; 1—muscle tissue loosened after pressing, easily separates); general appearance (5—no objections, surface moist, typical; 4—desired, surface slightly dried; 3—moderately desired, surface dry or slightly moist; 2—undesired, surface slimy, slightly sticky, colour changed in places; 1—very undesired, surface sticky from mucus). To ensure a proper assessment, the samples were coded and presented to the evaluators in white containers. The tests were carried out in appropriately prepared rooms free of foreign odours, at a temperature of 20 °C and with lighting eliminating any distracting factors, according to the relevant standards [57].

2.5. Statistical Analysis

The results obtained were statistically analysed using the analysis of variance (ANOVA) and Statistica 13.3 software package [58]. The research results are tabulated as the mean values, standard deviation, and range. The collected data were verified for normality using the Kolmogorov–Smirnov test. The impact of the effect of temperature (2 °C and 6 °C) and storage time (for 2, 4, 6, and 8 days) on the physicochemical and microbiological characteristics of the hen meat after the end of the laying period was evaluated with a Tukey's post hoc tests. The sensory evaluation was characteristics using non-parametric Kruksal–Wallis tests. Differences were considered significant at $p < 0.05$.

3. Results and Discussion

During storage, the physical–chemical characteristics of meat may undergo changes, including in pH, water absorption, colour, and tenderness. In our study, the acidity (pH) of raw hen breast meat on the first day after slaughter remained at the level of 5.62 ± 0.02 (Table 1). With increased storage time, both at a temperature of 2 °C and at a temperature of 6 °C, the degree of acidity of the meat increased. The increase in pH level of the meat during storage at a temperature of 6 °C was greater than at a temperature of 2 °C, and on the 8th day of storage reached the level of 6.89 ± 0.10. The production of lactic acid bacteria and the accumulation of alkaline components produced by psychrotrophic bacteria and the autolytic activity of the autochthonous enzymes may be the main reason for the change in pH during storage [59]. The increase in pH level of the meat during storage may be caused by the activity of microorganisms developing in the meat, which have the ability to produce enzymes catalysing the process of deamination of amino acids, which, consequently, causes the formation of nitrogenous bases and ammonia [26,59,60].

Table 1. Physicochemical characteristics of hen meat after the laying period depending on the temperature and period of time in cool storage (means ± standard deviation).

Tested Parameter	0 Day	Storage Temp. (°C)	Refrigerated Storage Time (Days)				p-Value
			2	4	6	8	
pH	5.62 [a] ± 0.02	2	5.66 [a] ± 0.03	5.77 [b] ± 0.03	5.82 [b] ± 0.02	5.85 [b] ± 0.02	<0.001
		6	5.67 [a] ± 0.04	6.20 [b] ± 0.24	6.25 [b] ± 0.08	6.89 [c] ± 0.10	<0.001
		p-value **	0.491	<0.001	<0.001	<0.001	
Drip loss (%)	-	2	0.93 [a] ± 0.15	1.39 [a] ± 0.15	2.01 [b] ± 0.12	2.32 [b] ± 0.10	<0.001
		6	1.07 [a] ± 0.10	2.39 [b] ± 0.25	2.92 [c] ± 0.26	3.15 [c] ± 0.23	<0.001
		p-value**	0.026	<0.001	<0.001	<0.001	
WHC (%)	24.39 [b] ± 2.34	2	31.81 [a] ± 2.58	25.00 [b] ± 2.61	23.33 [b] ± 2.80	28.53 [b] ± 3.10	0.041
		6	31.32 ± 2.32	30.41 ± 2.56	32.25 ± 2.40	32.041.12	0.066
		p-value **	0.232	<0.001	<0.001	<0.001	
Colour: L*—lightness	57.02 [a] ± 3.20	2	56.17 [a] ± 1.78	54.62 [b] ± 1.89	54.50 [b] ± 1.83	53.42 [b] ± 1.88	<0.001
		6	56.17 [a] ± 1.78	49.27 [b] ± 1.33	49.30 [b] ± 1.13	45.60 [b] ± 1.79	<0.001
		p-value **	0.023	<0.001	<0.001	<0.001	
a*—redness	3.69 [a] ± 0.62	2	3.73 [a] ± 0.66	3.78 [a] ± 0.28	3.90 [a] ± 0.36	4.22 [b] ± 0.45	0.044
		6	3.78 [a] ± 0.27	4.20 [b] ± 0.70	3.98 [ab] ± 0.25	4.28 [c] ± 0.39	0.003
		p-value **	0.051	0.030	0.535	0.902	
b*—yellowness	6.78 [a] ± 0.02	2	9.76 [b] ± 0.76	11.58 [c] ± 1.26	11.49 [c] ± 0.49	11.98 [c] ± 1.19	<0.001
		6	10.09 [b] ± 1.36	13.16 [d] ± 1.30	14.02 [d] ± 0.89	13.41 [d] ± 0.85	<0.001
		p-value **	0.504	0.004	<0.001	0.020	
Shear force (N)	31.20 [a] ± 4.50	2	31.00 [a] ± 2.70	30.25 [a] ± 3.40	28.90 [a] ± 3.62	25.60 [b] ± 3.80	0.038
		6	30.68 [a] ± 4.50	25.70 [b] ± 2.80	20.84 [c] ± 3.20	19.90 [c] ± 2.80	0.002
		p-value **	0.502	0.032	<0.001	<0.001	

** p-value, the level of significance when estimating the differences between the means of the parameters tested at storage temperatures of 2 and 6 °C; [a,b,c,d]—marked with different letters in the rows differ.

An increase in pH levels during the storage of poultry meat was also demonstrated by Surmei et al. [61], Ruíz-Cruz et al. [62], Nikmanesh et al. [59], Saleh et al. [25], and Katiyo et al. [12]. In the research of Sujiwo et al. [7], concerning the storage of broiler chicken meat at a temperature of 4 °C, no significant changes in the pH level were noted until the 9th day of storage. In our study, a high pH level (6.89) of meat on the 8th day of storage at a temperature of 6 °C was accompanied by the loss of suitability for consumption, which is concurrent with the results of Surmei and Usturoi [63], who stated that poultry meat is considered to be of very good quality with a pH not exceeding 6.2, whereas at a pH higher than 6.7, meat becomes unsuitable for consumption. In their study, Katiyo et al. [12] stated that changes in the pH level of meat have a significant influence on other meat characteristics, including colour and water absorption.

Our research demonstrated that the storage temperature had a considerable influence on the water holding capacity (WHC). Storage temperature and time, and microbiological growth are the main factors that influence the water retention ability of myofibrils in meat during storage under cold conditions. A significant decrease in water holding capacity was shown at the lower storage temperature. At a lower storage temperature, a significant fall in water absorption was observed on the 8th day of storage (Table 1). The research results obtained are consistent with the results of Sinhamahapatra et al. [64], Aziman et al. [65], and Hussein et al. [21] for the water absorption of broiler chicken meat during storage at a temperature of 4 °C. Most likely, the reduction in water contained in the meat with the passage of storage time is a result of greater leakage from mature meat, which in turn can increase the relative share of raw protein in stored meat [21]. The leakage of the cooling meat stored at temperatures of 2 °C and 6 °C increased along with the storage time at both temperatures of 2 °C and 6 °C (Table 1). The losses of cooled meat depend on the pH of the meat [66]. Poultry meat with a low pH has been associated with low water holding capacity (WHC), which results in increased cook loss, drip loss, and shelf life, and decreased tenderness. Similar results were noted in the research of Ruíz-Cruz et al. [62], and Chmiel and Słowiński [6], as well as Kondratowicz et al. [30].

Measuring the shear force is the most effective method of evaluating the tenderness of meat [62], including a raw meat. Our study indicated that the tenderness of the meat stored at a temperature of 6 °C, measured as a value of shear force, increased significantly ($p < 0.05$) from the 4th day of storage, while increased tenderness of the meat stored at a temperature of 2 °C was not noted until the 8th day of storage (Table 1). A reduction in the shear force during storage was observed in the studies of Sujiwo et al. [7] and Gratta et al. [32]. In the study by Gratta et al. [32], the test material was also raw meat. In the work of Ruíz-Cruz et al. [62], the shear force of broiler chicken fillets was reduced over the first 4 days of cool storage and increased in the next days. The tenderness of meat depends on myofibril proteolysis, which helps maintain the integrity of muscle fibre [63]. In the opinion of Kruk et al. [67], protein degradation in meat may be caused by bacterial or enzymatic processes occurring in meat in cold storage.

In our research it was indicated that along with the storage time and lightness (L^*) of colour, storage at a temperature of 2 °C did not present significant changes. However, while hen meat after the laying period stored at a temperature of 6 °C on the 4th, 6th, and 8th day of storage was characterised by a darker colour (lower indicator of lightness L^*) then on the first and second days of storage. The colour of the meat changed at 6 °C, where a higher proportion of red (parameter a^*) and yellow (parameter b^*) was shown compared to the meat stored at 2 °C (Table 1), regardless of storage time. The darkening of the chicken meats may be due to the reduction in the oxygen level in the surface tissue caused by microbial growth. This oxygen level reduction promotes the oxidation or denaturation of myoglobin and the formation of deoxymyoglobin, resulting in the degradation of the red colour within chicken meat [60]. At a temperature of 6 °C, the colour saturations of red (a^*) and yellow (b^*) also underwent changes. The increased colour saturation of yellow in the meat stored at the temperature of 6 °C was greater than at the temperature of 2 °C (Table 1). The results of the colour darkening during the 7-day storage period were obtained by Aziman et al. [65]. Changes in lightness during the storage period were also noted by Gratta et al. [32].

An approximative indicator of the microbiological quality of refrigerated meat is the overall number of microorganisms. This indicator assesses the total concentration of microorganisms in a sample, and helps determine the meat's level of hygiene and remaining shelf life [10,32,68]. However, *Pseudomonas* spp. bacteria are generally considered to be the microorganisms causing the spoiling of poultry meat stored in aerobic conditions [14,15,18,21,32,69]. Enterobacteriaceae are one of the potential bacterial spoilage groups in poultry meat [15,20,70]. However, the involvement of these bacteria and their role in poultry meat spoilage has not been fully investigated [71].

The research we conducted demonstrated that at a temperature of 2 °C, the overall number of bacteria of the family *Enterobacteriaceae* and the number of *Pseudomonas* spp. bacteria in the meat gradually increased, reaching the respective levels 4.64 log cfu/g,

4.16 log cfu/g, and 4.48 log cfu/g on the 8th day of cold storage (Table 2). In the meat stored at the temperature of 6 °C, the tempo of bacterial growth in all the assessment periods was significantly greater than at the temperature of 2 °C (Table 2). On the 6th day of storage at 6 °C, the overall number of bacteria in the meat exceeded 7 log cfu/g, considered to be the limit of microbiological safety [8]. On the 8th day of storage, microbiological evaluation of the meat stored at the temperature of 6 °C was ceased, as the sensory characteristics of the meat indicated its spoilage (Table 3) and loss of suitability for consumption.

Table 2. The microbiological quality of hen meat stored at temperatures of 2° and 6 °C (log cfu/g) (means ± standard deviation and range).

Parameter	0 Day	Storage Temp. (°C)	Refrigerated Storage Time (Days)				p-Value
			2	4	6	8	
Total microorganism count	3.48 [a] ± 0.39	2	3.50 [a] ± 0.05 3.43–3.57	3.72 [a] ± 0.44 3.30–4.30	3.80 [a] ± 0.54 3.30–4.43	4.64 [b] ± 0.38 4.34–5.34	<0.001
	2.85–3.87	6	4.11 [a] ± 0.45 3.63–4.66	6.29 [b] ± 0.15 6.13–6.47	7.07 [c] ± 0.37 6.62–7.62	nt	<0.001
		p-value **	0.007	<0.001	<0.001		
Family Enterobacteriaceae	<2	2	3.61 [a] ± 0.13 3.40–3.73	3.57 [a] ± 0.41 3.14–3.99	3.75 [a] ± 0.13 3.59–3.91	4.16 [b] ± 0.32 3.70–4.43	0.005
		6	3.70 [a] ± 0.10 3.53–3.80	5.70 [b] ± 0.54 5.09–6.24	6.76 [c] ± 0.09 6.62–6.88	nt	<0.001
		p-value **	0.248	<0.001	<0.001		
Pseudomonas spp.	2.87 [a] ± 0.35	2	2.85 [a] ± 0.77 2.20–3.32	3.69 [b] ± 0.13 3.59–3.83	3.70 [b] ± 0.14 3.51–3.91	4.48 [c] ± 0.25 4.00–4.66	<0.001
	2.41–3.32	6	3.70 [a] ± 0.04 3.62–3.73	6.32 [b] ± 0.12 6.18–6.51	7.11 [c] ± 0.23 6.73–7.38	nt	<0.001
		p-value **	0.023	0.020	0.007		
Salmonella spp.	nd	2	nd	nd	nd	nd	
		6	nd	nd	nd	nd	

nd—not detected; nt—not tested; ** p-value, the level of significance when estimating the differences between the means of the parameters tested at storage temperatures of 2 and 6 °C; [a,b,c]—marked with different letters in the rows differ.

Table 3. Evaluation of sensory characteristics of hen meat stored at temperatures of 2 °C and 6 °C (means ± standard deviation).

Parameter	0 Day	Storage Temp. (°C)	Refrigerated Storage Time (Days)			
			2	4	6	8
Intensity and desirability of the odour	4.90 [a] ± 0.30	2	4.62 [a] ± 0.30	[A] 4.10 [ab] ± 0.30	[A] 3.90 [ab] ± 0.32	[A] 3.30 [b] ± 0.52
		6	4.10 [a] ± 0.42	[B] 2.60 [b] ± 0.63	[B] 1.80 [c] ± 0.42	[B] 1.00 [c] ± 0.48
Outside colour	4.80 [a] ± 0.41	2	4.60 [a] ± 0.50	[A] 3.70 [ab] ± 0.51	[A] 3.80 [ab] ± 0.42	[A] 3.31 [b] ± 0.56
		6	4.30 [a] ± 0.60	[B] 2.50 [b] ± 0.42	[B] 2.60 [b] ± 0.69	[B] 1.40 [c] ± 0.51
Consistency	4.65 [a] ± 0.50	2	4.70 [a] ± 0.48	[A] 3.90 [b] ± 0.30	[A] 3.50 [b] ± 0.52	[A] 3.00 [c] ± 0.46
		6	3.50 [a] ± 0.36	[B] 2.20 [ab] ± 0.48	[B] 1.40 [c] ± 0.52	[B] 1.30 [c] ± 0.48
General appearance	4.85 [a] ± 0.36	2	4.60 [a] ± 0.52	[A] 3.90 [ab] ± 0.44	[A] 3.70 [ab] ± 0.48	[A] 3.00 [b] ± 0.52
		6	3.70 [a] ± 0.52	[B] 2.30 [b] ± 0.52	[B] 1.60 [bc] ± 0.52	[B] 1.50 [c] ± 0.52

[A,B]—marked with different letters in the columns differ at $p \leq 0.05$; [a,b,c]—marked with different letters in the rows differ at $p \leq 0.05$.

The increase in the overall number of bacteria in poultry meat stored in cooling conditions at a temperature of 4 °C was noted by Sujiwo et al. [7] and by Vergara-Figueroa et al. [20], and at the storage temperatures of 5 °C and 10 °C, as indicated by Ntzimani et al. [17]. Gratta et al. [32] demonstrated that both the overall number of microorganisms and of the *Pseudomonas* spp. bacteria increased along with the storage time of the poultry meat, normal and with myopathy, at a temperature of 4 °C, as well as Dourou et al. [18] at temperatures

0–5 °C in a refrigerator. The growth of Enterobacteriaceae in poultry meat in cold storage was indicated by Rouger et al. [15] and Vergara-Figueroa et al. [20]. Our research has demonstrated that in hen meat after the laying period, the presence of disease-causing bacteria (*Salmonella* spp.) was not noted, which reflects the good state of the health of hens from which the meat was obtained, as well as good sanitary conditions during the raising and slaughter of hens and proper storage conditions [3].

The study was conducted with the use of a MALDI-TOF MS Biotyper, allowing for an identification of bacteria in the meat stored at the temperatures of 2 °C and 6 °C. The results obtained in the current study have indicated that the initial microflora of hen meat after the laying period is consistent with published data for poultry meat [72–74]. From the test material obtained, 106 bacteria were isolated and correctly identified (score above 2000), of which the vast majority, 94%, were Gram-negative bacteria. The bacteria were clustered between nine families: Pseudomonadaceae (27%), Enterobacteriaceae (26%), Aeromonadaceae (18%), Staphylococcaceae (6%) and Moraxellaceae, Hafniaceae, Erwiniaceae, Comamonadaceae, and Yersiniaceae (less than 5%) (Figure 1). Based on the German Rules for Biological Agents #446, it was found that of the correctly identified bacteria, 34% were categorised as risk factor 1 (Biological agents that are unlikely to cause disease in an individual), while the remaining 66% were classified as risk factor 2 (biological agents that are likely to cause disease in humans and pose a risk to workers; spread in the community is unlikely; and effective prevention or treatment is usually possible). On day 0 of storage (24 h after cutting), bacteria belonging to four families were identified: Pseudomonadaceae (47%), Aeromonadaceae (21%), and Moraxellaceae (11%), as well as Staphylococcaceae, Enterobacteriaceae, Yersiniaceae, and Comamonadaceae (5% each). In the Pseudomonadaceae family, seven bacterial species were isolated: *Pseudomonas fluorescens* (11%), *Pseudomonas alcaligenes* (11%), *Pseudomonas koreensis* (5%), *Pseudomonas libanensis* (5%), *Pseudomonas synxantha* (5%), *Pseudomonas putida* (5%), and *Pseudomonas proteolytica* (5%). The Aeromonadaceae family was represented by *Aeromonas veronii* (21%), while the Moraxellaceae family by *Acinetobacter lwoffii* and *Acinetobacter calcoaceticus* (5% each). Among the Staphylococcaceae, the bacteria *Macrococcus caseolyticus* (5%) was isolated; in the Enterobacteriaceae family, the bacterium *Buttiauxella gaviniae* (5%) was isolated; among the Yersiniaceae, the bacterium *Serratia plymuthica* (5%) was isolated; and from the Comamonadaceae family, the bacterium *Comamonas aquatic* was identified. Bacteria from the family Moraxellaceae, Pseudomonadaceae, and members of the Vibrionaceae family were found in the meat of egg-laying poultry. *Pseudomonas* was found to be the predominant psychotropic spoilage-causing agent in meat [75–79]. On the 2nd day, the meat stored at 2 °C also had the highest percentage of bacteria from the Pseudomonadaceae family (31%), followed by Aeromonadaceae (23%), Enterobacteriaceae and Erwiniaceae (15% each), and the Hafniaceae and Staphylococcaceae families (8% each). Among the bacteria of the Pseudomonadaceae family, *Pseudomonas fragi* (23%) and *Pseudomonas anguilliseptica* (8%) were identified. *A. veronii* (15%) and *Aeromonas eucrenophila* (8%) were the bacteria correctly identified in the family Aeromonadaceae. The Enterobacteriaceae family was represented by *Escherichia coli* bacteria (15%), and Erwiniaceae by *Pantoea agglomerans* bacteria (15%). *Hafnia alvei* and *Macrococcus caseolyticus* bacteria (8% each) from the Hafniaceae and Staphylococcaceae families, respectively, were also identified.

On the 4th day of storage at 2 °C, all identified bacteria belonged to the Pseudomonadaceae family: *P. lundensis*, *P. alcaligenes*, *P. proteolytica*, and *P. teatrolens* in equal proportion. Additionally, on day six, bacteria from only one bacterial family were identified. Pseudomonadaceae was represented by *P. fragi* (50%), *P. alcaligenes* (20%), *P. putida* (10%), *P. proteolytica* (10%), and *P. teatrolens* (10%) (Figure 1). On the last eight days of storage of the meat sample at 2 °C, bacteria from the Pseudomonadaceae (50%), Moraxellaceae (25%), and Enterobacteriaceae (25%) families were identified. Among the bacteria of the Pseudomonadaceae family, *P. fragi* and *P. teatrolens* were identified (25% each), while for the Moraxellaceae family, *A. calcoaceticus* bacteria (25%) was identified, and among Enterobacteriaceae, *Enterobacter cloacae* bacteria (25%) was isolated.

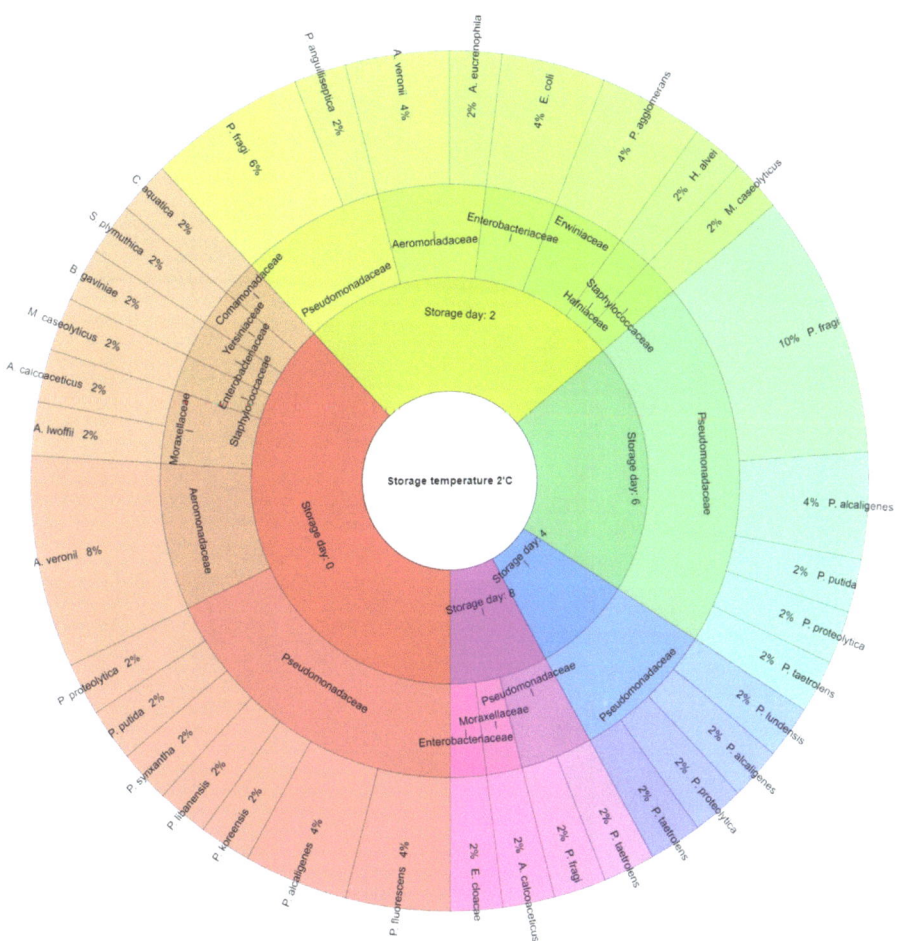

Figure 1. Isolated species of the bacteria from hen meat after the laying period storage at 2 °C.

On the 2nd day of meat storage at 6 °C, bacteria belonging to five different families were identified: Enterobacteriaceae (44%), Aeromonadaceae (38%), and Hafniaceae, Moraxellaceae, and Staphylococcaceae (6% each). In the Enterobacteriaceae family, the following were identified: *E. coli* (31%), *Kluyvera intermedia* (6%), and *L. amnigena* (6%). Among Aeromonadaceae, the following bacteria were isolated: *A. veronii* (19%) and *Aeromonas popoffii*, *Aeromonas hydrophila*, and *Aeromonas eucrenophila* (6% each). In addition, from the Hafniaceae family, the bacterium *H. alvei* (6%) was found; from the *Moraxellaceae* family, the bacterium *Acinetobacter pittii* (6%) was isolated; and from the *Staphylococcaceae* family, bacteria *S. pasteuri* (6%) were identified (Figure 2).

On the 4th day of storage at 6 °C, bacteria belonging to the following families were identified: Enterobacteriaceae (58%) and Aeromonadaceae (17%), as well as Staphylococcaceae, Hafniaceae, and Yersiniaceae (8% each). In the Enterobacteriaceae family, the bacteria identified were *Citrobacter freundii* (17%), as well as *Klebsiella oxytoca*, *E. cloacae*, *E. coli*, *Raoultella terrigena*, and *Enterobacter asburiae* (8% each); and in the Aeromonadaceae family, the bacteria *A. hydrophila* and *A. veronii* (8% each) were isolatead. In the case of bacteria from the families Staphylococcaceae, Hafniaceae, and Yersiniaceae, one genus each was identified, and these were *M. caseolyticus*, *H. Alvei*, and *Serratia liquefaciens* (8% each), respectively.

Figure 2. Isolated species of the bacteria from hen meat after the laying period storage at 6 °C.

On the 6th day of meat storage at 6 °C, bacteria belonging to seven families were identified: Enterobacteriaceae (36%), Yersiniaceae (25%), and Aeromonaidaceae (14%), as well as Staphylococcaceae, Erwiniaceae, Hafniiaceae (7% each), and Moraxellaceae (4%). In the Enterobacteriaceae family, the following bacteria were identified: *I. cloacae* (14%), *Raoultella planticola* (7%), *Cedecea* daivisae (7%), *C. freunidii* (4%), and *E. coli* (4%). In the family *Yersiniaceae*, the bacteria *S. liquefaciens* (25%) were identified; in the family Aeromonadiaceae, the bacteria *A. veronii* (11%) and *A. hydrophila* (4%) were isolated. On the other hand, from the family Staphylococcaceae, Erwiniaceae, Hafniaceae, and Moraxellaceae, the bacteria identified were *M. caseolyticus*, *P. agglomerans*, *H. alvei* (7% each), and *Acinetobacter calcoaceticus* (4%), respectively.

The composition of the bacterial microflora in the meat changed during storage, as confirmed by culture tests and identification via the MALDI-TOF MS Biotyper. The main microflora, regardless of the temperature used, were psychrophilic bacteria, which are characteristic and able to grow under refrigerated conditions. However, it was observed that during storage at 6 °C, the profile of the identified bacteria changed, with the majority of unfavourable microflora appearing, indicative of progressive spoilage processes. These bacteria may include, among others, *Aeromonas* spp., *Alcaligenes* spp., *Klebsiella* spp., and *Yersinia* spp. [20,75,80].

The effects of cool storage conditions on the sensory conditions of poultry meat were presented in the research of Chmiel and Słowiński [6], Ruíz-Cruz et al. [62], Yimenu et al. [36], Sujiwo et al. [7], and Kondratowicz et al. [30], as well as Garavito et al. [16]. In the study we conducted, over the storage time, all of the tested sensory characteristics of meat deteriorated, i.e., the intensity and desirability of odour, colour, texture, and general appearance (Table 3). The most obvious changes involved the odour of meat stored at a temperature of 6 °C. The intensity and desirability of the odour were reduced over the storage time.

Also, in the research of Katiyo et al. [12], the odour of chicken meat deteriorated more quickly than the colour or general appearance, and was strongly correlated with the growth in the number of microorganisms, which led the authors to state that the odour of raw meat may be a more reliable signal of rotting brought on by the development of microorganisms than the meat's appearance. The sensory quality of the breast meat of broiler chickens packed [36,81] and coated in an antibacterial coating [16] was reduced at a pace dependent on the temperature and increased storage time. In our research, the meat stored at a temperature of 2 °C was characterised by an acceptable, though not the highest, sensory quality until the 8th day of storage. On the other hand, the sensory characteristics of the meat stored at 6 °C on the 8th day of storage indicated its spoilage and loss of suitability for consumption (Table 3). According to Katiyo et al. [12], unpacked broiler chicken meat stored under aerobic conditions longer than 7 days was characterised by detrimental sensory characteristics and a total number of microorganisms higher than 8 log cfu/g.

4. Conclusions

The meat of hens of dual-purpose breeds after the laying period is over is now most often a waste product. In recent years, however, there has been growing interest in the possibility of using meat from the hens of such breeds. The results obtained from the evaluation of sensory, physicochemical, and microbiological characteristics showed that hen meat after the laying period stored under refrigerated conditions at 6 °C retains its shelf life only for up to 4 days, while storage at 2 °C retains shelf life up to 8 days of storage. The results obtained allow us to conclude that lowering the refrigerator temperature from 6 °C to 2 °C preserves the sensory characteristics and microbiological safety of post-lay hen meat at an acceptable level for longer, and therefore extends its shelf life.

The study provides new insights in terms of monitoring the breast muscle of laying hens after laying at different storage temperatures during different days, and gives an overview of how different bacterial species develop at different storage temperatures. The study provides new insights for practitioners in this field and identifies exactly which microorganisms develop during storage.

Author Contributions: Conceptualization, A.A.-P. and M.O.; methodology, A.A.-P., M.O., P.H. and M.K.; software, A.A.-P. and J.T.; validation, M.O., M.K. and P.H.; formal analysis, A.A.-P., P.H. and J.T.; investigation, A.A.-P.; resources, Z.S. and M.K.; data curation, A.A.-P. and P.H.; writing—original draft preparation, A.A.-P., M.O., Z.S., M.K. and P.H.; writing—review and editing, A.A.-P. and Z.S.; visualization, J.T.; supervision, Z.S.; project administration, J.T.; funding acquisition, J.T. All authors have read and agreed to the published version of the manuscript.

Funding: This project is financed by the program of the Minister of Education and Science named "Regional Initiative of Excellence" in the years 2019–2023, project number 026/RID/2018/19, and the amount of financing PLN 9 542 500.00.

Data Availability Statement: The data used to support the findings of this study can be made available by the corresponding author upon request.

Conflicts of Interest: The authors declare no conflict of interest.

References

1. Fan, H.; Wu, J. Conventional use and sustainable valorization of spent egg-laying hens as functional foods and biomaterials: A review. *Bioresour. Bioprocess.* **2022**, *9*, 43. [CrossRef] [PubMed]
2. Choe, J.; Kim, H.Y. Physicochemical characteristics of breast and thigh meats from old broiler breeder hen and old laying hen and their effects on quality properties of pressed ham. *Poult. Sci.* **2020**, *99*, 2230–2235. [CrossRef] [PubMed]
3. Sokołowicz, Z.; Augustyńska-Prejsnar, A.; Krawczyk, J.; Kačániová, M.; Kluz, M.; Hanus, P.; Topczewska, J. Technological and sensory quality and microbiological safety of RIR chicken breast meat marinated with fermented milk products. *Animals* **2021**, *1111*, 3282. [CrossRef] [PubMed]
4. D'Angelo, F. The European Union and the fight against food waste and losses: From policy to practice. In *Food Loss and Waste Policy*; Routledge: London, UK, 2022; pp. 92–105.
5. Laaninen, T. Reducing Food Waste in the European Union, EPRS: European Parliamentary Research Service. 2020. Available online: https://policycommons.net/artifacts/1426904/reducing-food-waste-in-the-european-union/2041486/ (accessed on 6 November 2023).
6. Chmiel, M.; Słowiński, M. Effect of Storage in Display Cases on the Sensory Quality of Chicken Breast Meat (M. Pectoralis). *Braz. J. Poult. Sci.* **2018**, *20*, 91–98. [CrossRef]
7. Sujiwo, J.; Kim, D.; Jang, A. Relation among quality traits of chicken breast meat during cold storage: Correlations between freshness traits and torrymeter values. *Poult. Sci.* **2018**, *97*, 2887–2894. [CrossRef] [PubMed]
8. Kaewthong, P.; Pomponio, L.; Carrascal, J.R.; Knøchel, S.; Wattanachant, S.; Karlsson, A.H. Changes in the Quality of Chicken Breast Meat due to Super chilling and Temperature Fluctuations during Storage. *J. Poult. Sci.* **2019**, *56*, 308–317. [CrossRef] [PubMed]
9. Kim, H.J.; Kim, H.J.; Jeon, J.; Nam, K.C.; Shim, K.S.; Jung, J.H.; Kim, K.S.; Choi, Y.; Kim, S.H.; Jang, A. Comparison of the quality characteristics of chicken breast meat from conventional and animal welfare farms under refrigerated storage. *Poult. Sci.* **2020**, *99*, 1788–1796. [CrossRef]
10. Marmion, M.; Ferone, M.T.; Whyte, P.; Scannell, A.G.M. The changing microbiome of poultry meat; from farm to fridge. *Food Microbiol.* **2021**, *99*, 103823. [CrossRef]
11. Fernández-Pan, I.; Carrión-Granda, X.; Mate, J.I. Antimicrobial efficiency of edible coatings on the preservation of chicken breast fillets. *Food Control* **2014**, *36*, 69–75. [CrossRef]
12. Katiyo, W.; de Kock, H.L.; Coorey, R.; Buys, E.M. Sensory implications of chicken meat spoilage in relation to microbial and physicochemical characteristics during refrigerated storage. *LWT-Food Sci. Technol.* **2020**, *128*, 109468. [CrossRef]
13. Arvanitoyannis, I.S.; Stratakos, A.C. Application of modified atmosphere packaging and active/smart technologies to red meat and poultry: A review. *Food Bioprocess Technol.* **2012**, *5*, 1423–1446. [CrossRef]
14. Hinton, A., Jr. Preventing spoilage of poultry meat. In *Achieving Sustainable Production of Poultry Meat*; Burleigh Dodds Science Publishing: Cambridge, UK, 2017; Volume 1, pp. 343–358.
15. Rouger, A.; Tresse, O.; Zagorec, M. Bacterial Contaminants of Poultry Meat: Sources, Species, and Dynamics. *Microorganisms* **2017**, *5*, 50. [CrossRef] [PubMed]
16. Garavito, J.; Moncayo-Martínez, D.; Castellanos, D.A. Evaluation of antimicrobial coatings on preservation and shelf life of fresh chicken breast fillets under cold storage. *Foods* **2020**, *9*, 1203. [CrossRef] [PubMed]
17. Ntzimani, A.; Kalamaras, A.; Tsironi, T.; Taoukis, P. Shelf Life Extension of Chicken Cuts Packed under Modified Atmospheres and Edible Antimicrobial Coatings. *Appl. Sci.* **2023**, *13*, 4025. [CrossRef]
18. Dourou, D.; Spyrelli, E.D.; Doulgeraki, A.I.; Argyri, A.A.; Grounta, A.; Nychas, G.J.E.; Chorianopoulos, N.G.; Tassou, C.C. Microbiota of chicken breast and thigh fillets stored under different refrigeration temperatures assessed by next-generation sequencing. *Foods* **2021**, *10*, 765. [CrossRef] [PubMed]
19. Sivarajan, M.; Lalithapriya, U.; Mariajenita, P.; Vajiha, B.A.; Harini, K.; Madhushalini, D.; Sukumar, M. Synergistic effect of spice extracts and modified atmospheric packaging towards non-thermal preservation of chicken meat under refrigerated storage. *Poult. Sci.* **2017**, *96*, 2839–2844. [CrossRef] [PubMed]
20. Vergara-Figueroa, J.; Cerda-Leal, F.; Alejandro-Martín, S.; Gacitúa, W. Evaluation of the PLA-nZH-Cu Nanocomposite Film on the Micro-Biological, Organoleptic and Physicochemical Qualities of Packed Chicken Meat. *Foods* **2022**, *11*, 546. [CrossRef] [PubMed]
21. Hussein, K.N.; Csehi, B.; József, S.; Ferenc, H.; Kiskó, G.; Dalmadi, I.; Friedrich, L. Effect of α-Terpineol on Chicken Meat Quality during Refrigerated Conditions. *Foods* **2021**, *10*, 1855. [CrossRef]
22. da Rocha, T.C.; Costa Filho, D.V.; de Carvalho, L.M.; de Carvalho, J.M.; Estevez, M.; Madruga, M.S. Effect of refrigeration and freezing on the oxidative stability of WB chicken breast. *LWT-Food Sci. Technol.* **2022**, *171*, 114108. [CrossRef]
23. Xu, Y.; Huang, J.C.; Huang, M.; Xu, B.C.; Zhou, G.H. The effects of different chilling methods on meat quality and calpain activity of pork muscle longissimus dorsi. *J. Food Sci.* **2012**, *77*, C27–C32. [CrossRef]
24. Latou, E.; Mexis, S.F.; Badeka, A.V.; Kontakos, S.; Kontominas, M.G. Combined effect of chitosan and modified atmosphere packaging for shelf life extension of chicken breast fillets. *LWT-Food Sci. Technol.* **2014**, *55*, 263–268. [CrossRef]
25. Saleh, E.; Morshdy, A.E.; El-Manakhly, E.; Al-Rashed, S.F.; Hetta, H.; Jeandet, P.; Yahia, R.; Batiha, G.; Ali, E. Effects of olive leaf extracts as natural preservative on retailed poultry meat quality. *Foods* **2020**, *9*, 1017. [CrossRef] [PubMed]

26. Al-jasser, M.S. Effect of cooling and freezing temperatures on microbial and chemical properties of chicken meat during storage. *J. Food Agric. Environ.* **2012**, *10*, 113–116.
27. Stonehouse, G.G.; Evans, J.A. The use of supercooling for fresh foods: A review. *J. Food Eng.* **2015**, *148*, 74–79. [CrossRef]
28. Bruckner, S.; Albrecht, A.; Petersen, B.; Kreyenschmidt, J. Characterization and comparison of spoilage processes in fresh pork and poultry. *J. Food Qual.* **2012**, *35*, 372–382. [CrossRef]
29. Andreani, N.A.; Fasolato, L. Pseudomonas and related genera. In *The Microbiological Quality of Food*; Woodhead Publishing: Cambridge, UK, 2017; pp. 25–59. [CrossRef]
30. Kondratowicz, J.; Chwastowska-Siwiecka, I.; Burczyk, E.; Piekarska, J.; Kuldo, Z. Sensory & microbiological assessment of turkey hens breast muscles depending on method and time of cold storage. *Żywność Nauka Technol. Jakość* **2011**, *18*, 143–152. [CrossRef]
31. Tuncer, B.; Sireli, U.T. Microbial growth on broiler carcasses stored at different temperatures after air- or water-chilling. *Poult. Sci.* **2008**, *87*, 793–799. [CrossRef]
32. Gratta, F.; Fasolato, L.; Birolo, M.; Zomeño, C.; Novelli, E.; Petracci, M.; Pascual, A.; Xiccato, G.; Trocino, A. Effect of breast myopathies on quality and microbial shelf life of broiler meat. *Poult. Sci.* **2019**, *98*, 2641–2651. [CrossRef]
33. Patsias, A.; Badeka, A.V.; Savvaidis, I.N.; Kontominas, M.G. Combined effect of freeze chilling and MAP on quality parameters of raw chicken fillets. *Food Microbiol.* **2008**, *25*, 575–581. [CrossRef]
34. Dawson, P.L.; Chaves, B.D.; Northcutt, J.K.; Han, I.Y. Quality and shelf life of fresh chicken breasts subjected to crust freezing with and without skin. *J. Food Qual.* **2013**, *36*, 361–368. [CrossRef]
35. Galarz, L.A.; Fonseca, G.G.; Prentice, C. Predicting bacterial growth in raw, salted, and cooked chicken breast fillets during storage. *Food Sci. Technol. Int.* **2016**, *22*, 461–474. [CrossRef] [PubMed]
36. Yimenu, S.M.; Koo, J.; Kim, B.S.; Kim, J.H.; Kim, J.Y. Freshness-based real-time shelf-life estimation of packaged chicken meat under dynamic storage conditions. *Poult. Sci.* **2019**, *98*, 6921–6930. [CrossRef]
37. Meredith, H.; Valdramidis, V.; Rotabakk, B.T.; Sivertsvik, M.; McDowell, D.; Bolton, D.J. Effect of different modified atmospheric packaging (MAP) gaseous combinations on Campylobacter and the shelf-life of chilled poultry fillets. *Food Microbiol.* **2014**, *44*, 196–203. [CrossRef] [PubMed]
38. Vaikousi, H.; Biliaderis, C.G.; Koutsoumanis, K.P. Applicability of a microbial time-temperature indicator (TTI) for monitoring spoilage of modified atmosphere packed minced meat. *Int. J. Food Microbiol.* **2009**, *133*, 272–278. [CrossRef] [PubMed]
39. da Silva, N.B.; Longhia, D.A.; Martinsa, W.F.; de Aragaoa, G.M.F.; Carciofia, B.A.M. Mathematical modeling of Lactobacillus viridescens growth in vacuum packed sliced ham under non isothermal conditions. *Procedia Food Sci.* **2016**, *7*, 33–36. [CrossRef]
40. Council Regulation (EC) No 834/2007 of 28 June 2007 on Organic Production and Labelling of Organic Products and Repealing Regulation (EEC) No 2092/91. 2007. Available online: https://eur-lex.europa.eu/eli/reg/2007/834/oj (accessed on 10 October 2023).
41. Commission Regulation (EC) No 889/2008 of 5 September 2008 Laying Down Detailed Rules for the Implementation of Council Regulation (EC) No 834/2007 on Organic Production and Labelling of Organic Products with Regard to Organic Production, Labelling and Control. 2008. Available online: https://eur-lex.europa.eu/eli/reg/2008/889/oj (accessed on 10 October 2023).
42. Law of June 23, 2022 on organic farming and production. *J. Laws* **2023**, *2023*, 1235.
43. Ziołecki, J.; Doruchowski, W. *Method of Assessing the Slaughter Value of the Poultry*; COBRD Publishing: Poznań, Poland, 1989.
44. Grau, R.; Hamm, R. Eine einfache Methode zur Bestimmung der Wasserbindung im Muskel. *Naturwiss* **1953**, *40*, 29–30. [CrossRef]
45. *PN-EN ISO 11133:2014-07*; Quality-Assured Culture Media for Food and Water Testing to Enhance Consumer Safety. ISO: Geneva, Switzerland, 2014.
46. *ISO 9308-1:2014*; Water Quality—Enumeration of *Escherichia coli* and Coliform Bacteria—Part 1: Membrane Filtration Method for Waters with Low Bacterial Background Flora. ISO: Geneva, Switzerland, 2014.
47. *ISO 13720:2010*; Meat and Meat Products—Enumeration of Presumptive *Pseudomonas* spp. ISO: Geneva, Switzerland, 2010.
48. *ISO 6579-1:2017*; Microbiology of the Food Chain—Horizontal Method for the Detection, Enumeration and Serotyping of Salmonella—Part 1: Detection of *Salmonella* spp. ISO: Geneva, Switzerland, 2017.
49. Shell, W.S.; Sayed, M.L.; Allah, F.M.G.; Gamal, F.E.M.; Khder, A.A.; Samy, A.A.; Ali, A.H.M. Matrix-assisted laser desorption-ionisation-time-of-flight mass spectrometry as a reliable proteomic method for characterisation of *Escherichia coli* and Salmonella isolates. *Vet. World* **2017**, *10*, 1083–1093. [CrossRef]
50. Kosikowska, U.; Stepień-Pysniak, D.; Pietras-Ozga, D.; Andrzejczuk, S.; Juda, M.; Malm, A. Application of MALDI-TOF MS for identification of clinical isolates of bacteria from humans and animals. *J. Lab. Diag.* **2015**, *51*, 23–30. [CrossRef]
51. Hess, C.; Alispahic, M.; Hess, M. Application of MALDI-TOF MS in veterinary and food microbiology. In *MALDI-TOF Mass Spectrometry in Microbiology*; Kostrzewa, M., Schubert, S., Eds.; Max Von Pettenkofer-Institut: Munich, Germany, 2016; pp. 109–126.
52. Peruzy, M.F.; Murru, N.; Yu, Z.; Cnockaert, M.; Joossens, M.; Proroga, Y.T.R.; Houf, K. Determination of the microbiological contamination in minced pork by culture dependent and 16S amplicon sequencing analysis. *Int. J. Food Microbiol.* **2019**, *290*, 27–35. [CrossRef]
53. Ramatla, T.; Ngoma, L.; Mwanza, M. The Utility of MALDI-TOF-Mass Spectrometry, Analytical Profile Index (API) and Conventional-PCR for the Detection of Foodborne Pathogens from Meat. *J. Food Nutr. Res.* **2021**, *9*, 442–448. [CrossRef]

54. Altakhis, M.; Pillidge, C.J.; Osborn, A.M.; Torley, P.J.; Kaur, M. Assessment of the potential use of MALDI-TOF MS for the identification of bacteria associated with chilled vacuum-packaged lamb meat. *Meat Sci.* **2021**, *177*, 108508. [CrossRef] [PubMed]
55. Kunová, S.; Sendra, E.; Haščík, P.; Vukovic, N.L.; Vukic, M.; Kačániová, M. Influence of essential oils on the microbiological quality of fish meat during storage. *Animals* **2021**, *11*, 3145. [CrossRef] [PubMed]
56. ISO 8586-2; Sensory Analysis—General Guidance for the Selection, Training and Monitoring of Assessors—Part 2: Expert Sensory Assessors. ISO: Geneva, Switzerland, 2008.
57. PN-EN ISO 8589:2010; General Guidelines for the Design of a Sensory Analysis Laboratory. iTeh Standards: Newark, DE, USA, 2010.
58. *StatSoft Electronic Statistics Textbook*; Data Analysis Software System, Version 13.3; StatSoft, Inc.: Kraków, Poland, 2018.
59. Nikmanesh, A.; Baghaei, H.; Mohammadi Nafchi, A. Development and Characterization of Antioxidant and Antibacterial Films Based on Potato Starch Incorporating Viola odorata Extract to Improve the Oxidative and Microbiological Quality of Chicken Fillets during Refrigerated Storage. *Foods* **2023**, *12*, 2955. [CrossRef] [PubMed]
60. Haghighatpanah, N.; Omar-Aziz, M.; Gharaghani, M.; Khodaiyan, F.; Hosseini, S.S.; Kennedy, J.F. Effect of mung bean protein isolate/pullulan films containing marjoram (*Origanum majorana* L.) essential oil on chemical and microbial properties of minced beef meat. *Int. J. Biol. Macromol.* **2022**, *201*, 318–329. [CrossRef] [PubMed]
61. Surmei, E.; Usturoi, M.G.; Albu, A. Studies on the Evolution of refrigerated poultry meat. *Lucr. Stiintifice Ser. Zootehni.* **2013**, *60*, 170–172.
62. Ruíz-Cruz, S.; Valenzuela-López, C.C.; Chaparro-Hernández, S.; Ornelas-Paz, J.D.J.; Toro-Sánchez, C.L.D.; Márquez-Ríos, E.; Valdez-Hurtado, S. Effects of chitosan-tomato plant extract edible coatings on the quality and shelf life of chicken fillets during refrigerated storage. *Food Sci. Technol.* **2018**, *39*, 103–111. [CrossRef]
63. Surmei, E.; Usturoi, M.G. Considerations regarding quality of Poultry Meat stored in refrigeration conditions. *Lucr. Științifice–Ser. Zooteh.* **2012**, *58*, 199–202.
64. Sinhamahapatra, M.; Biswas, S.; Das, A.K.; Battacharya, D. Comparative study of different surface decontaminants on chicken quality. *Br. Poult. Sci.* **2004**, *45*, 624–630. [CrossRef]
65. Aziman, N.; Jawaid, M.; Mutalib, N.A.A.; Yusof, N.L.; Nadrah, A.H.; Nazatul, U.K.; Tverezovskiy, V.V.; Tverezovskaya, O.A.; Fouad, H.; Braganca, R.M. Antimicrobial Potential of Plastic Films Incorporated with Sage Extract on Chicken Meat. *Foods* **2021**, *10*, 2812. [CrossRef]
66. Mushi, D.E.; Safari, J.; Mtenga, L.A.; Kifaro, G.C.; Eik, L.O. Effects of concentrate levels on fattening performance, carcass and quality attributes of Small East African × Norwegian crossbred goats fed low quality grass hay. *Livest. Sci.* **2009**, *124*, 148–155. [CrossRef]
67. Kruk, Z.A.; Yun, H.; Rutley, D.L.E.; Lee, J.; Kim, Y.J.; Jo, C. The effect of high pressure on microbial population, meat quality and sensory characteristics of chicken breast fillet. *Food Control* **2011**, *22*, 6–12. [CrossRef]
68. Yehia, H.M.; Elkhadragy, M.F.; Al-Megrin, W.A.; Al-Masoud, A.H. Citrox Improves the Quality and Shelf Life of Chicken Fillets Packed under Vacuum and Protects against Some Foodborne Pathogens. *Animals* **2019**, *9*, 1062. [CrossRef] [PubMed]
69. Bruckner, S.; Albrecht, A.; Petersen, B.; Kreyenschmidt, J. Influence of cold chain interruptions on the shelf life of fresh pork and poultry. *Int. J. Food Sci. Technol.* **2012**, *47*, 1639–1646. [CrossRef]
70. Rahkila, R.; Johansson, P.; Säde, E.; Björkroth, J. Identification of enterococci from broiler products and a broiler processing plant and description of *Enterococcus viikkiensis* sp. nov. *Appl. Environ. Microbiol.* **2011**, *77*, 1196–1203. [CrossRef] [PubMed]
71. Zeitoun, A.A.M.; Debevere, J.M.; Mossel, D.A.A. Significance of Enterobacteriaceae as index organisms for hygiene on fresh untreated poultry, poultry treated with lactic acid and poultry stored in a modified atmosphere. *Food Microbiol.* **1994**, *11*, 169–176. [CrossRef]
72. Bolton, D.J.; Meredith, H.; Walsh, D.; McDowell, D.A. The effect of chemical treatments in laboratory and broiler plant studies on the microbial status and shelf-life of poultry. *Food Control* **2014**, *36*, 230–237. [CrossRef]
73. Radha Krishnan, K.; Babuskin, S.; Azhagu Saravana Babu, P.; Sasikala, M.; Sabina, K.; Archana, G.; Sivarajan, M.; Sukumar, M. Antimicrobial and antioxidant effects of spice extracts on the shel life extension of raw chicken meat. *Int. J. Food Microbiol.* **2014**, *171*, 32–40. [CrossRef]
74. Rodríguez-Calleja, J.M.; Cruz-Romero, M.C.; O'Sullivan, M.G.; García-López, M.L.; Kerry, J.P. High-pressure-based hurdle strategy to extend the shelf-life of fresh chicken breast fillets. *Food Control* **2012**, *25*, 516–524. [CrossRef]
75. Doulgeraki, A.I.; Ercolini, D.; Villani, F.; Nychas, G.J.E. Spoilage microbiota associated to the storage of raw meat in different conditions. *Int. J. Food Microbiol.* **2012**, *157*, 130–141. [CrossRef]
76. Rouger, A.; Moriceau, N.; Prévost, H.; Remenant, B.; Zagorec, M. Diversity of bacterial communities in French chicken cuts stored under modified atmosphere packaging. *Food Microbiol.* **2018**, *70*, 7–16. [CrossRef]
77. Lee, H.S.; Kwon, M.; Heo, S.; Kim, M.G.; Kim, G.B. Characterization of the biodiversity of the spoilage microbiota in chicken meat using next generation sequencing and culture dependent approach. *Korean J. Food Sci. Anim. Resour.* **2017**, *37*, 535–541. [CrossRef]
78. Morales, P.A.; Aguirre, J.S.; Troncoso, M.R.; Figueroa, G.O. Phenotypic and genotypic characterization of Pseudomonas spp. present in spoiled poultry fillets sold in retail settings. *LWT-Food Sci. Technol.* **2016**, *73*, 609–614. [CrossRef]

79. Stellato, G.; Utter, D.R.; Voorhis, A.; De Angelis, M.; Murat Eren, A.; Ercolini, D. A few Pseudomonas oligotypes dominate in the meat and dairy processing environment. *Front. Microbiol.* **2017**, *8*, 264. [CrossRef]
80. Molenda, J. *Microbiology of Food of Animal Origin*; Wrocław University of Environmental and Life Sciences Publishing: Wrocław, Poland, 2010.
81. Herbert, U.; Albrecht, A.; Kreyenschmidt, J. Definition of predictor variables for MAP poultry filets stored under different temperature conditions. *Poult. Sci.* **2015**, *94*, 424–432. [CrossRef]

Disclaimer/Publisher's Note: The statements, opinions and data contained in all publications are solely those of the individual author(s) and contributor(s) and not of MDPI and/or the editor(s). MDPI and/or the editor(s) disclaim responsibility for any injury to people or property resulting from any ideas, methods, instructions or products referred to in the content.

Article

Green Extraction Strategy Using Bio-Based Aqueous Biphasic Systems for Polyphenol Valorization from Grape By-Product

Aleksandra Dimitrijević [1,*], Slađana Marić [1], Ana Jocić [1], Danijela Tekić [1], Jasmina Mušović [1] and Joana S. Amaral [2,3]

[1] Vinča Institute of Nuclear Sciences—National Institute of the Republic of Serbia, University of Belgrade, Mike Petrovica Alasa 12–14, 11000 Belgrade, Serbia; sladjana.maric@vin.bg.ac.rs (S.M.); ana.jocic@vin.bg.ac.rs (A.J.); danijela.tekic@vin.bg.ac.rs (D.T.); jasmina.musovic@vin.bg.ac.rs (J.M.)

[2] Centro de Investigação de Montanha (CIMO), Instituto Politécnico de Bragança, Campus de Santa Apolónia, 5300-253 Bragança, Portugal; jamaral@ipb.pt

[3] Laboratório Associado para a Sustentabilidade e Tecnologia em Regiões de Montanha (SusTEC), Instituto Politécnico de Bragança, Campus de Santa Apolónia, 5300-253 Bragança, Portugal

* Correspondence: daleksandra@vin.bg.ac.rs

Abstract: Polyphenols are natural compounds with enhanced antioxidant properties. They are present in relatively high concentrations in fruit/vegetable by-products. Therefore, there is a need for the development of efficient and cost-effective methods for the separation and purification of these valuable compounds. Traditional extraction with organic solvents needs to be switched to novel methods that are more efficient, with reduced extraction times and low consumption of organic solvents. Aiming at developing sustainable processes for the separation and purification of phenolic compounds, we used three model compounds, namely resveratrol, quercetin, and gallic acid, to investigate ionic liquid-based aqueous biphasic systems (IL-ABSs) formed by cholinium-based IL in combination with polypropylene glycol with a molecular mass of 400 g/mol (PPG400). The ABS composition in the two-phase region was selected according to a previously determined phase diagram. Extraction studies indicated the preferential partition of resveratrol and quercetin toward the hydrophobic PPG-rich phase that is mainly dominated by its hydrophobic nature and the strong salting-out effect of ILs. On the other hand, due to its considerably hydrophilic nature, gallic acid preferentially migrates toward the IL phase. The achieved results from grape stem extract demonstrated high extraction efficiencies of cholinium dihydrogen phosphate (~99% for resveratrol for the PPG phase and 78% for gallic acid for the IL phase), with considerable selectivity, demonstrating promising outcomes for potential applications.

Keywords: resveratrol; quercetin; gallic acid; ionic liquids; extraction; aqueous biphasic systems; polyphenols; grape by-products; food waste

Citation: Dimitrijević, A.; Marić, S.; Jocić, A.; Tekić, D.; Mušović, J.; Amaral, J.S. Green Extraction Strategy Using Bio-Based Aqueous Biphasic Systems for Polyphenol Valorization from Grape By-Product. *Foods* **2024**, *13*, 954. https://doi.org/10.3390/foods13060954

Academic Editor: Annalisa Tassoni

Received: 28 February 2024
Revised: 12 March 2024
Accepted: 19 March 2024
Published: 21 March 2024

Copyright: © 2024 by the authors. Licensee MDPI, Basel, Switzerland. This article is an open access article distributed under the terms and conditions of the Creative Commons Attribution (CC BY) license (https://creativecommons.org/licenses/by/4.0/).

1. Introduction

Food loss and food waste are currently major problems faced by the food supply chain globally. The Food and Agriculture Organization of the United Nations estimates that approximately one-third of food is either lost or wasted [1]. This poses significant challenges from environmental, economic, and social perspectives. Fruit and vegetable processing industries generate high amounts of "waste", such as pomace, peels, seeds, and stems [2]. Several purposes have been adopted for these by-product materials, including use as soil fertilizers or animal feed or as biomass to produce energy. However, these fruit and vegetable by-products are rich in a wide range of high-added-value bioactive phytochemicals, mainly polyphenolic compounds [3]. Bioactive polyphenols are on the top of the well-known value pyramid of biomass, which makes food by-products very attractive from a circular bioeconomy perspective [4]. Polyphenolic compounds are scavengers of free radicals, which are products that are harmful to aerobic metabolism, leading to oxidative stress in organisms. Multiple studies have shown that polyphenol compounds exhibit diverse biological activities, including antioxidant and anti-inflammatory properties, cardiovascular health benefits, potential anti-cancer effects, antimicrobial activity, metabolic improvements, and neuroprotective effects [5]. Moreover, dozens of studies have been published recently regarding the possible use of polyphenols to treat SARS-CoV-2 based on previous evidence of phenolic activity against different viruses [6]. Therefore, growing knowledge about polyphenols' health benefits and increased health awareness among consumers promote the use of these bioactive compounds as natural food additives over synthetic agents. According to Allied Market Research, the polyphenol market was valued at USD 1.6 billion in 2020, and it is projected to reach USD 2.7 billion by 2030, registering a compound annual growth rate of 5.2% from 2021 to 2030 [7]. Therefore, fruit and vegetable by-products are a prevailing source of these valuable compounds, making their valorization a core objective of much current research.

Grape by-products (grape stems and grape pomace) are among the most investigated agri-food matrices with the most potential due to their exceptional abundance of diverse polyphenols, along with their substantial residue fractions and waste volume [8]. The wine industry is the main contributor of grape by-products, generating 9 million tons of grape waste annually around the world, with the by-products retaining 45–65% of the total polyphenols [9]. Two main by-products obtained from wine cellars are grape pomace, accounting for 10–20% of the processed material, and stems, which constitute about 2–8% [10]. With this consideration, it is rational to view grape by-products as valuable sources of useful chemical compounds rather than mere waste. Polyphenols found in grape by-products include anthocyanins, phenolic acids, flavanols, flavonols, and stilbenes (Figure 1).

Figure 1. Main phenolic compounds present in selected by-products [11,12].

Anthocyanins and phenolic acids have a more hydrophilic character, i.e., they are more soluble in water compared to other polyphenolic compounds in by-product matrices. As can be seen from in Figure 1, grape stems are rich in phenolic acid, flavanols, flavonols, and stilbenes [13]. Recent concerns over the safety and adverse health effects of synthetic food additives, particularly on neurological functions and behavior [14], have ignited a strong interest in exploring natural biologically active alternatives. Therefore, by extracting and utilizing these bioactive phytoconstituents, we can reduce the need for synthetic additives or ingredients sourced from non-renewable resources. This promotes a more sustainable and diversified food system, enhancing its capacity to withstand challenges such as resource scarcity and disruptions in the global food supply chain. While grape pomace from wine processing is already industrially utilized to extract anthocyanins, there's a growing need to further valorize the residues from wine processing. This includes developing methods to extract high-value co-products like natural health remedies, food supplements, and innovative nutrifunctional food ingredients. However, there is still a long way to go until all these residues gain a factual recovery pathway, making the winemaking process a more sustainable activity.

Commercial methods such as solvent extraction (maceration) and Soxhlet extraction are commonly used to extract plant polyphenols. However, these techniques are not environmentally friendly due to their extensive use of organic solvents. Additionally, they are time-consuming and often yield lower extraction efficiencies [15,16]. Alternative methods like ultrasound-assisted extraction and microwave-assisted extraction have gained relevance in recent years, but they necessitate sophisticated equipment that can be challenging to scale up for industrial applications. High-pressure and high-temperature processes are generally avoided, as they can lead to the degradation of thermally unstable bio-active compounds during polyphenol extraction [17,18]. Polyphenols such as gallic acid, resveratrol, and quercetin can undergo significant degradation at higher temperatures, which makes high-temperature methods unsuitable for extraction. For example, degradation of GA begins from 60 °C, with 30% degradation occurring at 100 °C [19,20]. Generally all polyphenolic compounds are prone to pH and temperature-dependent degradation [17,21–23]. Moreover, a large fraction of phenolic compounds are left behind after extraction with traditional methods, as they are usually covalently bound to complex polysaccharides in the cell walls of food matrices [24]. In response to these challenges, it is of high importance to develop modern and efficient green solvent technology for extraction, separation, and concentration rooted in the principles of Green Chemistry, focusing on sustainable and environmentally friendly concepts for recovery of high-value polyphenols from selected food by-products. This entails employing efficient and benign solvents, and ideally, developed technology should integrate multiple steps, combining efficiency with sustainability and aligning with current societal and environmental standards.

Ionic liquids (ILs) are molten salts consisting entirely of ions (usually large asymmetric organic cations and organic or inorganic anions) that are liquid at ambient temperatures [25]. In the scientific literature, ILs are frequently designated as environmentally friendly alternatives to toxic organic solvents due to their stability and low vapor pressure, leading to a generalization of their benign nature [26,27]. In addition, it is well documented that ILs synthesized from carefully selected, safe, naturally-derived materials can be environmentally benign with enhanced biodegradability and biocompatibility [28–30]. Such ILs can be based on low-cost and non-toxic cations, thus opening doors for their application in bioactive molecule extraction and even their further preservation [31,32]. Ionic liquid-based aqueous biphasic systems (IL-ABSs) represent liquid–liquid extraction techniques that allow for the recovery of target products from complex samples [33,34]. IL-ABSs can be applied in both the extraction and purification steps of high-value compounds from food waste. Moreover, through the wise selection of the phase-forming components and their compositions, IL-ABSs may be biocompatible, efficient, and low-cost [30,35]. Briefly, IL-ABSs consist of two aqueous-rich phases formed by mixing aqueous solutions of ionic liquid and a polymer or salt at appropriate concentrations [36]. Recent investigations have demonstrated that ILs with cholinium cations paired with appropriate anions present outstanding biodegradable and low-toxicity properties [30,37]. Moreover, some cholinium ILs possess a remarkable biomass dissolution ability and cellular disruption potential, which could be crucial assets in designing extraction platforms for polyphenol recovery [33]. Saha et al., in their 2022 comprehensive review, highlighted that a considerable number of ABSs have been explored for polyphenol extraction. However, these systems predominantly involve polymer–polymer, polymer–inorganic salt, or imidazolium-based ABSs [35]. In polymer-based ABSs, a limited difference in polarities between the two phases often hampers their widespread application in extraction processes. Ionic liquids, well-known for their tunability, can span the entire spectrum of hydrophilicity to hydrophobicity. Previous reports focused mainly on the use of imidazolium IL-ABSs have shown promise in extracting phenolic acids; however, imidazolium ILs are not ideal due to their lack of biocompatibility and potential toxicity [38]. There is still a scarcity of publications related to the investigation of cholinium-based ABSs for polyphenol extraction [39–41]. Ribeiro and coworkers documented their use of choline chloride-based ABSs for the extraction of polyphenols and saponins from plant extracts, achieving extraction efficiencies of around 30% [42]. In a separate study, Wang et al. employed cholinium ILs featuring anions derived from amino acids to extract flavonoids and pectin from ponkan peels at room temperature. They then utilized an ABS for the subsequent separation of these two substances [43]. Moreover, a study by Neves et al. (2019) explored the use of ABSs formed by cholinium-derived ILs and carbohydrates for the simultaneous separation of antioxidants and carbohydrates from food waste. The process demonstrated high extraction efficiencies for both carbohydrate (89–92%) and antioxidant (65–75%) activities from an expired vanilla pudding sample. The methodology allowed for the recovery of antioxidants and the recycling of ILs, representing a promising approach for food waste valorization [41].

Considering the aforementioned motivations, our goal was to create a novel integrated approach combining environmentally friendly alternative solvents, biocompatible ionic liquids, and aqueous biphasic systems for selective separation of key polyphenolic compounds from grape by-products. Specifically, we focused on trans-resveratrol (RSV), quercetin-3-O-glucoside (QC), and gallic acid (GA), which are prominent representatives of their respective classes and have garnered significant interest due to their anti-cancer, cardiovascular protective, antioxidant, and anti-free radical properties. We investigated the partition potential of IL-ABSs formed by cholinium-based ILs with different anions (dihydrogen phosphate [DHP], dihydrogen citrate [DHCit], chloride, lactate [Lac], vanillate [Van], gallate [Gal], and nicotinate [Nic]) in combination with polypropylene glycol 400 for target compounds. [Ch][DHP], [Ch][DHCit], and [Ch]Cl were chosen as excellent salting-out agents among cholinium salts, while [Ch][Lac], [Ch][Gal], [Ch][Van], and [Ch][Nic] were chosen to obtain more customizable ABSs by introducing different anions derived from natural acids. The phase diagrams of the liquid–liquid equilibria of each ABS were determined at room temperature, and the appropriate composition was set up for subsequent partition experiments. Screening of these ABSs was conducted to extract polyphenols, and the logarithmic distribution coefficients were calculated to provide insights into the extraction mechanism. To optimize operational parameters and establish an experimental setup for real samples, [Ch][DHP] was selected as the ABS component due to its strong salting-out ability and selectivity, with the anticipation that this choice would enhance phase separation and extraction efficiency. The influence of ABS compositions on the extraction performance of [Ch][DHP]-based ABSs was investigated. Subsequently, a carefully tailored ABS was employed to extract polyphenols from grape stem extract.

2. Materials and Methods

2.1. Materials

Resveratrol (purity $\geq$ 99 wt%), quercetin (purity $\geq$ 99 wt%), and gallic acid (purity $\geq$ 99 wt%) were acquired from ExtraSynthese (Genay, France). Propylene glycol 400 (PPG400), choline hydroxide ([Ch][OH], purity $\geq$ 46 wt% in H_2O), cholinium dihydrogen citrate ([Ch][DHCit]), HPLC-grade acetonitrile, and HPLC-grade ethanol were purchased from Sigma-Aldrich (St. Louis, MA, USA). Nicotinic acid (purity $\geq$ 99%) and vanillic acid (purity $\geq$ 98 wt%) were procured from ThermoFisher (Dreieich, Germany), and lactic acid (purity $\geq$ 90%) was acquired from Fluka Chemie (Buchs, Switzerland). Cholinium chloride ([Ch]Cl, purity $\geq$ 98%) was obtained from Across Organic (Geel, Belgium), while cholinium dihydrogen phosphate ([Ch][DHP], purity $\geq$ 99%) was provided by Iolitec (Heilbronn, Germany). Cholinium vanillate ([Ch][Van]), cholinium gallate ([Ch][Gal]), cholinium lactate ([Ch][Lac]), and cholinium nicotinate ([Ch][Nic]) were synthesized in this work by following well-established protocols, via neutralization of cholinium hydroxide with the corresponding acid—vanillic, gallic, lactic, and nicotinic acid, respectively [30,44]. The water content was initially removed using a rotary evaporator (R-210 Rotavapor System, BÜCHI Labortechnik AG, Flawil, Switzerland) for 4 h, followed by an additional drying process at 70 °C under vacuum conditions for 36 h. ILs were subjected to Karl Fischer titration with a Metrohm 831 Karl Fischer coulometer, Herisau, Switzerland, revealing a water content of less than 400 ppm. The chemical structures of the synthesized ILs were confirmed by Fourier transform infrared spectroscopy (Nicolet iS5 spectrometer fitted with an iD7 ATR Accessory, Thermo Fisher Scientific, Waltham, MA, USA). The obtained spectra, along with peak assignments, are provided in the Supplementary Materials (Figure S1a–d). Figure 2 represents the chemical structures and abbreviations of the studied cholinium ILs/salts, as well as the structures and abbreviations of the studied polyphenolic compounds.

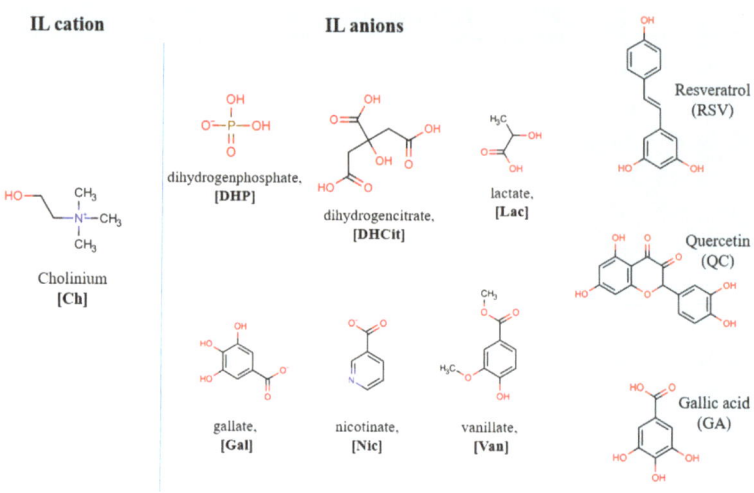

Figure 2. Chemical structures and abbreviations of the investigated cholinium ILs/salts and polyphenolic compounds.

2.2. Extraction of Polyphenolic Compounds

The capabilities of the investigated choline-based ABSs to extract RSV, QC, and GA were assessed. Considering previously established phase diagrams of ABSs, ternary mixtures within the biphasic region were prepared, containing 20 wt% salt/IL + 50 wt% PPG400 + 30 wt% of H_2O +10 μL of polyphenol solution. The polyphenol solutions were prepared in 80 vol% ethanol, with concentrations of circa 2500 mg L^{-1} for each compound. After screening of the partition behaviors of the studied polyphenols for the ABSs under study, the most selective system was selected for further optimization studies. This involved investigating the influence of diverse tie-line lengths (TLLs) of 56.88, 76.09, and 101.43 and initial compositions of ABSs along the chosen tie line, i.e., different phase ratios of the extraction parameters of [Ch][DHP]/PPG-based ABSs. The selected compositions of ABSs and corresponding TLLs are given in the Supplementary Materials.

Each ABS was prepared in 1.5 mL microcentrifuge tubes by adding the appropriate amount of constituents to achieve a final weight of 1 g. All systems were vigorously agitated (Reax Top, Heidolph, Schwabach, Germany) and left for 2 h at 25 °C in a thermo-shaker incubator (ALEMADR-MSC, Colo Lab Experts, Novo Mesto, Slovenia). Following 5 min of centrifugation at 5000× g rpm (LLG-uniCFUGE 5, Meckenheim, Germany), the phases were carefully separated, and their weights and volumes were measured. After appropriate dilution and filtration using syringe filters (0.45 μm), the contents of polyphenols in the phases were quantified by high-performance liquid chromatography (HPLC), as described in Section 2.5.

The logarithmic distribution coefficients (logD) of the polyphenols were calculated as the ratio of the equilibrium concentration of each polyphenol in the PPG400-rich and IL/salt-rich phases, as follows:

$$\log D = \log \frac{[\text{Polyphenol}]_{\text{PPG400-phase}}}{[\text{Polyphenol}]_{\text{IL/salt-phase}}}. \quad (1)$$

Recovery efficiencies of resveratrol and quercetin (RE, %) toward the PPG400-rich phase were calculated as follows:

$$\text{RE}\,(\%) = \frac{[\text{Polyphenol}]_{\text{PPG400-phase}} \cdot V_{\text{PPG400-phase}}}{[\text{Polyphenol}]_{\text{PPG400-phase}} \cdot V_{\text{PPG400-phase}} + [\text{Polyphenol}]_{\text{IL/salt-phase}} \cdot V_{\text{IL/salt-phase}}} \cdot 100 \quad (2)$$

while the recovery efficiency of gallic acid (RE%) toward the IL/salt-rich phase was calculated using following equation:

$$\text{RE}\,(\%) = \frac{[\text{Polyphenol}]_{\text{IL/salt-phase}} \cdot V_{\text{IL/salt-phase}}}{[\text{Polyphenol}]_{\text{PPG400-phase}} \cdot V_{\text{PPG400-phase}} + [\text{Polyphenol}]_{\text{IL/salt-phase}} \cdot V_{\text{IL/salt-phase}}} \cdot 100 \quad (3)$$

where $[\text{Polyphenol}]_{\text{PPG400-phase}}$ and $[\text{Polyphenol}]_{\text{IL/salt-phase}}$ represent the concentrations of RSV, QC, and GA, in the PPG400-rich and IL/salt-rich phases, respectively, and $V_{\text{PPG400-phase}}$ and $V_{\text{IL/salt-phase}}$ are the volumes of the PPG400-rich and IL/salt-rich phases, respectively.

The selectivities (S) with respect to gallic acid were calculated according to the following equation:

$$S_{\text{GA/Polyphenol}} = \frac{D_{\text{GA}}}{D_{\text{Polyphenol}}} \quad (4)$$

where the subscript "Polyphenol" stands for resveratrol or quercetin.

2.3. Recovery of Polyphenols from Grape Stems

Grapes collected from Quinta das Carvalhas, Pinhao (Regiao Demarcada do Douro, Portugal), were first separated into stems, seeds, and skins. All samples were freeze-dried, ground into powder, and stored in a desiccator. Two grams of the grape stem powder was mixed with 100 mL of a water:ethanol mixture (1:1, v:v), followed by stirring for 2 h and sonication for 5 min. The sample was centrifuged at 10,000× g rpm for 15 min, and the pellet was re-extracted. The supernatant was collected, and the solvent was evaporated under vacuum using a rotary evaporator at 40 °C. Using the HPLC analytical method, the concentrations of RSV and GA in grape stem extract were determined to be 5.2 and 8.9 µg g^{-1}, respectively, while quercetin was not detected (Figure S6 in Supplementary Materials). The resulting grape stem extract (5 mg) was added to the ternary ABS mixture with the composition indicated in the monophasic region of the phase diagram. This mixture was stirred continuously for at least two hours at a controlled temperature of 25 °C. Subsequently, an aqueous solution of [Ch][DHP] was introduced to the mixture to achieve a biphasic system composition of 40 wt% PPG400 and 25 wt% [Ch][DHP]. The solution was thoroughly mixed and allowed to equilibrate at room temperature for 2 h, followed by centrifugation at 5000× g rpm for 10 min and physical phase separation ([Ch][DHP]-rich and PPG400-rich phases). The phases were diluted, filtered through syringe filters (0.45 µm), and subjected to HPLC analysis. The experimental stages are summarized in Table 1.

Table 1. Experimental stages and appropriate system compositions (wt%).

Experimental Stage			IL	Polymer	Polyphenol Sample	T (°C)	System Composition	
							IL	PPG400
IL Synthesis		1	[Ch][Val]	/	/	25	/	/
		2	[Ch][Gal]					
		3	[Ch][Lac]					
		4	[Ch][Nic]					
Determination of ABS diagrams		1	[Ch][DHP]	PPG400	/	25	/	/
		2	[Ch][DHCit]					
		3	[Ch]Cl					
		4	[Ch][Lac]					
		5	[Ch][Gal]					
		6	[Ch][Nic]					
		7	[Ch][Van]					
Partition studies		1	[Ch][DHP]	PPG400	Standard ~2500 mg per L	25	~20	~50
		2	[Ch][DHCit]					
		3	[Ch]Cl					
		4	[Ch][Lac]					
		5	[Ch][Gal]					
		6	[Ch][Nic]					
		7	[Ch][Van]					
Optimization studies	TL	1	[Ch][DHP]	PPG400	Standard ~2500 mg per L	25	~20	~50
		2					~16	~40
		3					~12	~30
	ABS composition (phase ratio)	1	[Ch][DHP]	PPG400	Standard ~2500 mg per L	25	~20	~50
		2					~25	~40
		3					~40	~6
Recovery from grape stems		1	[Ch][DHP]	PPG400	5 mg stem extract per mL	25	~25	~40
Solubility		1	[Ch][DHP]	/	GA standard	25	~50	/
		2	/	PPG400	RSV standard	25	/	~80

2.4. Solubility Determination

The solubilities of RSV in a PPG400/water system with volume ratios of 80:20 and the solubility of GA in a [Ch][DHP]/water system with volume ratios of 50:50 were determined. Initially, RSV and GA were added to corresponding solutions in excess while continuously mixing at a constant temperature of 25 °C for 12 h. After centrifugation at 6000 rpm, the saturated solution was sampled, diluted, filtered, and quantified by UV spectroscopy using an LLG-uniSPEC2 spectrophotometer at wavelengths of 280 nm for GA and 307 nm for RSV. Quantification was carried out using calibration curves ($R^2 > 0.99$ in all cases) for each analyzed compound within the concentration range of 2.5 to 50 mg L^{-1}.

2.5. Quantification of Polyphenols

The quantification of polyphenols was carried out using HPLC (Knauer, Berlin, Germany), which includes a DAD detector and an Ascentis C18 (SUPELCO), Darmstadt, Germany, column with a particle size of 5 µm and dimensions 250 mm × 4.6 mm. Chromatographic separation of the target polyphenols occurred at a constant flow rate of 1 mL min^{-1}, the injection volume for all samples was 20 µL, and the temperature of the column was fixed at 45 °C, with 10% acetonitrile (v/v) as mobile phase A and pure acetonitrile as mobile phase B. For partition studies, isocratic conditions were used for the determination of each compound (A:B = 60:40). Quantification in a complex sample after extraction from the real sample was achieved using a gradient mode, varying the mobile phase from 100% A to 100% B for 30 min. The DAD detector was set to measure at 307 nm for RSV, 250 nm for QC, and 280 nm for GA, with retention times of 14.60, 12.50, and 2.60, respectively.

3. Results and Discussion

3.1. Liquid–Liquid Equilibria of Cholinium-Based ABSs

Seven extensively researched cholinium-based aqueous biphasic systems were selected to assess their efficacy in extracting polyphenols of varying polarities from aqueous solutions. These systems comprised three cholinium salts ([Ch][DHP], [Ch][DHCit], and [Ch]Cl) and four cholinium ionic liquids ([Ch][Lac], [Ch][Nic], [Ch][Van], and [Ch][Gal]), each combined with PPG400 and water to form biphasic systems for polyphenol extraction. Following the conventional definition of ionic liquids, cholinium-based compounds are categorized as such when their melting point is below 100 °C, as exemplified by [Ch][Gal], [Ch][Van], [Ch][Lac], and [Ch][Nic]. Conversely, cholinium salts like [Ch][DHP], [Ch][DHCit], and [Ch]Cl exhibit melting points above 100 °C [45]. Initial experiments involved precisely determining the mixture compositions capable of forming two phases suitable for extraction studies.

While there exists literature on liquid–liquid equilibria for selected cholinium-based ABSs with PPG400, there is a lack of comparative data regarding the ability of the studied series of ILs to form ABSs [30,37,46–48]. Therefore, we opted to investigate and compare solubility data to mitigate potential discrepancies stemming from temperature variations and variations in ionic liquid purities. In every ABS studied, the cholinium salt/IL and water predominantly form the lower phase, while the upper phase mainly consists of polypropylene glycol 400 (PPG400) and water. The respective phase diagrams are illustrated in Figure 3a, with the experimental data on weight fractions obtained via the cloud point method available in Table S1 of the Supplementary Materials. To further analyze the structural effects of the IL, the phase diagrams are also represented in molality units, as shown in Figure 3b, to account for the differences in molecular weights of the ILs. Regression parameters (A, B, and C) were calculated using the Merchuk equation [49], which is detailed as Equation (S1) in the Supplementary Materials, with values and their standard deviations listed in Table S2, indicating a high level of correlation. These parameters were applied to a [Ch][DHP]-based ABS to derive various tie-line (TL) data through a gravimetric method, as outlined in the Supplementary Materials. Figure S3 showcases a series of tie lines for this system, defined for different initial compositions and corresponding to increasing total tie-line lengths (56.88, 76.09, and 101.43). Additional information on these tie lines is available in Table S2 of the Supplementary Materials. An increase in hydrophobicity within the upper phase is observed, as highlighted by the rise in IL concentration at higher TLL values. For the tie line with a TLL of 101.43, two more ABSs were identified, each with different phase ratio volumes. These data on specific tie lines in the [Ch][DHP]-based ABS are crucial for the optimization studies presented in Section 3.3, providing insights into the system's performance under various conditions. In systems with larger biphasic regions, the ability of cholinium salts or ILs to undergo liquid–liquid separation is enhanced. Given that the cholinium cations remain constant across all salts/ILs, differences in phase behavior are mainly due to the distinct properties of the anions. This is in contrast to inorganic salt-based ABSs, where demixing is primarily driven by the salting-out effect of ions with high charge density. The demixing mechanism in IL polymer ABSs is recognized as more complex, stemming from the interaction dynamics among the three components, namely polymer, IL, and water [45,50]. There is a competitive interaction for water molecules between PPG400 and the IL ions, as well as among the ions themselves, with the IL anions more prone to forming hydration complexes compared to the IL cations [51].

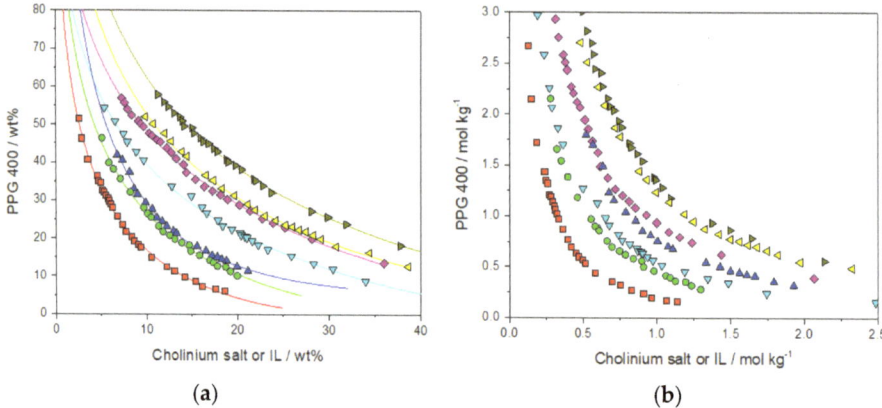

Figure 3. Ternary phase diagrams composed of cholinium salt or IL + PPG400 + H$_2$O at 25 °C. The compositions are given in (**a**) weight percents and (**b**) molality units (■, [Ch][DHP]; ●, [Ch][Lac]; ▽, [Ch][DHCit]; ▲, [Ch]Cl; ◆, [Ch][Gal]; ◁, [Ch][Nic]; ▶, [Ch][Van]).

According to the phase diagrams depicted in Figure 3b, the ability of cholinium salts/ILs to form ABSs decreases as follows: [Ch][DHP] > [Ch][Lac] > [Ch][DHCit] > [Ch]Cl > [Ch][Gal] [Ch][Nic] > [Ch][Van]. The ABS trend of high-melting-point cholinium salt is in accordance with data reported in existing literature [30,37]. [Ch][DHP] demonstrated the most prominent ability to expel PPG400. Earlier studies have highlighted that the primary distinction between cholinium salts and ILs with higher and lower melting points, which correlates with their capability to form ABSs and their affinity for water, lies in the values of their anion polar surfaces or logK$_{ow}$ values [45]. Anion polar surfaces represent the total surface area of all polar atoms, mainly oxygen and nitrogen, along with their bonded hydrogens. Typically, salts with higher melting points exhibit greater polar surface values due to a more localized charge. Although [DHCit]$^-$ has the highest polar surface charge among the studied cholinium compounds (132.13 Å^2), it is not a stronger salting-out agent than [Ch][DHP] (77.76 Å^2). As explained in the work of Pereira et al., [DHCit]$^-$ anions exhibit intramolecular hydrogen bonds between the hydroxyl hydrogen atoms and one of the oxygens of the central carboxyl group, which decreases their interaction with water and, consequently, the respective salting-out ability [32,45].

In the context of ILs with lower melting temperatures, the ability to form ABSs declines in the following order: [Ch][Lac] > [Ch][Gal] > [Ch][Nic] > [Ch][Van]. [Ch][Lac] demonstrates the strongest capacity among cholinium ILs to induce ABS formation, while [Ch][Van] has the lowest. This sequence does not align perfectly with the logK$_{ow}$ values of the ILs' anions, which are −3.74, −2.48, −3.03, and 1.50 for [Lac]$^-$, [Gal]$^-$, [Nic]$^-$, and [Van]$^-$, respectively [52]. The case of vanillate correlates well with its logK$_{ow}$ value, being the most hydrophobic and, thus, the least effective in ABS formation. However, despite nicotinate being more hydrophilic than gallate, it shows a lower tendency to form ABSs. Additionally, the polar surface areas of these anions (57.53, 97.99, 50.19, and 55.76 Å^2 for [Lac], [Gal], [Nic], and [Van], respectively) also do not mirror the decreasing ABS formation trend, as gallate has a higher value than lactate, while vanillate surpasses nicotinate. Nevertheless, [Gal] lags behind [Lac] in forming ABSs, and [Van] is weaker than [Nic]. This indicates that the formation of ABSs in these low-melting-point ILs with PPG is determined by a complex interplay of interactions with the polymer and water [30,53]. The gallate anion possesses four hydroxyl groups on its ring, providing four hydrogen donors and five hydrogen bond acceptors in its molecule. This contributes to its high polar surface area, which is even greater than that of the [DHP] anion, known as the strongest salting-out agent. The fact that [Ch][Gal] is a considerably weaker ABS phase former than [Ch][DHP] suggests that the solvation of IL anions may be largely affected by intermolecular interactions in ionic liquids, i.e., ion-pair binding energies. A multitude of ab initio studies have established that the binding energies between the cation and anion of an IL are closely related to the IL's melting point and its transport properties, like conductivity and viscosity [54–57]. Tot et al. showed that these binding energies play a significant role in determining the structure-making properties of ILs, i.e., their interactions with water [58]. Contrary to common expectations, they found that functionalization of an IL with a hydroxyl group does not enhance its structure-making properties, a key factor in the formation of ABSs, where the ability to influence structure is crucial. This could explain the lower ability of [Ch][Gal] to form ABS in comparison to [Ch][Lac]. A good indication of higher binding energy and more robust molecular packing of [Ch][Gal] is its higher melting temperature ([Ch][Gal] is solid, while [Ch][Lac] is liquid at room temperature; Figure S2). Additionally, even though [Ch][Nic] has a high logK_{ow} value, it acts as a weak salting-out agent. The presence of π electrons in [Ch][Nic], capable of forming strong hydrogen bond interactions with the ether and oxygen atoms of PPG, might significantly contribute to its reduced capability in facilitating the formation of ABSs [30].

3.2. Extraction of Polyphenolic Compounds

In selecting appropriate ILs to evaluate the recovery performance of selected polyphenols in this work, several factors were considered. Having in mind our goal, a strong focus was placed on selecting highly biocompatible ILs that are soluble in water. This led to the choice of ILs based on the quaternary ammonium cation, specifically the cholinium cation. Additionally, the selection of anions was also crucial in tailoring the physicochemical properties and interaction capabilities of the ILs to achieve adequate solvation ability for target natural compounds. Thus, the focus was also on anions derived from natural sources, such as from plant natural acids (gallic acid, vanillic acid, lactic acid, and nicotinic acid). Moreover, ILs with anions similar to those of the target compounds, such as gallate, vanillate, or nicotinate, allow for enhanced recovery selectivity [30]. We also considered more common commercial cholinium salts, such as [Ch][DHP], [Ch]Cl, and [Ch][DHCit], since they all have excellent salting-out ability among cholinium ILs. Therefore, by strategically considering both cations and anions, we aimed to select highly environmentally benign and biocompatible ABS constituents that possess desirable characteristics for our intended applications. To evaluate the effects of distinct ILs on the partitions of polyphenols in the proposed IL-based ABSs, the following appropriate composition of ternary ABS mixtures was selected: 20 wt% IL + 50 wt% PPG400 + 30 wt% water. This composition of the ternary mixtures was selected based on the previously determined phase diagrams with the aim of encompassing the biphasic range of each ABS, which is defined by the system with the lowest ability to undergo liquid−liquid demixing, namely the [Ch][Van]-based ABS. Moreover, when choosing the mixture compositions of the studied extraction ABSs, we endeavored to obtain similar phase ratios ($V_{IL\_phase}/V_{salt\_phase}$~0.5) of the systems to facilitate comparison. We evaluated two key parameters, namely the logarithmic distribution coefficients (logD) and recovery efficiency (RE), both crucial for discussing the performance of each ABS in extracting specific polyphenols. LogD values serve as an effective tool to delve into the partitioning mechanisms of polyphenols, revealing the complex interplay of interactions among ABS components—PPG400, the IL, the solute, and water. Meanwhile, recovery efficiency enabled us to identify the most efficient and selective systems, guiding the design of integrated processes for recovery of polyphenols from food by-products. The data presented as logarithmic distribution coefficients (logD values) in Figure 4 clearly demonstrate the varying affinities of the selected polyphenols for different phases in ABSs. LogD values greater than 0 signify a tendency of the polyphenol to favor the PPG-rich phase, whereas values less than 0 denote a predisposition towards the hydrophilic IL-rich phase.

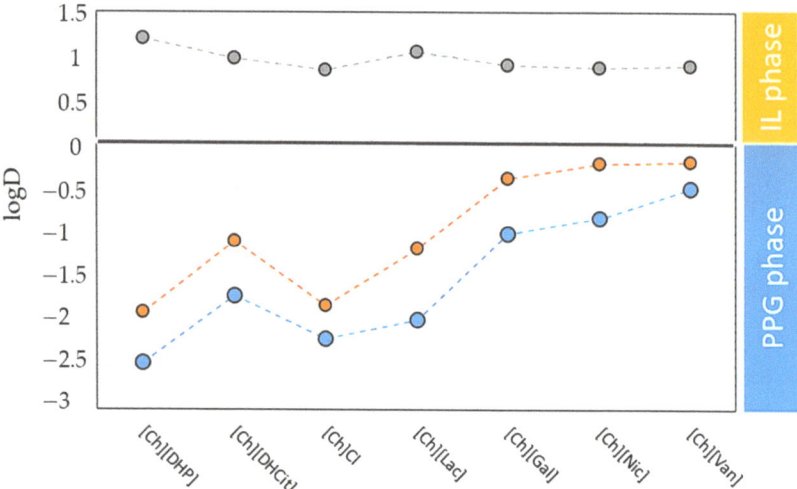

Figure 4. Distribution coefficients of polyphenols in cholinium-based ABSs at 25 °C and atmospheric pressure (RSV, blue circles; QC, orange circles; GA, gray circles).

The overall pattern in the partitioning of the analyzed polyphenols reveals that RSV and QC tend to migrate to the PPG-rich phase ($-0.49 < \log D_{RSV} < -2.47$ and $-0.19 < \log D_{QC} < -1.89$, respectively), while GA exhibits a preference for the IL-rich phase ($0.84 < \log D_{GA} < 1.2$). Among studied cholinium-based ABSs, the partitioning behavior of QC and RSV is aligned with the following sequence of ILs/salts: [Ch][DHP] > [Ch]Cl > [Ch][Lac] > [Ch][DHCit] > [Ch][Gal] > [Ch][Nic] > [Ch][Van]. This order is in close agreement with the ABS formation aptitude, with the exception of [Ch][DHCit]. At first glance, these results suggest that the distributions of hydrophobic polyphenols in cholinium-based ABSs are determined by the hydrophobic properties of the phases and the salting-out effect of cholinium salts/ILs (Figure S5). However, an exception is noted in systems using cholinium-based salts with melting temperatures above 100 °C. Specifically, in the [Ch][DHCit] ABS, the distribution coefficients are lower compared to those in the [Ch][DHP] and [Ch]Cl-based ABS, contradicting their expected behavior based on the phase diagram and the hydrophobicity of the PPG-rich phase. This odd behavior of [Ch][DHCit] might stem from specific intramolecular interactions within the [DHCit]$^-$ anions, which reduce its salting-out effect. A similar phenomenon was observed in a study by Dimitrijević et al., where [Ch][DHCit]/Pluronic PE6200 showed the lowest distribution coefficients for hydrophobic alkaloids like caffeine, theophylline, and theobromine [32]. The authors attributed this to the IL and water content in the Pluronic-rich phase. Considering PPG is a more hydrophobic polymer than Pluronic PE6200, it is plausible that the reduced salting-out capability of the [DHCit] anion is less marked than in Pluronic systems. This explains why logD values of [Ch][DHCit]/PPG ABSs are not lower in comparison to ABSs with lower-melting-point cholinium ILs. Additionally, the pronounced tendency of RSV in comparison to QC to migrate dominantly towards the upper PPG phase in each ABS is noticeable if comparing their determined distribution coefficients for each ABS. According to octanol–water partition coefficient values, RSV is the most lipophilic among the studied compounds ($\log K_{ow} = 3.40$ and $\log K_{ow} = 2.70$ for RSV and QC, respectively), which explains its high values of distribution coefficients (Table S4).

Contrary to stronger salting-out agents, the molecular mechanisms underlying the polyphenol partitions in [Ch][Lac], [Ch][Gal], [Ch][Nic], and [Ch][Van] ABSs are more complex, and specific interactions between components must be considered [32,44]. Therefore, the distribution pattern in these ABS is not only dominated by the hydrophobicity and salting out. Instead, it is largely a direct consequence of the favorable (or non-favorable) interactions that occur between the polymer, ILs, and polyphenols [20]. RSV and QC are more distributed toward the PPG-rich phase but with lower values of K than in the high-melting-point choline salt-based ABSs. Diminished K values in [Ch][Gal], [Ch][Nic], and [Ch][Van] ABSs can be attributed to the lower relative difference of hydrophobicity between the PPG and IL phases and π-π stacking interactions between IL anions (phenolic moieties) and aromatic structures of solutes.

Contrary to other compounds, GA predominantly partitions into the more hydrophilic IL-rich phases, as indicated by its negative octanol–water partition coefficients (logK_{ow} values ranging from 0.65 to −2.48 across pH 3.2–7.2). Furthermore, within the operational pH range of the ABSs, the fraction of negatively charged GA molecules reaches as high as 98.5% (Figure S4), emphasizing that the charged form of GA is prevalent, highlighting the significance of electrostatic interactions in its migration to the IL-rich phase. As shown in Figure 4, the logD values of GA do not completely mirror the hydrophobicity trend of ABSs outlined by their binodal curves. LogD values of GA align with the hydrophilicity of the IL-rich phase in cholinium salt ABSs in the following order: [Ch][DHP] > [Ch][DHCit] > [Ch]Cl. Intriguingly, [DHCit]-IL does not exhibit the same behavior as seen with non-charged, hydrophobic polyphenols, aligning with results from other studies [32,45]. In room-temperature IL systems, the distribution trends of GA align with their respective phase diagrams, showing a pattern of [Ch][Lac] > [Ch][Gal] > [Ch][Nic] > [Ch][Van]. However, the logD values for GA in these systems are not markedly lower compared to those in cholinium salt ABSs, even though their IL-rich phases are relatively less hydrophilic. This suggests that GA's distribution is influenced by a more intricate set of mechanisms, involving various interactions such as hydrogen bonding, π-π stacking, electrostatic forces, and van der Waals interactions. Notably, [Gal], [Nic], and [Van] anions establish additional π-π interactions with the aromatic part of GA that could compensate for their lower hydrophilicity.

The obtained recovery efficiencies (RE%) in the IL-rich phases for gallic acid and PPG-rich phases for quercetin and resveratrol are depicted in Figure 5 (detailed results are provided in the Supplementary Materials). The RE% values align more closely with the ability to form an ABS, as opposed to the trends observed in logD values, following the following sequence: [Ch][DHP] > [Ch][Lac] > [Ch][DHCit] > [Ch]Cl > [Ch][Gal] > [Ch][Nic] > [Ch][Van]. The highest REs for each polyphenol were obtained for the [Ch][DHP]-based ABS, namely 99.71 ± 0.60% for RSV and 98.92 ± 1.02% for QC toward the PPG-rich phase and 92.31 ± 2.31% for GA toward the IL-rich phase. Slightly lower values (reduced by less than 5%) were obtained for [Ch][Lac], [Ch][DHCit], and [Ch]Cl. Lower recovery efficiencies for QC and RSV were found with ILs containing anions from natural polyphenol acids, namely [Ch][Gal], [Ch][Nic], and [Ch][Van], ranging from 70.29 ± 4.89% to 92.27 ± 3.07% for RSV and from 56.04 ± 3.12% to 73.65 ± 3.07% for QC. This decrease in REs is due to the less hydrophobic nature of the PPG phases in these ABSs and the capacity of the anionic ILs to engage in π–electron interactions and hydrogen bonding with RSV and QC, which could be a significant factor in their reduced recovery efficiency for hydrophobic molecules. On the other hand, RE values of GA for different ABSs are generally steadier and higher, which is in accordance with their high distribution coefficients.

The ABS comprising [Ch][DHP]/PPG400 enables the highest partition coefficients for QC and RSV towards the PPG-rich phase and simultaneously yields the lowest coefficients for GA. This suggests that an effective separation of the examined polyphenols can be realized by selecting appropriate ABS components. Figure 6 illustrates the selectivity of the systems under study in separating resveratrol or quercetin from gallic acid. Among the evaluated ABSs, the [Ch][DHP]-based system stands out as the most selective in separating gallic acid from the other polyphenols, with a an SGA/RSV selectivity ratio exceeding 4500 and $S_{GA/QC}$ surpassing 1000.

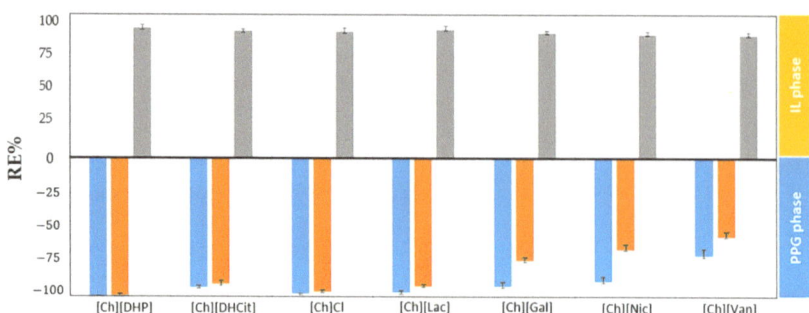

Figure 5. Recovery efficiencies (RE%) of GA (gray bar), RSV (blue bar), and QC (orange bar) in the studied ABSs at 25 °C.

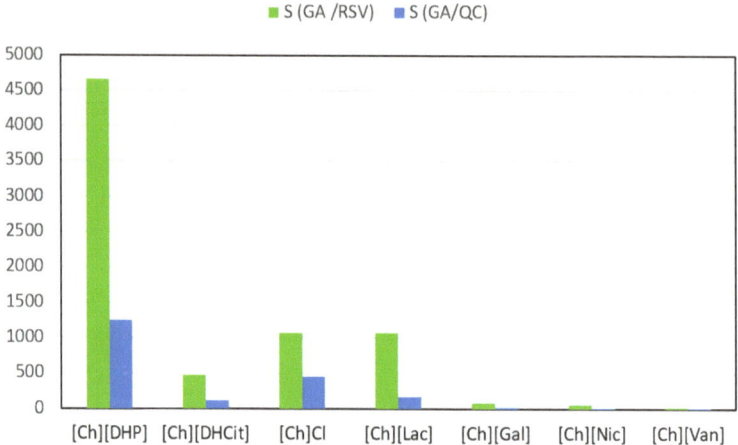

Figure 6. Selectivity parameters in studied ABSs at 25 °C.

Investigations of cholinium IL-based systems for the extraction and separation of polyphenols have been reported in the literature. For example, Xavier [39] employed ABSs based on [Ch]Cl and different surfactants to separate ferulic acid and p-coumaric acid from rice husk hydrolysate. Both compounds used in their study migrated to the surfactant-rich phase (with yields over 90%), which differs from the results we obtained for GA, which partitions dominantly in the IL-rich phase. Wang et al. [43] explored ABSs composed of cholinium-based ILs for the extraction of flavonoids and pectin from ponkan peel, achieving selective separation, with flavonoids favoring the IL-rich phase and pectin remaining in the salt-rich phase. While this study successfully separated polyphenols from polysaccharides, it did not accomplish selective separation among different polyphenols. Similarly, Neves et al. utilized cholinium-based ABSs for extraction and separation of antioxidants and carbohydrates from food waste [41]. The process demonstrated high efficiencies in separating antioxidants (65–75%) from an expired vanilla pudding sample. Several works have studied the extraction and purification of polyphenols using ABSs based on imidazolium ILs [33]. In these systems, polyphenols partitioned preferably to the IL-rich phase, so low selectivities were achieved. Additionally, imidazolium ILs are not considered ideal for polyphenol extraction due to concerns regarding their biocompatibility and potential toxicity. Generally, variations in the type of ionic liquids, polyphenols, samples, and experimental designs across different studies make direct comparisons of the obtained results with literature data challenging.

3.3. Recovery of Polyphenols from Grape Stems and Designing an Integrated Extraction and Separation Process

Our partition studies showed that cholinium IL/PPG ABSs hold considerable promise for the one-step extraction and separation of various polyphenols, including RSV, QC, and GA. By carefully choosing the anions in the IL's chemical structure, we were able to delve into the extraction mechanisms, enhance ABS performance, and fine-tune selectivity. To further evaluate the capability of cholinium-based ABSs for the direct extraction of polyphenols from grape stem extract, it was imperative to carefully choose the extraction system and fine-tune the operational parameters, notably the TLL and the composition mixture within the same tie line. The selection process for the optimal cholinium-based ABS as a polyphenol extraction/separation platform involved evaluating the extraction efficiency and selectivity of the systems. Among the evaluated cholinium salts, cholinium dihydrogen phosphate emerged as the ideal choice. The [Ch][DHP]-based ABS showcased exceptional performance in achieving the highest extraction efficiency and selectivity for each target compound in a single-step process. The salting-out effect, which is important in the formation of an ABS and the distribution of polyphenols, is a defining feature of [Ch][DHP]. By choosing an appropriate ABS composition, it is possible to establish considerably long tie lines. These longer tie lines contribute to improved recovery efficiencies and concentration factors while also reducing cross-contamination between the phases. This is achieved by ensuring that the PPG-rich phase contains minimal [Ch][DHP] and that the salt-rich phase is predominantly composed of water and salt.

The influence of the TLL on polyphenol partitions was assessed by varying the concentrations of ABS constituents as presented in Figure S3. Figure 7a depicts how the increase in TLL impacts RSV, QC, and GA extraction. By increasing the TLL under the tested conditions (polyphenol content = 2.5 g per L of ABS, pH~6, and T = 25 °C) increases of circa 7%, 9%, and 2% in RE_{RSV}, RE_{QC}, and RE_{GA} were obtained, respectively. The influence of the TLL on RE_{RSV} and RE_{QC} is primarily owing to the salting-out capability of [Ch][DHP] that also rises with the increase in the TLL. Adjusting the initial compositions of ABSs within the same tie line results in varying volume ratios of the coexisting phases, while their individual compositions remain unchanged. This variation affects the REs, as they are influenced by the phase volume ratios. Specifically, systems with different volume ratios of the PPG phase relative to the [Ch][DHP] phase, such as 0.50, 0.40, and 0.20, were examined. These systems, all with tie lines approximately 101 in length, are illustrated in Figure 7b. The findings show a decline in REs as the phase volume ratios decrease (for example, REs dropped from 99.71 ± 0.60% to 65.20 ± 3.89% for RSV, from 98.92 ± 1.02% to 55.70 ± 2.13% for QC, and from 92.31 ± 2.31% to 88.7 ± 3.91% for GA). Therefore, optimizing the volume ratios of the aqueous phases by reducing the volume of the PPG400 phase and minimizing the use of [Ch][DHP], without sacrificing extraction efficiency, is crucial.

Guided by the results from the optimization studies, polyphenols were extracted from grape stem extract with specific conditions in mind based on previously determined extraction parameters. These included the use of a [Ch][DHP] + PPG400 ABS, a temperature of 25 °C, a pH of approximately 6, a tie-line length of around 101, and grape stem extract quantities of about 5 mg per g of ABS, with the volume ratio of the salt phase relative to the PPG phase being approximately 0.4. In the process, mixtures of PPG, [Ch][DHP], and water in the monophasic region were directly employed in the extraction of valuable compounds from grape by-products, then stirred. Subsequently, an aqueous solution of [Ch][DHP] was added to the mix of PPG400, water, and grape stem extract to achieve an ABS composition of 40 wt% PPG400 and 25 wt% [Ch][DHP] at room temperature, as depicted in Figure 8. After stirring and equilibration, the ABS mixtures were centrifuged. Remarkably, RSV was completely extracted from the grape stem extract into the PPG-rich phase, achieving an exceptional 100% recovery efficiency. The RE for GA towards the [Ch][DHP] phase was 78.64%, and quercetin was not detected in the grape stem extract. Figure 9 shows the chromatograms of both ABS phases after extraction of the selected polyphenols using the [Ch][DHP]/PPG400 ABS, showing that RSV is dominantly presented in PPG-rich phase, while GA migrates toward the IL-rich phase. The application of optimized operational parameters for the extraction of polyphenols from grape stems resulted in an enhanced extraction of RSV, achieving a higher recovery efficiency than that observed during the optimization process. Conversely, the recovery efficiency for GA is lower than that found in our partition experiments. This could be due to the reduced hydrophilicity of the IL-rich phase, possibly because of the presence of additional polar compounds in the extract that compete with [Ch][DHP] for water molecules.

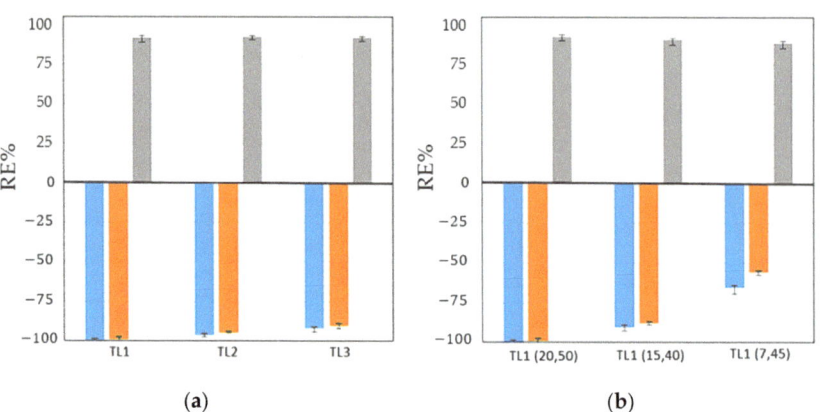

Figure 7. Impact of different TLLs (**a**) and different initial compositions along the same TL (**b**) on the extraction efficiencies of RSV (blue bar), QC (orange bar), and GA (gray bar) in the [Ch][DHP] + PPG400 + H_2O ABS at 25 °C.

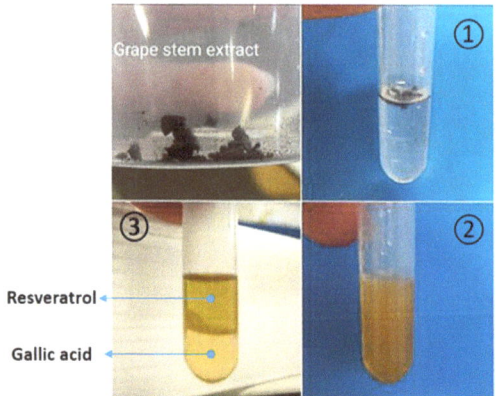

Figure 8. Extraction process from grape stem extract: 1—adding extract to IL + PPG400 aqueous solution; 2—mixing; 3—separation of the phases.

RSV is a highly hydrophobic compound and, thus, presents a low solubility in pure water (39 mg/L [59]). Considering that the upper PPG phase of the ABS mainly consists of PPG and water (with less than 0.1 wt% [Ch][DHP]), we aimed to determine the solubility of RSV in a PPG/water system matching the composition of the PPG phase with the purpose of estimating the capacity of the [Ch][DHP]/PPG ABS for RSV extraction. We achieved a solubility of 100 g/L for RSV, although this was not its maximum solubility. The notably high solubility of RSV in the PPG-rich phase suggests the feasibility of using this system to recover substantial amounts of hydrophobic compounds or for application in continuous processes before reaching saturation. Given that RSV content in grape stems ranges from 4.6 to 15.4 mg/kg [12] and based on the determined solubility of RSV in the PPG phase, it would theoretically be possible to process a minimum of 6.4 to 21.7 kg of plant material until RSV saturation if 1 L of the PPG rich phase is reached. Similarly, a high capacity for GA recovery is also expected. The maximal solubility of GA in the [Ch][DHP]-rich phase (around 50 wt% [Ch][DHP]) was found to be 139.8 g/L, which is significantly higher than its solubility in water (14.38 g/L). Claudio et al. corroborated the exceptional ability of ILs to act as hydrotropes through the formation of ionic liquid–biomolecule aggregates [60]. This characteristic is crucial from the perspective of developing new strategies for polyphenol recovery. Specifically, GA from approximately 12 kg of material could potentially be recovered in 1 L of the [Ch][DHP] phase, showcasing the efficacy of this method.

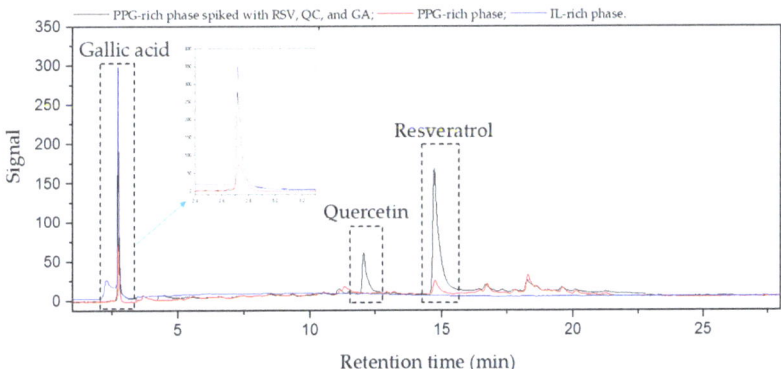

Figure 9. Chromatograms of IL-rich (blue line) and PPG-rich (red line) phases after solid–liquid extraction and liquid–liquid separation from grape stem using a [Ch][DHP]/PPG400 ABS. The black line represents a PPG phase spiked with a standard solution of RSV, QC, and GA.

The evaluated ABS is able to remarkably recover, concentrate, and separate value-added compounds directly from grape stem samples and largely contributes to the green concept of the designed integrated separation platform. All components applied as ABS phase formers in the proposed integrated concept are benign from a toxicological point of view. PPG400 is a harmless and highly biocompatible polymer approved as a food additive by the FDA. Cholinium-based salts and ILs are safe, generally non-toxic, and biodegradable ionic compounds. A conceptual design of the developed integrated platform to extract and recover value-added compounds from food by-products is schematically shown in Figure 10.

After the extraction and separation step, bioactive polyphenols need to be isolated from the PPG and IL matrix. Although the isolation step is outside of the scope of the experimental study reported here, some suggestions are provided with respect to how to isolate and reuse the ABS components (dashed lines in Figure 10) to further emphasize the sustainable character of the process. RSV could be easily recovered from the PPG-rich phase by antisolvent crystallization, whereby adding water as antisolvent can significantly lower the solubility of hydrophobic polyphenol [61]. Isolation from the IL matrix could be achieved using affinity resins. For example, Rosa et al. achieved pigment isolation from the IL and PPG matrix by a solid-phase extraction approach using affinity resins, with high recoveries of over 96% for betalains and 98% for chlorophylls [34]. Only by addressing and guaranteeing the recycling of phase-forming components will it be possible to be economically competitive with other well-established techniques.

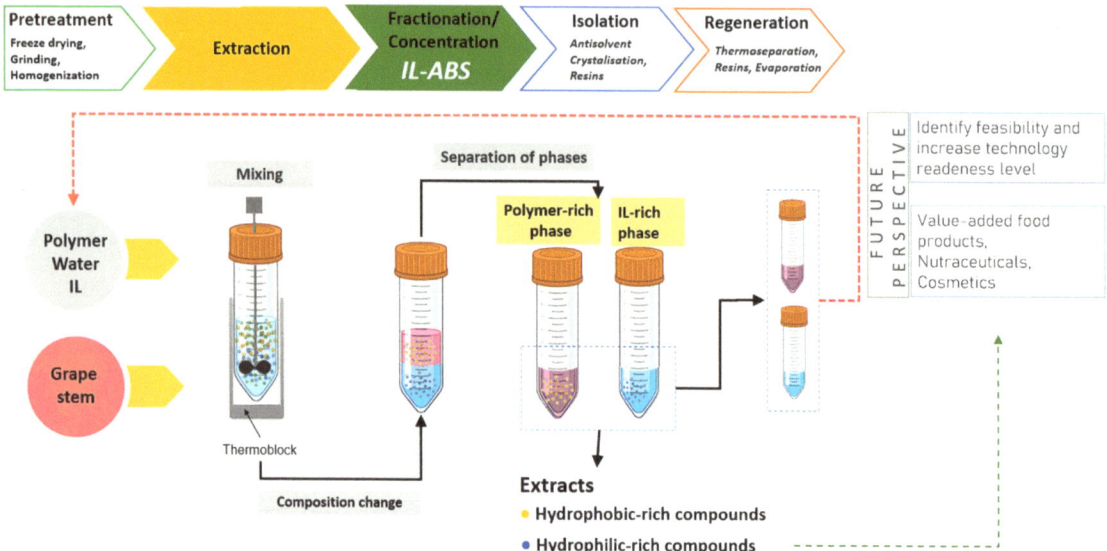

Figure 10. Schematic representation of the proposed integrated platform to extract and recover value-added compounds from grape by-products using ABSs composed of cholinium-based ILs and PPG400.

4. Conclusions

In this work, we proposed an integrated approach based on ABSs to simultaneously extract, separate, and concentrate selected polyphenols (resveratrol, quercetin, and gallic acid) from grape stem extract. For that purpose, phase diagrams composed of PPG400 and seven different salting-out agents ([Ch][DHP], [Ch]Cl, [Ch][DHCit], [Ch][Lac], [Ch][Gal], [Ch][Nic], and [Ch][Van]) were determined at 25 °C and under atmospheric pressure. We established the aptitude of salts to generate ABS with PPG400 in the following order: [Ch][DHP] > [Ch]Cl > [Ch][Lac] > [Ch][DHCit] > [Ch][Gal] > [Ch][Nic] > [Ch][Van]. While the aptitude of higher-melting-point cholinium salts for ABSs follows the well-established ability of their anions to be solvated by water and act as salting-out species, lower-melting-point cholinium ILs completely follow neither $\log K_{ow}$ nor polar anion surface parameters, as described earlier in the literature. As discussed, IL ion-pair binding energies play an important role in determining the interactions of IL anions with water and with the polymer, which are crucial in ABS formation. Next, extraction experiments were performed to determine the partition behavior of each polyphenol in the investigated ABSs, and different distribution patterns were observed for the investigated polyphenols. [Ch][DHP] exhibited the highest extraction efficiencies of above 95% for RSV and QC toward the PPG phase and 90% for GA in the IL phase. Also, this system stands out as the most selective in separating GA from the other polyphenols, with a $S_{GA/RSV}$ selectivity ratio exceeding 4500 and $S_{GA/QC}$ surpassing 1000. The distribution patterns in other ABSs are not solely governed by hydrophobicity and salting-out effects but are also a result of the specific interaction between the polymer, ILs, and polyphenols. After screening the extraction ability of each ABS, key operational conditions, specifically the TLL and mixture composition along the same tie line, were established to optimize extraction performance from a real sample. It was shown that RSV was completely extracted from the grape stem extract into the PPG-rich phase, achieving an exceptional 100% recovery efficiency. The RE for GA towards the [Ch][DHP] phase was 78.64%, and quercetin was not detected in the grape stem extract. The lower recovery of GA could be due to the reduced hydrophilicity of the IL-rich phase. The evaluated ABS is able to remarkably recover, concentrate, and separate value-added compounds directly from grape stem samples and largely contributes to the green concept of the designed integrated separation platform.

Supplementary Materials: The following supporting information can be downloaded at: https://www.mdpi.com/article/10.3390/foods13060954/s1, Figure S1a–d: FTIR spectra of the synthesized ionic liquids; Figure S2: Synthesized [Ch][Gal] (left) and [Ch][Lac] (right); Figure S3: Ternary phase diagram composed of [Ch][DHP] + PPG 400 + H2O at 25 °C with illustrated TLs; Figure S4: Speciation diagram of gallic acid; Figure S5: Chromatograms of investigated polyphenolic in [Ch][DHP] or [Ch][Nic] -rich phase (black line) and PPG400-rich phase (red line) at 280 nm; Table S1: Experimental binodal mass fraction data for the salt/IL (X) + PPG400 (Y) + H$_2$O ABS at 25 °C and at p = 0.1 MPa.; Table S2: Correlation parameters, standard deviations and determination coefficients (R2) of the salt/IL (X) + PPG400 (Y) + H$_2$O ABS obtained by the Merchuk equation at 25 °C and at p = 0.1 MPa); Table S3: Experimental tie-lines data in percentage weight fraction for the ABS composed of ([Ch][DHP] + PPG400 (Y) + H$_2$O (Z)) at 25 °C, pH = 6, and 0.1 MPa and volume ratios (Vr).; Table S4: The properties of the polyphenolic compounds. Figure S6: Chromatograms of grape stem extract (red line) and spiked grape stem extract (black line) in ethanol at 280 nm. References [49,62] are cited in the supplementary materials.

Author Contributions: Conceptualization, A.D. and J.S.A.; methodology, A.D., S.M., A.J., D.T., J.M. and J.S.A.; validation, A.D. and J.S.A.; formal analysis, A.D., S.M. and A.J.; investigation, A.D., D.T. and J.M.; writing—original draft preparation, A.D., S.M. and A.J.; writing—review and editing, A.D. and J.S.A.; visualization, A.D. and A.J.; supervision, A.D. and J.S.A.; funding acquisition, A.D. and J.S.A. All authors have read and agreed to the published version of the manuscript.

Funding: This research received support from the Ministry of Science, Technological Development and Innovation of the Republic of Serbia under Contract number 451-03-66/2024-03/200017. Additionally, it was supported by a Short-Term Scientific Mission (STSM) Grant (Reference: E-COST-GRANT-CA20133-975d1689) from the COST Action CA20133 (Fullrec4us—Cross Border Transfer and Development of Sustainable Resource Recovery Strategies Towards Zero Waste). J.S. Amaral is thankful to FCT/MCTES (PIDDAC) for the funding of CIMO, UIDB/00690/2020 (DOI: 10.54499/UIDB/00690/2020), UIDP/00690/2020 (DOI: 10.54499/UIDP/00690/2020), and SusTEC, LA/P/0007/2020 (DOI: 10.54499/LA/P/0007/2020). J.S.Amaral also thanks the financing of the project "BacchusTech–Integrated Approach for the Valorization of Winemaking Residues" (POCI-01-0247-FEDER-069583), supported by the Competitiveness and Internationalization Operational Program (COMPETE 2020), under the PORTUGAL 2020 Partnership Agreement, through the European Regional Development Fund (ERDF).

Institutional Review Board Statement: Not applicable.

Informed Consent Statement: Not applicable.

Data Availability Statement: The original contributions presented in the study are included in the article/Supplementary Material, further inquiries can be directed to the corresponding author.

Conflicts of Interest: The authors declare no conflicts of interest.

References

1. Tackling Food Loss and Waste: A Triple Win Opportunity, Newsroom (n.d.). Available online: https://www.fao.org/newsroom/detail/FAO-UNEP-agriculture-environment-food-loss-waste-day-2022/en (accessed on 21 February 2024).
2. Torres-León, C.; Ramírez-Guzman, N.; Londoño-Hernandez, L.; Martinez-Medina, G.A.; Díaz-Herrera, R.; Navarro-Macias, V.; Alvarez-Pérez, O.B.; Picazo, B.; Villarreal-Vázquez, M.; Ascacio-Valdes, J.; et al. Food Waste and Byproducts: An Opportunity to Minimize Malnutrition and Hunger in Developing Countries. *Front. Sustain. Food Syst.* **2018**, *2*, 52. [CrossRef]
3. Trigo, J.P.; Alexandre, E.M.C.; Saraiva, J.A.; Pintado, M.E. High Value-Added Compounds from Fruit and Vegetable by-Products–Characterization, Bioactivities, and Application in the Development of Novel Food Products. *Crit. Rev. Food Sci. Nutr.* **2020**, *60*, 1388–1416. [CrossRef] [PubMed]
4. Mazzutti, S.; Pedrosa, R.C.; Ferreira, S.R.S. Green Processes in Foodomics. Supercritical Fluid Extraction of Bioactives. In *Comprehensive Foodomics*; Elsevier: Amsterdam, The Netherlands, 2020; pp. 725–743. ISBN 9780128163955.
5. De la Rosa, L.A.; Alvarez-Parrilla, E.; González-Aguilar, G.A. *Fruit and Vegetable Phytochemicals*; de la Rosa, L.A., Alvarez-Parrilla, E., González-Aguilar, G.A., Eds.; Wiley: Hoboken, NJ, USA, 2009; ISBN 9780813803203.
6. Montenegro-Landívar, M.F.; Tapia-Quirós, P.; Vecino, X.; Reig, M.; Valderrama, C.; Granados, M.; Cortina, J.L.; Saurina, J. Polyphenols and Their Potential Role to Fight Viral Diseases: An Overview. *Sci. Total Environ.* **2021**, *801*, 149719. [CrossRef]

7. Allied Market Research. Polyphenol Market Size, Share & Growth—Industry Report, 2030. Available online: https://www.alliedmarketresearch.com/polyphenol-market (accessed on 24 December 2023).
8. Mir-Cerdà, A.; Nuñez, O.; Granados, M.; Sentellas, S.; Saurina, J. An Overview of the Extraction and Characterization of Bioactive Phenolic Compounds from Agri-Food Waste within the Framework of Circular Bioeconomy. *TrAC—Trends Anal. Chem.* **2023**, *161*, 116994. [CrossRef]
9. Yang, C.; Han, Y.; Tian, X.; Sajid, M.; Mehmood, S.; Wang, H.; Li, H. Phenolic Composition of Grape Pomace and Its Metabolism. *Crit. Rev. Food Sci. Nutr.* **2022**. ahead of print. [CrossRef]
10. Panouillé, M.; Ralet, M.C.; Bonnin, E.; Thibault, J.F. Recovery and Reuse of Trimmings and Pulps from Fruit and Vegetable Processing. In *Handbook of Waste Management and Co-Product Recovery in Food Processing*; Elsevier: Amsterdam, The Netherlands, 2007; Volume 1, pp. 417–447. ISBN 9781845692520.
11. Kumar, M.; Barbhai, M.D.; Hasan, M.; Punia, S.; Dhumal, S.; Radha; Rais, N.; Chandran, D.; Pandiselvam, R.; Kothakota, A.; et al. Onion (*Allium cepa* L.) Peels: A Review on Bioactive Compounds and Biomedical Activities. *Biomed. Pharmacother.* **2022**, *146*, 112498. [CrossRef]
12. Silva, V.; Igrejas, G.; Falco, V.; Santos, T.P.; Torres, C.; Oliveira, A.M.P.; Pereira, J.E.; Amaral, J.S.; Poeta, P. Chemical Composition, Antioxidant and Antimicrobial Activity of Phenolic Compounds Extracted from Wine Industry by-Products. *Food Control* **2018**, *92*, 516–522. [CrossRef]
13. Samota, M.K.; Sharma, M.; Kaur, K.; Sarita; Yadav, D.K.; Pandey, A.K.; Tak, Y.; Rawat, M.; Thakur, J.; Rani, H. Onion Anthocyanins: Extraction, Stability, Bioavailability, Dietary Effect, and Health Implications. *Front. Nutr.* **2022**, *9*, 917617. [CrossRef]
14. Miller, M.D.; Steinmaus, C.; Golub, M.S.; Castorina, R.; Thilakartne, R.; Bradman, A.; Marty, M.A. Potential Impacts of Synthetic Food Dyes on Activity and Attention in Children: A Review of the Human and Animal Evidence. *Environ. Health A Glob. Access Sci. Source* **2022**, *21*, 45. [CrossRef] [PubMed]
15. Sridhar, A.; Ponnuchamy, M.; Kumar, P.S.; Kapoor, A.; Vo, D.V.N.; Prabhakar, S. Techniques and Modeling of Polyphenol Extraction from Food: A Review. *Environ. Chem. Lett.* **2021**, *19*, 3409–3443. [CrossRef]
16. Chuo, S.C.; Nasir, H.M.; Mohd-Setapar, S.H.; Mohamed, S.F.; Ahmad, A.; Wani, W.A.; Muddassir, M.; Alarifi, A. A Glimpse into the Extraction Methods of Active Compounds from Plants. *Crit. Rev. Anal. Chem.* **2022**, *52*, 667–696. [CrossRef] [PubMed]
17. Antony, A.; Farid, M. Effect of Temperatures on Polyphenols during Extraction. *Appl. Sci.* **2022**, *12*, 2107. [CrossRef]
18. Sólyom, K.; Solá, R.; Cocero, M.J.; Mato, R.B. Thermal Degradation of Grape Marc Polyphenols. *Food Chem.* **2014**, *159*, 361–366. [CrossRef] [PubMed]
19. Volf, I.; Ignat, I.; Neamtu, M.; Popa, V.I. Thermal Stability, Antioxidant Activity, and Photo-Oxidation of Natural Polyphenols. *Chem. Pap.* **2014**, *68*, 121–129. [CrossRef]
20. Boles, J.S.; Crerar, D.A.; Grissom, G.; Key, T.C. Aqueous thermal degradation of gallic acid. *Geochim. Cosmochim. Acta* **1988**, *52*, 341–344. [CrossRef]
21. Nadja, B.; Angelika, K.; Rohn, S.; Kroh, L.W. Effect of Thermal Processing on the Flavonols Rutin and Quercetin. *Rapid Commun. Mass Spectrom.* **2010**, *24*, 1457–1466. [CrossRef]
22. Zupančič, Š.; Lavrič, Z.; Kristl, J. Stability and Solubility of Trans-Resveratrol Are Strongly Influenced by PH and Temperature. *Eur. J. Pharm. Biopharm.* **2015**, *93*, 196–204. [CrossRef]
23. de Lima Marsiglia, W.I.M.; Oliveira, L.D.S.C.; Almeida, R.L.J.; Santos, N.C.; da Silva Neto, J.M.; Santiago, Â.M.; de Melo, B.C.A.; da Silva, F.L.H. Thermal Stability of Total Phenolic Compounds and Antioxidant Activities of Jaboticaba Peel: Effect of Solvents and Extraction Methods. *J. Indian Chem. Soc.* **2023**, *100*, 100995. [CrossRef]
24. Anson, N.M.; Selinheimo, E.; Havenaar, R.; Aura, A.M.; Mattila, I.; Lehtinen, P.; Bast, A.; Poutanen, K.; Haenen, G.R.M.M. Bioprocessing of Wheat Bran Improves in Vitro Bioaccessibility and Colonic Metabolism of Phenolic Compounds. *J. Agric. Food Chem.* **2009**, *57*, 6148–6155. [CrossRef] [PubMed]
25. Rogers, R.D.; Seddon, K.R. Ionic Liquids—Solvents of the Future? *Science* **2003**, *302*, 792–793. [CrossRef]
26. Dinodia, M. Ionic Liquids: Environment-Friendly Greener Solvents for Organic Synthesis. *Curr. Org. Synth.* **2022**, *19*, 543–557. [CrossRef] [PubMed]
27. Dimitrijević, A.; Jocić, A. Ionic Liquids as Promising Media in (Pre)Analytical Treatments and Degradation of Organophosphate Pesticides. In *Organophosphates: Detection, Exposure and Occurrence. Volume 1: Impact on Health and the Natural Environment*; Nova Science Publishers: New York, NY, USA, 2022; pp. 181–214. ISBN 978-1-68507-652-5.
28. Weaver, K.D.; Kim, H.J.; Sun, J.; MacFarlane, D.R.; Elliott, G.D. Cyto-Toxicity and Biocompatibility of a Family of Choline Phosphate Ionic Liquids Designed for Pharmaceutical Applications. *Green Chem.* **2010**, *12*, 507–551. [CrossRef]
29. Pavez, P.; Figueroa, R.; Medina, M.; Millán, D.; Falcone, R.D.; Tapia, R.A. Choline [Amino Acid] Ionic Liquid/Water Mixtures: A Triple Effect for the Degradation of an Organophosphorus Pesticide. *ACS Omega* **2020**, *5*, 26562–26572. [CrossRef] [PubMed]
30. Dimitrijević, A.; Tavares, A.P.M.M.; Almeida, M.R.; Vraneš, M.; Sousa, A.C.A.A.; Cristóvão, A.C.; Trtić-Petrović, T.; Gadžurić, S.; Freire, M.G. Valorization of Expired Energy Drinks by Designed and Integrated Ionic Liquid-Based Aqueous Biphasic Systems. *ACS Sustain. Chem. Eng.* **2020**, *8*, 5683–5692. [CrossRef]
31. Le Donne, A.; Bodo, E. Cholinium Amino Acid-Based Ionic Liquids. *Biophys. Rev.* **2021**, *13*, 147–160. [CrossRef] [PubMed]
32. Dimitrijević, A.; Tavares, A.P.M.; Jocić, A.; Marić, S.; Trtić-Petrović, T.; Gadžurić, S.; Freire, M.G. Aqueous Biphasic Systems Comprising Copolymers and Cholinium-Based Salts or Ionic Liquids: Insights on the Mechanisms Responsible for Their Creation. *Sep. Purif. Technol.* **2020**, *248*, 117050. [CrossRef]

33. Ventura, S.P.M.; e Silva, F.A.; Quental, M.V.; Mondal, D.; Freire, M.G.; Coutinho, J.A.P. Ionic-Liquid-Mediated Extraction and Separation Processes for Bioactive Compounds: Past, Present, and Future Trends. *Chem. Rev.* **2017**, *117*, 6984–7052. [CrossRef] [PubMed]
34. Rosa, M.E.; Ferreira, A.M.; Neves, C.M.S.S.; Almeida, M.R.; Barros, R.; Cristovão, A.C.; Sousa, A.C.A.; Reis, P.M.; Rebelo, L.P.N.; Esperança, J.M.S.S.; et al. Valorisation of Red Beet Waste: One-Step Extraction and Separation of Betalains and Chlorophylls Using Thermoreversible Aqueous Biphasic Systems. *Green Chem.* **2023**, *25*, 1852–1864. [CrossRef]
35. Saha, N.; Sarkar, B.; Sen, K. Aqueous Biphasic Systems: A Robust Platform for Green Extraction of Biomolecules. *J. Mol. Liq.* **2022**, *363*, 119882. [CrossRef]
36. Freire, M.G.; Cláudio, A.F.M.; Araújo, J.M.M.; Coutinho, J.A.P.; Marrucho, I.M.; Lopes, J.N.C.; Rebelo, L.P.N. Aqueous Biphasic Systems: A Boost Brought about by Using Ionic Liquids. *Chem. Soc. Rev.* **2012**, *41*, 4966. [CrossRef]
37. Mondal, D.; Sharma, M.; Quental, M.V.; Tavares, A.P.M.M.; Prasad, K.; Freire, M.G. Suitability of Bio-Based Ionic Liquids for the Extraction and Purification of IgG Antibodies. *Green Chem.* **2016**, *18*, 6071–6081. [CrossRef] [PubMed]
38. Romero, A.; Santos, A.; Tojo, J.; Rodríguez, A. Toxicity and Biodegradability of Imidazolium Ionic Liquids. *J. Hazard. Mater.* **2008**, *151*, 268–273. [CrossRef] [PubMed]
39. Xavier, L.; Rocha, M.; Pisani, J.; Zecchi, B. Aqueous Two-Phase Systems Based on Cholinium Ionic Liquids for the Recovery of Ferulic and p-Coumaric Acids from Rice Husk Hydrolysate. *Appl. Food Res.* **2024**, *4*, 100381. [CrossRef]
40. Mourão, T.; Tomé, L.C.; Florindo, C.; Rebelo, L.P.N.; Marrucho, I.M. Understanding the Role of Cholinium Carboxylate Ionic Liquids in PEG-Based Aqueous Biphasic Systems. *ACS Sustain. Chem. Eng.* **2014**, *2*, 2426–2434. [CrossRef]
41. Neves, C.M.S.S.; Figueiredo, M.; Reis, P.M.; Sousa, A.C.A.; Cristóvão, A.C.; Fiadeiro, M.B.; Rebelo, L.P.N.; Coutinho, J.A.P.; Esperança, J.M.S.S.; Freire, M.G. Simultaneous Separation of Antioxidants and Carbohydrates from Food Wastes Using Aqueous Biphasic Systems Formed by Cholinium-Derived Ionic Liquids. *Front. Chem.* **2019**, *7*, 454370. [CrossRef] [PubMed]
42. Ribeiro, B.D.; Coelho, M.A.Z.; Rebelo, L.P.N.; Marrucho, I.M. Ionic Liquids as Additives for Extraction of Saponins and Polyphenols from Mate (*Ilex paraguariensis*) and Tea (*Camellia sinensis*). *Ind. Eng. Chem. Res.* **2013**, *52*, 12146–12153. [CrossRef]
43. Wang, R.; Chang, Y.; Tan, Z.; Li, F. Applications of Choline Amino Acid Ionic Liquid in Extraction and Separation of Flavonoids and Pectin from Ponkan Peels. *Sep. Sci. Technol.* **2016**, *51*, 1093–1102. [CrossRef]
44. Sintra, T.E.; Luís, A.; Rocha, S.N.; Ferreira, A.I.M.C.L.; Gonçalves, F.; Santos, L.M.N.B.F.; Neves, B.M.; Freire, M.G.; Ventura, S.P.M.; Coutinho, J.A.P. Enhancing the Antioxidant Characteristics of Phenolic Acids by Their Conversion into Cholinium Salts. *ACS Sustain. Chem. Eng.* **2015**, *3*, 2558–2565. [CrossRef]
45. Pereira, J.F.; Kurnia, K.A.; Cojocaru, O.A.; Gurau, G.; Rebelo, L.P.N.; Rogers, R.D.; Freire, M.G.; Coutinho, J.A. Molecular Interactions in Aqueous Biphasic Systems Composed of Polyethylene Glycol and Crystalline vs. Liquid Cholinium-Based Salts. *Phys. Chem. Chem. Phys.* **2014**, *16*, 5723–5731. [CrossRef] [PubMed]
46. Li, Z.; Liu, X.; Pei, Y.; Wang, J.; He, M. Design of Environmentally Friendly Ionic Liquid Aqueous Two-Phase Systems for the Efficient and High Activity Extraction of Proteins. *Green Chem.* **2012**, *14*, 2941. [CrossRef]
47. Ramalho, C.C.; Neves, C.M.S.S.; Quental, M.V.; Coutinho, J.A.P.; Freire, M.G. Separation of Immunoglobulin G Using Aqueous Biphasic Systems Composed of Cholinium-Based Ionic Liquids and Poly(Propylene Glycol). *J. Chem. Technol. Biotechnol.* **2018**, *93*, 1931–1939. [CrossRef]
48. Marić, S.; Jocić, A.; Tekić, D.; Mušović, J.; Milićević, J.; Dimitrijević, A. Customizable Cholinium-Based Aqueous Biphasic Systems as Ecofriendly Extraction Platform for Removal of Pesticide from Wastewaters. *Sep. Purif. Technol.* **2024**, *340*, 126609. [CrossRef]
49. Merchuk, J.C.; Andrews, B.A.; Asenjo, J.A. Aqueous Two-Phase Systems for Protein Separation. *J. Chromatogr. B Biomed. Sci. Appl.* **1998**, *711*, 285–293. [CrossRef] [PubMed]
50. Pereira, J.F.B.; Magri, A.; Quental, M.V.; Gonzalez-Miquel, M.; Freire, M.G.; Coutinho, J.A.P. Alkaloids as Alternative Probes to Characterize the Relative Hydrophobicity of Aqueous Biphasic Systems. *ACS Sustain. Chem. Eng.* **2016**, *4*, 1512–1520. [CrossRef]
51. Neves, C.M.S.S.; Shahriari, S.; Lemus, J.; Pereira, J.F.; Freire, M.G.; Coutinho, J.A.P. Aqueous Biphasic Systems Composed of Ionic Liquids and Polypropylene Glycol: Insights into Their Liquid-Liquid Demixing Mechanisms. *Phys. Chem. Chem. Phys.* **2016**, *18*, 20571–20582. [CrossRef] [PubMed]
52. Chemicalize Was Used for Predicting ILs Properties. Developed by ChemAxon. Available online: http://www.chemaxon.com (accessed on 28 December 2023).
53. Shaker Shiran, H.; Baghbanbashi, M.; Ghazizadeh Ahsaie, F.; Pazuki, G. Study of Curcumin Partitioning in Polymer-Salt Aqueous Two Phase Systems. *J. Mol. Liq.* **2020**, *303*, 112629. [CrossRef]
54. Armaković, S.; Armaković, S.J.; Vraneš, M.; Tot, A.; Gadžurić, S. DFT Study of 1-Butyl-3-Methylimidazolium Salicylate: A Third-Generation Ionic Liquid. *J. Mol. Model.* **2015**, *21*, 246. [CrossRef]
55. Gadžurić, S.; Tot, A.; Armaković, S.; Armaković, S.; Panić, J.; Jović, B.; Vraneš, M. Uncommon Structure Making/Breaking Behaviour of Cholinium Taurate in Water. *J. Chem. Thermodyn.* **2017**, *107*, 58–64. [CrossRef]
56. Bernard, U.L.; Izgorodina, E.I.; MacFarlane, D.R. New Insights into the Relationship between Ion-Pair Binding Energy and Thermodynamic and Transport Properties of Ionic Liquids. *J. Phys. Chem. C* **2010**, *114*, 20472–20478. [CrossRef]
57. Demurtas, M.; Onnis, V.; Zucca, P.; Rescigno, A.; Lachowicz, J.I.; De Villiers Engelbrecht, L.; Nieddu, M.; Ennas, G.; Scano, A.; Mocci, F.; et al. Cholinium-Based Ionic Liquids from Hydroxycinnamic Acids as New Promising Bioactive Agents: A Combined Experimental and Theoretical Investigation. *ACS Sustain. Chem. Eng.* **2021**, *9*, 2975–2986. [CrossRef]

58. Tot, A.; Podlipnik, Č.; Bešter-Rogač, M.; Gadžurić, S.; Vraneš, M. Influence of Oxygen Functionalization on Physico-Chemical Properties of Imidazolium Based Ionic Liquids—Experimental and Computational Study. *Arab. J. Chem.* **2017**, *13*, 1598–1611. [CrossRef]
59. Ghazwani, M.; Alam, P.; Alqarni, M.H.; Yusufoglu, H.S.; Shakeel, F. Various Propylene Glycol + Water Mixtures. *Molecules* **2021**, *26*, 3091. [CrossRef] [PubMed]
60. Cláudio, A.F.M.; Neves, M.C.; Shimizu, K.; Canongia Lopes, J.N.; Freire, M.G.; Coutinho, J.A.P. The Magic of Aqueous Solutions of Ionic Liquids: Ionic Liquids as a Powerful Class of Catanionic Hydrotropes. *Green Chem.* **2015**, *17*, 3948–3963. [CrossRef] [PubMed]
61. Das, L.; Paik, S.P.; Sen, K. Poly(Propylene Glycol) vs. Sugar Alcohol-Based Aqueous Biphasic System to Extract Drugs and Subsequent Recovery of the Polymer. *J. Chem. Eng. Data* **2021**, *66*, 4629–4638. [CrossRef]
62. Gutowski, K.E.; Broker, G.A.; Willauer, H.D.; Huddleston, J.G.; Swatloski, R.P.; Holbrey, J.D.; Rogers, R.D. Controlling the aqueous miscibility of ionic liquids: Aqueous biphasic systems of water-miscible ionic liquids and water-structuring salts for recycle, metathesis, and separations. *J. Am. Chem. Soc.* **2003**, *125*, 6632. [CrossRef] [PubMed]

Disclaimer/Publisher's Note: The statements, opinions and data contained in all publications are solely those of the individual author(s) and contributor(s) and not of MDPI and/or the editor(s). MDPI and/or the editor(s) disclaim responsibility for any injury to people or property resulting from any ideas, methods, instructions or products referred to in the content.

Article

Assessment of the Content of Glycoalkaloids in Potato Snacks Made from Colored Potatoes, Resulting from the Action of Organic Acids and Thermal Processing

Anna Pęksa, Agnieszka Tajner-Czopek *, Artur Gryszkin, Joanna Miedzianka, Elżbieta Rytel and Szymon Wolny

Department of Food Storage and Technology, Faculty of Biotechnology and Food Science, Wrocław University of Environmental and Life Sciences, Chełmońskiego St. 37, 51-630 Wrocław, Poland; anna.peksa@upwr.edu.pl (A.P.); artur.gryszkin@upwr.edu.pl (A.G.); joanna.miedzianka@upwr.edu.pl (J.M.); elzbieta.rytel@upwr.edu.pl (E.R.); szymon.wolny@upwr.edu.pl (S.W.)

* Correspondence: agnieszka.tajner-czopek@upwr.edu.pl; Tel.: +48-071-320-7761

Abstract: Glycoalkaloids (TGAs, total glycoalkaloids), toxic secondary metabolites, are found in potatoes (110–335 mg·kg^{-1} DW), mainly in the peel. Colorful, unpeeled potatoes are an innovative raw material for the production of snacks which are poorly tested in terms of their glycoalkaloid content. Third-generation snacks and French fries made from red-fleshed Mulberry Beauty (MB) and purple-fleshed Double Fun (DF) potatoes were produced with the use of 1% solutions of ascorbic, citric, lactic, malic, and tartaric acids to stabilize the structure of anthocyanins in the raw material and maintain their color in obtained products. The influence of the type of acid and thermal processes, like frying, microwaving, and baking, on the content of glycoalkaloids in ready-made products was examined. Only 0.45–1.26 mg·100 g^{-1} of TGA was found in pellet snacks and 1.32–1.71 mg·100 g^{-1} in French fries. Soaking blanched potatoes in organic acid solution reduced the α-chaconine content by 91–97% in snacks and by 57–93% in French fries in relation to the raw material to the greatest extent after the use of malic acid and the DF variety. The effect of lactic and citric acid was also beneficial, especially in the production of baked French fries from MB potatoes.

Keywords: red- and purple-fleshed potatoes; snacks; French fries; glycoalkaloids; organic acids; microwaving; frying; baking

1. Introduction

Colored potatoes, as shown by various authors all over the world [1–4], are characterized by increased attractiveness due to the unusual color that distinguishes them from the traditional yellow, cream, or white varieties that is given to the products obtained from them. They are also a potential source of health-promoting ingredients, including anthocyanins, representing numerous polyphenolic compounds found in potatoes [1,2,5–8]. Many studies are being carried out on increasing the amount or stabilizing the structure and activity of these ingredients in potatoes and their products, both for health reasons and sensory attractiveness. The importance of anthocyanins as active substances with antioxidant and free radical scavenging properties is emphasized, including their importance in inhibiting fat oxidation processes in fried potato products [2,9].

Potatoes of varieties with colored flesh, like those with traditional flesh color, contain glycoalkaloids produced in tubers as secondary metabolites. These compounds are represented in potatoes mainly by α-solanine and α-chaconine and exhibit toxic effects when their total amount reaches approximately 200 mg·kg^{-1} fresh weight FW, causing neurological disorders such as apathy, drowsiness, and disorientation, and may even be fatal [10–13]. However, few studies also indicate the beneficial effects of small amounts of glycoalkaloids, including antipyretic and anti-inflammatory effects [2,10,14]. Research by these authors also shows that α-chaconine is a more toxic compound. Its content in potato

tubers is 2–5 times higher than that of α-solanine. The total content of glycoalkaloids in potatoes ranges from 85 to 182 mg·kg^{-1} DM [2,10,11,14], and they are usually located at a distance of 1–1.5 mm from the outer part of the tuber [15]. The literature on the subject indicates that potatoes intended for food should not contain more than 12 mg·100 g^{-1} FW of total glycoalkaloids (TGA), as they are considered bitter and unacceptable to consumers. The result of research conducted by various research centers are potato varieties in which the accumulation of these compounds does not exceed 10 mg of glycoalkaloids in 100 g of tuber. Varieties with low TGA content also include, to a large extent, varieties with colored flesh [10,16–18].

Potato products, such as selected assortments of French fries and crisps, obtained from unpeeled tubers, contain more polyphenolic compounds, proteins and dietary fiber compared to products from peeled tubers, and the production process is more efficient and profitable. In addition, leaving the skin on the surface of tubers with colored skin significantly increases the content of anthocyanins in the products obtained from them [15]. However, it should be taken into account that attractive for consumers, products with a skin are also characterized by an increased dry matter, and thus a potentially higher share of glycoalkaloids in the product [2,7,15,17].

The peeling and blanching steps during potato processing allows for the removal of approximately 90% of the glycoalkaloids found in the tubers [15,19]. The blanching procedure, commonly used in potato processing, is supposed to have a positive effect on the color of the products and its uniformity, but also on the consistency, especially of French fries. According to science and practice, inorganic acids, such as sulfurous and organic acids, such as citric and ascorbic acids, used in the production of potato products, prevent the browning of potato tissues, i.e., the effect of the Maillard reaction [20,21]. When processing potatoes with red and purple flesh, the color of the finished products is also related to the transformation of anthocyanins that occur under the influence of high temperature [13,20,21]. In that conditions anthocyanins transform into colourless chalcones, which when oxidized, can form high-molecular coloured compounds [22]. Due to the large number of factors, the mechanism of these transformations is not fully known. Anthocyanins have a high ability to attach numerous hydroxyl and methoxy groups in the flavylium cation ring. The degree of glycosylation and acylation of anthocyanidin with phenolic acids or organic acids improves the stability of anthocyanin pigments [23].

The literature on the subject lacks information on the influence of various organic acids on the degree and direction of stabilization of the structure and color of various acylated anthocyanins found in colored potatoes, stable in a wide pH range. Potato varieties with colored flesh are attracting the interest of an increasingly wide group of consumers in various countries around the world. Therefore, they are increasingly the subject of detailed research in terms of the possibility of their industrial processing. Due to the need to develop research on the properties of such potatoes an attempt was made to determine the effect of treating tubers with skin with selected food organic acids on the content of glycoalkaloids in dried potato grits, third-generation snacks made from it and in French fries in the process of frying, microwave oven or baking. The research used ascorbic, citric, lactic, malic, and tartaric acids, which in pilot studies had a positive effect on the color of potato products with purple and red flesh.

Therefore, the aim of the research was to determine the influence of the type of organic acid, the method of preparing products for consumption and the type of raw material, i.e., unpeeled potatoes of 2 varieties, differing in the color of the flesh, and thus in the structure and content of anthocyanins, on the content of glycoalkaloids in dried potatoes and third-generation snacks, and fries.

2. Materials and Methods

2.1. Raw Materials Characteristic

Samples of tubers of two potato varieties with colored flesh, red Mulberry Beauty (MB) and purple Double Fun (DF), from the 2021/2022 growing season, purchased directly from the producer in Poland constituted the research material. The tested potatoes were harvested in full maturity, without mechanical damage, greening, and infestation and without sprouts. The size of the potato tubers of MB variety was closed to 60 × 80 mm (width × length), and the tubers of the DF variety to 60 × 90 mm (Supplementary Figure S1). In total, 20 kg of raw material of each variety was taken for the experiment potatoes collected for the study and then divided. Potatoes collected for the studies were divided into raw materials intended for obtaining freeze-dried potatoes (experimental grits), used in the course of research as a component of pellet snacks, and intended for the production of French fries. Five organic acids of analytical grade were used in the research: citric acid, lactic acid, L-ascorbic acid, malic acid, and L-tartaric acid. In the production of pellet snacks, apart from the experimental dried potatoes, potato starch obtained from a starch factory in Niechlów, as well as corn grits produced by Sante, Poland; and salt (NaCl) by Kłodawa, Poland, were also used.

2.2. Basic Analyses of Raw Materials and Ready-to-Eat Products

The basic chemical composition and a sum of anthocyanins were determined in the raw material. Determinations of the dry matter of fresh, unpeeled potato, freeze-dried raw material, ready French fries, and pellet snacks were carried out by the reduced weight until constant weight was achieved [24]. The starch content was determined in raw potato tubers indirectly by measuring the specific gravity of tubers and reading the starch value from Maercer's tables, according to the methodology described by Houghland [25]. The reducing sugar was determined by the colorimetric method described by Lindsay [26]. The anthocyanin content was determined using the HPLC method, as described below. Freeze-dried potato samples were extracted with 70% aqueous acetone (acidified with 0.1% acetic acid). The homogenized mixture was left for 2 h at room temperature, and then the acetone–water mixture was separated with chloroform to remove lipophilic compounds. The acetone–water fraction was collected, and the remaining acetone was evaporated on a Büchi rotary evaporator. The obtained aqueous extract was brought to a known volume with distilled water and stored at 20 °C until analysis. Samples filtered through 0.2 μm filters were used for HPLC analysis of anthocyanins. Anthocyanins were determined using a Dionex HPLC system (Waltham, MA, USA) equipped with an Ultimate 3000 model diode detector, an LPG-3400A quaternary pump, an EWPS-3000SI automatic sampler, and a TCC-3000SD thermostabilized column chamber, controlled by Chromeleon v.6.8 software. An Atlantic T3 reversed-phase column (250 mm × 4.6 i.d., 5 μm) (Waters, Wexford, Ireland) and an Atlantis T3 guard column (20 × 4.6 i.d., 5 μm) (Waters Corp., Milford, MA, USA) were used. The following solvents were used as the mobile phase: A/4.5% formic acid and B/acetonitrile. The following elution conditions were used: 0–1 min, 5% B isocratic; 1–6 min, linear gradient from 5 and 10% B; 6–26 min, linear gradient 10–20% B; 26–33 min, linear gradient from 20 to 100% B; and then the initial conditions. The flow rate was 1 mL·min^{-1}, and the injection volume was 40 μL. The column was operated at 30 °C. Anthocyanins were monitored at a wavelength of 520 nm.

2.3. Conditions for Producing Experimental Coloured Potato Grits

Washed, unpeeled potatoes intended for the preparation of experimental dried potatoes (in the form of grits of 0.5–1.0 mm) were cut into 1.0 cm slices using a mechanical slicer (Robot coupe CL 50, Vincennes, France). Then, they were blanched in water under conditions typical for French fries' production, i.e., using water at a temperature of 75 °C and in 10 min. The hot slices were cooled in the ice water (2 min.), and then samples were separated and were immersed in parallel in water (control sample) and in 1% solutions of five organic acids, namely citric acid, lactic acid, L-ascorbic acid, malic acid, and L-tartaric acid, for 5 min. The potato slices obtained in this way were dried using the freeze-drying method, and the dried samples were grounded in a laboratory grinder (Retsch GM 200, Hann, Germany) and then passed through a sieve with a mesh size of 1 × 1 mm. The obtained dried potatoes in the form of grits were used as pellet ingredients. These were stored in the freezer before taking into the experiments. The scheme for preparing experimental potato grits, intermediate products, ready-made snacks, and French fries is shown in Figure 1.

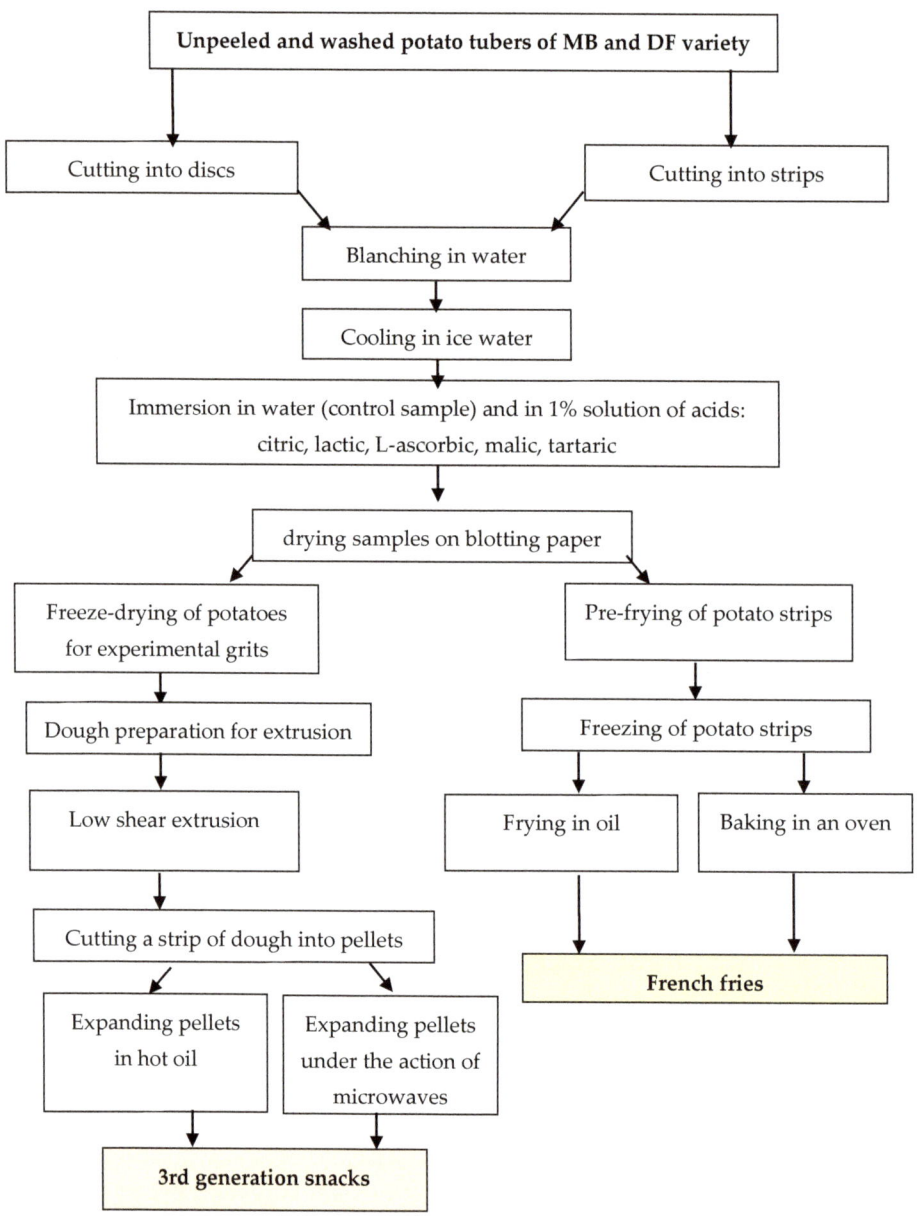

Figure 1. Scheme of laboratory preparation of third-generation snacks and French fries from potatoes of 2 varieties with colored flesh.

2.4. Conditions for Obtaining Extruded Pellets and Colorful Snacks

To the mixture of potato starch (650 g·kg^{-1} of the mixture), experimental dried potato (260 g·kg^{-1} of the mixture), corn grits (50 g·kg^{-1} of the mixture), and salt (10 g·kg^{-1} of the mixture) water was added in an amount that the moisture of the obtained dough was around 35%. The proportions of the mixture for the extruded pellets were elaborated previously by the authors Pęksa et al. [27]. Then, the dough was passed through a sieve with a mesh size of 1×1 mm, packed in polyethylene bags, and kept at a temperature of $20 \pm 2\ °C$ for 24 h. After this time, the samples were rubbed to obtain uniform granulation and moisture, and then they were subjected to the extrusion process in a Brabender laboratory extruder, type 20 DN, using a screw with a compression ratio of 1:1, screw rotation speed of 120 rpm, a head with a die size of 80×0.5 mm, and the process temperature in three successive sections as 50–60–80 °C. The extruded product in the form of a strip was cut into pieces of 30×15 mm and dried at 20–22 °C to a moisture content of about 11%, i.e., for about 14–16 h. Ready semi-products (pellets) were stored for 1–2 days in tightly closed polyethylene bags at the room temperature until expanded snacks were obtained from them.

The pellet samples were divided into two batches, one for the preparation of snacks that expand during frying in oil, and the other for expansion under the influence of microwaves at 750 W. The microwave exposure time was set at 25 s through pilot studies based on sensory evaluation (results not included). The process of frying the pellets was carried out in rapeseed oil heated to the temperature of 180 °C, measuring the period of 3 s after the snacks floated to the surface of the oil. Samples of fried snacks were drained of excess oil on filter paper.

2.5. Conditions for Obtaining Coloured Potato French Fries

Unpeeled potato samples of the tested varieties, intended for the preparation of French fries, were washed, cut using a cutting machine (Robot Coupe CL50, Haas, Germany) into strips with a cross-section of 1×1 cm, and blanched in water at 75 °C for 10 min. Blanched cut potatoes were cooled in ice water for 2 min, and then the whole sample was divided into six parts and immersed for 5 min in parallel in water (control sample) and 1% solutions of five organic acids, i.e., citric acid, lactic acid, L-ascorbic acid, malic acid, and L-tartaric acid. Then, strips of experimental potatoes were drained on paper and subjected to two methods of thermal treatment (frying or baking). French fries were prepared using a two-step method. In the first stage, experimental samples of potato strips were fried in rapeseed oil at 175 °C for 1 min [28], and then they were cooled, frozen, and packed in polyethylene bags. These pre-fried frozen French fried potato samples were divided into two parts for final processing. One part was fried in rapeseed oil at a temperature of 175 °C for about 5 min (the proportion of oil to raw material was: 200 $g \cdot L^{-1}$ of oil), while the other part was thermally treated in a convection oven (Type SCC61 WE, Rational, Landsberg, Germany). For baking, samples of pre-fried French fries were placed on special baking trays and baked at 190 °C for 25 min. Samples of fried and baked ready-made French fries were degreased using the Soxhlet method and intended for determining the content of glycoalkaloids. The time and temperature of the thermal processing of potato strips in a convection oven were determined experimentally in pilot studies, based on the measurement of the dry weight of ready French fries and sensory evaluation (results not included).

2.6. Sample Preparation for the Chromatographic Analysis of α-Solanine and α-Chaconine

The freeze-dried samples of raw material and ready products, after grinding in a laboratory electric mill, constituted a fixed material in which the content of α-solanine and α-chaconine and the sum of these compounds (TGA) were determined. Samples of fried snacks and French fries were previously degreased in a Soxhlet apparatus. These were used for determining the content of glycoalkaloids based on the methodology provided by Saito et al. [29] and Pęksa et al. [30], with some modifications. A sample 5 g of freeze-dried material was mixed with 25 mL of methanol and placed in an ultrasound bath for 30 min, followed by filtration. The filtrate was brought to a final volume of 50 mL with methanol. An aliquot of 5 mL was mixed with 8 mL of water and cleaned up on the SPE column—Chromabond C18 ec.; 500 mg; 6.0 mL (Macherey-Nagel, Dueren, Germany). The eluate was evaporated to dryness under vacuum at a temperature of 50 °C, and dry residue was dissolved in 1 mL of methanol. The whole solution was filtered through syringe filter (Pureland, hydrophobic PTFE, 0.22 µm). Standard solutions (1 mg·mL^{-1}) were prepared by dissolving 10 mg of α-solanine (Sigma-Aldrich, Poznań, Germany) in 10 mL of methanol and 5 mg of α-chaconine (Sigma-Aldrich, Germany) in 5 mL of methanol. A calibration curve was prepared in the range from 1 to 50 µg·mL^{-1} for both analytes. Twenty microliters were injected into the column.

2.7. Apparatus and Conditions of the Glycoalkaloids Separation

To determine the content of α-solanine and α-chaconine, a high-pressure liquid chromatography HPLC (Prominence-i LC-2030C Plus) was used, made by Shimadzu Corporation (Kyoto, Japan), equipped with LC-2030 UV detector, Supelcosil LC-18 (25 cm × 4.6 mm, 5 µm) analytical column, (Supelco Inc., Bellefonte, PA, USA) and a computer system monitoring the chromatograph (Shimadzu LabSolutions, Darmstadt, Germany). As a mobile phase, 0.1 M KH2PO$_4$ and acetonitrile (70:30 v/v) was used. The separation was performed at 70 °C with a 1 mL·min^{-1} flow rate, applying the light wavelength of 200 nm and injection volume of 20 µL.

2.8. Statistical Analysis

The results obtained in the experiment were subjected to statistical calculations of the Statistica v. 13.1 software StatSoft: Tulsa, OK, USA [31]. ANOVA/MANOVA analysis of variance of the data. The Duncan test was performed for comparing the means, and homogeneous groups were determined to prove the significance of the observed differences, using multiple comparisons, and the standard deviations (±SDs) were estimated. All experiments were performed in two technological replications, and the present results show the average values obtained in this investigation.

3. Results and Discussion

3.1. Raw Material Characteristics

The two analyzed varieties of potatoes with colored flesh, namely Mulberry Beauty (MB) with red flesh and Double Fun (DF) with purple flesh, slightly differed only in terms of dry matter and the content of starch and reducing sugars (Table 1). The analyzed samples contained 21.75–22.24 g·100^{-1} g dry weight (d.w.), 16.08–16.66 g·100^{-1} g starch and 0.18–0.22 g·100^{-1} g reducing sugars, and for this reason, they were a suitable raw material for the production of fried snacks [32]. However, they varied to a large extent in content of α-solanine, α-chaconine, and the sum of these compounds (TGA), as well as the ratio of both these of glycoalkaloids forms. Mulberry Beauty potatoes contained more of both forms of glycoalkaloids. The share of α-chaconine in potatoes of the Double Fun variety was higher (2.8) than in the Mulberry Beauty variety (2.4). Urban et.al. [12] determined the content of glycoalkaloids in fourteen new potato cultivars with purple and red flesh in comparison with yellow- and white-fleshed control potatoes; the TGA levels in tubers' flesh ranged from 33.69 to 167.77 mg·kg^{-1} fresh matter (FM), and the ratio of α-chaconine to α-solanine ranged from 1.18 to 3.78. Friedman and Levin [19] report that the TGA content of tubers of potatoes with different color flesh ranges from 8.0 to 63.1 mg·kg^{-1} FW.

Table 1. Characteristics of studied potatoes of two varieties.

Potato Variety	Compound	Raw Unpeeled Potatoes	Dried Unpeeled Potatoes
(MB)	Dry matter [g·100 g^{-1}]	22.24 ± 0.23 [A]	92.88 ± 0.14 [A]
	Starch [g·100 g^{-1}]	16.66 ± 0.11 [A]	69.58 ± 0.07 [A]
	Reducing sugars [g·100 g^{-1}]	0.22 ± 0.08 [A]	0.92 ± 0.05 [A]
	α-solanine [mg·100 g^{-1}]	1.78 ± 0.02 [A]	7.42 ± 0.01 [A]
	α-chaconine [mg·100 g^{-1}]	4.33 ± 0.04 [A]	18.09 ± 0.02 [A]
	TGA [mg·100 g^{-1}]	6.04 ± 0.11 [A]	25.21 ± 0.13 [A]
	α-chaconine/α-solanine	2.4	2.4
	A sum of anthocyanins [mg·100 g^{-1}]	24.95 ± 0.13 [B]	104.2 ± 0.15 [B]
(DF)	Dry matter [g·100 g^{-1}]	21.75 ± 0.19 [B]	92.68 ± 0.25 [A]
	Starch [g·100 g^{-1}]	16.06 ± 0.15 [A]	68.43 ± 0.03 [B]
	Reducing sugars [g·100 g^{-1}]	0.18 ± 0.02 [A]	0.77 ± 0.01 [A]
	α-solanine [mg·100 g^{-1}]	0.61 ± 0.03 [B]	2.60 ± 0.02 [B]
	α-chaconine [mg·100 g^{-1}]	1.73 ± 0.01 [B]	7.36 ± 0.02 [B]
	TGA [mg·100 g^{-1}]	2.34 ± 0.03 [B]	9.96 ± 0.02 [B]
	α-chaconine/α-solanine	2.8	2.8
	A sum of anthocyanins [mg·100 g^{-1}]	50.38 ± 0.15 [A]	214.7 ± 0.11 [A]

[A,B]—the same capital letters within the same column were not significantly different at ($p < 0.05$); according to Duncan's least significant difference test; mean values ($n = 6$); ±SD—standard deviation; MB—Mulberry Beauty; DF—Double Fun.

The authors Ieri et al. [3], Urban et al. [12], and Friedman and Levin [19] also stated that a decisive influence on TGA content had the cultivar genotype, not flesh color. Jansen and Flamme [11] showed the presence of total glycoalkaloids in raw potatoes of the red-fleshed Red Cardinal (2.86 mg·100 g^{-1} FW) and in the blue-fleshed varieties Blaue Zimmerli and Violettfleischige 2.21 mg·100 g^{-1} and 7.5 mg·100 g^{-1} FW, respectively.

In the conducted research, the TGA content in potatoes of the MB variety on average stated about 6.04 mg·100 g^{-1} FM, more than found in the tubers of the Double Fun variety, i.e., 2.34 mg·100 g^{-1} FM; however, both of them were within the range of the permitted quantity for health reasons, which should not be higher than 12 mg·100 g^{-1} of raw tubers [10]. These amounts were lower than those obtained by other authors, such as Urban et al. [12], and analyzing the content of glycoalkaloids in potatoes of varieties with colored flesh, i.e., mostly in the range from 4.10 to 11.0 mg·100 g^{-1} FM. Lachman et al. [33] also found that the TGA levels in red-fleshed potato varieties were higher compared to blue-fleshed varieties.

The tested samples of tubers of both potato varieties differed in terms of their total anthocyanin content. Potatoes of the DF variety with purple flesh contained twice as much of these compounds, i.e., on average, 50.38 mg·100 g^{-1} FW (Table 1). Lachman et al. [33] reports that potatoes with red flesh contained an average of 231.9 mg·100 g^{-1} DM of anthocyanins, while those with purple flesh contained 146.8 mg·100 g^{-1} DM.

The conducted research assumed a beneficial effect of the potato blanching process on reducing the content of glycoalkaloids, as confirmed by the authors Tian et al. [7], Singh et al. [17], Lachman et al. [33], Omayio et al. [34], and Nie et al. [35], who noted that microwaving and baking processes reduced the amount of glycoalkaloids in unpeeled potatoes by 45% and 51%, respectively, compared to the raw material. The data in Table 2 show that immersing blanched potato slices in 1% solutions of five organic acids contributed to reducing the amount of glycoalkaloids in the resulting experimental droughts, compared to the raw material—dried, unpeeled tubers (Table 1). The decrease in the amount of glycoalkaloids depended on both the potato variety and the type of organic acid in which the blanched potato slices were soaked. The TGA content in experimental grits decreased in relation to the raw material by 49–79% when MB variety potatoes were used and by 19–57% when DF variety potatoes were used. Experimental potato grits made from potatoes of the MB variety contained, apart from samples soaked in citric and tartaric acids, more glycoalkaloids than grits obtained from tubers of the DF variety.

Table 2. Characteristics of experimental potato grits depending on the type of organic acid used.

Potato Variety	Compound	Experimental Dried Potato					
		Water/Control	Citric Acid	Lactic Acid	Ascorbic Acid	Malic Acid	Tartaric Acid
(MB)	Dry matter [g·100 g^{-1}]	92.21 ± 0.23 [B]	93.05 ± 0.17 [A]	93.10 ± 0.16 [A]	92.75 ± 0.20 [A]	93.08 ± 0.14 [A]	92.99 ± 0.19 [A]
	α-solanine [mg·100 g^{-1}]	2.32 ± 0.03 [bA]	1.50 ± 0.06 [cB]	3.82 ± 0.01 [aA]	2.66 ± 0.02 [bA]	3.96 ± 0.03 [aA]	1.52 ± 0.02 [cB]
	α-chaconine [mg·100 g^{-1}]	7.46 ± 0.02 [bA]	4.44 ± 0.02 [cdB]	9.12 ± 0.03 [aA]	5.25 ± 0.03 [cA]	8.20 ± 0.01 [abA]	3.82 ± 0.03 [dB]
	TGA [mg·100 g^{-1}]	9.78 ± 0.08 [bA]	5.94 ± 0.01 [dB]	12.94 ± 0.02 [aA]	7.91 ± 0.02 [cA]	12.16 ± 0.02 [aA]	5.34 ± 0.01 [dB]
	α-chaconine/α-solanine	3.1	2.9	2.4	2.0	2.1	2.5
(DF)	Dry matter [g·100 g^{-1}]	93.39 ± 0.18 [A]	92.91 ± 0.17 [B]	93.54 ± 0.15 [A]	92.77 ± 0.19 [A]	92.30 ± 0.21 [B]	93.01 ± 0.14 [A]
	α-solanine [mg·100 g^{-1}]	1.54 ± 0.01 [abB]	1.58 ± 0.05 [abA]	1.29 ± 0.06 [cB]	0.92 ± 0.02 [cB]	1.70 ± 0.01 [aB]	1.66 ± 0.04 [aA]
	α-chaconine [mg·100 g^{-1}]	5.51 ± 0.03 [abB]	6.10 ± 0.06 [aA]	4.69 ± 0.08 [abB]	3.37 ± 0.02 [bB]	3.88 ± 0.02 [bB]	6.44 ± 0.03 [aA]
	TGA [mg·100 g^{-1}]	7.04 ± 0.02 [abB]	7.68 ± 0.04 [abA]	5.98 ± 0.07 [bcB]	4.29 ± 0.02 [cB]	5.58 ± 0.01 [bcB]	8.10 ± 0.03 [aA]
	α-chaconine/α-solanine	3.6	3.9	3.6	3.7	2.3	3.9

[a–d] The same lower letters within the same rows were not significantly different at $p < 0.05$; [A,B] the same capital letters within the same column were not significantly different at ($p < 0.05$); according to Duncan's least significant difference test; mean values ($n = 6$); ±SD—standard deviation; MB—Mulberry Beauty; DF—Double Fun.

The lowest TGA content in experimental grits was found in samples obtained from tubers of the MB variety treated with tartaric and citric acid, and from potatoes of the DF variety treated with ascorbic acid. With regard to the raw material, in experimental grits, especially those obtained from potatoes of the DF variety, the content of α-solanine decreased under the influence of blanching and immersion in acid solutions to a greater extent than α-chaconine, with the exception of samples immersed in malic acid (Tables 1 and 2). As a result of the effect of technological factors, including the type of organic acid used, the ratio of the amount of α-chaconine to α-solanine changed. In most experimental grits from tubers of the DF variety, this parameter increased to 3.6–3.9, to the greatest extent after the use of citric and tartaric acids. In samples of grits from (MB) tubers treated with citric and tartaric acids was recorded an increase in the proportion of α-chaconine in TGA to 2.5–2.9 and in the control sample to 3.1.

When analyzing the obtained results, it can be concluded that immersing blanched potato slices in solutions of appropriately selected low-concentration organic acids may contribute to a significant reduction in the amount of toxic glycoalkaloids in products obtained from varieties of potatoes of colored flesh.

Research by various authors, such as Ieri et al. [3], Tian et al. [7], D'Amelia et al. [13], Rytel et al. [15], Lachman et al. [33], Nie et al. [35], and Rytel et al. [36], shows that glycoalkaloids contained in potatoes can be largely removed during their processing, as a result of rinsing, but also as a result of the degradation of these compounds at raised temperatures, e.g., during frying or baking processes. From the other side, information in the literature on the influence of various organic acids on changes in the glycoalkaloid content in color-fleshed potatoes during processing and in ready products is scarce.

3.2. Influence of Different Factors in the Production of Pellet Snacks on the Content of Glycoalkaloids in Ready Products

The potato grits obtained during the experimental research were, next to potato starch, the basic ingredient of extruded pellets, which in turn were a semi-product for obtaining expanded, crispy snacks of the third generation. A highly significant effect of the factors used in the research on the content of glycoalkaloids in ready-made snacks was found (Table 3). Products obtained with grits from DF variety tubers contained more α-chaconine and TGA than those made with grits from MB variety potatoes; however, regardless of the origin of the grits, the amount of glycoalkaloids in the snacks remained at a low level, α-solanine did not exceed 0.47 mg·100 g^{-1}, and α-chaconine 0.79 mg·100 g^{-1}, so there was a total of 1.26 mg·100 g^{-1} TGA.

Table 3. Results of ANOVA and LSD multiple range tests of the influence of the pellet expansion method, the type of organic acid used, and the potato variety on the content of glycoalkaloids in the obtained third-generation snacks.

Factor	α-Solanine	α-Chaconine [mg·100 g^{-1}]	TGA
ANOVA Test			
Potato variety	NS	***	***
Expanding method	***	***	***
Acid type	***	***	***
Potato variety x acid type	***	***	***
Expanding method x acid type	***	***	***
Potato variety x expanding method	***	***	***
LSD multiple range test			
Expanding method			
Frying	0.15 b	0.30 b	0.45 b
Microwaving	0.47 a	0.79 a	1.26 a
Type of acid			
Control	0.30 cd	0.52 bc	0.82 c
Citric acid	0.29 c	0.53 b	0.81 cd
Lactic acid	0.32 ab	0.54 b	0.86 bc
Ascorbic acid	0.33 ab	0.58 ab	0.91 ab
Malic acid	0.27 c	0.46 c	0.73 d
Tartaric acid	0.35 a	0.64 a	0.99 a
Potato variety			
Mulberry Beauty	0.31 a	0.51 b	0.82 b
Double Fun	0.31 a	0.57 a	0.88 a

NS = not significant at $p > 0.05$; *** significant at $p < 0.001$, respectively; ($n = 6$), a–d—values followed by the same letter, within the same column, were not significantly different ($p > 0.05$), according to Duncan's least significant difference test.

The statistical analysis of the one-way ANOVA data (Table 3) and interactions (Figure 2A; Table S1) showed that fried snacks obtained from pellets based on grits from tubers of both potato varieties contained 62–64% less glycoalkaloids than those expanded under the influence of microwaves, and, accordingly, the content of glycoalkaloids in fried snacks from MB variety potatoes reached 0.38 mg·100 g^{-1} TGA, and from DF variety potato tubers, 0.52 mg·100 g^{-1} TGA, while in samples expanded under the influence of microwaves, these values were 1.27 and 1.24 mg·100 g^{-1} TGA, respectively. Regardless of the snack expansion method, they contained more α-chaconine than α-solanine. In snacks made from MB potatoes, the proportion of α-chaconine to α-solanine was 1.59, and in products made from DF potatoes, it was 1.82 (Figure 2; Supplementary Table S1).

These differences between fried and microwaved snacks were probably related to the presence of fat in fried snacks and a lower share of the dry matter of potato tubers and other ingredients of the crisp recipe in the weight of these products. Fried potato or snacks of the third generation usually contain about 20–30% fat [32,37].

Organic acids used in the process of obtaining experimental potato grits had a different impact on the content of glycoalkaloids in the snacks obtained with them. The use of tartaric or ascorbic acid contributed to a higher content of both forms of glycoalkaloids in the snacks: on average, they contained 0.35 mg·100 g^{-1} α-solanine, 0.64 mg·100 g^{-1} α-chaconine, and a total of TGA 0.99 mg·100 g^{-1} (Table 3). On the other side, the use of malic acid in the process of obtaining snacks, as well as citric acid, resulted in a reduction in the glycoalkaloid content. The products of the control sample, in which no organic acid was used—only water—were similar in terms of glycoalkaloid content to the remaining crisp samples, except for the samples obtained with tartaric acid.

Regardless of the pellet expansion method, there was a significant impact of the potato variety and the type of organic acid used on the glycoalkaloid content in ready-to-eat snacks (Figure 2B; Supplementary Table S1).

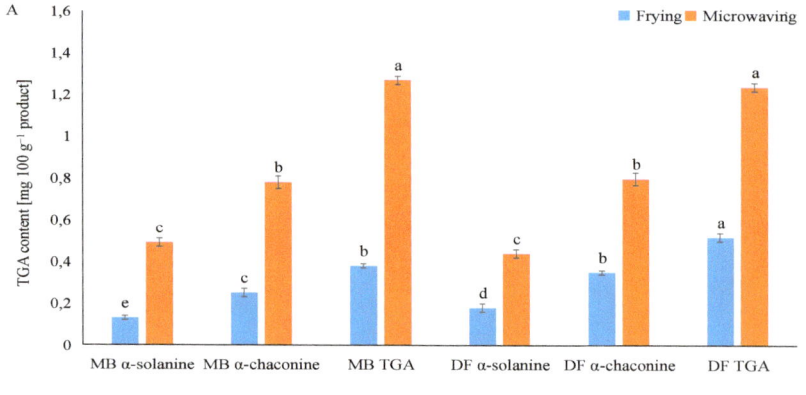
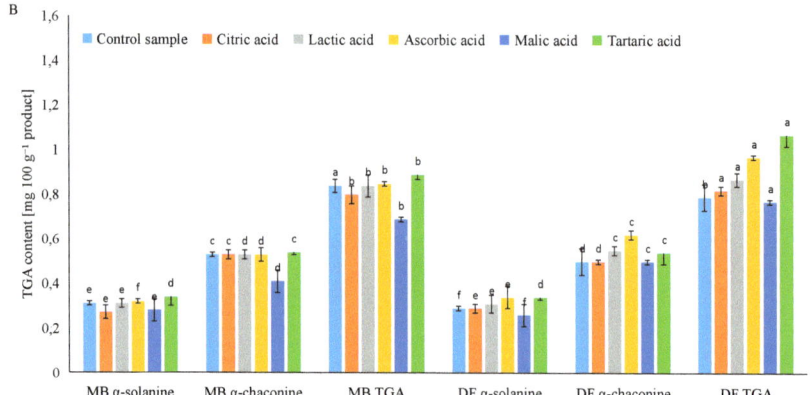
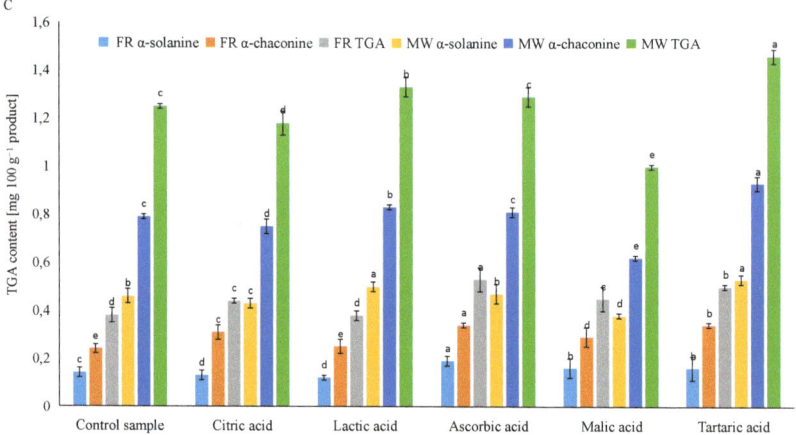

Figure 2. The influence of (**A**) potato variety and expanding method, (**B**) potato variety and acid type, and (**C**) acid type and expansion method on the glycoalkaloids content in snacks. MB—Mulberry Beauty; DF—Double Fun; TGAs—total glycoalkaloids; FR—frying; MW—microwave. Values are represented as mean (±SD) standard deviation ($n = 6$); $^{a-f}$—indicate significant differences. (**A**) Potato variety and expanding method, (**B**) potato variety and acid type, and (**C**) acid type and expansion method (Duncan's test, $p \leq 0.05$).

Tartaric acid contributed to a reduction in the content of α-chaconine in products based on grits from tubers of both varieties, while ascorbic acid caused a greater increase in the content of α-chaconine in snacks made from potatoes of the DF variety. Compared to the control sample, the use of organic acids, apart from malic acid, contributed to a slight increase in the α-solanine content in the ready products, especially those made with grits from DF variety tubers. The content of total glycoalkaloids in snacks made with organic acids differed only slightly from the control sample when the raw material was MB variety potatoes. However, it was shown that the use of organic acids in the production of grits from potatoes of the DF variety with purple flesh, apart from malic acid, resulted in an increase in the TGA content in the analyzed products, compared to the control sample.

As shown from the data present in Figure 2C and Supplementary Table S1, the content of glycoalkaloids in the analyzed snacks depended more on the method of expanding the pellets than on the type of organic acid used for their production. Fried products from the control sample contained over 3.3 times less TGA than microwaved snacks. Fried products made with organic acids contained 2.2 (malic acid) to 3.5 (lactic acid) times less TGA compared to microwave samples. The greatest differences in the content of individual forms of glycoalkaloids and their sum between the fried and microwaved snacks samples occurred in the samples in which lactic acid was used. The differences in the content of α-solanine, α-chaconine and TGA in these snack products were 4.2, 3.3, and 3.5 times lower, respectively, in fried products than in microwaved products. The smallest differences were observed in snack samples prepared with malic acid, which contained 2.4, 2.1, and 2.2 times less glycoalkaloids in the fried samples than in the samples heated in a microwave oven, respectively. Thus, malic acid turned out to be the most preferred organic acid in terms of the glycoalkaloids content in snacks made from potatoes of colored flesh varieties, but it was particularly beneficial in the production of microwaved snacks. The available literature lacks information on the impact of the use of organic acids in the production of fried products from potatoes with red or purple flesh. Research involving acids primarily examines the specific color of the resulting products and their texture [38]. In contrast, some authors, such as Huang et al. [39] and Negoiță et al. [40], used the addition of blanching acids, i.e., citric, acetic, and ascorbic, during the production process of fried potato snacks in order to reduce the amount of acrylamide.

By comparing the content of glycoalkaloids in the tested fried and microwaved snacks with their content in the raw material, i.e., in the experimental grits, at the same level of dry matter, it was found that, in the fried crisps made with grits from Mulberry Beauty tubers, from 2.95 to 5.86% α-solanine remained, from 3.17 to 4.68% α-chaconine, and from 3.31 to 4.81% TGA; and in crisps containing Double Fun potato grits, from 9.32 to 15.9% α-solanine, from 5.78 to 8.98% α-chaconine, and from 6.14 to 9.27% TGA. Higher glycoalkaloid residues were found in microwaved crisps than in fried snacks. Depending on the type of acid used, glycoalkaloid residues in potato snacks ranged from 11.5 to 15.0% TGA when experimental grits from Mulberry Beauty potato tubers were used and from 22.6 to 23.7% TGA when it was Double Fun potato grits. However, more α-solanine than α-chaconine remained in the snacks.

Research conducted by Rytel et al. [15] showed that the French fries they obtained contained 3%, crisps 16%, and dried potatoes 17% of the initial amount of total glycoalkaloids (TGA) contained in the raw material, and the reduction in the content of α-chaconine and α-solanine was at a similar level.

3.3. Influence of Different Factors in the Production of French Fries on the Content of Glycoalkaloids in Ready Products

The content of both forms of glycoalkaloids in French fries obtained during the research from potatoes with colored flesh depended primarily on the potato variety and the method of preparing the fries for consumption, and only to a small extent on the type of organic acid used at the stage of blanching the potato strips (Table 4). The interactions of the factors used had a significant impact on the content of glycoalkaloids in ready-made French fries (Figure 3A–C; Supplementary Table S2).

French fries obtained from tubers of the purple-colored DF variety contained, on average, 2 times less α-solanine compared to French fries prepared from tubers of the MB variety with red flesh, while the differences between French fries from potatoes of these varieties in terms of α-chaconine and TGA content were minor. A more advantageous method of preparing French fries for consumption in terms of the glycoalkaloid content in the final product was baking a frozen semi-finished product. This concerned the content of α-solanine to a greater extent. Baked French fries contained on average approximately 3 times less α-solanine than fried French fries. Regardless of the type of acid used, the content of α-solanine in the tested French fries decreased compared to the control sample (immersion in water), while the content of α-chaconine and TGA increased, especially after the use of ascorbic acid (Table 4).

Table 4. Results of ANOVA and LSD multiple range tests of the influence of French fries' preparing method, the type of organic acid used, and the potato variety on the content of glycoalkaloids in ready products.

Factor	α-Solanine	α-Chaconine [mg·100 g^{-1}]	TGA
ANOVA Test			
Potato variety	***	***	***
Preparing method	***	***	***
Acid type	*	*	NS
Potato variety × preparing method	***	***	***
Preparing method × acid type	***	**	**
Potato variety × acid type	***	***	***
LSD multiple range test			
Preparing method			
Frying	0.20 a	1.51 a	1.71 a
Baking	0.08 b	1.24 b	1.32 b
Type of acid			
Control	0.20 a	1.23 b	1.43 a
Citric acid	0.14 ab	1.22 b	1.36 a
Lactic acid	0.17 ab	1.26 b	1.43 a
Ascorbic acid	0.13 ab	1.67 a	1.80 a
Malic acid	0.09 b	1.40 ab	1.50 a
Tartaric acid	0.10 ab	1.49 ab	1.59 a
Potato variety			
Mulberry Beauty	0.18 a	1.31 b	1.49 a
Double Fun	0.09 b	1.45 a	1.54 b

NS = not significant at $p > 0.05$; *, **, and *** significant at $p < 0.05$, 0.01, and 0.001, respectively ($n = 6$); a,b—values followed by the same letter, within the same column, were not significantly different ($p > 0.05$), according to Duncan's least significant difference test.

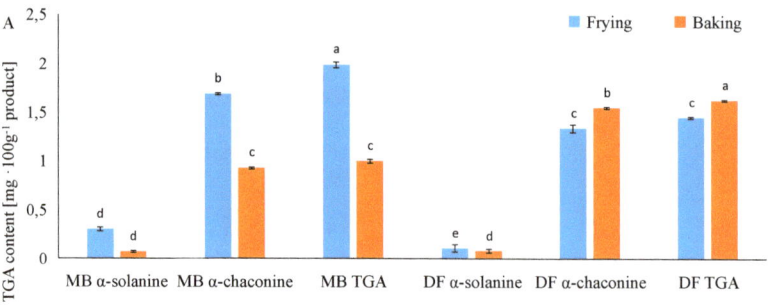

Figure 3. Cont.

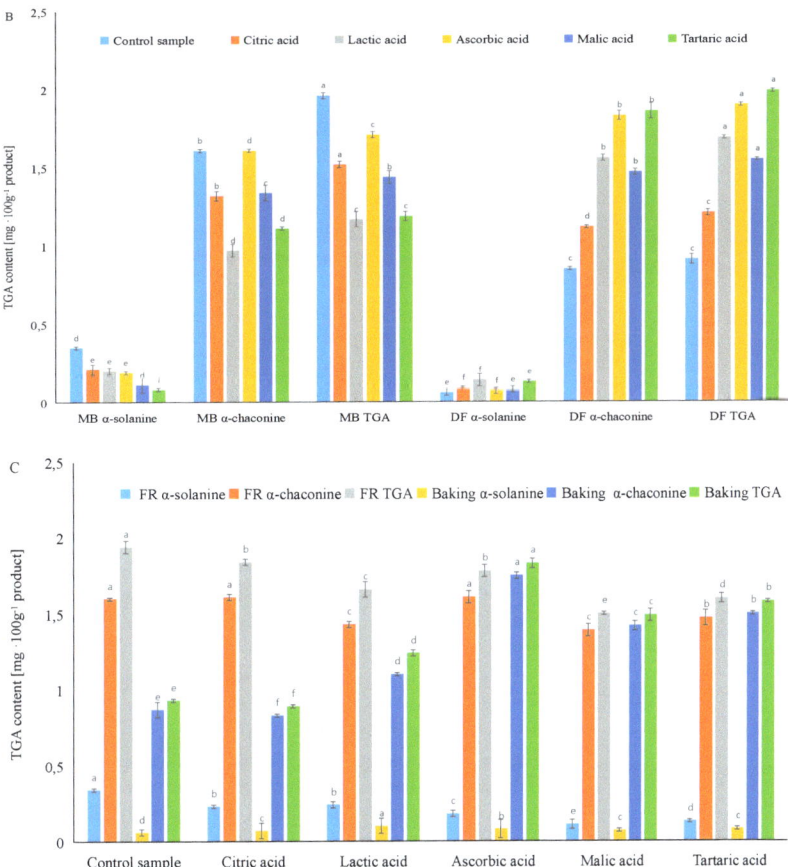

Figure 3. The influence of the (**A**) preparing method and potato variety, (**B**) acid type and potato variety, and (**C**) preparing method and acid type on the glycoalkaloids content in French fries made from colored potatoes of two varieties. MB—Mulberry Beauty; DF—Double Fun; FR—frying. Values are represented as mean (± SD) standard deviation, ($n = 6$). $^{a–f}$ Significant differences. (**A**) Preparing method and potato variety, (**B**) acid type and potato variety, and (**C**) preparing method and acid type (Duncan's test, $p \leq 0.05$).

A more pronounced influence of the method of preparing French fries on the content of glycoalkaloids was found in products obtained from potatoes of the MB variety (Figure 3A; Supplementary Table S2). French fries fried in the second stage, obtained from tubers of this variety, contained more of both forms of glycoalkaloids than baked ones. However, such relationship was not found in French fries obtained from potatoes of the DF variety, with purple flesh. The process of soaking potato strips in solutions of various organic acids turned out to be particularly beneficial in terms of glycoalkaloid content for French fries obtained from potatoes of the MB variety. These samples contained fewer glycoalkaloids as compared to the control products, regardless of the type of organic acid used (except ascorbic acid). Potatoes of the DF variety reacted in the opposite way to the use of organic acids. The lactic and tartaric acids were the most beneficial acids in the production of French fries from MB variety tubers, while the citric acid was found to be the best in the production of French fries from DF variety potatoes (Figure 3B; Supplementary Table S2). Ascorbic acid turned out to be unfavorable acid in the production of French fries when the MB variety was used, and ascorbic and tartaric acids when the DF variety was used.

The method of preparing the French fries to consumption had a significant influence on the content of both forms of glycoalkaloids and their sum (TGA). The data in Figure 3C and Supplementary Table S2 show that, regardless of the method of preparing French fries, i.e., frying or baking, the α-solanine content in the ready-to-eat French fries was at a low level: in fried samples, it was found from 0.13 to 0.34 mg·100 g^{-1} of product, and in baked samples, from 0.05 to 0.10 mg·100 g^{-1} of product. The use of organic acids in the production of deep-fried French fries contributed to a reduction in the α-solanine content compared to the control sample, while in the samples of baked French fries, there was no effect of the type of acid on the content of this form of glycoalkaloid.

The use of malic and tartaric acids turned out to be particularly beneficial. In the French fries obtained in the study, regardless of the method of their preparation for consumption, the α-chaconine content predominated. Fried samples of ready products contained from 1.39 to 1.61 mg·100 g^{-1} of this alkaloid, while baked samples contained from 0.83 to 1.75 mg·100 g^{-1} (Figure 3C; Supplementary Table S2). The content of total glycoalkaloids ranged from 1.50 to 1.94 mg·100 g^{-1} of fried French fries and from 0.89 to 1.83 mg·100 g^{-1} of baked ones. In fried French fries, the use of lactic, malic, and tartaric acids contributed to a greater reduction in the content of α-chaconine and TGA than when other acids were used.

Baked products from the control sample and samples in which citric and lactic acid were used contained approximately 37–53% less α-chaconine and approximately 32–51% less total glycoalkaloids as compared to French fries obtained with ascorbic acid. Thus, in the production of baked French fries, the use of citric and lactic acids turned out to be beneficial due to the content of glycoalkaloids, and in the production of deep-fried fries, the use of lactic, malic, and tartaric acids turned out to be favorable. In a study conducted by Liu et al. [41], soaking potato strips in clean water for 8 h had a similar effect on reducing the glycoalkaloid content in raw potatoes as immersing them in acetic acid solutions for the same time. As a result, less than 8.6% of α-solanine and α-chaconine remained in the raw potato samples, both in the control sample and after acid application, especially after longer soaking times, i.e., after 12 and 24 h. In our experiment, a significant effect in reducing the content of glycoalkaloids in products from tubers of varieties with colored flesh was achieved after just 5 min of soaking in 1% solutions of the five acids used. The analysis of the glycoalkaloid content in ready-made French fries in relation to the amount of glycoalkaloids present in raw, unpeeled tubers intended for the production of French fries, carried out at the same DM content, showed that α-solanine residues in fried French fries ranged from 1.65 to 14.6%, α-chaconine from 11.7 to 42.7% and TGA from 8.84 to 34.9%, with larger amounts of glycoalkaloids coming from tubers of the Double Fun variety. Baked fries contained from 1.2 to 9.71% of α-solanine derived from the raw material, from 6.78 to 12.5% of α-chaconine, and from 5.15 to 44.2% of TGA, with higher values for fries obtained from Double Fun potatoes. Liu et al. [41] obtained an over 90% reduction in the content of glycoalkaloids in fried French fries in relation to raw tubers by soaking potato pieces in acetic acid solutions for a period of 1 to 8 h. These authors examined potatoes of a light-fleshed variety and found that the reduction in the content of glycoalkaloids in fries was mainly influenced by the soaking time and not the concentration of the solution of this acid. It should also be noted that the heat treatment of the potatoes themselves can have an impact on reducing the amount of TGA in the ready products. D'Amelia et al. [13] report that cooking potatoes had a reduction effect on the amounts for α-solanine and α-chaconine of about 80% and 65%, respectively, while the frying process reduced the amount of total TGA by about 90% on average. According to Nie et al. [35], the frying process of chips had an effect of reducing the amount of TGA by about 94% compared to tubers. Authors D'Amelia et al. [13] also found that the use of microwave and oven heat treatments had an effect on reducing the amount of glycoalkaloids (α-solanine and α-chaconine), but less than the frying process.

4. Conclusions

Processing potatoes containing anthocyanins requires maintaining the structure and color of these phenolic compounds and their health-promoting effects. Processing tubers with skin increases the amounts of anthocyanins in the products obtained from them, but it is possible to introduce other compounds found in potato skin into the finished products, including glycoalkaloids, which are toxic in larger amounts.

The research showed a varied impact of the use of 1% solutions of ascorbic, citric, lactic, malic, and tartaric acids in the production of dried potato groats and third-generation snacks and French fries obtained from them on the content of α-solanine and α-chaconine. The use of tartaric acid in the production of third-generation snacks contributed to an increase in the content of TGA and α-chaconine, regardless of the potato variety. Malic acid turned out to be the most beneficial in their production, regardless of the potato variety and the method of expanding the pellets. Due to the content of glycoalkaloids, a more favorable method of expanding the pellets was deep-frying in hot oil. Third-generation fried snacks contained, on average, three times less glycoalkaloids than those from the microwave, which may be related to the higher dry matter content of potato snacks that came from the raw material.

In the production of French fries, the use of organic acids had little or no effect on the glycoalkaloid content in the finished products and depended on the potato variety. In this respect, it was more advantageous to use potatoes of the MB variety with red flesh and the use of lactic acid and baking fried and frozen semi-finished products rather than deep-frying them in oil. Regardless of the potato variety, the process of baking fried French fries turned out to be more beneficial than frying them due to the three times lower content of glycoalkaloids in the finished products. This was probably related to the greater degradation of glycoalkaloids in a longer process and higher temperature. Studies have shown a greater impact of technological treatments and organic acids on reducing the content of less toxic α-solanine in finished products. With regard to the raw material, 3.31–9.27% TGA remained in third-generation fried snacks, from 11.5 to 23.7% TGA in microwaved snacks, from 8.84 to 34.9% TGA in fried French fries, and from 5.15 to 44.2% TGA in baked fries.

Supplementary Materials: The following supporting information can be downloaded at https://www.mdpi.com/article/10.3390/foods13111712/s1, Figure S1. Photo of potato tubers of the Mulberry Beauty variety with red flesh and the Double Fun variety with purple flesh, including the size of the tubers (own photos). Table S1. The content of glycoalkaloids in samples of pellet snacks obtained from dried potatoes of two varieties, depending on the type of organic acid and the method of expanding used for their production. Table S2. The content of glycoalkaloids in samples of French fries obtained from two varieties, depending on the type of organic acid used in their production and the method of their preparing for eating.

Author Contributions: Conceptualization, A.P. and A.T.-C.; methodology, A.P. and A.T.-C.; software, J.M.; validation, S.W., A.G. and J.M.; investigation, A.P., A.T.-C., S.W. and A.G.; data curation, A.P. and A.T.-C.; writing—original draft preparation A.P. and A.T.-C.; writing—review and editing A.P. and A.T.-C.; visualization, A.T.-C. and J.M.; supervision, E.R. and J.M. All authors have read and agreed to the published version of the manuscript.

Funding: This research received no external funding.

Institutional Review Board Statement: Not applicable.

Informed Consent Statement: Not applicable.

Data Availability Statement: The original contributions presented in the study are included in the article/Supplementary Materials, further inquiries can be directed to the corresponding author.

Conflicts of Interest: The authors declare no conflict of interest.

References

1. Wrolstad, R.E.; Durst, R.W.; Lee, J. Tracking color and pigment changes in anthocyanin products. *Trends Food Sci. Technol.* **2005**, *16*, 423–428. [CrossRef]
2. Lachman, J.; Hamouz, K.; Orsák, M. Coloured potatoes. In *Advances in Potato Chemistry and Technology*, 2nd ed.; Singh, J., Kaur, L., Eds.; Academic Press: London, UK, 2016; Chapter 9; pp. 249–281.
3. Ieri, F.; Innocenti, M.; Andrenelli, L.; Vecchio, V.; Mulinacci, N. Rapid HPLC/DAD/MS method to determine phenolic acids, glycoalkaloids and anthocyanins in pigmented potatoes (*Solanum tuberosum* L.) and correlations with variety and geographical origin. *Food Chem.* **2011**, *125*, 750–759. [CrossRef]
4. Ru, W.; Pang, Y.; Gan, Y.; Liu, Q.; Bao, J. Phenolic Compounds and Antioxidant Activities of Potato Cultivars with White, Yellow, Red and Purple Flesh. *Antioxidants* **2019**, *8*, 419. [CrossRef] [PubMed]
5. Camire, M.E.; Kubow, S.; Donnelly, D.J. Potatoes and Human Health. *Crit. Rev. Food Sci. Nutr.* **2009**, *49*, 823–840. [CrossRef] [PubMed]
6. Nayak, B.; Berrios, J.D.J.; Powers, J.R.; Tang, J.; Ji, Y. Colored potatoes (*Solanum tuberosum* L.) dried for antioxidant-rich value-added foods. *J. Food Process. Preserv.* **2011**, *35*, 571–580. [CrossRef]
7. Tian, J.; Chen, J.; Lv, F.; Chen, S.; Chen, J.; Liu, D.; Ye, X. Domestic cooking methods affect the phytochemical composition and antioxidant activity of purple-fleshed potatoes. *Food Chem.* **2016**, *197*, 1264–1270. [CrossRef]
8. Rasheed, H.; Ahmad, D.; Bao, J. Genetic Diversity and Health Properties of Polyphenols in Potato. *Antioxidants* **2022**, *11*, 603. [CrossRef] [PubMed]
9. Harada, K.; Kano, M.; Takayanagi, T.; Yamacawa, O.; Ishikawa, F. Absorption of acylated antocyanins in rats and humans after ingesting an extract of Ipomoea batatas purple sweet potato tuber. *Biosci. Biotechnol. Biochem.* **2004**, *68*, 1500–1507. [CrossRef] [PubMed]
10. Friedman, M. Potato glycoalkaloids and metabolites: Roles in the plant and in the diet. *J. Agric. Food Chem.* **2006**, *15*, 8655–8681. [CrossRef] [PubMed]
11. Jansen, G.; Flamme, W. Coloured potatoes (*Solanum tuberosum* L.)–anthocyanin content and tuber quality. *Genet. Resour. Crop Evol.* **2006**, *53*, 1321–1331. [CrossRef]
12. Urban, J.; Hamouz, K.; Lachman, J.; Pulkrábek, J.; Pazderů, K. Effect of genotype, flesh colour and environment on the glycoalkaloid content in potato tubers from integrated agriculture. *Plant Soil Environ.* **2018**, *64*, 186–191. [CrossRef]
13. D'Amelia, V.; Sarais, G.; Fais, G.; Dessì, D.; Giannini, V.; Garramone, R.; Carputo, D.; Melito, S. Biochemical Characterization and Effects of Cooking Methods on Main Phytochemicals of Red and Purple Potato Tubers, a Natural Functional Food. *Foods* **2022**, *11*, 384. [CrossRef] [PubMed]
14. Deußer, H.; Guignard, C.; Hoffmann, L.; Evers, D. Polyphenol and glycoalkaloid contents in potato cultivars grown in Luxembourg. *Food Chem.* **2012**, *135*, 2814–2824. [CrossRef] [PubMed]
15. Rytel, E.; Tajner-Czopek, A.; Kita, A.; Kucharska, A.Z.; Sokół-Łętowska, A.; Hamouz, K. Content of anthocyanins and glycoalkaloids in blue-fleshed potatoes and changes in the content of α-solanine and α-chaconine during manufacture of fried and dried products. *Int. J. Food Sci. Technol.* **2018**, *53*, 719–727. [CrossRef]
16. Friedman, M.; Kozukue, N.; Kim, H.-J.; Choi, S.-H.; Mizuno, M. Glycoalkaloid, phenolic, and flavonoid content and antioxidative activities of conventional nonorganic and organic potato peel powders from commercial gold, red, and Russet potatoes. *J. Food Compos. Anal.* **2017**, *62*, 69–75. [CrossRef]
17. Singh, B.; Dutt, S.; Raigond, P. Potato glycoalkaloids. In *Potato: Nutrition and Food Security*; Springer: Singapore, 2020; pp. 191–212. ISBN 9789811576621.
18. Martinez-Garcia, I.; Gaona-Scheytt, C.; Morante-Zarcero, S.; Sierra, I. Development of a Green, Quick, and Efficient Method Based on Ultrasound-Assisted Extraction Followed by HPLC-DAD for the Analysis of Bioactive Glycoalkaloids in Potato Peel Waste. *Foods* **2024**, *13*, 651. [CrossRef] [PubMed]
19. Friedman, M.; Levin, C.E. Glycoalkaloids and Calystegine Alkaloids in Potatoes. In *Advances in Potato Chemistry and Technology*, 2nd ed.; Singh, J., Kaur, L., Eds.; Academic Press: London, UK, 2016; Chapter 7; pp. 167–194.
20. Vinci, R.M.; Mestdagh, F.; Van Poucke, C.; Kerkaert, B.; De Muer, N.; Denon, Q.; Van Peteghem, C.; De Meulenaer, B. Implementation of acrylamide mitigation strategies on industrial production of French fires: Challenges and Pitfalls. *J. Agric. Food Chem.* **2011**, *59*, 898–906. [CrossRef] [PubMed]
21. Amaral, R.D.A.; Benedetti, B.C.; Pujolà, M.; Achaerandio, I. Effect of citric acid on browning of fresh-cut potatoes and on texture after frying. *Acta Hortic.* **2018**, *1209*, 259–264. [CrossRef]
22. Oszmiański, J.; Bąkowska, A.; Piacente, S. Thermodynamic characteristics of copigmentation reaction of acylated anthocyanin isolated from blue flowers of *Scutellaria baicalensis* Georgi with copigments. *J. Sci. Food Agric.* **2004**, *84*, 1500–1506. [CrossRef]
23. Brownmiller, C.; Howard, L.R.; Prior, R.L. Processing and storage on monomeric anthocyanins percent polymeric color, and antioxidant capacity of processed blueberry products. *J. Food Sci.* **2008**, *5*, 72–79. [CrossRef]
24. Thiex, N.J.; Van Erem, T. Determination of water (moisture) and dry matter in animal feed, grain, and forage (plant tissue) by Karl Fischer titration: Collaborative study. *J. AOAC Int.* **2002**, *85*, 318–327. [CrossRef] [PubMed]
25. Houghland, G.V.C. New conversion table for specific gravity, dry matter and starch in potatoes. *Am. Pot. J.* **1966**, *43*, 138. [CrossRef]

26. Lindsay, H.A. A colorimetric estimation of reducing sugars in potatoes with 3,5-dinitrosalicylic acid. *Pot. Res.* **1973**, *16*, 176–179. [CrossRef]
27. Pęksa, A.; Miedzianka, J.; Kita, A.; Tajner-Czopek, A.; Rytel, E. The quality of fried snacks fortified with fiber and protein supplements. *Poravinárstvo* **2010**, *4*, 59–64. [CrossRef]
28. Tajner-Czopek, A.; Kita, A.; Rytel, E. Characteristics of French Fries and Potato Chips in Aspect of Acrylamide Content—Methods of Reducing the Toxic Compound Content in Ready Potato Snacks. *Appl. Sci.* **2021**, *11*, 3943. [CrossRef]
29. Saito, K.; Horie, M.; Hoshino, Y.; Nose, N. High-performance liquid chromatographic determination of glycoalkaloids in potato products. *J. Chromatogr.* **1990**, *508*, 141–147. [CrossRef]
30. Pęksa, A.; Gołubowska, G.; Rytel, E.; Lisińska, G.; Aniołowski, K. Influence of harvest date on glycoalkaloid contents of three potato varieties. *Food Chem.* **2002**, *78*, 313–317. [CrossRef]
31. StatSoft, Inc. *Electronic Statistics Textbook*; StatSoft: Tulsa, OK, USA, 2013.
32. Lisińska, G.; Pęksa, A.; Kita, A.; Rytel, E.; Tajner-Czopek, A. The quality of potato for processing and consumption. In *Potato III. Food, Global Science Books*; Yee, N., Bussel, W.T., Eds.; Unitec: Auckland, New Zeland, 2009; pp. 99–104.
33. Lachman, J.; Hamouz, K.; Musilová, J.; Hejtmánková, K.; Kotíková, Z.; Pazderů, K.; Domkářová, J.; Pivec, V.; Cimr, J. Effect of peeling and three cooking methods on the content of selected phytochemicals in potato tubers with various colour of flesh. *Food Chem.* **2013**, *138*, 1189–1197. [CrossRef] [PubMed]
34. Omayio, O.G.; Abong, G.O.; Okoth, M.W. A Review of Occurrence of Glycoalkaloids in Potato and Potato Products. *Curr. Res. Nutr. Food Sci.* **2016**, *4*, 195–202. [CrossRef]
35. Nie, X.; Zhanga, G.; Lvc, S.; Guo, H. Steroidal glycoalkaloids in potato foods as affected by cooking methods. *Int. J. Food Prop.* **2018**, *21*, 1875–1887. [CrossRef]
36. Rytel, E.; Kułakowska, K.; Nemś, A. The effect of chips processing on the content of toxic compounds. *It. J. Food Sci.* **2012**, *24*, 376–383. Available online: https://www.proquest.com/docview/1266221529/fulltextPDF/5545A6D922314383PQ/1?accountid=48845&sourcetype=Scholarly%20Journals (accessed on 3 May 2024).
37. Kita, A.; Lisińska, G.; Tajner-Czopek, A.; Pęksa, A.; Rytel, E. The properties of Potato Snacks Influenced by the Frying Medium. In *Potato IV. Food, Global Science Books*; Yee, N., Bussel, W.T., Eds.; Unitec: Auckland, New Zeland, 2009; pp. 93–98.
38. Liua, J.; Wena, C.; Wang, M.; Wang, S.; Dong, N.; Lei, Z.; Lin, S.; Zhu, B. Enhancing the hardness of potato slices after boiling by combined treatment with lactic acid and calcium chloride: Mechanism and optimization. *Food Chem.* **2020**, *308*, 124832. [CrossRef] [PubMed]
39. Huang, Y.; Lu, J.; Li, M.; Li, C.; Wang, Y.; Shen, M.; Chen, Y.; Nie, S.; Zeng, M.; Che, J.; et al. Effect of acidity regulators on acrylamide and 5-hydroxymethylfurfural formation in French fries: The dual role of pH and acid radical ion. *Food Chem.* **2022**, *371*, 131154. [CrossRef] [PubMed]
40. Negoiţă, M.; Mihai, A.L.; Hornet, G.A. Influence of Water, NaCl and Citric Acid Soaking Pre-Treatments on Acrylamide Content in French Fries Prepared in Domestic Conditions. *Foods* **2022**, *11*, 1204. [CrossRef]
41. Liu, H.; Roasa, J.; Mats, L.; Zhu, H.; Shao, S. Effect of acid on glycoalkaloids and acrylamide in French fries. *Food Addit. Contam. Part A* **2020**, *37*, 938–945. [CrossRef]

Disclaimer/Publisher's Note: The statements, opinions and data contained in all publications are solely those of the individual author(s) and contributor(s) and not of MDPI and/or the editor(s). MDPI and/or the editor(s) disclaim responsibility for any injury to people or property resulting from any ideas, methods, instructions or products referred to in the content.

Disclaimer/Publisher's Note: The title and front matter of this reprint are at the discretion of the Guest Editors. The publisher is not responsible for their content or any associated concerns. The statements, opinions and data contained in all individual articles are solely those of the individual Editors and contributors and not of MDPI. MDPI disclaims responsibility for any injury to people or property resulting from any ideas, methods, instructions or products referred to in the content.

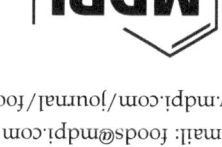

MDPI AG
Grosspeteranlage 5
4052 Basel
Switzerland
Tel.: +41 61 683 77 34

Foods Editorial Office
E-mail: foods@mdpi.com
www.mdpi.com/journal/foods